T0258229

**IET CONTROL ENGINEERING SERIES 70**

Series Editors: Professor D.P. Atherton
Professor G.W. Irwin
Professor S. Spurgeon

# Intelligent Control Systems using Computational Intelligence Techniques

# Other volumes in this series:

# Intelligent Control Systems using Computational Intelligence Techniques

### Edited by A.E. Ruano

The Institution of Engineering and Technology

Published by The Institution of Engineering and Technology, London, United Kingdom

First edition © 2005 The Institution of Electrical Engineers
New cover © 2008 The Institution of Engineering and Technology

First published 2005

The Institution of Engineering and Technology
Michael Faraday House
Six Hills Way, Stevenage
Herts, SG1 2AY, United Kingdom

www.theiet.org

**British Library Cataloguing in Publication Data**
Intelligent control systems using computational intelligence techniques
  1. Intelligent control systems 2. Computational intelligence
  I. Ruano, A.E. (Antonio E.) II. Institution of Electrical Engineers
  629.8'9

**ISBN (10 digit) 0 86341 489 3**
**ISBN (13 digit) 978-0-86341-489-3**

Typeset in India by Newgen Imaging Systems (P) Ltd, Chennai

To
Graça, Carina and João

# Contents

**2    An overview of nonlinear identification and control with
    neural networks                                                37**
    *A. E. Ruano, P. M. Ferreira and C. M. Fonseca*

**3    Multi-objective evolutionary computing solutions for control and
    system identification                                         89**
    *Ian Griffin and Peter John Fleming*

**9    Reinforcement learning for online control and optimisation    293**
*James J. Govindhasamy, Seán F. McLoone, George W. Irwin,*
*John J. French and Richard P. Doyle*

**10   Reinforcement learning and multi-agent control within an internet environment    327**
*P. R. J. Tillotson, Q. H. Wu and P. M. Hughes*

# Preface

Intelligent Control techniques are nowadays recognised tools in both academia and industry. Methodologies coming from the field of computational intelligence, such as neural networks, fuzzy systems and evolutionary computation, can lead to accommodation of more complex processes, improved performance and considerable time savings and cost reductions. This book is neither intended to be an introductory text in the subject, nor a book specialised in a specific topic. Rather, it covers most of the active research topics in the subject, and gives a comprehensive overview of the present state of intelligent control using computational intelligence techniques.

The level of this book makes it suitable for practising control engineers, final year students and both new and advanced researchers in this field. The contents have been chosen to present the different topics with a sound theoretical basis, but without unnecessary mathematical details. In addition to the description of the topic in itself, in each chapter an exhaustive set of important references in the subject has been incorporated, so that the interested reader can complement his or her reading. Also, for each topic covered in this book illustrative examples are included, mainly industrially based.

The first three chapters are introductory chapters on the use of fuzzy systems, neural networks and evolutionary algorithms for nonlinear identification and control. Chapter 1 introduces fuzzy systems, providing also a suitable example of identification and predictive control of an air-conditioned system. Chapter 2 gives an overview of artificial neural networks, in the context of identification and control. An example, of greenhouse climate models, illustrates a multi-objective genetic algorithm (MOGA) approach to the design of neural models. Evolutionary algorithms are introduced in Chapter 3, in the context of multi-objective optimisation. Case studies are incorporated in Chapter 3 illustrating how complex control design problems can be mapped into the MOGA framework, as well as its use for system identification.

Chapters 4–7 discuss different approaches in the important topics of local linear models and neuro-fuzzy systems. In Chapter 4 local linear models are constructed with clustering methods (self-organising maps and probability density mixture models), for input space partitioning. The use of these models for systems identification is demonstrated in chaotic times-series prediction and in a synthetic problem. Employing an inverse error dynamics control scheme, the use of these models is demonstrated for the control of an unmanned aerial vehicle and a transonic wind tunnel. Chapter 5 discusses neuro-fuzzy models of the Takagi–Sugeno type for nonlinear systems identification, using a construction algorithm, LOLIMOT, based on axis orthogonal partitioning. The integration of prior process knowledge in model and identification signals design is demonstrated by the modelling of a cross-flow heat exchanger. A different approach is described in Chapter 6, where linear local Gaussian process networks are discussed. These are non-parametric models, which offer, besides their approximation capabilities, information about the confidence in the prediction. The use of these networks for model predictive control and internal model control is addressed there. Takagi–Sugeno neuro-fuzzy models based on an input grid structure, and employing a decomposition of multivariate functions in additive functions of smaller dimensionality (ANOVA decomposition) are introduced in Chapter 7. The ASMOD algorithm, a constructive algorithm used with this type of model is first reviewed, and subsequently extended for fuzzy rule regularisation and subspace-based information extraction, by combining a locally regularised orthogonal least-squares algorithm and a D-optimality criterion for subspace rule selection.

The design of control systems requires the consideration of various types of constraints (actuator saturation, constraints in the states and others) as well as the use of different performance measures, such as minimum energy, minimum time and others. Chapter 8 discusses a framework to find nearly optimal controllers in state feedback for general nonlinear systems with constraints, with non-quadratic performance functionals. Neural networks are used here to solve for the value function of the Hamilton–Jacobi–Bellman (HJB) equation. Numerical examples are discussed and simulated.

There are many examples in industrial control where conventional automatic control systems (e.g. self-tuning controllers) are not yet sufficiently advanced to cater for nonlinear dynamics across different operating regions or to predict the effect of current controller changes in the long term. In these situations, an intelligent approach for evaluating possible control alternatives can be of value. One such framework, called the adaptive critic design, is currently attracting much renewed interest in the academic community, and it is the subject of Chapters 9 and 10. In the former, the model-free, action dependent adaptive critic design of Si and Wang is extended to produce a fully online neurocontroller without the necessity to store plant data during a successful run. The performance of this scheme for reinforcement learning is validated using simulation results from an inverted pendulum control task, and using data from an actual industrial grinding process used in the manufacture of substrates for disk drives. Adaptive heuristic critics and Q-learning are discussed in Chapter 10, in the context of multi-agent control. An application of these techniques to improve the routing control within an internet is described. Chapter 11 outlines some of the recent research on fault detection and isolation and fault diagnosis for dynamic systems, using an approach where analytical and soft-computing techniques are combined to achieve good global and nonlinear modelling and robustness. Illustrative examples are given throughout the chapter.

The book concludes with two more application-oriented chapters. Chapter 12 presents an intelligent system for autonomous parking of vehicles, including intelligent navigation and searching of parking place. Soft-computing techniques are here used to take advantage of knowledge of expert drivers, to manage imprecise information from the sensors, to navigate searching for a parking place and to select and perform a parking manoeuvre. Chapter 13 discusses applications of intelligent control in medicine. The first part of this chapter describes two surveys on the use of fuzzy technology and on the use of smart and adaptive systems in the whole field of medicine. The merits and limitations of current work are shown and signposts for promising future developments in the field are given. The second part of the chapter describes two relevant case studies: the first concerning intelligent systems in anaesthesia monitoring and control, and the second on the development of nonlinear models based on neural networks and neuro-fuzzy techniques for both classification and prediction in cancer survival studies.

In total, this book provides a strong coverage of state-of-the-art methods in intelligent control using computational intelligence techniques, together with guidelines for future research. It is possible to use this book as a complete text, or in terms of individual chapters, detailing specific areas of interest. I hope that this book will be of use to readers wishing to increase their knowledge in this field, as well as control engineers who wish to know about the advantages that this technology might bring to their applications.

The editor would like to express his gratitude to all the authors for their contributions and cooperation during the production process. Thanks also to Wendy Hiles, Michelle Williams and Phil Sergeant from the IEE for their continuous assistance, and to Dr Pedro Frazão for his help in making incompatible word-processing formats compatible.

António Ruano
November, 2004

# Contributors

**M. F. Abbod**
Department of Automatic Control
and Systems Engineering
University of Sheffield
Mappin Street, Sheffield
S1 3JD, UK
m.f.abbod@sheffield.ac.uk


**Murad Abu-Khalaf**
Automation and Robotics Research
Institute
The University of Texas at Arlington
7300 Jack Newell Blvd. S.
Fort Worth, Texas 76118, USA
abukhalaf@arri.uta.edu


**Robert Babuška**
Delft Center for Systems and Control
Faculty of Mechanical Engineering
Delft University of Technology
Mekelweg 2, 2628 CD Delft
The Netherlands
r.babuska@dcsc.tudelft.nl

**Jeongho Cho**
Computational NeuroEngineering
Laboratory
NEB 451, Bldg #33
University of Florida
Gainesville, FL 32611, USA


**F. Cuesta**
Dept. Ingeniería de Sistemas y
Automática
Escuela Superior de Ingenieros
Univ. de Sevilla
Camino de los Descubrimientos
E-41092 Sevilla, Spain
fede@cartuja.us.es


**Richard P. Doyle**
Seagate Technology Media Ltd.
99 Dowland Road
Aghanloo Industrial Estate
Limavady BT49 OHR
Northern Ireland, UK
Richard.P.Doyle@seagate.com

**Deniz Erdogmus**
Oregon Graduate Institute (OHSU
West Campus)
Department of CSEE
20000 NW Walker Road
Beaverton, OR 97006, USA
derdogmus@iee.org

**P. M. Ferreira**
Centre for Intelligent Systems
Faculty of Science and Technology
University of Algarve
Campus de Gambelas
8000 Faro, Portugal
pfrazao@ualg.pt

**Alexander Fink**
Institute of Automatic Control
Darmstadt University of Technology
Landgraf-Georg-Strasse 4
64283 Darmstadt, Germany
alexander.fink@siemens.com

**Peter John Fleming**
Department of Automatic Control and
Systems Engineering
The University of Sheffield
Mappin Street, Sheffield S1 3JD, UK
P.Fleming@sheffield.ac.uk

**C. M. Fonseca**
Centre for Intelligent Systems
Faculty of Science and Technology
University of Algarve
Campus de Gambelas
8000 Faro, Portugal
cmfonsec@ualg.pt

**John J. French**
Seagate Technology Media Ltd.
99 Dowland Road
Aghanloo Industrial Estate
Limavady BT49 OHR
Northern Ireland, UK
jjf@surftechintl.co.uk

**F. Gómez-Bravo**
Dept. Ingeniería Electrónica
Sistemas Informáticos y Automática
Esc. Superior Politécnica de la Rábida
Universidad de Huelva, Huelva, Spain
fernando.gomez@diesia.uhu.es

**James J. Govindhasamy**
Intelligent Systems and Control
Research Group
Queen's University Belfast
Belfast BT9 5AH, Northern Ireland
UK
j.govindasamy@ee.qub.ac.uk

**Gregor Gregorčič**
Department of Electrical Engineering
and Microelectronics
University College Cork, Ireland
gregorg@rennes.ucc.ie

**Ian Griffin**
Department of Automatic Control
and Systems Engineering
The University of Sheffield
Mappin Street, Sheffield S1 3JD, UK
griffin@acse.shef.ac.uk

**C. J. Harris**
School of Electronics and
Computer Science
University of Southampton
Southampton SO17 1BJ, UK
cjh@ecs.soton.ac.uk

**X. Hong**
Department of Cybernetics
University of Reading, RG6 6AY, UK
x.hong@reading.ac.uk

**P. M. Hughes**
BT Laboratories, Martlesham Heath
Ipswich, Suffolk IP5 7RE, UK

**George W. Irwin**
Intelligent Systems and Control
Research Group
Queen's University Belfast
Belfast BT9 5AH, Northern Ireland
UK
g.irwin@ee.qub.ac.uk

**Rolf Isermann**
Institute of Automatic Control
Darmstadt University of Technology
Landgraf-Georg-Strasse 4
64283 Darmstadt, Germany
risermann@iat.tu-darmstadt.de

**Józef Korbicz**
Institute of Control and Computation
Engineering
University of Zielona Góra
ul. Podgórna 50, 65-246 Zielona Góra
Poland
J.Korbicz@issi.uz.zgora.pl

**Jing Lan**
Computational NeuroEngineering
Laboratory
NEB 451, Bldg #33
University of Florida
Gainesville, FL 32611, USA

**Frank L. Lewis**
Automation and Robotics Research
Institute
The University of Texas at Arlington
7300 Jack Newell Blvd. S.
Fort Worth, Texas 76118, USA
flewis@arri.uta.edu

**Gordon Lightbody**
Department of Electrical Engineering
and Microelectronics
University College Cork, Ireland
gordon@rennes.ucc.ie

**D. A. Linkens**
Department of Automatic Control
and Systems Engineering
University of Sheffield
Mappin Street
Sheffield S1 3JD, UK
D.Linkens@sheffield.ac.uk

**Seán F. McLoone**
Department of Electronic
Engineering
National University of
Ireland Maynooth
Maynooth, Co. Kildare
Ireland
sean.mcloone@eeng.may.ie

**Mark Motter**
Computational NeuroEngineering
Laboratory
NEB 451, Bldg #33
University of Florida
Gainesville, FL 32611, USA

**A. Ollero**
Dept. Ingeniería de Sistemas
y Automática
Escuela Superior de Ingenieros
Univ. de Sevilla
Camino de los Descubrimientos
E-41092 Sevilla, Spain
aollero@cartuja.us.es

**Ronald J. Patton**
Department of Engineering
University of Hull
HU6 7RX Kingston upon Hull
East Yorkshire, UK
R.J.Patton@hull.ac.uk

**Jose C. Principe**
Computational NeuroEngineering
Laboratory
NEB 451, Bldg #33
University of Florida
Gainesville, FL 32611, USA
principe@cnel.ufl.edu

**A. E. Ruano**
Centre for Intelligent Systems
Faculty of Science and Technology
University of Algarve
Campus de Gambelas
8000 Faro, Portugal
aruano@ualg.pt

**P. R. J. Tillotson**
Department of Electrical
Engineering and Electronics
The University of Liverpool
Liverpool, L69 3GJ, UK
peter.tillotson@roke.co.uk

**Faisel Uppal**
Department of Engineering
University of Hull
HU6 7RX Kingston upon Hull
East Yorkshire, UK
f.uppal@hull.ac.uk

**Michael Vogt**
Institute of Automatic Control
Darmstadt University of Technology
Landgraf-Georg-Strasse 4
64283 Darmstadt, Germany

**Martin Witczak**
Institute of Control and Computation
Engineering
University of Zielona Góra
ul. Podgórna 50, 65-246 Zielona Góra
Poland
M.Witczak@issi.uz.zgora.pl

**Q. H. Wu**
Department of Electrical
Engineering and Electronics
The University of Liverpool
Liverpool, L69 3GJ, UK
q.h.wu@liverpool.ac.uk

**Ralf Zimmerschied**
Institute of Automatic Control
Darmstadt University of Technology
Landgraf-Georg-Strasse 4
64283 Darmstadt, Germany
rzimmerschied@iat.tu-darmstadt.de

*Chapter 1*

# An overview of nonlinear identification and control with fuzzy systems

*Robert Babuška*

## 1.1 Introduction

The design of modern control systems is characterised by stringent performance and robustness requirements and therefore relies on model-based design methods. This introduces a strong need for effective modelling techniques. Many systems are not amenable to conventional modelling approaches due to the lack of precise knowledge and strong nonlinear behaviour of the process under study. Nonlinear identification is therefore becoming an important tool which can lead to improved control systems along with considerable time saving and cost reduction. Among the different nonlinear identification techniques, methods based on fuzzy sets are gradually becoming established not only in the academia but also in industrial applications [1–3].

Fuzzy modelling techniques can be regarded as grey-box methods on the boundary between nonlinear black-box techniques and qualitative models or expert systems. Their main advantage over purely numerical methods such as neural networks is the transparent representation of knowledge in the form of fuzzy if–then rules. Linguistic interpretability and transparency are therefore important aspects in fuzzy modelling [4–8]. The tools for building fuzzy systems are based on algorithms from the fields of fuzzy logic, approximate reasoning, neural networks, pattern recognition, statistics and regression analysis [1,2,9].

This chapter gives an overview of system identification techniques for fuzzy models and some selected techniques for model-based fuzzy control. It starts with a brief discussion of the position of fuzzy modelling within the general nonlinear identification setting. The two most commonly used fuzzy models are the Mamdani model and the Takagi–Sugeno model. An overview of techniques for the data-driven construction of the latter model is given. We discuss both structure selection (input variables, representation of dynamics, number and type of membership functions) and parameter estimation (local and global estimation techniques, weighted least squares, multi-objective optimisation).

Further, we discuss control design based on a fuzzy model of the process. As the model is assumed to be obtained through identification from sampled data, we focus on discrete-time methods, including: gain-scheduling and state-feedback design, model-inverse control and predictive control. A real-world application example is given.

## 1.2    Nonlinear system identification

Consider a multiple-input, single-output (MISO) nonlinear time-invariant dynamic system:

$$y(t) = S(\mathbf{u}(t)), \qquad y(t) \in \mathbb{R}, \quad \mathbf{u}(t) \in \mathbb{R}^p, \tag{1.1}$$

where $p$ is the number of inputs, $t$ denotes continuous time and $S$ is an operator relating the input signals $\mathbf{u}(t)$ to the output signal $y(t)$. The mathematical description of the system itself is unknown, but we assume that input–output data are available. The inputs and the output are sampled at a constant rate, resulting in discrete-time signals denoted by $\mathbf{u}(k)$ and $y(k)$. System (1.1) can then be approximated by a MISO nonlinear auto-regressive model with exogenous input (NARX):

$$\hat{y}(k + 1) = f(\mathbf{x}(k)), \tag{1.2}$$

where the hat denotes approximation and $\mathbf{x}(k) \in X \subset \mathbb{R}^n$ is the regression vector defined as the collection of previous process inputs and outputs:

$$\mathbf{x}(k) = [y_1(k), \dots, y_1(k - n_1 + 1), u_1(k), \dots,$$
$$u_1(k - m_1 + 1), \dots, u_p(k), \dots, u_p(k - m_p + 1)]^T. \tag{1.3}$$

The parameters $n_1$ and $m_1, \dots, m_p$ are integers related to the dynamic order of system (1.1). Further, denote $n = n_1 + \sum_{j=1}^p m_j$ the dimension of the regression vector. To properly account for noise disturbances (e.g. sensor noise, process noise, etc.), more complicated model structures can be chosen. Some common examples are the nonlinear output error (NOE) model, which involves the past model predictions instead of the process output data:

$$\mathbf{x}(k) = [\hat{y}_1(k), \dots, \hat{y}_1(k - n_1 + 1), u_1(k), \dots,$$
$$u_1(k - m_1 + 1), \dots, u_p(k), \dots, u_p(k - m_p + 1)]^T, \tag{1.4}$$

or the 'innovations' forms with the additional freedom to describe the effect of disturbances. In the NARMAX model, for instance, the prediction error $e(k) = y(k) - \hat{y}(k)$ and its past values are included in the regression vector as well (stated for single-input–single-output (SISO) systems for brevity):

$$\mathbf{x}(k) = [y(k), \dots, y(k - n_1 + 1), u(k), \dots,$$
$$u(k - m_1 + 1), e(k), \dots, e(k - n_e)]^T. \tag{1.5}$$

The above structures directly apply to systems with pure transport delays. The discussion of fuzzy models in the subsequent sections is restricted to the NARX structure (1.2) for $p = 1$. In the sequel, we also drop the hat denoting predictions (the model output thus becomes $y$). When the physical state $\xi$ of the system is measurable, one usually prefers the nonlinear state–space description:

$$\xi(k + 1) = g(\xi(k), \mathbf{u}(k)),$$

$$\mathbf{y}(k) = h(\xi(k)). \tag{1.6}$$

The problem of nonlinear system identification is to infer the (partly) unknown function $f$ in (1.2) or the functions $g$ and $h$ in (1.6) from the sampled data sequences $\{(u(k), y(k)) \mid k = 1, 2, \ldots, N\}$. One can use general function approximators such as neural networks, fuzzy systems, splines, interpolated look-up tables, etc. If the aim of modelling is only to obtain an accurate predictor for $y$, there is not much difference between these models, as they all can approximate smooth nonlinear systems arbitrarily well. However, besides accurate predictions, one often wants to have a model that can be used to learn something about the underlying system and analyse its properties. From this point of view, fuzzy and neuro-fuzzy systems are more transparent and flexible than most other black-box techniques.

## 1.3  Fuzzy models

The introduction of fuzzy systems was originally motivated by the need to represent (ambiguous) human knowledge and the corresponding deductive processes [10]. In system identification, rule-based fuzzy models are usually applied. In these models, relations among variables are described by means of if–then rules with fuzzy predicates, such as 'if heating valve is open then temperature is high'. The ambiguity in the definition of the linguistic terms (e.g. high temperature) is represented by using fuzzy sets, which are sets with overlapping boundaries.

Fuzzy sets are defined through their membership functions (denoted by $\mu$) which map the elements of the considered universe to the unit interval [0, 1]. The extreme values 1 and 0 denote complete membership and non-membership, respectively, while a degree between 0 and 1 means partial membership in the fuzzy set. A particular domain element can simultaneously belong to several sets (with different degrees of membership). In Figure 1.1, for instance, $t = 20°\text{C}$ belongs to the set of high temperatures with membership 0.4 and to the set of medium temperatures with membership 0.2. This gradual transition from membership to non-membership facilitates a smooth outcome of the reasoning (deduction) with fuzzy if–then rules; in fact a kind of interpolation.

Depending on the structure of the if–then rules, two main types of fuzzy models can be distinguished: the Mamdani (or linguistic) model and the Takagi–Sugeno (TS) model.

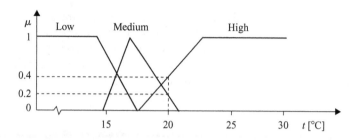

*Figure 1.1    Partitioning of the temperature domain into three fuzzy sets*

### 1.3.1    Mamdani model

In this model, the antecedent (if-part of the rule) and the consequent (then-part of the rule) are fuzzy propositions:

$$\mathcal{R}_i: \quad \text{If } \mathbf{x} \text{ is } A_i \text{ then } y \text{ is } B_i, \quad i = 1, 2, \dots, K. \tag{1.7}$$

Here $A_i$ and $B_i$ are the antecedent and consequent linguistic terms (such as 'small', 'large', etc.), represented by fuzzy sets, and $K$ is the number of rules in the model. The linguistic fuzzy model is typically used to represent qualitative expert knowledge, although data-driven techniques can also be used to optimise a given rule base and the membership function parameters [11,12]. In this chapter we focus on the TS model described in Section 1.3.2.

### 1.3.2    Takagi–Sugeno model

In data-driven identification, the model due to Takagi and Sugeno [13] has been extensively used. In this model, the antecedent is fuzzy, while the consequent is an affine linear function of the input variables:

$$\mathcal{R}_i: \quad \text{If } \mathbf{x} \text{ is } A_i \text{ then } y_i = \mathbf{a}_i^{\mathsf{T}} \mathbf{x} + b_i, \quad i = 1, 2, \dots, K, \tag{1.8}$$

where $\mathbf{a}_i$ is the consequent parameter vector, $b_i$ is a scalar offset and $i = 1, \dots, K$. This model combines a linguistic description with standard functional regression; the antecedents describe fuzzy regions in the input space in which the consequent functions are valid. The output $y$ is computed as the weighted average of the individual rules' contributions:

$$y = \frac{\sum_{i=1}^{K} \beta_i(\mathbf{x}) y_i}{\sum_{i=1}^{K} \beta_i(\mathbf{x})} = \frac{\sum_{i=1}^{K} \beta_i(\mathbf{x})(\mathbf{a}_i^{\mathsf{T}} \mathbf{x} + b_i)}{\sum_{i=1}^{K} \beta_i(\mathbf{x})}, \tag{1.9}$$

where $\beta_i(\mathbf{x})$ is the degree of fulfilment of the $i$th rule. For the rule (1.8), it is simply the membership degree, $\beta_i(\mathbf{x}) = \mu_{A_i}(\mathbf{x})$, but it can also be a more complicated expression, as shown later. For notational convenience, we define the normalised degree of fulfilment $\gamma_i(\mathbf{x})$ as follows:

$$\gamma_i(\mathbf{x}) = \frac{\beta_i(\mathbf{x})}{\sum_{j=1}^{K} \beta_j(\mathbf{x})}. \tag{1.10}$$

This allows us to write the output as a weighted sum of the individual rules' consequents: $y = \sum_{i=1}^{K} \gamma_i(\mathbf{x}) y_i$. The antecedent fuzzy sets are usually defined to describe distinct, partly overlapping regions in the input space. The parameters $\mathbf{a}_i$ can then represent approximate local linear models of the considered nonlinear system. The TS model can thus be regarded as a smooth piece-wise linear approximation of a nonlinear function or a parameter-scheduling model. Note, however, that the antecedent and consequent may generally contain different sets of variables.

As the consequent parameters are first-order polynomials in the input variables, model (1.8) is also called the first-order TS model. This is to distinguish it from the zero-order TS model whose consequents are constants (zero-order polynomials):

$$\mathcal{R}_i: \text{ If } \mathbf{x} \text{ is } A_i \text{ then } y_i = b_i, \quad i = 1, 2, \ldots, K. \tag{1.11}$$

The input–output equation of this model is obtained as a special case of (1.9) for $\mathbf{a}_i = 0$, $\forall i$, and also as a special case of the Mamdani system (1.7) in which the consequent fuzzy sets $B_i$ degenerate to singletons (real numbers):

$$\mu_{B_i}(y) = \begin{cases} 1, & \text{if } y = b_i, \\ 0, & \text{otherwise.} \end{cases} \tag{1.12}$$

In fuzzy systems with multiple inputs, the antecedent proposition is usually represented as a combination of terms with univariate membership functions, using logic operators 'and' (conjunction), 'or' (disjunction) and 'not' (complement). In fuzzy set theory, several families of operators have been introduced for these logical connectives. Table 1.1 shows the two most common ones.

As an example, consider the conjunctive form of the antecedent, which is given by:

$$\mathcal{R}_i: \text{ If } x_1 \text{ is } A_{i1} \text{ and }, \ldots, \text{ and } x_n \text{ is } A_{in} \text{ then } y_i = \mathbf{a}_i^T \mathbf{x} + b_i \tag{1.13}$$

with the degree of fulfilment

$$\beta_i(\mathbf{x}) = \min(\mu_{A_{i1}}(x_1), \ldots, \mu_{A_{in}}(x_n)) \quad \text{or} \quad \beta_i(\mathbf{x}) = \mu_{A_{i1}}(x_1) \cdots \mu_{A_{in}}(x_n) \tag{1.14}$$

for the minimum and product conjunction operators, respectively. The complete set of rules (1.13) divide the input domain into a lattice of overlapping axis-parallel hyperboxes. Each of these hyperboxes is a Cartesian-product intersection of the corresponding univariate fuzzy sets.

*Table 1.1  Common fuzzy logic operators*

|  | A and B | A or B | not A |
|---|---|---|---|
| Zadeh | $\min(\mu_A, \mu_B)$ | $\max(\mu_A, \mu_B)$ | $1 - \mu_A$ |
| Probabilistic | $\mu_A \cdot \mu_B$ | $\mu_A + \mu_B - \mu_A \cdot \mu_B$ | $1 - \mu_A$ |

### 1.3.3   Dynamic fuzzy models

In the modelling of dynamic systems, fuzzy models are used to parameterise the nonlinear functions $f$ in (1.2) or $g$ and $h$ in (1.6). An example is the following TS NARX model:

$$\mathcal{R}_i: \quad \text{If } \mathbf{x}(k) \text{ is } A_i \text{ then } y_i(k+1) = \sum_{j=1}^{n_1} a_{ij} y(k-j+1)$$

$$+ \sum_{j=1}^{m_1} b_{ij} u(k-j+1) + c_i, \qquad (1.15)$$

where the antecedent regressor $\mathbf{x}(k)$ is generally given by (1.3), but it may of course contain only some of the past inputs and outputs or even variables other than $u$ and $y$. Similarly, state–space models can be represented in the TS framework:

$$\mathcal{R}_i: \quad \text{If } \boldsymbol{\xi}(k) \text{ is } A_i \text{ and } \mathbf{u}(k) \text{ is } B_i \text{ then } \begin{cases} \boldsymbol{\xi}_i(k+1) = \boldsymbol{\Phi}_i \boldsymbol{\xi}(k) + \boldsymbol{\Gamma}_i \mathbf{u}(k) + \mathbf{a}_i, \\ \mathbf{y}_i(k) = \mathbf{C}_i \boldsymbol{\xi}(k) + \mathbf{c}_i. \end{cases}$$

$$(1.16)$$

An advantage of the state–space modelling approach is that the structure of the model can be easily related to the physical structure of the real system, and, consequently, the estimated model parameters are physically relevant. This is usually not the case with the input–output models (1.15).

The TS model possesses many degrees of freedom, which can be exploited by the user in order to make the model suit the problem at hand. For instance, the local models in the rule consequents can have different dynamic orders or delays, to represent systems with operating regime dependent structural parameters. Examples are processes where transportation lags depend on state variables (such as the speed or flow-rate of some transported medium), systems switching between different modes (such as forced dynamics and free-run), models of systems that operate in nominal situations but also in the presence of faults, etc.

The antecedent and consequent variables can be selected to match the physical structure of the process under study. Well-known types of block-oriented models are special cases of the dynamic TS models (1.15) and (1.16). For instance, by including only the current input $\mathbf{u}(k)$ in the antecedent of (1.16) and by restricting the $\boldsymbol{\Phi}$ and $\mathbf{C}$ matrices to be constant for all the rules, a Hammerstein fuzzy model (a static input nonlinearity in series with linear dynamics) is obtained:

$$\mathcal{R}_i: \quad \text{If } \mathbf{u}(k) \text{ is } B_i \text{ then } \begin{cases} \boldsymbol{\xi}_i(k+1) = \boldsymbol{\Phi} \boldsymbol{\xi}(k) + \boldsymbol{\Gamma}_i \mathbf{u}(k) + \mathbf{a}_i, \\ \mathbf{y}_i(k) = \mathbf{C} \boldsymbol{\xi}(k) + \mathbf{c}. \end{cases}$$

In a similar way, a Wiener fuzzy model can be obtained, which consists of linear dynamics in series with an output nonlinearity. Another example is the input-affine model resulting from (1.15) by excluding $u(k)$ from the antecedent vector:

$$\mathbf{x}(k) = [y(k), y(k-1), \ldots, y(k-n_1+1), u(k-1), \ldots, u(k-m_1+1)]^{\mathrm{T}}.$$

Substitute $\mathbf{x}(k)$ in equation (1.9) to see that this model is indeed affine linear in $u(k)$. To conclude this section, let us summarise that the TS model structure provides a flexible framework for approximating nonlinear processes with both smooth and abrupt nonlinearities (through the choice of the membership functions), specific structure (e.g. Wiener and Hammerstein systems) and for modelling processes where the dynamic structure varies with some known variables (switching systems, failure modes, etc.).

## 1.4   Constructing fuzzy models

Expert knowledge and process data are typically combined in the construction of fuzzy systems. Two main approaches can be distinguished:

1.  Expert knowledge is stated as a collection of if–then rules to form an initial model. The parameters of this model (membership functions, consequent parameters) are then fine-tuned by using process data.
2.  Fuzzy rules are constructed from scratch by using data. The resulting rule base can then be analysed and interpreted by an expert (which is not possible with truly black-box structures like neural networks). An expert can compare the information stored in the rule base with his own knowledge, can modify the rules, or supply additional ones to extend the validity of the model, etc.

### 1.4.1   Structure and parameters

The two basic steps in system identification are structure identification and parameter estimation. The choice of the model's structure is very important, as it determines the flexibility of the model in the approximation of (unknown) systems. A model with a rich structure can approximate more complicated functions, but, at the same time, will have worse generalisation properties.[1] The structure selection process involves the following main choices:

*Selection of input variables.* This involves not only the physical inputs **u** but also the dynamic regressors, defined by the input and output lags in equation (1.3). Prior knowledge, insight in the process behaviour and the purpose of the modelling exercise are the typical sources of information for the choice of an initial set of possible inputs. Automatic data-driven selection can then be used to compare different structures in terms of some specified performance criteria.

*Number and type of membership functions, number of rules.* These two structural parameters are mutually related (for more membership functions more rules must be defined) and determine the level of detail, called the granularity, of the model. The purpose of modelling and the amount of available information (knowledge and data) will determine this choice. Automated methods can be used to add or remove membership functions and rules.

After the structure is fixed, the fuzzy model can be fitted to the available data by estimating or fine-tuning its antecedent and consequent parameters. The antecedent parameters are the parameters of the membership functions $A_i$. The type of membership function can be freely chosen as, for instance, the trapezoidal function

$$\mu(x; \alpha_1, \alpha_2, \alpha_3, \alpha_4) = \max\left(0, \min\left(\frac{x - \alpha_1}{\alpha_2 - \alpha_1}, 1, \frac{\alpha_4 - x}{\alpha_4 - \alpha_3}\right)\right), \tag{1.17}$$

where $\alpha_j$ are the coordinates of the trapezoid apexes. Often, smooth membership functions are employed, such as the exponential function:

$$\mu(x; c, \sigma) = \exp\left(-\frac{(x - c)^2}{2\sigma^2}\right), \tag{1.18}$$

whose parameters are the centre $c$ and the width $\sigma$. We now combine equations (1.9) and (1.10) into:

$$y = \sum_{i=1}^{K} \gamma_i(\mathbf{x}; \boldsymbol{\alpha}_i)(\mathbf{x}^{\mathrm{T}}\mathbf{a}_i + b_i) = \sum_{i=1}^{K} \gamma_i(\mathbf{x}; \boldsymbol{\alpha}_i)[\mathbf{x}^{\mathrm{T}}, 1]\boldsymbol{\theta}_i, \tag{1.19}$$

where $\boldsymbol{\theta}_i^{\mathrm{T}} = [\mathbf{a}_i^{\mathrm{T}}, b_i]$ and $\boldsymbol{\alpha}_i$ are the consequent and antecedent parameters vectors, respectively. The available set of $N$ input–output data pairs $\{(\mathbf{x}_k, y_k) \mid k = 1, 2, \ldots, N\}$ is represented as a matrix $\mathbf{X} \in \mathbb{R}^{N \times n}$, having the vectors $\mathbf{x}_k^{\mathrm{T}}$ in its rows, and a column vector $\mathbf{y} \in \mathbb{R}^N$, containing the outputs $y_k$:

$$\mathbf{X} = [\mathbf{x}_1, \ldots, \mathbf{x}_N]^{\mathrm{T}}, \qquad \mathbf{y} = [y_1, \ldots, y_N]^{\mathrm{T}}. \tag{1.20}$$

The complete parameter estimation problem in model (1.19) can now be formulated as the minimisation of the following nonlinear least-square criterion:

$$\{\boldsymbol{\alpha}_1, \ldots, \boldsymbol{\alpha}_K, \boldsymbol{\theta}_1, \ldots, \boldsymbol{\theta}_K\} = \arg\min \sum_{k=1}^{N}\left(y_k - \sum_{i=1}^{K} \gamma_i(\mathbf{x}_k; \boldsymbol{\alpha}_i)[\mathbf{x}_k^{\mathrm{T}}, 1]\boldsymbol{\theta}_i\right)^2.$$

$$\tag{1.21}$$

The commonly used optimisation techniques can be divided into two main categories:

1. Methods based on global nonlinear optimisation of all the parameters, such as genetic algorithms, neuro-fuzzy learning techniques (backpropagation), product-space fuzzy clustering, etc.
2. Methods that exploit the fact that equation (1.19) is nonlinear in $\boldsymbol{\alpha}_i$ (due to inherently nonlinear parameterisation of the membership functions), while it is linear in $\boldsymbol{\theta}_i$. Typically, the linear estimation problem is solved as a local problem within one iteration of the antecedent parameter optimisation problem.

In Section 1.4.3, we give examples of several selected methods, while other important algorithms can be found in the references. First, however, the linear least-squares estimation of the consequent parameters is discussed. This issue is a little more complicated than it might seem.

### 1.4.2    Consequent parameter estimation

Assume for the moment that the antecedent membership functions are known. As the output in equation (1.19) is linear in the consequent parameters, it is straightforward to estimate them such that an optimal fit to the data is obtained. However, extra care must be taken, as for many problems, the TS model tends to be over parameterised, which may lead to numerical problems, over fitting and meaningless parameter estimates. Let us illustrate this by the following example.

**Example 1**    Assume we wish to approximate a second-order polynomial $y = f_s(u) = 3u^2 - 5u + 6$ by a TS model with two rules. Choose two operating points $t_1$ and $t_2$ and define triangular membership functions for $t_1 \leq u < t_2$:

$$\mu_{A_1}(u) = \frac{u - t_1}{t_2 - t_1}, \qquad \mu_{A_2} = 1 - \mu_{A_1}. \tag{1.22}$$

The TS model rules are:

$$\mathcal{R}_i: \text{If } u \text{ is } A_i \text{ then } \hat{y}_i = a_i u + b_i, \quad i = 1, 2.$$

By substituting the membership functions (1.22) into (1.9), the output of the TS model is obtained (after some elementary algebra):

$$\hat{y} = \frac{a_1 - a_2}{t_2 - t_1} u^2 + \frac{t_2 a_2 - t_1 a_1 + b_1 - b_2}{t_2 - t_1} u + \frac{t_2 b_2 - t_1 b_1}{t_2 - t_1}.$$

As this is a second-order polynomial in $u$, the TS model can perfectly represent the given nonlinear system. However, the TS model has four free parameters $(a_1, a_2, b_1, b_2)$ while three are sufficient to fit the polynomial – it is thus over parameterised. This is a very simple example, but the essence of the over parameterisation problem remains the same when approximating complex unknown systems.

To circumvent over parameterisation, the global least-squares criterion (1.21) can be combined with additional criteria for local fit or with constraints on the parameter values, as discussed in Section 1.4.2.1.

#### 1.4.2.1    Global least-squares estimation

The global least-squares estimation method yields parameters that minimise the global prediction error criterion:

$$\theta = \arg \min \sum_{k=1}^{N} \left( y_k - \sum_{i=1}^{K} \gamma_i(\mathbf{x}_k)[\mathbf{x}_k^{\mathsf{T}}, 1]\theta_i \right)^2, \tag{1.23}$$

where $\theta$ is the concatenation of all the individual rules' parameter vectors $\theta_i$. For the data matrices (1.20), this criterion can be rewritten in a matrix form:

$$\theta = \arg \min(\mathbf{y} - \mathbf{\Gamma}_\varphi \theta)^{\mathsf{T}}(\mathbf{y} - \mathbf{\Gamma}_\varphi \theta) \tag{1.24}$$

with $\mathbf{\Gamma}_\varphi = [\mathbf{\Gamma}_1 \varphi \cdots \mathbf{\Gamma}_K \varphi]$ where $\varphi = [\mathbf{X}, 1]$ and $\mathbf{\Gamma}_i = \mathrm{diag}(\gamma_i(\mathbf{x}_1) \cdots \gamma_i(\mathbf{x}_N))$, i.e. a diagonal matrix having $\gamma_i(\mathbf{x}_k)$ as its $k$th diagonal element. The optimal solution of (1.24) is then directly obtained by using matrix pseudo-inverse:

$$\boldsymbol{\theta} = (\boldsymbol{\Gamma}_\varphi^\mathrm{T}\boldsymbol{\Gamma}_\varphi)^{-1}\boldsymbol{\Gamma}_\varphi^\mathrm{T}\mathbf{y}. \tag{1.25}$$

### 1.4.2.2   Local least-squares estimation

While the global solution gives the minimal prediction error, it may bias the estimates of the consequents as parameters of local models. If locally relevant model parameters are required, a weighted least-squares approach applied per rule should be used. This is done by minimising the following set of $K$ weighted local least-square criteria:

$$\boldsymbol{\theta}_i = \arg\min(\mathbf{y} - \boldsymbol{\varphi}\boldsymbol{\theta}_i)^\mathrm{T}\boldsymbol{\Gamma}_i(\mathbf{y} - \boldsymbol{\varphi}\boldsymbol{\theta}_i), \quad i = 1, 2, \ldots, K \tag{1.26}$$

for which the solutions are

$$\boldsymbol{\theta}_i = (\boldsymbol{\varphi}^\mathrm{T}\boldsymbol{\Gamma}_i\boldsymbol{\varphi})^{-1}\boldsymbol{\varphi}^\mathrm{T}\boldsymbol{\Gamma}_i\mathbf{y}, \quad i = 1, 2, \ldots, K. \tag{1.27}$$

In this case, the consequent parameters of the individual rules are estimated independently, and therefore the result is not influenced by the interactions of the rules. At the same time, however, a larger prediction error is obtained with global least squares.

When validating and interpreting fuzzy models obtained from data, one has to be aware of the trade-offs between local and global estimation. Constrained and multicriteria optimisation can also be applied to restrict the freedom in the parameters.

### 1.4.2.3   Constrained estimation

Realise that the affine TS model (1.8) can be regarded as one quasi-linear system

$$y = \left(\sum_{i=1}^{K}\gamma_i(\mathbf{x})\mathbf{a}_i^\mathrm{T}\right)\mathbf{x} + \sum_{i=1}^{K}\gamma_i(\mathbf{x})b_i = \mathbf{a}^\mathrm{T}(\mathbf{x})\mathbf{x} + b(\mathbf{x})$$

with input-dependent 'parameters' $\mathbf{a}(\mathbf{x})$, $b(\mathbf{x})$ which are convex linear combinations of the individual consequent parameters $\mathbf{a}_i$ and $b_i$:

$$\mathbf{a}(\mathbf{x}) = \sum_{i=1}^{K}\gamma_i(\mathbf{x})\mathbf{a}_i, \qquad b(\mathbf{x}) = \sum_{i=1}^{K}\gamma_i(\mathbf{x})b_i. \tag{1.28}$$

Knowledge about the dynamic system such as its stability, minimal or maximal static gain or its settling time can be translated into convex constraints on the consequent parameters. Using input–output data, optimal parameter values are found by means of quadratic programming, instead of least squares. There are two types of constraints, global and local. Local constraints represent detailed knowledge pertaining to each specific rule, while global constraints apply to the entire model and should thus refer to some global system properties such as the overall stability.

As an example, assume that we have the prior knowledge that the process is stable. It is well known that the poles of a stable discrete-time model of a properly sampled, continuous-time system cannot be situated in the left half of the complex plane. This knowledge on sampling can be translated into inequality constraints on the parameters of the model parameters:

$$(-1)^i a_i \geq 0, \quad 1 \leq i \leq n_1.$$

Additional constraints can be derived from stability considerations. Let $C(m, R)$ denote the set of complex numbers within or at a circle with a real-valued centre $m$ and radius $R$,

$$C(m, R) = \{z \in \mathbb{C} \mid \|z - m\| \leq R; \ m \in \mathbb{R}; \ R \in \mathbb{R}^+\}.$$

In Reference 14 it is shown that for the poles of a linear discrete-time system to be in $C(m, R)$, the following linear constraints on the model parameters must be satisfied:

$$\mathfrak{R} \cdot [1, a_1, a_2, \ldots, a_{n_y}]^T \geq 0, \tag{1.29}$$

$$\mathfrak{R} \cdot \mathfrak{I} \cdot [1, a_1, a_2, \ldots, a_{n_y}]^T \geq 0, \tag{1.30}$$

where the non-zero elements of the right and left triangular matrices $\mathfrak{R}$ and $\mathfrak{I}$ are defined by

$$[\mathfrak{R}]_{ij} = (-1)^i \binom{j}{i}, \quad i = 0, 1, \ldots, n_y, \quad j = 0, 1, \ldots, n_y,$$

$$[\mathfrak{I}]_{ij} = (m - R)^{i-j} \binom{n_y - j}{i - j} (2R)^{n_y - i}.$$

If $m = 0$ and $R = 1$, the previous equation constitutes the smallest convex hull of the admissible parameter region corresponding to the stable system model having all poles in the unit circle, $C(0, 1)$, in the complex plane. These are therefore necessary conditions for asymptotic stability.

**Example 2**    Consider the following second-order consequent model: $y(k + 1) = a_1 y(k) + a_2 y(k - 1) + b_1 u(k) + c$. In this case $\mathfrak{R} \cdot \mathfrak{I}$ becomes:

$$\begin{bmatrix} 1 & 1 & 1 \\ 0 & -1 & -2 \\ 0 & 0 & 1 \end{bmatrix} \begin{bmatrix} 4 & 0 & 0 \\ -4 & 2 & 0 \\ 1 & -1 & 1 \end{bmatrix} = \begin{bmatrix} 1 & 1 & 1 \\ 2 & 0 & -2 \\ 1 & -1 & 1 \end{bmatrix}. \tag{1.31}$$

From equations (1.30) and (1.31), the following inequality constraints are obtained:

$$1 + a_1 + a_2 \geq 0,$$

$$2 - 2a_2 \geq 0,$$

$$1 - a_1 + a_2 \geq 0.$$

Often, not only the stability of the plant is known beforehand, but also the admissible intervals or at least the signs of the static gains are known. One may also have information on the form of the nonlinearity (e.g. monotonically increasing, decreasing, etc.).

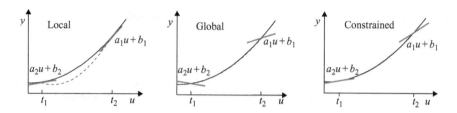

*Figure 1.2    The result of different methods in estimating the consequent parameters. The solid line is the function being approximated and the dashed line is the output of the model*

These pieces of information can also be formalised as constraints on the consequent parameters and can considerably improve the quality of the model especially when one only has poor data [15].

**Example 3**    The application of local, global and constrained estimation to the TS model from Example 1 results in the consequent models given in Figure 1.2. Note that the consequents estimated by local least squares describe properly the local behaviour of the function, but the model does not have a good global fit. For global least squares, the opposite holds – a perfect fit is obtained, but the consequents are not relevant for the local behaviour of the system. Constrained estimation results in a compromise: the consequent parameters do not represent perfect local models, but they are not completely meaningless either. The only constraints taken into account here are $a_1 > 0$, $a_2 > 0$.

#### 1.4.2.4    Multi-objective optimisation

Another possibility is to regularise the estimation by penalising undesired local behaviour of the model. This can dramatically improve the robustness of the construction algorithm, eventually leading to more relevant (interpretable) parameter estimates. One way is to minimise the weighted sum of the global and local identification criteria (1.24) and (1.26):

$$\theta = \arg \min \left\{ (\mathbf{y} - \boldsymbol{\Gamma}_\varphi \theta)^\mathrm{T} (\mathbf{y} - \boldsymbol{\Gamma}_\varphi \theta) + \sum_{i=1}^{K} \delta_i (\mathbf{y} - \varphi \theta_i)^\mathrm{T} \boldsymbol{\Gamma}_i (\mathbf{y} - \varphi \theta_i) \right\}.$$

The weighting parameters $\delta_i \geq 0$ parameterise the set of Pareto-optimal solutions of the underlying multi-objective optimisation problem and thus determine the trade-off between the possibly conflicting objectives of global model accuracy and local interpretability of the parameters [3,16,17].

### 1.4.3    Construction of antecedent membership functions

For the successful application of consequent estimation methods, good initialisation of the antecedent membership functions is important. As the optimisation problem (1.21) is nonlinear in the antecedent parameters $\alpha_i$, a variety of methods have been introduced, but none of them is really effective in general. Several frequently used methods are briefly reviewed in this section.

### 1.4.3.1  Template-based membership functions

With this method, the domains of the antecedent variables are *a priori* partitioned by a number of user-defined membership functions (usually evenly spaced and shaped). The rule base is then established to cover all the combinations of the antecedent terms. A severe drawback of this approach is that the number of rules in the model grows exponentially. Furthermore, if no knowledge is available as to which variables involve the nonlinearity of the system, all the antecedent variables are usually partitioned uniformly. However, the complexity of the system's behaviour is typically not uniform. In some relatively large operating regions, the system can be well approximated by a local linear model, while other regions require a rather fine partitioning. In order to obtain an efficient representation with as few rules as possible, the membership functions must be placed such that they capture the non-uniform behaviour of the system. This can be achieved by automated construction methods, as described below.

### 1.4.3.2  Discrete search methods

Tree-search algorithms can be applied to successively decompose the antecedent space into hyper-rectangles by axis-orthogonal splits. In each step, the quality of the model is evaluated and the region with the worst local error measure is divided into two subsets. Splits in all dimensions of the input are tested and the one with the largest performance improvement is chosen. This successive partitioning stops when a specified error goal is met or when the desired number of rules is reached.

The first three steps of such an algorithm are illustrated in Figure 1.3. In this example, the algorithm starts with a single TS rule, i.e. a global linear model. In step 1,

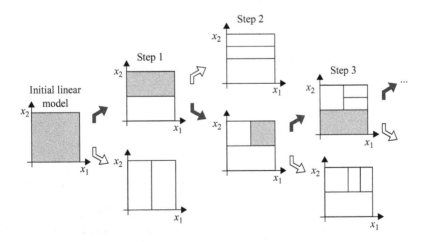

*Figure 1.3*    *Antecedent space decomposition by a heuristic search algorithm. The dark areas represent TS rules with the worst local fit in the given step*

the two possible splits are tested. The splitting of the $x_2$ domain gives a better result and is therefore used for further partitioning. The model performance is tested and the upper region is found to have the worst local fit to the data. In step 2, it is therefore further split into smaller subregions, etc.

An advantage of this approach is its effectiveness for high-dimensional data and the transparency of the obtained partition. A drawback is that the tree building procedure is suboptimal (greedy) and hence the number of rules obtained can be quite large [18].

### 1.4.3.3   Neuro-fuzzy learning

At the computational level, a fuzzy system can be seen as a layered structure (network), similar to artificial neural networks of the RBF type [19]. In order to optimise parameters in a fuzzy system, gradient-descent training algorithms known from the area of neural networks can be employed. Hence, this approach is usually referred to as neuro-fuzzy modelling [20–22]. As an example, consider a zero-order TS fuzzy model with the following two rules:

If $x_1$ is $A_{11}$ and $x_2$ is $A_{21}$ then $y = b_1$,

If $x_1$ is $A_{12}$ and $x_2$ is $A_{22}$ then $y = b_2$.

Figure 1.4 shows a network representation of these two rules. The nodes in the first layer compute the membership degree of the inputs in the antecedent fuzzy sets.

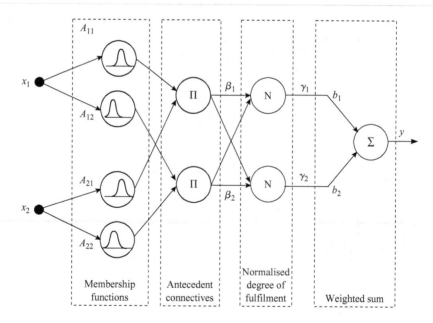

*Figure 1.4    An example of a zero-order TS fuzzy model with two rules represented as a neuro-fuzzy network*

The product nodes $\Pi$ in the second layer represent the antecedent connective (here the 'and' operator). The normalisation node N and the summation node $\sum$ realise the fuzzy-mean operator (1.9). This system is called ANFIS – adaptive neuro-fuzzy inference system [20].

Typically, smooth antecedent membership functions are used, such as the Gaussian functions:

$$\mu_{A_{ij}}(x_j; c_{ij}, \sigma_{ij}) = \exp\left(-\frac{(x_j - c_{ij})^2}{2\sigma_{ij}^2}\right),\tag{1.32}$$

for which the input–output equation (1.9) becomes:

$$y = \frac{\sum_{i=1}^{K} b_i \prod_{j=1}^{n} \exp(-(x_j - c_{ij})^2/2\sigma_{ij}^2)}{\sum_{i=1}^{K} \prod_{j=1}^{n} \exp(-(x_j - c_{ij})^2/2\sigma_{ij}^2)}.\tag{1.33}$$

The first-order TS fuzzy model can be represented in a similar fashion. Consider again an example with two rules

If $x_1$ is $A_{11}$ and $x_2$ is $A_{21}$ then $y_1 = a_{11}x_1 + a_{12}x_2 + b_1$,

If $x_1$ is $A_{12}$ and $x_2$ is $A_{22}$ then $y_2 = a_{21}x_1 + a_{22}x_2 + b_2$,

for which the corresponding network is given in Figure 1.5.

The input–output equation of this first-order TS model is:

$$y = \frac{\sum_{i=1}^{K} (\mathbf{a}_i^{\mathsf{T}}\mathbf{x} + b_i) \prod_{j=1}^{n} \exp(-(x_j - c_{ij})^2/2\sigma_{ij}^2)}{\sum_{i=1}^{K} \prod_{j=1}^{n} \exp(-(x_j - c_{ij})^2/2\sigma_{ij}^2)}.$$

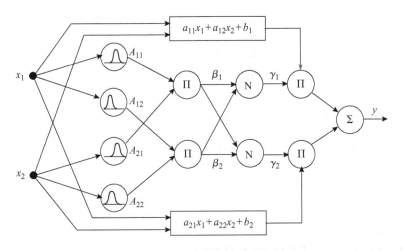

*Figure 1.5    An example of a first-order TS fuzzy model with two rules represented as a neuro-fuzzy network*

It is quite straightforward to derive the gradient-descent learning rule for the $b_i c_{ij}$ and $\sigma_{ij}$ parameters. The procedure is identical to the derivation of the backpropagation formulas for neural nets. Consider the zero-order ANFIS model, given by equation (1.33) and the quadratic cost function of the type (1.21). For the consequent parameters $b_i$, we have the Jacobian:

$$\frac{\partial J}{\partial b_i} = \frac{\partial J}{\partial y} \cdot \frac{\partial y}{\partial b_i} = -\gamma_i (y - \hat{y}) \tag{1.34}$$

and the gradient-descent update law:

$$b_i(n+1) = b_i(n) + \alpha(n)\gamma_i(y - \hat{y}), \tag{1.35}$$

where $\alpha(n)$ is the learning rate and $\hat{y}$ is the model output (using the hat to distinguish from the data $y$). For the centres and widths of the Gaussian membership functions (1.32) we apply the chain rule for differentiation and after some algebra, the following update formulas are obtained:

$$c_{ij}(n+1) = c_{ij}(n) + 2\alpha(n)\gamma_i(b_i - \hat{y})\frac{x_j - c_{ij}}{\sigma_{ij}^2}(y - \hat{y})$$

and

$$\sigma_{ij}(n+1) = \sigma_{ij}(n) + 2\alpha(n)\gamma_i(b_i - \hat{y})\frac{(x_j - c_{ij})^2}{\sigma_{ij}^3}(y - \hat{y}).$$

The parameter-update equations for the first-order ANFIS model can be derived in a similar fashion.

### 1.4.3.4   Fuzzy clustering

Construction methods based on fuzzy clustering originate from data analysis and pattern recognition, where the concept of fuzzy membership is used to represent the degree to which a given data object is similar to some prototypical object. The degree of similarity can be calculated by using a suitable distance measure. Based on the similarity, data vectors are clustered such that the data within a cluster are as similar as possible, and data from different clusters are as dissimilar as possible.

To identify TS models, the prototypes can be defined as linear subspaces or the clusters are ellipsoids with adaptively determined shape. Fuzzy clustering is applied in the Cartesian product space of the regressor and the output: $X \times Y$ to partition the data into subsets, which can be approximated by local linear models [9]. The data set $\mathbf{Z}$ to be clustered is formed by appending $\mathbf{y}$ to $\mathbf{X}$:

$$\mathbf{Z} = [\mathbf{X}, \mathbf{y}]^T. \tag{1.36}$$

Given $\mathbf{Z}$ and the desired number of clusters $K$, the Gustafson–Kessel (GK) algorithm [23] computes the fuzzy partition matrix, $\mathbf{U}$, the prototype matrix of cluster means, $\mathbf{V}$, and a set of cluster covariance matrices $\mathbf{F}$:

$$(\mathbf{Z}, K) \xrightarrow{\text{clustering}} (\mathbf{U}, \mathbf{V}, \mathbf{F}). \tag{1.37}$$

Given $(\mathbf{U}, \mathbf{V}, \mathbf{F})$, the antecedent membership functions $A_i$ and the consequent parameters $\mathbf{a}_i$, $b_i$ and $c_i$ are computed. From each cluster, one TS fuzzy rule is extracted. The membership functions $A_i$ can be computed analytically by using the distance of $\mathbf{x}(k)$ from the projection of the cluster centre $\mathbf{v}_i$ onto $X$, and then computing the membership degree in an inverse proportion to the distance. Denote $\mathbf{F}_i^x = [f_{jl}]$, $1 \leq j, l \leq n$ the sub-matrix of the cluster covariance matrix $\mathbf{F}_i$, to describe the shape of the cluster in the antecedent space $X$. Let $\mathbf{v}_i^x = [v_{1i}, \ldots, v_{ni}]^{\mathrm{T}}$ denote the projection of the cluster centre onto the antecedent space $X$. The inner-product distance norm,

$$d_{ik} = (\mathbf{x} - \mathbf{v}_i^x)^{\mathrm{T}} \det(\mathbf{F}_i^x)^{1/n} (\mathbf{x} - \mathbf{v}_i^x) \tag{1.38}$$

is converted into the membership degree by:

$$\mu_{A_i}(\mathbf{x}) = \frac{1}{\sum_{j=1}^{K} \left(d_{ik}/d_{jk}\right)^{2/(m-1)}}, \tag{1.39}$$

where $m$ is the fuzziness parameter of the GK algorithm. The antecedent membership functions can also be obtained by projecting the clusters point-wise onto the individual regressor variables [9].

The number of clusters directly determines the number of rules in the fuzzy model obtained. It can either be defined *a priori* or sought automatically. Two main approaches to automatically determine the appropriate number of clusters can be distinguished [9]:

1. Cluster the data for different values of $K$ and then use validity measures to assess the goodness of the obtained partitions. Different validity measures have been proposed in connection with adaptive distance clustering techniques [24].
2. Start with a sufficiently large number of clusters, and successively reduce this number by merging clusters that are similar (compatible) with respect to some predefined criteria [25,26]. One can also adopt an opposite approach, i.e. start with a small number of clusters and iteratively insert clusters in the regions where the data points have low degree of membership in the existing clusters [24].

Models obtained by fuzzy clustering may have some redundancy in terms of similar membership functions. Similarity-based reduction techniques have been developed to simplify the models [27,28].

## 1.5 Example: nonlinear identification of an air-conditioning system

Identification is used to obtain an accurate and compact fuzzy model of a heating, ventilating and air-conditioning (HVAC) system [29]. The purpose of this modelling exercise is to get a model with a high predictive power, suitable for model-based control. The considered air-conditioning system consists of a fan-coil unit placed inside a room (test cell), see Figure 1.6. Hot water at 65°C is supplied to the coil, which exchanges the heat between the hot water and the surrounding air. In the fan-coil unit, the air coming from the outside (primary air) is mixed with the return air

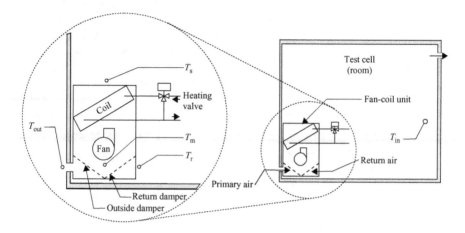

*Figure 1.6    The considered heating, ventilation and air-conditioning system*

from the room. The air flows are controlled by dampers and by the velocity of the fan, which forces the air to pass through the coil.

The control goal is to keep the temperature of the test cell $T_{in}$ at a certain reference value, ensuring that enough ventilation air is supplied to the room. In this example, the fan is kept at low velocity in order to minimise the noise level. Both dampers are half-open, allowing ventilation from the outside, and only the heating valve is used as a control input. As shown in Figure 1.6, temperatures can be measured at different parts of the fan-coil. The supply temperature $T_s$, measured just after the coil, is chosen as the most relevant temperature to control.

An initial TS fuzzy model of the system was constructed from process data. The inputs were selected on the basis of correlation analysis and a physical understanding of the process. The model predicts the supply air temperature $T_s$ based on its present and previous value, the mixed air temperature $T_m$ and the heating valve position $u$:

$$\mathbf{x}(k) = [T_s(k), T_s(k-1), u(k-1), T_m(k)]^T, \qquad y(k) = T_s(k+1). \qquad (1.40)$$

The model consists of ten rules, each with four antecedent fuzzy sets, of the form:

$$\mathcal{R}_i: \text{If } T_s(k) \text{ is } A_{i1} \text{ and } T_s(k-1) \text{ is } A_{i2} \text{ and } u(k-1) \text{ is } A_{i3} \text{ and } T_m(k) \text{ is } A_{i4}$$

$$\text{then } T_s(k+1) = \mathbf{a}_i \mathbf{x} + b_i.$$

The antecedent membership functions and the consequent parameters were estimated from a set of input–output measurements by using product-space clustering equation (1.37) and global least-square estimation equation (1.25), respectively. The identification data set contains $N = 800$ samples with a sampling period of 30 s. A separate data set was recorded on another day for validation purposes. The number of clusters was selected to be 10, based on cluster validity measures. The initial model thus contains 10 rules and a total of 40 antecedent fuzzy sets, shown in Figure 1.7(a).

Inspecting the initial HVAC process model, we notice that there are a lot of overlapping and similar fuzzy sets in the rule base (see Figure 1.7(a)). In order to

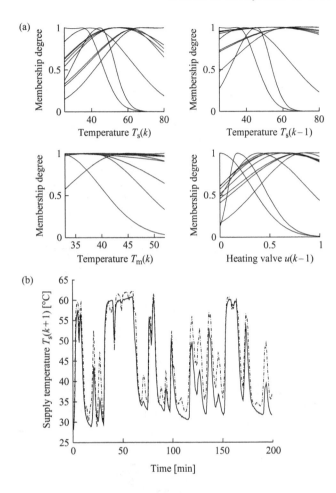

*Figure 1.7    Fuzzy sets used in the initial model (a) and the simplified model M2 (c), and their corresponding prediction of the validation data in a recursive simulation on a validation data set (b) and (d)*

reduce the complexity of the model, and thereby also the computation time, we apply a rule base simplification method [28].

From Figure 1.7 we can see that the accuracy of model M2 is just slightly lower than that of the initial model, even though, it is significantly reduced compared to the initial model. Model M2 consists of only four rules and nine different fuzzy sets. The antecedent of model M2 is given in Table 1.2.

The significant simplification of the rule base has made model M2 easy to inspect. When looking at the antecedent of the rules, we notice an interesting result for the antecedent variables $T_s(k)$ and $T_s(k-1)$. The partitioning of their domains are almost equal, as seen from the membership functions $A_1, A_2$ and $B_1, B_2$ in Figure 1.7(c). This suggests that one of the two variables could be removed from the model. If we

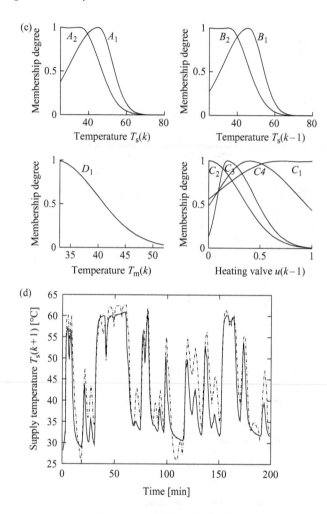

Figure 1.7    Continued

Table 1.2    *Antecedent of the simplified model M2*

| IF | $T_s(k)$ | $T_s(k-1)$ | $u(k-1)$ | $T_m(k)$ |
|---|---|---|---|---|
| $R_1$: | — | — | $C_1$ | — |
| $R_2$: | $A_1$ | $B_1$ | $C_2$ | — |
| $R_3$: | $A_2$ | $B_2$ | $C_3$ | $D_1$ |
| $R_4$: | — | — | $C_4$ | — |

remove the variable $T_s(k-1)$ from the antecedent of model M2 and re-estimate the consequent parameters using global least squares, we end up with the strongly reduced model M3. The antecedent of this model is given in Table 1.3.

The antecedent fuzzy sets of model M3 are the same as for model M2, shown in Figure 1.7(c), except that the variable $T_s(k-1)$ is no longer in the model. The validation of model M3 on predicting the validation data in a recursive simulation is shown in Figure 1.8(a). The error of model M3 is lower than that of the initial model. This indicates that the antecedent structure of the initial model was redundant.

To appreciate the performance of the fuzzy model, we compare its predictive power to a linear state–space model

$$\boldsymbol{\xi}(k+1) = \boldsymbol{\Phi}\boldsymbol{\xi}(k) + \boldsymbol{\Gamma}\mathbf{u}(k),$$
$$\mathbf{y}(k) = \mathbf{C}\boldsymbol{\xi}(k), \tag{1.41}$$

obtained through subspace identification [30]. This fifth-order model has one input, the heating valve, and two outputs, the mixed air temperature $T_m$ and the supply temperature $T_s$. Figure 1.8(b) compares the supply temperature measured and

*Table 1.3    Antecedent of the simplified model M3*

| IF | $T_s(k)$ | $u(k-1)$ | $T_m(k)$ |
|---|---|---|---|
| $R_1$: | — | $C_1$ | — |
| $R_2$: | $A_1$ | $C_2$ | — |
| $R_3$: | $A_2$ | $C_3$ | $D_1$ |
| $R_4$: | — | $C_4$ | — |

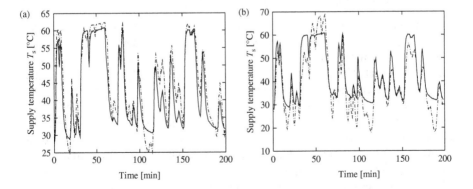

*Figure 1.8    Comparison with a linear model (solid line – measured output, dashed line – model output). (a) TS fuzzy model M3. (b) Linear state–space model*

predicted by the linear model. One can see that the performance of the linear model is considerably worse than in the case of the fuzzy model.

## 1.6    Control based on fuzzy models

An available fuzzy or neuro-fuzzy model can be used in the design of a controller or can become a part of a model-based control scheme. The main distinction can be made between feedback methods (such as state-feedback control or gain-scheduled PID control) and techniques that primarily rely on a feedforward component (such as model-inverse or predictive control). In this section, we briefly address examples of methods from each class.

### 1.6.1    Fuzzy gain-scheduled control

Even with the introduction of powerful nonlinear control strategies such as feedback linearisation or model predictive control, gain scheduling remains attractive mainly because of its conceptual simplicity and practical usefulness. In classical gain scheduling, slow varying scheduling variables are used to capture nonlinearities and parameter dependencies. The control law is obtained by interpolating a number of locally valid linear controllers. In the context of fuzzy systems, gain-scheduled control is obtained when using the TS fuzzy controller, usually designed on the basis of a TS model of the plant, represented by the following set of rules:

$$\text{If } \mathbf{z}(k) \text{ is } A_i \text{ then } y(k) = H_i u(k), \quad i = 1, 2, \ldots, K, \tag{1.42}$$

where $H_i$ denotes a locally valid linear time-invariant model, e.g. a linear transfer function (1.15) or a state–space model (1.16), and $\mathbf{z}(k) \in D \subset \mathbb{R}^{n_z}$ is the vector of scheduling variables. The corresponding fuzzy gain-scheduled controller consists of a similar set of rules:

$$\text{If } \mathbf{z}(k) \text{ is } B_i \text{ then } u(k) = C_i y(k), \quad i = 1, 2, \ldots, K, \tag{1.43}$$

where $C_i$ is a linear time-invariant controller. Once the structures of $H_i$ and $C_i$ are given, for gain-scheduling one can represent these dynamic systems as vectors of their real-valued parameters. It is reasonable to assume that the fuzzy model is in some kind of 'minimal' form, i.e. no rules are redundant or linearly dependent on other rules. The number of controller rules that are needed for the accurate parameterisation of the controller's nonlinearity is then equal to the number of rules in the model.

In the vast majority of fuzzy gain-scheduled control design approaches, it is assumed that the model membership functions are also used in the controller, i.e. $A_i = B_i, \forall i$. The control design then boils down to the computation of the local controller parameters by using some design method (e.g. pole placement, robust control design, LMI optimisation, etc.). However, sharing the model membership functions results in suboptimal closed-loop behaviour, as demonstrated in Reference 31, where a method to the design of optimal controller's membership functions is also given.

## 1.6.2   Parallel distributed compensation

Recently, much attention has been focused on the development of design methods for fuzzy control rules that guarantee closed-loop stability and robustness, regardless of the rate of change of the scheduling variables $\mathbf{z}$. This framework was initially termed 'parallel distributed compensation' (PDC) [32]. An offset-free case of the state–space TS fuzzy model (1.16) is considered

$$\mathcal{R}_i: \text{ If } \mathbf{z}(k) \text{ is } A_i \text{ then } \begin{cases} \boldsymbol{\xi}(k+1) = \boldsymbol{\Phi}_i \boldsymbol{\xi}(k) + \boldsymbol{\Gamma}_i \mathbf{u}(k), \\ \mathbf{y}(k) = \mathbf{C}_i \boldsymbol{\xi}(k). \end{cases} \quad i = 1, 2, \ldots, K,$$

$$(1.44)$$

along with the following state-feedback controller:

$$\mathcal{R}_i: \text{ If } \mathbf{z}(k) \text{ is } A_i \text{ then } \mathbf{u}(k) = -\mathbf{L}_i \boldsymbol{\xi}(k), \quad i = 1, 2, \ldots, K. \tag{1.45}$$

The antecedent variables $\mathbf{z}$ are assumed to be independent of the control input $\mathbf{u}$. Given the standard weighted mean interpolation (1.9), the closed-loop system is given by:

$$\boldsymbol{\xi}(k+1) = \sum_{i=1}^{K} \gamma_i(\mathbf{z}(k))\gamma_i(\mathbf{z}(k))\mathbf{G}_{ii}\boldsymbol{\xi}(k)$$

$$+ \frac{1}{2} \sum_{i=1}^{K} \sum_{j>i}^{K} \gamma_i(\mathbf{z}(k))\gamma_j(\mathbf{z}(k))(\mathbf{G}_{ij} + \mathbf{G}_{ji})\boldsymbol{\xi}(k)$$

with $\mathbf{G}_{ij} = \boldsymbol{\Phi}_i - \boldsymbol{\Gamma}_i \mathbf{L}_j$. The equilibrium of the closed-loop system (1.46) is asymptotically stable in the large if there exists a common positive definite matrix $\mathbf{P}$ such that

$$\mathbf{G}_{ii}^{\mathrm{T}} \mathbf{P} \mathbf{G}_{ii} - \mathbf{P} < 0, \tag{1.46}$$

$$\left( \frac{\mathbf{G}_{ij} + \mathbf{G}_{ji}}{2} \right)^{\mathrm{T}} \mathbf{P} \left( \frac{\mathbf{G}_{ij} + \mathbf{G}_{ji}}{2} \right) - \mathbf{P} \leq 0, \quad i < j, \tag{1.47}$$

for all $i$ and $j$ except the pairs $(i, j)$ such that $\gamma_i(\mathbf{z}(k))\gamma_i(\mathbf{z}(k)) = 0$, $\forall k$. The above inequalities can be checked or a stable fuzzy controller can be automatically designed by using LMI (linear matrix inequality) techniques [33,34]. In this framework, the scheduling variables $\mathbf{z}(k)$ are no longer required to vary slowly with respect to the plant dynamics (as opposed to classical gain scheduling). The price one has to pay is the conservatism of the approach (i.e. stability often cannot be proven for systems that actually are stable). Consequently, much research effort is currently devoted to reducing the conservatism, see, e.g. References 34–36 among many other references. Similar developments can also be found in the classical automatic control literature [37,38].

### 1.6.3    Inverse model control

A straightforward approach to model-based design of a controller for a nonlinear process is inverse control. It can be applied to a class of systems that are open-loop stable (or that are stabilisable by feedback) and whose inverse is stable as well, i.e. the system does not exhibit nonminimum phase behaviour. The approach here is explained for SISO models without transport delays from the input to the output:

$$y(k+1) = f(\mathbf{x}(k), u(k)). \tag{1.48}$$

In this section, the vector $\mathbf{x}(k) = [y(k), \ldots, y(k - n_1 + 1), u(k - 1), \ldots, u(k - m_1 + 1)]^{\mathrm{T}}$ denotes the actual state and thus it does not include the current input $u(k)$. The objective of inverse control is to compute for the current state $\mathbf{x}(k)$ the control input $u(k)$, such that the system's output at the next sampling instant is equal to the desired (reference) output $r(k + 1)$. This can be achieved if the process model (1.48) can be inverted according to:

$$u(k) = f^{-1}(\mathbf{x}(k), r(k + 1)). \tag{1.49}$$

Generally, it is difficult to find the inverse function $f^{-1}$ in an analytical form. It can, however, always be found by numerical optimisation, using the following objective function:

$$J(u(k)) = [r(k + 1) - f(\mathbf{x}(k), u(k))]^2. \tag{1.50}$$

The minimisation of $J$ with respect to $u(k)$ gives the control corresponding to the inverse function (1.49), if it exists, or the best least-square approximation of it otherwise. A wide variety of optimisation techniques can be applied (such as Newton or Levenberg–Marquardt). This approach directly extends to MIMO systems. Its main drawback, however, is the computational complexity due to the numerical optimisation that must be carried out online.

Some special forms of (1.48) can be inverted analytically. Examples are the input-affine TS model and a singleton model with triangular membership functions for $u(k)$, as discussed below.

#### 1.6.3.1    Inverse of the affine TS model

Consider the following input–output TS fuzzy model:

$\mathcal{R}_i$: If $y(k)$ is $A_{i1}$ and, $\ldots$, and $y(k - n_1 + 1)$ is $A_{in_1}$ and

$\quad u(k - 1)$ is $B_{i2}$ and, $\ldots$, and $u(k - m_1 + 1)$ is $B_{im_1}$ then

$$y_i(k+1) = \sum_{j=1}^{n_1} a_{ij} y(k - j + 1) + \sum_{j=1}^{m_1} b_{ij} u(k - j + 1) + c_i, \quad i = 1, \ldots, K.$$

$$\tag{1.51}$$

As the antecedent does not include the input term $u(k)$, the model output $y(k+1)$ is affine linear in the input $u(k)$:

$$y(k+1) = \sum_{i=1}^{K} \gamma_i(\mathbf{x}(k)) \left[ \sum_{j=1}^{n_1} a_{ij} y(k-j+1) + \sum_{j=2}^{m_1} b_{ij} u(k-j+1) + c_i \right]$$
$$+ \sum_{i=1}^{K} \gamma_i(\mathbf{x}(k)) b_{i1} u(k). \tag{1.52}$$

This is a nonlinear input-affine system which can in general terms be written as:

$$y(k+1) = g(\mathbf{x}(k)) + h(\mathbf{x}(k))u(k). \tag{1.53}$$

Given the goal that the model output at time step $k+1$ should equal the reference output, $y(k+1) = r(k+1)$, the corresponding input, $u(k)$, is obtained:

$$u(k) = \frac{r(k+1) - g(\mathbf{x}(k))}{h(\mathbf{x}(k))}. \tag{1.54}$$

In terms of equation (1.52) we have the eventual inverse-model control law:

$$u(k) = \left( r(k+1) - \sum_{i=1}^{K} \gamma_i(\mathbf{x}(k)) \left[ \sum_{j=1}^{n_1} a_{ij} y(k-j+1) \right. \right.$$
$$\left. \left. + \sum_{j=2}^{m_1} b_{ij} u(k-j+1) + c_i \right] \right) \left( \sum_{i=1}^{K} \gamma_i(\mathbf{x}(k)) b_{i1} \right)^{-1}. \tag{1.55}$$

### 1.6.3.2 Inverse of the singleton model

Consider a SISO singleton fuzzy model with the rules:

$$\text{If } \mathbf{x}(k) \text{ is } A_i \text{ and } u(k) \text{ is } B_i \text{ then } y(k+1) \text{ is } c_i. \tag{1.56}$$

Denote by $M$ the number of fuzzy sets $A_i$ defined for the state $\mathbf{x}(k)$ and $N$ the number of fuzzy sets $B_j$ defined for the input $u(k)$. Assume that the rule base consists of all possible combinations of sets $A_i$ and $B_j$, the total number of rules is then $K = MN$. The entire rule base can be represented as a table:

| $\mathbf{x}(k)$ | $u(k)$ | | | |
|---|---|---|---|---|
| | $B_1$ | $B_2$ | $\cdots$ | $B_N$ |
| $A_1$ | $c_{11}$ | $c_{12}$ | $\cdots$ | $c_{1N}$ |
| $A_2$ | $c_{21}$ | $c_{22}$ | $\cdots$ | $c_{2N}$ |
| $\vdots$ | $\vdots$ | $\vdots$ | $\vdots$ | $\vdots$ |
| $A_M$ | $c_{M1}$ | $c_{M2}$ | $\cdots$ | $c_{MN}$ |

(1.57)

The degree of fulfilment $\beta_{ij}(k)$ is computed by using the product operator:

$$\beta_{ij}(k) = \mu_{A_i}(\mathbf{x}(k)) \cdot \mu_{B_j}(u(k)). \tag{1.58}$$

The output $y(k+1)$ of the model is computed by the weighted mean:

$$
\begin{aligned}
y(k+1) &= \frac{\sum_{i=1}^{M} \sum_{j=1}^{N} \beta_{ij}(k) \cdot c_{ij}}{\sum_{i=1}^{M} \sum_{j=1}^{N} \beta_{ij}(k)} \\
&= \frac{\sum_{i=1}^{M} \sum_{j=1}^{N} \mu_{A_i}(\mathbf{x}(k)) \cdot \mu_{B_j}(u(k)) \cdot c_{ij}}{\sum_{i=1}^{M} \sum_{j=1}^{N} \mu_{A_i}(\mathbf{x}(k)) \cdot \mu_{B_j}(u(k))}.
\end{aligned}
\tag{1.59}
$$

The inversion method requires that the antecedent membership functions $\mu_{B_j}(u(k))$ are triangular and form a partition, i.e. fulfil:

$$\sum_{j=1}^{N} \mu_{B_j}(u(k)) = 1. \tag{1.60}$$

The basic idea is that for each particular state $\mathbf{x}(k)$, the multivariate mapping equation (1.48) is reduced to a univariate mapping $y(k+1) = f_x(u(k))$, where the subscript $x$ denotes that $f_x$ is obtained for the particular state $\mathbf{x}$. From this mapping, which is piece-wise linear, the inverse mapping $u(k) = f_x^{-1}(r(k+1))$ can be easily found, provided the model is invertible, which is easy to check, see Reference 9. Using equation (1.60), the output equation of the model (1.59) simplifies to:

$$
\begin{aligned}
y(k+1) &= \frac{\sum_{i=1}^{M} \sum_{j=1}^{N} \mu_{A_i}(\mathbf{x}(k)) \cdot \mu_{B_j}(u(k)) \cdot c_{ij}}{\sum_{i=1}^{M} \sum_{j=1}^{N} \mu_{A_i}(\mathbf{x}(k)) \mu_{B_j}(u(k))} \\
&= \sum_{j=1}^{N} \mu_{B_j}(u(k)) \sum_{i=1}^{M} \gamma_i(\mathbf{x}(k)) \cdot c_{ij},
\end{aligned}
\tag{1.61}
$$

where $\gamma_i(\mathbf{x}(k))$ is the normalised degree of fulfilment of the state part of the antecedent:

$$\gamma_i(\mathbf{x}(k)) = \frac{\mu_{A_i}(\mathbf{x}(k))}{\sum_{j=1}^{K} \mu_{A_j}(\mathbf{x}(k))}. \tag{1.62}$$

As the state $\mathbf{x}(k)$ is available, the latter summation in equation (1.61) can be evaluated, yielding:

$$y(k+1) = \sum_{j=1}^{N} \mu_{B_j}(u(k)) c_j, \tag{1.63}$$

where

$$c_j = \sum_{i=1}^{M} \gamma_i(\mathbf{x}(k)) \cdot c_{ij}. \tag{1.64}$$

This is an equation of a singleton model with one input $u(k)$ and one output $y(k + 1)$:

$$\text{If } u(k) \text{ is } B_j \text{ then } y(k + 1) \text{ is } c_j(k), \quad j = 1, \ldots, N. \tag{1.65}$$

Each of the above rules is inverted by exchanging the antecedent and the consequent:

$$\text{If } r(k + 1) \text{ is } c_j(k) \text{ then } u(k) \text{ is } B_j, \quad j = 1, \ldots, N. \tag{1.66}$$

Here, the reference $r(k+1)$ was substituted for $y(k+1)$. Since $c_j(k)$ are singletons, it is necessary to interpolate between the consequents $c_j(k)$ in order to obtain $u(k)$. This interpolation is accomplished by fuzzy sets $C_j$ with triangular membership functions:

$$\mu_{C_1}(r) = \max\left(0, \min\left(1, \frac{c_2 - r}{c_2 - c_1}\right)\right), \tag{1.67}$$

$$\mu_{C_j}(r) = \max\left(0, \min\left(\frac{r - c_{j-1}}{c_j - c_{j-1}}, \frac{c_{j+1} - r}{c_{j+1} - c_j}\right)\right), \quad 1 < j < N, \tag{1.68}$$

$$\mu_{C_N}(r) = \max\left(0, \min\left(\frac{r - c_{N-1}}{c_N - c_{N-1}}, 1\right)\right). \tag{1.69}$$

The output of the inverse controller is thus given by:

$$u(k) = \sum_{j=1}^{N} \mu_{C_j}(r(k + 1))b_j, \tag{1.70}$$

where $b_j$ are the cores of $B_j$. The inversion is thus given by equations (1.62), (1.69) and (1.70). It can be verified that the series connection of the controller and the inverse model gives an identity mapping (perfect control)

$$y(k + 1) = f_x(u(k)) = f_x(f_x^{-1}(r(k + 1))) = r(k + 1), \tag{1.71}$$

when $u(k)$ exists such that $r(k + 1) = f(\mathbf{x}(k), u(k))$. When no such $u(k)$ exists, the difference

$$|r(k + 1) - f_x(f_x^{-1}(r(k + 1)))|$$

is the least possible. The proof can be found in Reference 9. Apart from the computation of the membership degrees, both the model and the controller can be implemented using standard matrix operations and linear interpolations, which makes the algorithm suitable for real-time implementation.

The invertibility of the fuzzy model can be checked in run-time, by checking the monotonicity of the aggregated consequents $c_j$ with respect to the cores of the input fuzzy sets $b_j$. This is useful, since nonlinear models can be noninvertible only locally, resulting in a kind of exception in the inversion algorithm. Moreover, for models adapted on line, this check is necessary.

For a noninvertible rule base, a set of possible control commands can be found by splitting the rule base into two or more invertible parts. For each part, a control action is found by inversion. Among these control actions, only one has to be selected, which requires some additional criteria, such as minimal control effort (minimal $u(k)$ or $|u(k) - u(k - 1)|$, for instance).

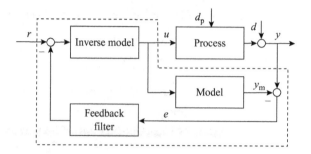

*Figure 1.9    IMC scheme*

### 1.6.4    Internal model control

Disturbances acting on the process, measurement noise and model-plant mismatch cause differences in the behaviour of the process and of the model. In open-loop inverse-model control, this results in an error between the reference and the process output. The internal model control (IMC) scheme [39] is one way of compensating for this error. Figure 1.9 depicts the IMC scheme, which consists of three parts: the controller based on an inverse of the process model, the model itself and a feedback filter. The control system (dashed box) has two inputs, the reference and the measurement of the process output, and one output, the control action.

The purpose of the process model working in parallel with the process is to subtract the effect of the control action from the process output. If the predicted and the measured process outputs are equal, the error $e$ is zero and the controller works in an open-loop configuration. If a disturbance $d$ acts on the process output, the feedback signal $e$ is equal to the influence of the disturbance and is not affected by the effects of the control action. This signal is subtracted from the reference. With a perfect process model, the IMC scheme is hence able to cancel the effect of unmeasured output-additive disturbances. The feedback filter is introduced in order to filter out the measurement noise and to stabilise the loop by reducing the loop gain for higher frequencies.

### 1.6.5    Model-based predictive control

Model-based predictive control (MBPC) is a general methodology for solving control problems in the time domain. It is based on three main concepts:

1. A model is used to predict the process output at future discrete-time instants, over a prediction horizon.
2. A sequence of future control actions is computed over a control horizon by minimising a given objective function.
3. Only the first control action in the sequence is applied, the horizons are moved towards the future and optimisation is repeated. This is called the receding horizon principle.

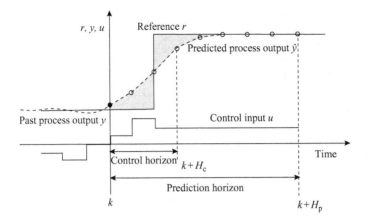

*Figure 1.10     The basic principle of MBPC*

Because of the optimisation approach and the explicit use of the process model, MBPC can realise multivariable optimal control, deal with nonlinear processes and can efficiently handle constraints. A predictive controller can be directly used in the IMC scheme Figure 1.9, instead of the inverse controller.

### 1.6.5.1   Prediction and control horizons

The future process outputs are predicted over the prediction horizon $H_p$ using a model of the process. The predicted output values, denoted $\hat{y}(k+i)$ for $i = 1, \ldots, H_p$, depend on the state of the process at the current time $k$ and on the future control signals $u(k+i)$ for $i = 0, \ldots, H_c - 1$, where $H_c \leq H_p$ is the control horizon. The control signal is manipulated only within the control horizon and remains constant afterwards, i.e. $u(k+i) = u(k + H_c - 1)$ for $i = H_c, \ldots, H_p - 1$, see Figure 1.10.

### 1.6.5.2   Objective function and optimisation algorithms

The sequence of future control signals $\mathbf{u}(k+i)$ for $i = 0, 1, \ldots, H_c - 1$ is usually computed by optimising the following quadratic cost function [40]:

$$J = \sum_{i=1}^{H_p} \|(\mathbf{r}(k+i) - \hat{\mathbf{y}}(k+i))\|_{\mathbf{P}_i}^2 + \sum_{i=1}^{H_c} \|(\Delta\mathbf{u}(k+i-1))\|_{\mathbf{Q}_i}^2. \tag{1.72}$$

The first term accounts for minimising the variance of the process output from the reference, while the second term represents a penalty on the control effort (related, for instance, to energy). The latter term can also be expressed by using $u$ itself. $\mathbf{P}_i$ and $\mathbf{Q}_i$ are positive definite weighting matrices that specify the importance of two terms in equation (1.72) relative to each other and to the prediction step. Additional terms can be included in the cost function to account for other control criteria.

For systems with a dead time of $n_d$ samples, only outputs from time $k + n_d$ are considered in the objective function, because outputs before this time cannot be influenced by the control signal $u(k)$. Similar reasoning holds for nonminimum phase systems.

Hard constraints, e.g. level and rate constraints of the control input, process output, or other process variables can be specified as a part of the optimisation problem. The minimisation of (1.72) generally requires nonlinear (non-convex) optimisation for which no general effective methods are available. Various solutions have been proposed in the literature:

*Iterative optimisation algorithms.* These include Nelder–Mead, sequential quadratic programming, etc. For longer control horizons ($H_c$), these algorithms usually converge to local minima, which results in a poorly performing predictive controller. A partial remedy is to find a good initial solution, for instance, by grid search [41]. This, however, is only efficient for small-size problems.

*Discrete search techniques.* These include dynamic programming, branch-and-bound methods, genetic algorithms (GAs) and other randomised algorithms. A drawback of these methods is that the control signal domain typically has to be discretised and that the computation time can become rather long.

*Taylor linearisation.* The nonlinear model is linearised at each sampling instant and used in a standard linear predictive control scheme [42,43]. Linearisation can be applied at the current time step $k$ or along the predicted trajectory. These methods yield approximate solutions.

*Feedback linearisation.* This method yields an exact linearised model with time-invariant dynamics, for which the predictive controller is easy to tune. However, feedback linearisation transforms input constraints in a nonlinear way, which rules out a direct application of quadratic programming. Some solutions to this problem have been suggested [44,45].

### 1.6.5.3    Receding horizon principle

Only the control signal $u(k)$ is applied to the process. At the next sampling instant, the process output $y(k + 1)$ is available and the optimisation and prediction can be repeated with the updated values. This is called the receding horizon principle. The control action $u(k + 1)$ computed at time step $k + 1$ will be generally different from the one calculated at time step $k$, since more up-to-date information about the process is available.

## 1.7    Example: nonlinear predictive control of an air-conditioning system

A nonlinear predictive controller for the air-conditioning system introduced in Section 1.5 is described. Through nonlinear identification, a reasonably accurate TS fuzzy model was constructed. The predictive controller employs a branch-and-bound optimisation method [29] and is incorporated in the IMC scheme of Figure 1.11 to compensate for modelling errors and disturbances.

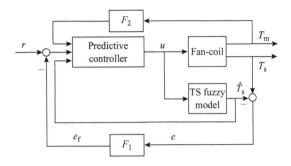

*Figure 1.11    The fuzzy predictive control scheme for the fan-coil unit*

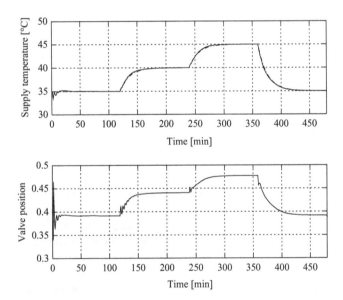

*Figure 1.12    Simulated response of the predictive controller based on the TS model.
Solid line is the measured output, dashed line is the reference*

The controller's inputs are the set-point, the predicted supply temperature $\hat{T}_s$, and the filtered mixed-air temperature $T_m$. The error signal, $e(k) = T_s(k) - \hat{T}_s(k)$, is passed through a first-order low-pass digital filter $F_1$. A similar filter $F_2$ is used to filter $T_m$. Both filters were designed as Butterworth filters, the cut-off frequency was adjusted empirically, based on simulations, in order to reliably filter out the measurement noise, and to provide a fast response.

The results obtained by using the TS fuzzy model and the linear model within the IMC scheme are first compared in simulation. The TS fuzzy model simulates the process. Prediction and control horizons are set to $H_c = H_p = 5$. The upper graph in Figure 1.12 presents a response of the supply temperature to several steps in the reference for the fuzzy predictive controller. The corresponding control actions are

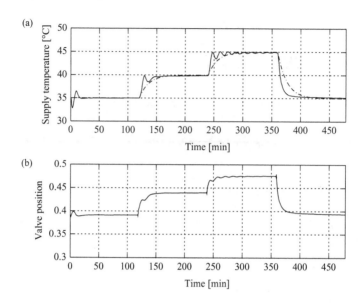

*Figure 1.13*    *Simulated response of the predictive controller based on a linear model. Solid line is the measured output, dashed line is the reference*

given in the lower graph. Figure 1.13 shows the same signals for the predictive controller based on the linear model.

Both controllers are able to follow the reference without steady-state errors, but the response of the fuzzy controller is much smoother than that of the controller based on the linear model. The reason is that in the latter case, the IMC scheme has to compensate severe model-plant mismatches, which causes the oscillations observed in Figure 1.13.

The fuzzy controller was also tested in real time in the test cell. Figure 1.14 shows some of the real-time results obtained for $H_c = 2$ and $H_p = 4$. Contrary to the simulations, this setting gave the best results for the experimental process. This can be attributed to a model-plant mismatch which leads to inaccurate predictions for a longer horizon.

## 1.8    Concluding remarks

Fuzzy modelling is a flexible framework in which different paradigms can be combined, providing, on the one hand, a transparent interface with the designer and, on the other hand, a tool for accurate nonlinear modelling and control. The rule-based character of neuro-fuzzy models allows for the analysis and interpretation of the result. Conventional methods for numerical validation can be complemented by human expertise, which often involves heuristic knowledge and intuition.

A drawback of fuzzy modelling is that the current techniques for constructing and tuning fuzzy models are rather complex, and their use requires specific skills

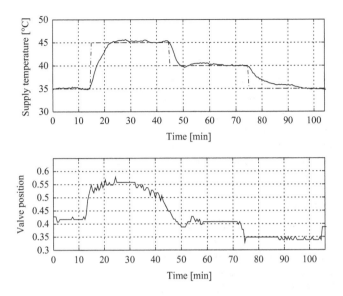

*Figure 1.14    Real-time response of the air-conditioning system. Solid line is the measured output, dashed line is the reference*

and knowledge. In this sense, neuro-fuzzy modelling will probably never become a 'one-button', fully automated identification technique. It should rather be seen as an interactive method, facilitating the active participation of the user in a computer-assisted modelling session. This holds, to a certain degree, also for linear identification methods. Modelling of complex systems will always remain an interactive approach.

Techniques are available to develop nonlinear controllers based on a fuzzy model of the process under consideration. They can be classified into feedback methods (such as state-feedback control or gain-scheduled control) and techniques that primarily rely on feedforward compensation (such as model-inverse or predictive control).

## Further Reading

More details on the different methods and tools can be found in References 1,21,22 and 46, among others. A large number of works are being regularly published in fuzzy systems oriented journals (*IEEE Transactions on Fuzzy Systems, Fuzzy Sets and Systems*) and also *IEEE Transactions on Systems Man and Cybernetics*. MATLAB tools (Fuzzy Modelling and Identification Toolbox) used to generate some of the results in this chapter can be found at http://www.dcsc.tudelft.nl/~babuska.

## Notes

1.   Good generalisation means that a model fitted to one data set will also perform well on another data set from the same process.

## References

1  HELLENDOORN, H., and DRIANKOV, D. (Eds): 'Fuzzy model identification: selected approaches' (Springer, Berlin, Germany, 1997)
2  NELLES, O.: 'Nonlinear system identification' (Springer, New York, 2000)
3  MAERTENS, K., JOHANSEN, T. A., and BABUŠKA, R.: 'Engine load prediction in off-road vehicles using multi-objective nonlinear identification', *Control Engineering Practice*, 2004, **12** (5), pp. 615–24
4  SETNES, M., BABUŠKA, R., and VERBRUGGEN, H. B.: 'Rule-based modeling: precision and transparency', *IEEE Transactions on Systems, Man, and Cybernetics, Part C: Applications and Reviews*, 1998, **28** (1), pp. 165–9
5  de OLIVEIRA, J. V., and VALENTE, J.: 'Semantic constraints for membership function optimisation', *IEEE Transactions on Fuzzy Systems*, 1999, **19** (1), pp. 128–38
6  POMARES, H., ROJAS, I., ORTEGA, J., GONZALEZ, J., and PRIETO, A.: 'A systematic approach to a self-generating fuzzy rule-table for function approximation', *IEEE Transactions on Systems, Man, and Cybernetics – Part B: Cybernetics*, 2000, **30** (3), pp. 431–47
7  JIN, Y.: 'Fuzzy modeling of high-dimensional systems: complexity reduction and interpretability improvement', *IEEE Transactions on Fuzzy Systems*, 2000, **8**, pp. 212–22
8  JOHANSEN, T. A., SHORTEN, R., and MURRAY-SMITH, R.: 'On the interpretation and identification of dynamic Takagi–Sugeno fuzzy models', *IEEE Transactions on Fuzzy Systems*, 2000, **8**, pp. 297–313
9  BABUŠKA, R.: 'Fuzzy modeling for control' (Kluwer Academic Publishers, Boston, MA, 1998)
10 ZADEH, L. A.: 'Outline of a new approach to the analysis of complex systems and decision processes', *IEEE Transactions on Systems, Man and Cybernetics*, 1973, **1**, pp. 28–44
11 de OLIVEIRA, J. V., and VALENTE, J.: 'Toward neuro–linguistic modeling constraint for optimization of membership functions', *Fuzzy Sets and Systems*, 1999, **106** (3), pp. 357–80
12 CORDÓN, O., and HERRERA, F.: 'A proposal for improving the accuracy of linguistic modeling', *IEEE Transactions on Fuzzy Systems*, 2000, **8** (3), pp. 335–44
13 TAKAGI, T., and SUGENO, M.: 'Fuzzy identification of systems and its application to modeling and control', *IEEE Transactions on Systems, Man and Cybernetics*, 1985, **15** (1), pp. 116–32
14 TULLEKEN, H. J. A. F.: 'Gray-box modelling and identification using physical knowledge and Bayesian techniques', *Automatica*, 1993, **29**, pp. 285–308
15 ABONYI, J., BABUŠKA, R., VERBRUGGEN, H. B., and SZEIFERT, F.: 'Incorporating prior knowledge in fuzzy model identification', *International Journal of Systems Science*, 2000, **31** (5), pp. 657–67

16 YEN, J., WANG, L., and GILLESPIE, C. W.: 'Improving the interpretability of TSK fuzzy models by combining global learning and local learning', *IEEE Transactions on Fuzzy Systems*, 1998, **6**, pp. 530–7

17 JOHANSEN, T. A., and BABUŠKA, R.: 'Multi-objective identification of Takagi–Sugeno fuzzy models', *IEEE Transactions on Fuzzy Systems*, 2003, **11** (6), pp. 847–60

18 NELLES, O., FINK, A., BABUŠKA, R., and SETNES, M.: 'Comparison of two construction algorithms for Takagi–Sugeno fuzzy models', *International Journal of Applied Mathematics and Computer Science*, 2000, **10** (4), pp. 835–55

19 JANG, J.-S. R., and SUN, C.-T.: 'Functional equivalence between radial basis function networks and fuzzy inference systems', *IEEE Transactions on Neural Networks*, 1993, **4** (1), pp. 156–9

20 JANG, J.-S. R.: 'ANFIS: adaptive-network-based fuzzy inference systems', *IEEE Transactions on Systems, Man and Cybernetics*, 1993, **23** (3), pp. 665–85

21 BROWN, M., and HARRIS, C.: 'Neurofuzzy adaptive modelling and control' (Prentice Hall, New York, 1994)

22 JANG, J.-S. R., SUN, C.-T., and MIZUTANI, E.: 'Neuro-fuzzy and soft computing; a computational approach to learning and machine intelligence' (Prentice-Hall, Upper Saddle River, NJ, 1997)

23 GUSTAFSON, D. E., and KESSEL, W. C.: 'Fuzzy clustering with a fuzzy covariance matrix'. Proceedings of IEEE CDC, San Diego, CA, 1979, pp. 761–6

24 GATH, I., and GEVA, A. B.: 'Unsupervised optimal fuzzy clustering', *IEEE Transaction of Pattern Analysis and Machine Intelligence*, 1989, **7**, pp. 773–81

25 KRISHNAPURAM, R., and FREG, C.-P.: 'Fitting an unknown number of lines and planes to image data through compatible cluster merging', *Pattern Recognition*, 1992, **25** (4), pp. 385–400

26 KAYMAK, U., and BABUŠKA, R.: 'Compatible cluster merging for fuzzy modeling'. Proceedings of FUZZ-IEEE/IFES'95, Yokohama, Japan, 1995, pp. 897–904

27 SETNES, M., BABUŠKA, R., and VERBRUGGEN, H. B.: 'Transparent fuzzy modeling: data exploration and fuzzy sets aggregation', *International Journal of Human Computer Studies*, 1998, **49** (2), pp. 159–79

28 SETNES, M., BABUŠKA, R., KAYMAK, U., and van NAUTA LEMKE, H. R.: 'Similarity measures in fuzzy rule base simplification', *IEEE Transactions on Systems, Man, and Cybernetics*, 1998, **28** (3), pp. 376–86

29 SOUSA, J. M., BABUŠKA, R., and VERBRUGGEN, H. B.: 'Fuzzy predictive control applied to an air-conditioning system', *Control Engineering Practice*, 1997, **5** (10), pp. 1395–406

30 VERHAEGEN, M., and DEWILDE, P.: 'Subspace model identification. Part I: the output-error state space model identification class of algorithms', *International Journal of Control*, 1992, **56**, pp. 1187–210

31 BABUŠKA, R., and OOSTEROM, M.: 'Design of optimal membership functions for fuzzy gain-scheduled control'. Proceedings of IEEE International Conference on *Fuzzy systems*, St. Louis, MO, 2003, pp. 476–81

32  WANG, H. O., TANAKA, K., and GRIFFIN, M. F.: 'An approach to fuzzy control of nonlinear systems: stability and design issues', *IEEE Transactions on Fuzzy Systems*, 1996, **4**, pp. 14–23

33  BOYD, S. P., GHAOUI, L. E., and FERON, E.: 'Linear matrix inequalities in systems and control theory' (SIAM, Philadelphia, PA, 1994)

34  TANAKA, K., IKEDA, T., and WANG, H. O.: 'Fuzzy regulators and fuzzy observers: relaxed stability conditions and LMI-based designs', *IEEE Transactions on Fuzzy Systems*, 1998, **6** (2), pp. 250–65

35  KIM, E., and LEE, H.: 'New approaches to relaxed quadratic stability condition of fuzzy control systems', *IEEE Transactions on Fuzzy Systems*, 2000, **8** (5), pp. 523–34

36  GUERRA, T. M., and VERMEIREN, L.: 'Control laws for Takagi–Sugeno fuzzy models', *Fuzzy Sets and Systems*, 2001, **120**, pp. 95–108

37  APKARIAN, P., GAHINET, P., and BECKERS, G.: 'Self-scheduled $H_\infty$ control of linear parameter-varying systems: a design example', *Automatica*, 1995, **31**, pp. 1251–61

38  APKARIAN, P., and ADAMS, R. J.: 'Advanced gain-scheduling techniques for uncertain systems', in El GHAOUI, L., and NICULESCU, S.-I. (Eds): 'Advances in linear matrix inequality methods in control' (SIAM, Philadelphia, PA, 2000)

39  ECONOMOU, C. G., MORARI, M., and PALSSON, B. O.: 'Internal model control. 5. Extension to nonlinear systems', *Industrial & Engineering Chemistry Process Design & Development*, 1986, **25**, pp. 403–11

40  CLARKE, D. W., MOHTADI, C., and TUFFS, P. S.: 'Generalised predictive control. Part 1: The basic algorithm. Part 2: Extensions and interpretations', *Automatica*, 1987, **23** (2), pp. 137–60

41  FISCHER, M., and ISERMANN, R.: 'Inverse fuzzy process models for robust hybrid control', in DRIANKOV, D., and PALM, R. (Eds): 'Advances in fuzzy control' (Springer, Heidelberg, Germany, 1998), pp. 103–27

42  MUTHA, R. K., CLUETT, W. R., and PENLIDIS, A.: 'Nonlinear model-based predictive control of nonaffine systems', *Automatica*, 1997, **33** (5), pp. 907–13

43  ROUBOS, J. A., MOLLOV, S., BABUŠKA, R., and VERBRUGGEN, H. B.: 'Fuzzy model based predictive control by using Takagi–Sugeno fuzzy models', *International Journal of Approximate Reasoning*, 1999, **22** (1/2), pp. 3–30

44  OLIVEIRA, S. de, NEVISTIĆ, V., and MORARI, M.: 'Control of nonlinear systems subject to input constraints'. Preprints of NOLCOS'95 vol. 1, Tahoe City, CA, 1995, pp. 15–20

45  BOTTO, M. A., van den BOOM, T., KRIJGSMAN, A., and Sá da COSTA, J.: 'Constrained nonlinear predictive control based on input–output linearization using a neural network'. Preprints 13th IFAC World Congress, San Francisco, CA, 1996

46  DRIANKOV, D., and PALM, R. (Eds): 'Advances in fuzzy control' (Springer, Heidelberg, Germany, 1998)

*Chapter 2*

# An overview of nonlinear identification and control with neural networks

*A. E. Ruano, P. M. Ferreira and C. M. Fonseca*

## 2.1 Introduction

The aim of this chapter is to introduce background concepts in nonlinear systems identification and control with artificial neural networks. As this chapter is just an overview, with a limited page space, only the basic ideas will be explained here. The reader is encouraged, for a more detailed explanation of a specific topic of interest, to consult the references given throughout the text. Additionally, as general books in the field of neural networks, the books by Haykin [1] and Principe *et al.* [2] are suggested. Regarding nonlinear systems identification, covering both classical and neural and neuro-fuzzy methodologies, Reference 3 is recommended. References 4 and 5 should be used in the context of B-spline networks.

The main goal of systems identification is to determine models from experimental data. These models can afterwards be employed for different objectives, such as prediction, simulation, optimisation, analysis, control, fault detection, etc. Neural networks, in the context of system identification, are black-box models, meaning that both the model parameters and the model structure are determined from data.

Model construction is an iterative procedure, often done in an *ad hoc* fashion. However, a sequence of steps should be followed, in order to decrease the number of iterations needed to obtain, in the end, a satisfactory model [3]. As a first step, with the help of any prior knowledge available, the relevant inputs for the model must be identified. Afterwards, as these are data-driven models, data must be acquired and pre-processed. Then, the model architecture must be chosen. If the process has dynamics, then a suitable structure must be chosen, and a suitable model order determined. Afterwards, the model structure must be identified, and the model parameters estimated. At the end of this iteration, the results of the latter steps are evaluated using fresh data, and an analysis of these results constitutes prior knowledge to be incorporated for future iterations.

Only some of these steps will be discussed in this chapter. Section 2.2 will introduce three different neural models, multilayer perceptrons, radial basis function networks and B-spline networks, in a unified framework. Parameter estimation will be discussed next, where both offline and online learning will be addressed. Structure selection techniques, with a view of determining well-performing models with the smallest possible complexity, will be treated in Section 2.4. The last task in systems identification to be addressed in this chapter is dynamics representation and model order selection, which will be covered in Section 2.5. Control structures based on neuro-fuzzy systems and neural networks do not differ very much. As the main control schemes have been introduced in Section 1.6 of Chapter 1, Section 2.6 will refer to them, and important references for each different control scheme, in the context of neural networks, will be given. Section 2.7 will illustrate an application of neural networks for nonlinear system identification, particularly in the design of predictive models for greenhouse environmental control, with the aid of multi-objective genetic algorithms. Conclusions will be drawn in Section 2.8.

## 2.2   Model architectures

For system identification applications, artificial neural networks (Figure 2.1) can be envisaged as a model that performs a nonlinear transformation between a $p$-dimensional input space and a one-dimensional output space.

The model is, in a first instance, characterised by $w$ parameters and a certain structure, both of which need to be determined. In all neural networks described in this chapter, according to the type of dependence of the output in the model parameters, these can be decomposed into two classes: linear and nonlinear.

In Figure 2.2, $u$ is a vector of $l$ linear weights, $\Psi$ is a vector of $l$ basis functions and $v$ is a vector of nonlinear weights. In a neural network, all the basis functions

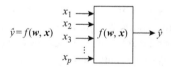

$$\hat{y} = f(w, x)$$

*Figure 2.1    Neural model*

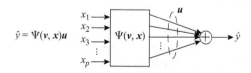

$$\hat{y} = \Psi(v, x)u$$

*Figure 2.2    Model with parameter separability*

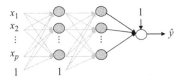

*Figure 2.3    MLP network*

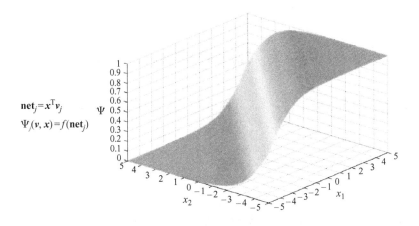

$$net_j = x^T v_j$$
$$\Psi_j(v, x) = f(net_j)$$

*Figure 2.4    Nonlinear neuron of an MLP network*

are of the same type, or can all be obtained from the same elementary function. Each basis function is $p$-dimensional, and is constructed using a unidimensional function. According to the construction method and the function used, the different model types are obtained.

### 2.2.1    Multilayer perceptrons

In multilayer perceptrons (MLPs) [6] (Figure 2.3) the nonlinear neurons are grouped into one or two hidden layers.

The nonlinear neurons use a sigmoidal-type unidimensional function, and the construction method employed consists in projecting the input vector over the nonlinear weight vector associated with each basis function (Figure 2.4).

It should be noticed that, as the employed functions are of the sigmoidal type, with infinite support, any variation in the linear parameters associated with the basis functions influences the output in the overall range of the input space, that is, the network has a global behaviour. MLPs are universal approximators [7,8].

### 2.2.2    Radial basis function networks

The radial basis function networks (RBFs) [9] employ, as the name suggests, a function with a radial behaviour around one point (centre), and use, as the

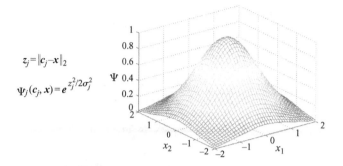

$z_j = \|c_j - x\|_2$

$\Psi_j(c_j, x) = e^{z_j^2/2\sigma_j^2}$

*Figure 2.5    Nonlinear neuron of a RBF network (Gaussian function and Euclidean distance)*

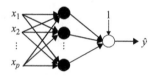

*Figure 2.6    RBF network*

construction method, a distance, usually a Euclidean distance, although other metrics can be employed (Figure 2.5).

In this network, the nonlinear parameters are the centres ($c$) and the spreads ($\sigma$), and the nonlinear neurons are grouped into only one hidden layer (Figure 2.6).

The RBF basis functions, in contrast to the function employed in MLPs, have a local behaviour, although not strictly local as it depends on the $\sigma$ parameter. Besides being universal approximators, RBF networks possess the best approximation property [10], which does not happen with MLPs.

### 2.2.3    B-spline networks

In these networks [4] the basis function is a polynomial function, with a pre-defined order ($k$), and the construction method is a tensorial product. For each input, its range is divided into intervals ($I$), and over each interval there are exactly $k$ active functions.

In order to define the intervals, vectors of knots ($\lambda$), the nonlinear parameters of this network, must be defined, one for each input axis. There is usually a different number of knots for each dimension, and they are generally placed at different positions (Figure 2.7).

The $j$th interval of the $i$th input is denoted as $I_{i,j}$ and is defined as:

$$I_{i,j} = \begin{cases} [\lambda_{i,j-1}, \lambda_{i,j}) & \text{for } j = 1, \ldots, r_i, \\ [\lambda_{i,j-1}, \lambda_{i,j}] & \text{if } j = r_i + 1. \end{cases} \tag{2.1}$$

The application of the knot vectors produces a lattice in the input space. In each lattice cell, there are exactly $\prod_{i=1}^{n} k_i$ active functions.

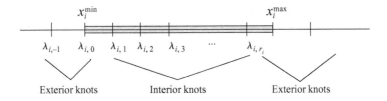

*Figure 2.7    A knot vector in one dimension, with $r_i$ interior knots and $k_i = 2$*

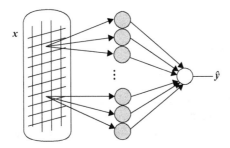

*Figure 2.8    B-spline network*

The $j$th univariate basis function of order $k$ will be denoted here as $\Psi_k^j(x)$, and is defined by the following recurrence relationships:

$$\Psi_k^j(x) = \left( \frac{x - \lambda_{j-k}}{\lambda_{j-1} - v_{j-k}} \right) \Psi_{k-1}^{j-1}(x) + \left( \frac{\lambda_j - x}{\lambda_j - \lambda_{j-k+1}} \right) \Psi_{k-1}^j(x),$$

$$\Psi_1^j(x) = \begin{cases} 1 & \text{if } x \in I_j, \\ 0 & \text{otherwise.} \end{cases} \tag{2.2}$$

Multidimensional basis functions are obtained applying a tensorial product – $\Psi_k^j(x) = \prod_{i=1}^p \Psi_{k_i,i}^j(x_i)$ – over the input dimension (Figure 2.9).

As the basis functions have a compact support, the network has a strictly local behaviour. Associated with these networks there is the problem of curse of dimensionality, which states that the complexity of these networks increases exponentially with the input dimension. These neural networks can also be envisaged as fuzzy systems satisfying certain assumptions (please see Section 7.2 of Chapter 7 for details) which means that the parameters of these fuzzy systems can be indirectly obtained employing neural network learning algorithms.

## 2.3    Parameter estimation

Once the model architecture is chosen, its structure and its parameters must be determined. The present section will focus on the latter problem, and the former will be dealt within the next section.

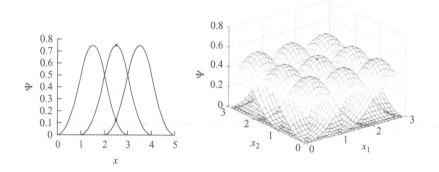

_Figure 2.9   B-spline quadratic functions ($k = 3$). Left – unidimensional function; right – bidimensional function_

As the models are nonlinear, parameter estimation is an iterative process. If the parameter update is performed after the presentation of a data set, which will be used in every iteration of the learning algorithm, the process is denominated offline learning, batch learning, or simply training. On the other hand, if the update is performed in a pattern-by-pattern basis, or after the presentation of a data set that changes from iteration to iteration, the learning process is called online learning or adaptation.

## 2.3.1   Model training

Training a neural model means to determine the model parameters in such a way that the sum of the squared errors ($e$), between the process output ($y$) and the model output ($\hat{y}$), subject to the same inputs ($x$), is minimised:

$$\Omega(w, x) = \sum_{i=1}^{n} \frac{(y(x_i) - \hat{y}(w, x_i))^2}{2} = \frac{\|y - \hat{y}(w)\|_2^2}{2} = \frac{\|e\|_2^2}{2}. \tag{2.3}$$

Training methods will be divided into two classes: methods that can be applied to all the architectures referred above, and those specific to each architecture.

### 2.3.1.1   Applicable to all architectures

There are two classes of algorithms which are applicable to all architectures described in the last section: gradient-based methods and evolutionary algorithms. The former will be detailed first.

#### 2.3.1.1.1   Gradient-based algorithms

All the models presented in the last section are differentiable in their parameters and, as a consequence, gradient-based methods are applicable to them. The most widely known class of algorithms is based on the steepest descent technique and its variants. For MLPs, this algorithm is known as the error-back-propagation (BP) algorithm [6,11]. This method uses the following parameter update:

$$w[k+1] = w[k] - \eta g[k], \tag{2.4}$$

where $\eta$ is called the learning rate, and $g = d\Omega/dw$ is the gradient vector. The gradient is computed using a clever and local application of the derivative chain rule. A description of the BP algorithm can be found in any standard textbook on neural networks, and will not be detailed here. Several variants of this algorithm exist, from the inclusion of a momentum term [6], the incorporation of line-search algorithms [12] and the use of adaptive learning rates [13].

As neural network training can be envisaged as a nonlinear optimisation problem, different methods employed for unconstrained nonlinear optimisation have been proposed. Usually, the performance of these methods, in terms of convergence rate, is inversely proportional to their complexity, measured both in terms of memory requirements and computational complexity, per iteration. This way, and in an increasing order of complexity, we have:

*Conjugate gradient.* This method is similar to the steepest descent method with the incorporation of a momentum term, with the difference that, instead of choosing the momentum parameter in an *ad hoc* fashion, here it is chosen to produce, in each iteration, conjugate search directions [14], that is:

$$\Delta w[k] = -\eta g[k] + \beta[k]\Delta w[k-1]. \tag{2.5}$$

Different strategies are available to determine $\beta[k]$. Probably, the most popular is the Fletcher–Reeves approach [15]:

$$\beta[k] = \frac{g^{\mathrm{T}}[k]g[k]}{g^{\mathrm{T}}[k-1]g[k-1]}. \tag{2.6}$$

*Quasi-Newton.* In this case, an approximation of the inverse ($H$) of the Hessian (matrix of the second derivatives of $\Omega$) is employed, the parameter update being obtained as:

$$w[k+1] = w[k] - \eta[k]H[k]g[k]. \tag{2.7}$$

The most used approximation is the Broyden, Fletcher, Goldfarb and Shanno (BFGS) [15] approach:

$$H_{\mathrm{BFGS}}[k+1] = H[k] + \left(1 + \frac{q^{\mathrm{T}}[k]H[k]q[k]}{s^{\mathrm{T}}[k]q[k]}\right)\frac{s[k]s^{\mathrm{T}}[k]}{s^{\mathrm{T}}[k]q[k]}$$

$$- \left(\frac{s[k]q^{\mathrm{T}}[k]H[k] + H[k]q[k]s^{\mathrm{T}}[k]}{s^{\mathrm{T}}[k]q[k]}\right), \tag{2.8}$$

where

$$s[k] = w[k] - w[k-1],$$
$$q[k] = g[k] - g[k-1]. \tag{2.9}$$

Usually the quasi-Newton method employs a line-search algorithm.

*Levenberg–Marquardt.* This method specifically exploits the nature of the problem at hand, which is a nonlinear least-squares problem [16,17]. In this class of optimisation problems the Hessian matrix ($G$) contains both information of the first derivative of the output vector, $J$, and of its second derivative. The following approximation is used:

$$G \approx J^T J, \tag{2.10}$$

where $J$ is called the Jacobian matrix. The parameter update is given as the solution of:

$$(J^T[k]J[k] + v[k]I)\Delta w[k] = -J^T[k]e[k]. \tag{2.11}$$

The Levenberg–Marquardt (LM) algorithm is a trust region or restricted step algorithm. Convergence to a local minimum is guaranteed even in situations where the Jacobian matrix is not full-rank. The $v$ parameter is called the regularisation parameter and controls both the search direction and its magnitude. The update of this parameter should reflect how well the actual function is approximated by a quadratic function. Using this approach, the predicted reduction is given by:

$$\Delta\Omega^P[k] = \Omega(w[k]) - \frac{(e^P[k])^T(e^P[k])}{2}, \tag{2.12}$$

$$e^P[k] = e[k] - J[k]\Delta w[k],$$

where $e$ represents the error vector. As the actual reduction is given by

$$\Delta\Omega[k] = \Omega(w[k]) - \Omega(w[k] + \Delta w[k]), \tag{2.13}$$

the ratio $r[k] = \Delta\Omega[k]/\Delta\Omega^P[k]$ is employed to control the update of $v$. A commonly used heuristic is:

$$v[k+1] = \begin{cases} \dfrac{v[k]}{2}, & r[k] > \dfrac{3}{4}, \\ 4v[k], & r[k] < \dfrac{1}{4}, \\ v[k], & \text{others.} \end{cases} \tag{2.14}$$

If $r[k] < 0$ only $v$ is updated, the model parameters being kept unchanged. This algorithm was proposed for MLP training [18], subsequently for RBFs [19] and for B-splines and neuro-fuzzy systems [20].

In the same way that the LM algorithm exploits the specific nature of the optimisation problem at hand, the separability of the model parameters into linear and nonlinear can also be exploited to improve the performance of the training algorithms [21]. The output of the models can be expressed as:

$$\hat{y} = \Phi u, \tag{2.15}$$

where $\Phi$ represents the output matrix of the last nonlinear hidden layer (possibly with a column of ones to account for the model output bias). When (2.15) is replaced in (2.3), we have:

$$\Omega(v, u) = \frac{\|y - \Phi(v)u\|_2^2}{2}. \tag{2.16}$$

For any value of $v$, the minimum of $\Omega$ with respect to $u$ can be obtained using the least squares solution, here determined with the application of a pseudo-inverse:

$$\hat{u}(v) = \Phi^+ y. \tag{2.17}$$

If (2.17) is replaced in (2.16), a new criterion is obtained:

$$\psi(v) = \frac{\|y - \Phi(v)\Phi(v)^+ y\|_2^2}{2}, \tag{2.18}$$

which just depends on the nonlinear parameters. To minimise this criterion with any of the methods described before, the gradient of (2.18) with respect to $v$ must be computed, or, for the case of the LM method, the Jacobian matrix. It can be proved [21,22] that the gradient of $\psi$ can be determined, computing first the optimal value of the linear parameters, using (2.17), replacing this in the model, and subsequently performing the usual calculation of the gradient (only for the partition related with the nonlinear parameters). The same procedure can be applied for the computation of a suitable Jacobian for $\psi$:

$$\begin{bmatrix} 0 \\ g_\psi \end{bmatrix} = g_\Omega|_{u=\hat{u}} = -\begin{bmatrix} \Phi^T \\ J_\psi^T \end{bmatrix} e_\psi. \tag{2.19}$$

The use of this criterion presents some advantages, comparing with the use of criterion (2.16):

- it lowers the dimensionality of the problem;
- when the LM is used, each iteration is computationally cheaper;
- usually a smaller number of iterations is needed for convergence to a local minimum, as:
  - the initial value of (2.18) is much lower than the one obtained with (2.16);
  - equation (2.18) usually achieves a faster rate of convergence than (2.16).

Derivative-based algorithms converge, at most, for a local minimum. As the training process can be envisaged as the evolution of a nonlinear difference equation, its final value will depend on the initial conditions, or, in other words, on the initial value of the parameters. This way, there are two problems associated with this class of algorithms: where to start the training and when to finish it. The first problem depends on the chosen architecture. As to the second question, training is normally interrupted when a certain user-specified resolution in the approximation is achieved or when a user-specified number of iterations is met. However, the aim in the training is to converge to a local minimum criteria that detect if this has been achieved and can be employed. A termination criterion that is commonly used in nonlinear optimisation is [15]:

$$\Omega[k-1] - \Omega[k] < \theta[k], \tag{2.20}$$

$$\|w[k-1] - w[k]\| < \sqrt{\tau_f} \cdot (1 + \|w[k]\|), \tag{2.21}$$

$$\|g[k]\| \leq \sqrt[3]{\tau_f} \cdot (1 + |\Omega[k]|), \tag{2.22}$$

where

$$\theta[k] = \tau_f \cdot (1 + \Omega[k]), \tag{2.23}$$

$\tau_f$ being a measure of the desired number of correct digits in the training criterion. If (2.18) is employed, $\Omega$ and $w$ should be replaced by $\psi$ and $v$ in (2.20)–(2.23). When the three conditions are simultaneously met, the training terminates.

Although the use of a termination criterion such as the one described before enables to detect when a local minimum is attained, as a consequence of the bias-variance dilemma, if a small value for $\tau_f$ is specified, overtraining can occur. This can be avoided using the early stopping technique (also called implicit regularisation). Data is divided into training and test or generalisation data, the latter being used to determine when the best performance (minimum error) is achieved, terminating the training at that point. Obviously this technique can be used together with the criterion given in (2.20)–(2.23), terminating the training as soon as the first of the two criteria is met.

#### 2.3.1.1.2    Evolutionary algorithms

Evolutionary algorithms are powerful optimisation tools (please see Chapter 3). As such, they can be employed for model training (please see references on Section 2 of Reference 23), with some advantages compared with gradient-based algorithms. As they evolve a population and not a single point, they perform a more global search. As they do not employ gradients, their calculation is not needed. However, if their performance for parameter estimation of neural models is compared with the fast gradient-based methods, they normally present slower rates of convergence, and do not possess mechanisms for local (and global) minimum detection. As a consequence, fast gradient algorithms are usually a more popular choice.

### 2.3.1.2    Training methods specific for each architecture

#### 2.3.1.2.1    Multilayer perceptrons

The gradient-based methods described before were initially introduced for MLPs and are essentially the training methods used for this architecture. In what concerns parameter initialisation, it is not possible to incorporate a prior knowledge. Admitting that the training data is normalised, the nonlinear parameters are usually initialised with small random values, with heuristics that try to assure that all neurons are in an active region and that the network is not particularly badly conditioned [24,25].

### 2.3.1.2.2 *RBF networks*

One of the advantages associated with RBFs is the possibility of employing hybrid methods, composed of unsupervised methods for the determination of the nonlinear parameters (centres and spreads), and the use of the least-squares solution for the linear parameters. The centres are determined using data clustering methods, the $k$-means algorithm [26] being the most popular technique (see Algorithm 2.1).

---

**Algorithm 2.1** $k$-means clustering

1. Initialization – Choose random values for the centres; they must be all different; $j = 1$
2. While *go_on*

   2.1. Sampling – Find a sample vector $x(j)$ from the input matrix
   2.2. Similarity matching – Find the centre (out of $m_1$) closest to $x(j)$. Let its index be $k(x)$:

$$k(x) = \arg \min_i \|x(j) - c_i\|_2, \quad i = 1, \ldots, m_1.$$

   2.3. Updating – Adjust the centres of the radial basis functions according to:

$$c_i[j + 1] = \begin{cases} c_i[j] + \eta(x(k) - c_i), & i = k(x) \\ c_i[j], & \text{otherwise} \end{cases}$$

   2.4. $j = j + 1$

   end

---

In Reference 27 a change to this algorithm has been proposed, incorporating for each pattern a weight relative to the error, resulting in the error-sensitive clustering method. An adaptive version of the $k$-means algorithm, the optimal adaptive $k$-means (OAKM) algorithm, has been introduced in Reference 28. In what concerns the spreads, the $k$-nearest neighbours heuristic, $\sigma_i = \left( \sum_{j=1}^{k} \|c_i - c_j\|_2 \right) / k\sqrt{2}$, $k$ being a user-defined percentage of the closest centres to $c_i$ is the most used technique, although alternatives such as the maximum distance between patterns – $\sigma = (\max_{i,j=1,\ldots,m} \|x_i - x_j\|) / 2\sqrt{2}$, $m$ being the total number of patterns, the empirical standard deviation – $\sigma_i = \sum_{j=1}^{n} \sqrt{\|c_i - x_j\|_2^2 / n}$, $n$ being the total number of patterns assigned to cluster $i$, and others are also employed [1]. In practice, the joint use of clustering algorithms and heuristics for the spreads cannot guarantee a complete cover of the input space. In low dimensionality problems, the centres are sometimes located in a lattice, which additionally enables the interpretation of an RBF network as a fuzzy system.

Radial basis function networks can also be derived from regularisation theory [10]. RBF networks trained using this approach are usually called generalised RBF networks [1].

If gradient-based algorithms are used to determine the nonlinear parameters, the centres are usually initialised as random values, or, alternatively, a random selection of training patterns can be performed.

### 2.3.1.2.3    B-spline networks

For this type of networks heuristics (or meta-heuristics such as evolutionary algorithms) are usually employed, which besides determining the structure iteratively, change the position of the input knots. Therefore, algorithms such as the ASMOD algorithm (see Section 7.3 of Chapter 7) perform structure selection and parameter estimation simultaneously.

If gradient methods are used, the initial position of the input knots is usually evenly distributed across the range of each variable. If previous knowledge about the process to be modelled is available, the knot distribution can reflect this knowledge.

## 2.3.2    Model adaptation

When the parameters of a nonlinear model are adapted, there are two phenomena that should be taken into consideration: the first one, parameter shadowing, refers to the fact that a continuously applied parameter adaptation scheme, minimising the error between the target and the model output, may cause the parameters not to converge to their optimal values (in the offline sense). Parameter interference reflects the fact that adapting the model in a region of the input space may cause deterioration of the mapping already learned for other regions of the space. Obviously these problems become more severe for networks with global behaviour, such as MLPs [29].

Essentially, there are two types of adaptive algorithms: those that adapt only the linear parameters, keeping the nonlinear parameters unchanged, and those that adapt all model parameters. For this last class, there is a subclass of algorithms that exploits the linear–nonlinear separability of parameters.

### 2.3.2.1    Adaptation of the linear parameters

In this case it is assumed that the nonlinear parameters have been previously determined by an offline method such as the ones described in Section 2.3.1, and are kept fixed, while the linear parameters change. Any linear adaptive algorithm can be used in this case.

Relatively to first-order methods, the least-means square (LMS) can be employed [25]:

$$u[k] = u[k-1] + \eta e[k] \Psi[k], \tag{2.24}$$

or, preferably, its normalised version (NLMS) [4]:

$$u[k] = u[k-1] + \frac{\delta e[k] \Psi[k]}{\|\Psi[k]\|_2^2}, \tag{2.25}$$

stable adaptation being achieved for $0 \leq \delta \leq 2$. In the same way as their offline versions, the convergence of these algorithms is usually very slow and highly dependent on the correlation of the output data of the last hidden layer. When modelling error or measurement errors are present, these algorithms do not converge to a unique point, but to a minimum capture zone [4]. In these cases dead-zone algorithms can be employed, where $e[k]$ in (2.24) or (2.25) should be replaced by:

$$e^d[k] = \begin{cases} 0 & \text{if } (|e[k]\| \leq \varsigma), \\ e[k] + \varsigma & \text{if } (e[k] < -\varsigma), \\ e[k] - \varsigma & \text{if } (e[k] > \varsigma), \end{cases} \tag{2.26}$$

$\varsigma$ being the dead-zone parameter. Alternatively, algorithms with adaptive learning rates can be employed, where, for the case of the NLMS, $\delta$ in (2.25) is replaced by $\delta_i[k] = \delta_1/(1 + k_i/\delta_2), \delta_1, \delta_2 > 0$, $k_i$ being the number of times the $i$th basis function has been updated.

Alternatives to the first-order methods described above are the use of sliding-window or second-order methods.

Sliding-window adaptation [30] consists in the use of training methods that employ $L$ samples previously stored. These methods include [31]: moving average of LMS/NLMS search directions [32]

$$\Delta \boldsymbol{u}[k] = \frac{\eta}{L} \sum_{i=1}^{L} e_i \boldsymbol{x}_i + (1 - \eta)e[k]\,\boldsymbol{x}[k], \quad \text{for the case of LMS}, \tag{2.27}$$

maximum error method [33]

$$\Delta \boldsymbol{u}[k] = \eta e_b\,\boldsymbol{x}_b, \quad b = \arg\{\max_i(|e_i|)\} \tag{2.28}$$

and Gram–Schmidt orthogonalisation [33]. In (2.27) and (2.28) the variables indicated by $e_i$ and $\boldsymbol{x}_i$ are variables in the store. Associated with the use of sliding windows, there are different strategies for the management of the stored data [31]: the simplest and the most obvious is to use a FIFO policy, but other strategies based on the sample correlation, where the sample to be removed from the store is the one that is most highly correlated with the current sample, or based on the distance between samples, where the sample to be removed is the one closest (in Euclidean distance) to the current sample, can be used.

In what concerns second-order methods, the well-known recursive least-squares algorithm and Kalman filters and their variants [34] are popular solutions. An alternative is the use of a recursive least-squares algorithm with regularisation [35].

### 2.3.2.2 Adapting all the parameters

In the same way as in the former algorithms, the methods that adapt all the network parameters can be divided into recursive implementations of offline algorithms and sliding-window-based methods.

#### 2.3.2.2.1 Recursive implementations

The most known and used method is the stochastic error BP [6], which is identical to the LMS algorithm applied to all model parameters, and where the update equation employs the input data instead of the output of the last hidden layer:

$$w[k] = w[k-1] + \delta e[k] x[k]. \tag{2.29}$$

In the same way as the LMS, a momentum term can be incorporated in (2.29). Several heuristics have been proposed for the selection of the learning rate, as, for instance: the search, then converge [36] technique

$$\delta[k] = \frac{1 + (c/\delta_0) \cdot (k/\tau)}{1 + (c/\delta_0) \cdot (k/\tau) + \tau(k/\tau)^2}, \tag{2.30}$$

where $\delta_0$ is the initial learning rate, $\tau$ is the reset parameter and $c$ determines the rate of decay; the Chan and Fallside procedure [37]

$$\delta[k] = \delta[k-1](1 + 0.5\cos(2\theta[k])),$$
$$\cos(2\theta[k]) = \frac{-g[k]\Delta w[k]}{(g^{\mathrm{T}}[k]g[k])(\Delta w^{\mathrm{T}}[k]\Delta w[k])}, \tag{2.31}$$

and the bold driving [38] method, as well as methods that assign different learning rates for different parameters, such as the delta-bar-delta rule [13]:

$$\Delta\delta_i[k] = \begin{cases} \alpha & \text{if } (s_i[k-1]g_i[k] > 0), \\ -\beta\delta_i[k-1] & \text{if } (s_i[k-1]g_i[k] < 0), \\ 0 & \text{otherwise,} \end{cases} \tag{2.32}$$

where $s_i$ measures the average of the previous gradients, for weight $i$:

$$s_i[k] = (1-\gamma)g_i[k] + \gamma s_i[k-1], \tag{2.33}$$

$\alpha, \beta$ and $\gamma$ having to be defined by the user.

Relative to second-order methods, the recursive least-squares algorithm can be adapted to nonlinear models by employing a linear approximation of the output at each sampling instant. Using this approach the recursive prediction error (RPE) algorithm is obtained [39]. On incorporating a regularisation term in the RPE algorithm, a recursive implementation of the LM [40] is achieved. As these methods are computationally expensive, for real-time applications decompositions of the covariance matrix, ignoring the dependencies between neurons, have been proposed [41]. The separability of the parameters (the use of criterion (2.18)) can also be employed in recursive implementations, obtaining better results than the conventional algorithms [31].

### 2.3.2.2.2 Sliding-window algorithms

These algorithms are basically offline methods applied to a time-varying data set. Any gradient-based method referred in Section 2.3.1.1 can be employed, together with the window management strategies mentioned in Section 2.3.2.1. As examples, in Reference 42 the error-BP algorithm was used with a FIFO management strategy, in Reference 31 the conjugate gradient algorithm is employed with different management strategies and in Reference 43 the LM algorithm, minimising (2.16) and (2.18) is used with a FIFO management strategy.

## 2.4    Structure identification

In the previous section the model structure was assumed to be fixed, and the problem at hand was to estimate the model parameters. Here, the goal is to determine the 'best' model structure, using only the experimental data. Each structure has a different complexity that here denotes the number of parameters to be estimated.

The aim here is to find parsimonious models that have a satisfactory performance for the given data, with the smallest possible complexity. It is known that, as the model complexity increases, for the same amount of training data, the model performance in this data set improves but, above a certain limit in the complexity, the model performance in the test data is deteriorated. There is a compromise between the approximation obtained in the training set and the generalisation in data not seen by the model. The final goal of any structure selection method is to find this compromise value. If the model complexity is below it, we are in the presence of underfitting or undermodelling; if, on the other hand the model complexity is above that value, there is overfitting or overmodelling. The determination of this compromise value is a difficult problem as, besides assuming that one of the methods described in the last section is able to find the optimal model (with the additional problem of the existence of local minima), this compromise value in practice depends on the amount and on the quality of the training data.

Before specifying methods for structure selection, it is necessary to define the criteria used to assess the model quality. Different criteria are usually applied, which are given below.

1.  If a test data is available, the performance (normally measured with the mean-square error) of models with different complexity is compared in the test set; in this case, the 'best' model is the one with the best performance in the test set.
2.  If only a training set is available, information criteria, not only dependent on the quality of the approximation obtained, but also on the amount of training data and on the model complexity, can be used [44]. Examples of these criteria are:

    *Akaike information criterion* – $\text{AIC} = n \cdot \ln(\text{MSE}) + 2k$,
    *Baysean information criterion* – $\text{BIC} = n \cdot \ln(\text{MSE}) + k \cdot \ln(k)$,
    *Final prediction error* – $\text{FPE} = n \cdot \ln(\text{MSE}) + n \cdot \ln((n+k)/(n-k))$,

    where $n$ denotes the amount of training data and $k$ the model complexity. These criteria have different sensitivities to an increase in complexity, leading therefore to different results. Using this perspective, the selected model is the one that has the minimum value of the information criterion employed.
3.  Using again only a training set, by defining an appropriate kernel product, and optimising the empirical risk [45]:

$$R_{\text{emp}} = \frac{1}{n} \sum_{i=1}^{n} L_\varepsilon(y_i, \hat{y}_i), \tag{2.34}$$

where

$$L_\varepsilon(y_i, \hat{y}_i) = \begin{cases} |y_i - \hat{y}| - \varepsilon, & |y_i - \hat{y}_i| \geq \varepsilon, \\ 0, & \text{others,} \end{cases} \tag{2.35}$$

using restricted quadratic optimisation. Having defined the kernel, the insensitivity parameter $\varepsilon$, and a parameter which weights the quality of the approximation against the generalisation capacity, the construction method is automatic, resulting in the support vector machines.

4.  The criteria introduced above weight, in a different way, the quality of the approximation in the training set against the performance in the test set or the model complexity. An alternative is to simultaneously optimise these conflicting criteria, using multi-objective evolutionary algorithms [46,47]. Using this approach, the result is not a single model, but a set of non-dominated or Pareto model solutions (please see Chapter 3 and the example at the end of this chapter). These solutions can be ranked if preferences and/or goals are associated with the different objectives [46].

Having defined the criteria used to compare the quality of the models, the large number of structure selection methods will be now very briefly summarised. They can be broadly classified into two large classes of techniques: iterative methods, which explicitly optimise the model structure, and those that aim to do that in one step.

## 2.4.1   Single step methods

In this class of methods the nonlinear structure of an overmodelled network is assumed as given. These techniques have user-selected control parameters, whose value is usually tuned by the model performance in the test set.

### 2.4.1.1   Structure regularisation

In this case, in reality the structure in itself does not vary, it is the number of effective parameters, or degrees of freedom of the model that is reduced. This is translated, in terms of model error, by a large decrease in the variance error and by a smaller increase in the bias error [3,5].

Regularisation can be obtained by incorporating a penalty term in the training criterion (2.16):

$$\phi = \frac{\|y - \hat{y}\|_2^2}{2} + \alpha V(w), \tag{2.36}$$

where $\alpha$ is the regularisation parameter. In practice two types of penalty terms are used:

$$V(w) = w^{\mathrm{T}} K w, \tag{2.37}$$

$$V(w) = \sum_{k=1}^{n} \left| \frac{\mathrm{d}^2}{\mathrm{d} x(k)^2} \hat{y}(x(k), w) \right|^2. \tag{2.38}$$

When $K = I$ in (2.37) this is called zero-order regularisation, while the use of (2.38) is usually known as second-order regularisation. The latter is usually used when sparse or noisy training data is available, resulting in smoother outputs, while the use of the former produces a smaller parameter vector norm.

While the optimal values of the nonlinear parameters are implicitly limited by the range of the input data (usually normalised), the same thing does not happen with the linear parameters which, in badly conditioned problems, might assume very high values. Therefore, it is justifiable to replace in (2.37) the total parameters vector $w$ by the linear parameters vector $u$ only. In this case, as (2.36), with zero-order regularisation, can be expressed as:

$$\phi = \frac{\left\| \begin{bmatrix} y \\ 0 \end{bmatrix} - \begin{bmatrix} \Phi u \\ \sqrt{\alpha} K u \end{bmatrix} \right\|_2^2}{2} = \frac{\left\| \begin{bmatrix} y \\ 0 \end{bmatrix} - \begin{bmatrix} \Phi \\ \sqrt{\alpha} k \end{bmatrix} u \right\|_2^2}{2} = \frac{\| \bar{y} - \bar{\Phi} u \|_2^2}{2}, \tag{2.39}$$

all the gradient-based methods introduced in Section 2.3.1.1 can be applied directly to minimise $\phi$. If the separability of the parameters is exploited, one must use (2.40) to determine the optimal value of the linear parameters, instead of (2.17).

$$\hat{u} = (\Phi^T \Phi + \alpha K I)^{-1} \Phi^T y = \bar{\Phi}^+ \bar{y}. \tag{2.40}$$

### 2.4.1.2   Support vector machines

This approach has solid roots in statistical learning theory [45,48], and was initially introduced for classification problems. The models generated by this technique are non-parametric models, and the mapping shown in Figure 2.2 is represented here by:

$$y = \sum_{j=0}^{m_1} \Psi_j(x, x_j) u_i = \Phi(x) u, \tag{2.41}$$

where $x_j$ represents the $j$th training pattern, $\Psi_0(x, x_0) = 1$, and $m_1$ represents the number of basis functions, chosen by the construction method, among the $n$ possible basis functions, $n$ being the number of training patterns.

The nonlinear basis functions should satisfy the Mercer theorem [1,5]. Functions that can be used are Gaussian functions (the $\sigma$ parameter must be defined by the user and be constant for all neurons), even order B-splines, and perceptron-like functions (satisfying some restrictions).

This methodology employs a pre-defined nonlinear mapping between the input space and a large-dimensional space (denoted as the feature space), the support vectors being determined in this space. Essentially the functional (2.42) is minimised, by applying constrained quadratic optimisation, using the dual formulation of the Lagrange multipliers method (for more details, please see, for instance, Reference 1).

$$c \sum_{i=1}^{n} L_\varepsilon(y_i, \hat{y}_i) + \frac{1}{2} u^T u. \tag{2.42}$$

Clearly (2.42) may be interpreted as a regularisation problem, the first term being related with the desired approximation in the training set (function of the $\varepsilon$ parameter), and the second term with model generalisation. The regularisation parameter $C$ is also user-defined, and must be tuned simultaneously with $\varepsilon$. The solution to this problem is:

$$u_{1,...,m_1} = \alpha_i - \alpha_i^*, \tag{2.43}$$

where $\alpha_i$, $\alpha_i^*$ are the Lagrange multipliers associated with the training patterns for which $|y_i - \hat{y}_i| \geq \varepsilon$ (the support vectors) and $u_0$ is determined using the data that satisfy the equality $|y_i - \hat{y}_i| = 0$.

## 2.4.2 Explicit structure optimisation

For this large class of methods, the number of the model parameters, or the number of structures (neurons, sub-models) varies from iteration to iteration. There are essentially four classes of methods [3]:

*Generic.* This class of computationally heavy methods envisage this problem as a combinatorial problem. In this class we have a pure inspection of a pre-defined set of models, or evolutionary algorithms are used to perform a search in the model space. The latter has been a hot topic of research, and different approaches can be found in Section 3 of Reference 23. An illustrative example of the application of evolutionary algorithms for neural network design will be given in the end of this chapter.

*Constructive.* These methods start with a simple model, which is iteratively refined, by adding more parameters or sub-structures. Examples of these methods, applicable to architectures not covered in this chapter are the projection pursuit algorithm [49,50] and the cascade correlation networks [51].

*Pruning.* In contrast with constructive methods, pruning methods start with a very complex model, and remove, in each iteration, parameters or sub-structures. They are usually slower than constructive algorithms.

*Mixed.* In this class of methods, in each iteration there are steps that increase the model complexity, and others that decrease it.

Depending on the selected architecture, different heuristics can be used.

### 2.4.2.1   Multilayer perceptrons

In the class of constructive algorithms, the method of sequential construction [49] can be mentioned. The basic idea is, in each iteration, to add neurons whose parameters, randomly initialised, are trained, in a first phase, by keeping the parameters of the original neurons fixed. Subsequently, all the parameters (the original and the new ones) are trained.

There are several pruning algorithms for MLPs [52]. Some use regularisation techniques, such as the technique of weight decay, where parameters which have little influence (denoted as excess weights) on the model are forced to have small values, the technique of weight elimination [53], where excess weights are eliminated, and the smoothing technique [54], essentially a weight decay approach where the influence of linear and nonlinear weights is differentiated. Other methods are based in the computation of the sensitivity of the different parameters, after the network has been trained. This is the case of the optimal brain damage procedure [55] and its posterior version, the optimal brain surgeon [56].

### 2.4.2.2  RBF networks

The most used structure selection method for RBFs is, without doubt, the orthogonal least-squares (OLS) technique [57,58]. The basic idea is to orthogonalise, iteratively (using for instance, Gram–Schmidt orthogonalisation), a subset of basis functions among a larger set of RBFs previously determined (using for instance data clustering techniques). In each iteration, among the columns not yet orthogonalised, the one chosen to orthogonalise is the one that corresponds to the largest error-reduction rate. The method is terminated when a pre-defined number of regressors is obtained, or when the model output variance unexplained by the model is lower than a pre-defined value. There are variants of this technique, such as the incorporation of a regularisation term [59].

Another frequently used alternative is the resource allocation network (RAN) method, proposed in Reference 60. A new neuron is added to the network when the error or the distance between the current sample and its nearest neighbour exceeds pre-defined values. This method was improved in Reference 61 by the incorporation of an extended Kalman filter. The possibility of network pruning was proposed in Reference 62, and a comparison between the different alternatives was done in Reference 63. As the model performance assessment of these methods is done in a pattern-by-pattern basis, they are suitable for online applications.

### 2.4.2.3  B-spline networks

As has been already mentioned, these networks suffer from the curse of dimensionality. The main technique used for these networks is a mixed algorithm, ASMOD – adaptive spline modelling of observational data [64] and its variants, which are based on the decomposition of a high dimensionality function into a sum of lower dimensionality functions. These algorithms will be addressed in Chapter 7 and therefore will not be mentioned here.

As an alternative to the ASMOD algorithms, evolutionary algorithms were proposed for the design of B-spline networks. In Reference 65 single and multi-objective genetic programming have been compared with the ASMOD algorithm and in Reference 66 bacterial algorithms were employed, with success, to the same problem.

## 2.5  Dynamics representation and model order selection

When the process to be modelled has dynamics, the model should reflect this fact. There are essentially two approaches for that: to use models with external dynamics or to employ models with internal dynamics.

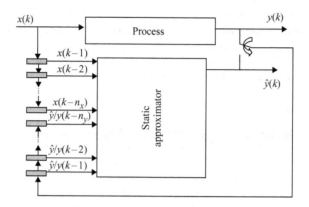

*Figure 2.10    Static model with external dynamics*

### 2.5.1    Models with external dynamics

Models with external dynamics can be considered as static nonlinear approxima-tors, whose inputs are obtained through the application of time-delay lines (see Figure 2.10). If the aim is to use the model for simulation, then the switch is con-nected to the model output. In the context of neural networks, this is called a parallel configuration. If the model is used for prediction, then the switch is connected to the process output, and this is called a series–parallel configuration [67].

As with linear dynamic models, there are several nonlinear model structures available. For a detailed description of the latter, the reader can refer to Reference 68. Concerning the former, Reference 34 is recommended.

All these models can be expressed as:

$$\hat{y}[k] = f(\varphi[k]), \qquad (2.44)$$

where the regression vector $\varphi[k]$ contains the current and past inputs, possibly past (model or process) outputs, and eventually past prediction errors.

In the case when the regression values only contain inputs, we have models without output feedback. To obtain a good model, a large number of regressors is needed, which is translated into a greater model complexity. For these reasons they are seldom used.

The most common nonlinear structures are:

$$\text{NARX: } \varphi[k] = [u[k-1] \cdots u[k-n_u] \, y[k-1] \cdots y[k-n_y]]^{\mathrm{T}}, \qquad (2.45)$$

$$\text{NARMAX: } \varphi[k] = [u[k-1] \cdots u[k-n_u] \, y[k-1] \cdots y[k-n_y]$$

$$e[k-1] \cdots e[k-n_y]]^{\mathrm{T}}, \qquad (2.46)$$

$$\text{NOE: } \varphi[k] = [u[k-1] \cdots u[k-n_u] \, \hat{y}[k-1] \cdots \hat{y}[k-n_y]]^{\mathrm{T}}. \qquad (2.47)$$

The NARX model is trained in a series–parallel configuration, while the NOE is trained in a parallel configuration. The NARMAX model uses more inputs (the past errors), which is translated into a larger complexity and for this reason it is not usually employed. The NARX structure is the most employed, as the NOE (as well as the NARMAX) needs, for training, the computation of dynamic gradients, as it uses the model output, and not the process output.

## 2.5.2   Models with internal dynamics

These can be described by the following state–space representation:

$$\hat{\xi}[k + 1] = h(\hat{\xi}[k], x[k]),$$
$$\hat{y}[k] = g(\hat{\xi}[k]),$$
(2.48)

where $\hat{\xi}[k]$ represents the model state vector which, as a general rule, does not bear any relation with the process state vector. Basically, there are four types of internal dynamic models:

1. Fully recurrent models, introduced in Reference 69 for sequence recognition. Due to the slow convergence of the training algorithms and to stability problems, they are rarely used for system identification purposes.
2. Models with partial recurrence, proposed by Elman [70] and Jordan [71]. In this case, the network has a multilayer structure, with an additional set of neurons (called context neurons), which receive feedback from the hidden neurons or from the model output. In comparison with fully recurrent models, they possess better convergence and stability properties.
3. Models with state recurrence, proposed by Schenker [72]. They are the most direct implementation of a nonlinear state–space model, but where the internal states are not known. These models offer some advantages over the two last types, but still training can become unstable, and the trained model can be unstable.
4. Locally recurrent globally feedforward (LRGF) networks, proposed in Reference 73. These are extensions of multilayer networks, where the dynamic is implemented using linear FIR or IIR filters as connections between the neurons, applied to the net input, or providing feedback from the output to the input. An advantage of these networks over the others is that their stability is much easier to check. With the exception of the local output feedback, if all the filters are stable, then the network is stable.

One of the problems associated with internal dynamic models lies in the complexity of the training phase (this problem is common to the NARMAX and NOE structures). Independent of the architecture being linear or nonlinear in its parameters, training is always a nonlinear optimisation problem. A more serious problem is the fact that the gradient depends on the past model states, and, therefore, it requires a more complex calculation. There are essentially two classes of training algorithms for these networks:

1. The backpropagation-through-time (BPTT) algorithm, introduced in Reference 6. This is basically an extension of the standard BP algorithm, for a structure composed of copies of the original model, unfolded in time.

2. The real-time recurrent learning (RTRL) algorithm, proposed in Reference 69. This is a much more efficient algorithm, not needing to unfold the model in time. It is, however, more computationally demanding than the static BP.

Due to the greater computational complexity of the training algorithms, and to the possibility of instability associated with dynamic neural networks, the models with external dynamics are, by far, the most used. Among them, the NARX structure is the most commonly employed, for the reasons given above.

### 2.5.3    Model order selection

For models with internal dynamics, the model order is somehow related with the number of internal states, although this is not an immediate relation (with the exception of the LRGF networks). When there is no prior knowledge of the process order, these models can be envisaged as a black-box, in the sense that their internal dynamic can efficiently approximate a large variety of process dynamic orders.

In what concerns models with external dynamics, the model order selection can be viewed as an input selection problem. There are essentially four different approaches, the first three being also applicable for a generic input selection problem:

1. To compare the performance of models with different orders in a test set.
2. To use information criteria such as the ones introduced in Section 2.4 with models of different orders [74].
3. To compute the Lipschitz indices, as proposed in Reference 75, as a function of an increasing number of inputs. The correct number is determined as the smallest number of inputs for which the minimum value of the indices is obtained.
4. To extend the correlation tests used in linear systems [76] to nonlinear systems. Essentially, for models with correctly captured dynamics, the error is not correlated with the model inputs. As in practice the correlation is not zero, statistical tests are employed to verify if this is met with a certain probability. For nonlinear models, Billings and co-authors [77,78] have proposed tests involving higher order correlations.

## 2.6    Control based on neural networks

Different control schemes were already introduced in Section 1.6, in the context of fuzzy systems. In fact, as fuzzy systems and neural networks are both universal approximators, they are used in the same control structures. Therefore, in this section, rather than introducing them again, the reader is referred to the corresponding section in Chapter 1, and important references, in terms of neural networks, are given. Good surveys on the application of neural networks for control can be found in, for instance, References 79 and 80. Also, please see the papers in the special issue [81].

### 2.6.1    Gain scheduling

Gain scheduling control is covered in Section 1.6.1. Neural network applications of this approach can be found in, for instance, Reference 82, where RBFs are used as a gain-scheduling controller for the lateral motion of a propulsion controlled aircraft, and in Reference 83, where a regularised neural network is used to improve the performance of a classic continuous-parameter gain-scheduling controller.

## 2.6.2   Inverse control

The most straightforward application of neural networks is adaptive inverse control. This topic is covered in Section 1.6.3, and important neural network applications can be found in References 84–87.

## 2.6.3   Internal model control

Internal model control with fuzzy systems is introduced in Section 1.6.4. This approach was introduced, in the context of neural networks in Reference 88. This technique was used for the control of a bioreactor in Reference 89, for pH neutralisation, using local model networks in Reference 90, for air temperature control of a heat exchanger in Reference 91, and in Reference 92 a design procedure of neural internal controllers for stable processes with delay is discussed.

## 2.6.4   Controller tuning

Neural networks have been also used for parameter tuning of conventional controllers of a given known structure. For PID autotuning, please see, for instance, References 93–95. For other controller types, please see Reference 96.

## 2.6.5   Model-based predictive control

Model-based predictive control is one of the techniques that has found more applications in industry. The topic is covered in Section 1.6.5. A few of a large number of references in this topic are, for instance, References 97–99 for chemical process control applications, and where local linear neuro-fuzzy networks were employed [100,101].

## 2.6.6   Feedback linearisation

Feedback linearisation [102] is a nonlinear control technique, whose objective is to cancel the model nonlinearities using feedback, in such a way that the resulting model is linear. In the continuous case, Lyapunov theory is used to establish stability and convergence results. RFFs and MLPs have been used with the objective of approximating the Lie derivatives, in such a way that a controller can subsequently be designed using feedback linearisation. States are assumed accessible [103–107], or they are estimated with an observer [108]. Recurrent networks, to approximate the process dynamics, have also been used in the continuous case. In the works by Rovithakis and Christodoulou [109] and Ponzyak *et al.* [110], the controller uses the available process states, while in References 111–113 the states are estimated by the recurrent network.

In the discrete case, there are also two approaches. Considering an affine input–output model:

$$y[k + 1] = f(x[k]) + g(x[k])u[k], \tag{2.49}$$

MLPs or RBFs are used to approximate the functions $f(\cdot)$ and $g(\cdot)$, in a procedure similar to the approximation of the Lie derivatives, in the continuous case [114–118]. Recurrent state–space neural models, whose states are used for the determination of the control action, have also been proposed. Examples are Reference 119, in a non-adaptive context, and Reference 120 in an adaptive scheme.

## 2.7    Example

Greenhouses are mainly used to improve the environmental conditions in which plants are grown. Greenhouse environmental control (GEC) provides a means to further improve these conditions in order to optimise the plant production process. GEC methods aimed at efficiently controlling the greenhouse climate environment and optimising the crop production must take into account the influences of the outside weather, the actuators and the crop, which is achieved by the use of models. The production of most crops is primarily affected by the greenhouse air temperature, humidity and $CO_2$ concentration.

This section presents an example application of two of the most widely used soft-computing approaches (neural computation and evolutionary computation) for the identification of predictive greenhouse climate models and outside weather models.

### 2.7.1    Data pre-processing

Relevant outside weather variables available for this system identification problem are the air temperature $(t_w)$, global solar radiation $(sr_w)$, wind speed $(ws_w)$ and wind direction $(wd_w)$. Greenhouse climate variables available are the air temperature $(t_c)$, relative humidity $(h_c)$ and global solar radiation $(sr_c)$. Throughout the rest of this chapter, relative humidity will be simply denoted as humidity.

The data was acquired at a 1 min sampling rate in a plastic covered greenhouse during 12 days. Considering that climate variables change slowly over time, the amount of data is reduced by changing the sampling rate from 1 to 5 min, with the benefits of reducing the computational effort in the identification methods, and, for this particular data set, of allowing better model excitation. Instead of directly subsampling the original data, a five sample average is applied to the entire data set, which also filters high frequencies that may be present. All the data is scaled to the [0;1] interval due to the different scales of the variables.

### 2.7.2    Predictive model structure and arrangement

The models are required to predict 3 h (36 steps) in the future (multistep prediction). As the predictions of $1, 2, \ldots, l$ steps ahead are also required, models that directly predict $l$ steps into the future become inappropriate. Given these requirements, the obvious solution is to build one-step-ahead (OSA) predictive models of the general form,

$$\hat{y}(k+1) = f(y(k), \ldots, y(k-n_y), u_1(k), \ldots, u_1(k-n_{u_1}), \ldots,$$
$$u_m(k), \ldots, u_m(k-n_{u_m})) \tag{2.50}$$

and when predicting a further step ahead, substituting $k \rightarrow k+1$ and using $\hat{y}(k+1)$ from the previous prediction step. These operations are repeated until the prediction horizon $l$ is reached.

It is known from previous work [121] that the variables $ws_w$, $wd_w$ and $sr_c$ are not as relevant as the remaining ones when modelling the greenhouse air temperature. Given the fact that the greenhouse temperature and humidity are strongly coupled quantities, it is assumed that those three variables can also be discarded for humidity modelling. $ws_w$ and $wd_w$ could be of interest for greenhouse humidity models if outside humidity measurements and the greenhouse openings state were available, which is not the case. This leaves us with an input universe $X = \{t_w, sr_w, t_c, h_c\}$.

Taking into account the composition of $X$ and the structure of equation (2.50), an NARX structure, the greenhouse temperature and humidity models have the following general structure:

$$\hat{h}_c(k+1) = f_{h_c}(h_c(k), \ldots, h_c(k-n_{h_c}), t_c(k), \ldots, t_c(k-n_{t_c})), \tag{2.51}$$

$$\hat{t}_c(k+1) = f_{t_c}(t_c(k), \ldots, t_c(k-n_{t_c}), h_c(k), \ldots, h_c(k-n_{h_c}),$$
$$t_w(k), \ldots, t_w(k-n_{t_w}), sr_w(k), \ldots, sr_w(k-n_{sr_w})). \tag{2.52}$$

The dependence of $f_{t_c}$ on $t_w$ and $sr_w$, and the requirement of a prediction horizon $l = 36$, make predictive models for $t_w$ and $sr_w$ necessary for a real application. Since we are in the context of black-box modelling, and no other relevant input quantities measurements are available, $sr_w$ can be modelled with a nonlinear auto-regressive (NAR) structure, which corresponds to (2.50) without the exogenous $(u_1, \ldots, u_m)$ inputs. Regarding the outside temperature, although it is affected by solar radiation, there are many other local and global climate factors that strongly affect it, for which measurements are also unavailable and that $sr_w$ cannot account for. Again, the NAR structure can be used for the $t_w$ predictive model. Both these models have the following general form:

$$\hat{sr}_w(k+1) = f_{sr_w}(sr_w(k), \ldots, sr_w(k-n_{sr_w})), \tag{2.53}$$

$$\hat{t}_w(k+1) = f_{t_w}(t_w(k), \ldots, t_w(k-n_{t_w})). \tag{2.54}$$

The interconnections between climate and weather models are depicted in Figure 2.11. Thick lines indicate paths for predicted values as needed by the models, and explicitly show that for $l > 1$ both measured and predicted values are used as inputs. Model identification occurs exclusively along thin lines. When performing multistep

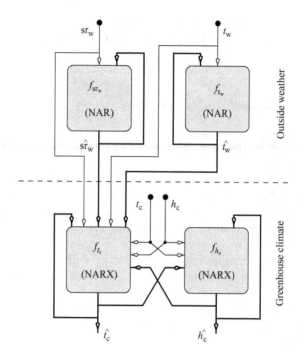

*Figure 2.11    Predictive models interconnection*

prediction and $l$ is sufficiently large, thin lines cease to be used and the model can be denoted as a simulation model (Figure 2.11).

### 2.7.3    Identification framework

#### 2.7.3.1    RBF neural networks

The authors have previously worked on greenhouse climate modelling problems, namely inside air temperature [43,121–126], and also on outside solar radiation prediction [127]. Work focused on the application of RBFs to these modelling problems. As pointed out before, this rather simple feedforward neural network architecture is characterised by having $n$ neurons in the hidden layer, each defined by a radial function $(\varphi_i)$ (usually Gaussian), whose parameters are its centre location $(c_i)$ on input space and its spread around the centre $(\sigma_i)$. At the output it has a linear combiner assigning a weight $(u_i)$ to the output of each neuron, plus an optional bias weight. The RBF neural network architecture is used in the four models, $f_{h_c}$, $f_{t_c}$, $f_{sr_w}$ and $f_{t_w}$.

The training method used to determine the neural network parameters $\{c_i, \sigma_i, u_i\}$, $i = 1, \ldots, n$ is the Levenberg–Marquardt method minimising criterion (2.18). The nonlinear parameters are initialised with the OAKM method and the termination criterion is the early stopping technique. This training procedure only computes the network parameters for a given model structure. The number of neurons and lagged input-variable terms need to be given beforehand, but there is no straightforward way to determine them for a given problem. In fact, this constitutes a combinatorial optimisation problem by itself, which becomes even harder if the optimisation of the neural network structure involves multiple, possibly conflicting, goals. A typical formulation consists of searching for combinations of the number of neurons and input terms, minimising both prediction errors and the neural network number of parameters. This formulation adds a multi-objective character to the nonlinear optimisation problem. Although trial and error can provide some insight, the number of possibilities is often enormous, and it may result in the execution of many trials without obtaining the desired performance and/or neural network complexity. Moreover, the results from the trials easily become misleading, as the shape of the search space is unknown, and may induce the designer into some poor local minima.

### 2.7.3.2   Multi-objective genetic algorithm

Multi-objective evolutionary algorithms (MOEA) are one type of nonlinear optimisers that mimic the human trial and error approach over the entire search space, and based on a set of performance measures try to evolve satisfactory solutions that meet certain pre-specified goals. One of the advantages of MOEA techniques over other techniques is that they can provide a diverse set of non-dominated solutions to problems involving a number of possibly conflicting objectives, in a single run of the algorithm. They have been successfully applied [128,129] to this type of system identification problem. In particular the multi-objective genetic algorithm (MOGA) [46] has been applied by the authors to greenhouse inside air temperature prediction [121]. It is used in this example to select the RBF structure of the models discussed in Section 2.7.2, equations (2.51)–(2.54). The MOGA is an evolutionary computing approach, inspired by the theory of natural selection and the notion of survival of the fittest, which performs a population-based search by employing operators, such as selection, mating and mutation. One run of a MOGA algorithm consists of a sufficiently large number of generations to allow the evolution of individuals meeting certain pre-specified requirements.

### 2.7.3.3   RBF neural network representation and the MOGA

The structure of a multiple-input single-output RBF neural network can be represented by the number of neurons in the hidden layer and the number of inputs. The chromosome for one such network can be represented by a string of integers, the first of which corresponds to the number of neurons, $n$, and the remaining represent a variable number of distinct input terms. Input terms are represented by their position in a lookup table of the lagged terms considered for each available variable. Figure 2.12 illustrates the chromosome representation and the lookup table method.

After the individuals in a generation are evaluated, the population is ranked using the preferability relation [46] and then the individuals selected are mated to produce two offsprings from each pair of parents. Parent recombination is done in such a way that the offspring respect the maximum model length with no loss of input terms [128]. The resulting offspring may be longer, shorter or equally sized as their parents.

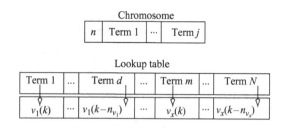

*Figure 2.12     Chromosome and terms lookup table*

The mutation operator is implemented by three basic operations: substitution, deletion and addition of one element. The number of neurons is mutated, with a given probability, by adding or subtracting one neuron to the model, verifying boundary conditions that no neural network can have fewer or more neurons than pre-specified values. Each model input term in the chromosome is tested and, with a given probability, is either replaced by a new term not in the model, or deleted. Finally a new term may be appended to the chromosome.

#### 2.7.3.4   Model design cycle

Globally, the model structure optimisation problem can be viewed as a sequence of actions undertaken by the model designer, which should be repeated until pre-specified design goals are achieved. These actions can be grouped into three major categories: problem definition, solution(s) generation and analysis of results. In the context of this identification framework, the procedure is executed as depicted in Figure 2.13. In summary, the problem definition is carried out by choosing a number of hypothetically relevant variables and corresponding lagged terms; this affects the size of the search space. Another aspect to be defined is the set of objectives and goals to be attained; this affects the quantity, quality and class of the resulting solutions. When the analysis of the solutions provided by the MOGA requires the process to be repeated, the problem definition steps should be revised. In this case, three major actions can be carried out: input space reduction by removing one or more variables, reducing the number of input terms by choosing more appropriate ones from the set of solutions already obtained and restricting the trade-off surface coverage by changing objectives or redefining goals. This cycle of actions can be iterated until a refined set of satisfactory solutions is obtained.

### 2.7.4   Methodology

#### 2.7.4.1   Data sets

The input universe, $X$, described in Section 2.7.2 and pre-processed as discussed in Section 2.7.1 is split up into three equally sized (4 days) data sets, $DS_t$, $DS_g$ and $DS_v$, for model training, generalisation and validation, respectively. For each variable in the data sets, lagged terms up to 24 h are computed, resulting in 288 possible input terms per variable. In order to reduce the size of the search space, a subset of all

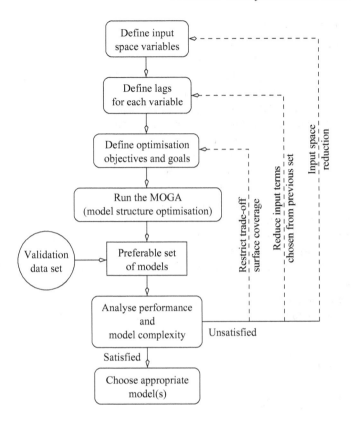

*Figure 2.13    Model design cycle*

possible input terms is selected according to the following reasoning. Consider the fact that climate values at a given time instant are strongly related to their most recent values, and also, to a certain extent, to their values 24 h before. So that recent values predominate and a smaller number of not so recent values, including those from 1 day before, are present, per-variable lagged term subsets are chosen as follows:

$$LT_i = \text{round}(lt_i), \quad i = 1, \dots, N,$$
$$lt_i = lt_{i-1}h, \quad i = 2, \dots, N, \tag{2.55}$$
$$lt_1 = 1,$$

where $N$ is the desired number of terms and the value of $h$ is such that $LT_N = 288$. As an example, for $N = 10$, the lagged terms considered would be:

$$LT = [1, 2, 4, 7, 12, 23, 44, 82, 154, 288],$$

and $h = 1.8761$. In this work, for each variable considered, $N$ is set to 144 and $h = 1.0404$.

### 2.7.4.2   Model evaluation

In order to rank the individuals in each generation, the MOGA requires these to be evaluated. The objectives used in the experiments can be classified into two groups: model complexity and model performance. For the former, the 2-norm of the RBF neural network output linear parameters, $\|u\|$, is used. Model performance objectives are the sum of the OSA training error (OSATE), the sum of the OSA testing or generalisation error (OSAGE) and a long-term prediction error measure, $R(E^{\mathrm{T}})$, described below. Another class of objectives which are sometimes used [121] in the MOGA optimisation, are model validation objectives. These are usually implemented by means of correlation-based model validity tests [77,78], consisting of cross-correlations involving the model residuals, model output and model inputs. As the models are trained as OSA predictors, and the validity tests efficiency depends essentially on the OSA training error, preliminary work, not reported here, indicated that the tests bias the search towards better OSA predictors. For this reason, correlation based model validation tests are not employed here.

The long-term prediction performance of a model is assessed using the DS$_g$ data set and a horizon of 36 steps (3 h). To reduce computational time, prediction is computed starting at 1 h intervals, resulting in 96 prediction horizons. For each starting point, the prediction error $e[k]$ is taken over the prediction horizon. Consider the matrix,

$$
E = \begin{bmatrix} e_{(1,1)} & \cdots & e_{(1,36)} \\ \cdots & \cdots & \cdots \\ e_{(96,1)} & \cdots & e_{(96,36)} \end{bmatrix},
$$

where each row corresponds to the errors obtained over each prediction horizon. Let $R(\cdot)$ be the root mean square error (RMSE) function operating over the rows of a matrix. In this way, $R(E)$ and $R(E^{\mathrm{T}})$ are vectors denoting the RMSE over the full predicted horizon for every time instant, and the RMSE over all the horizons, for each prediction instant within the considered horizon, respectively.

### 2.7.4.3   Weather models: $f_{\mathrm{sr_w}}$ and $f_{t_w}$

Both outside weather models are selected by MOGA algorithms in two independent experiments where the main goal is to select good predictive models with the lowest possible complexity. Each individual is represented by the RBF neural network chromosome described in Section 2.7.3.3. The number of neurons, $n$, is required to be between 2 and 15 and the maximum number of input terms allowed is 15, resulting in a chromosome length of 16. The population size is set to 150 individuals and each MOGA run is composed of 250 generations. The selective pressure, cross-over rate and mutation survival rate [46] are, respectively, 2, 0.7 and 0.5. Table 2.1 shows how the MOGA is set up in terms of objectives for the two weather models. Values are presented unnormalised, temperature in °C and solar radiation in Wm$^{-2}$. OSATE and $\|u\|$ are set up as restrictions with equal priority. Four MOGA runs were executed for each model.

Table 2.1  MOGA goals and objectives for the weather models

| Objective | Solar radiation | Temperature |
|---|---|---|
| OSATE | $\leq 50$ | $\leq 0.5$ |
| OSAGE | Minimise | Minimise |
| $\max(R(E^{\mathrm{T}}))$ | Minimise | Minimise |
| $\|u\|$ | $\leq 4000$ | $\leq 80$ |

Table 2.2  MOGA runs objectives for the climate models

| Objective | Case 1 | Case 2 |
|---|---|---|
| $\mathrm{OSATE}_{(t_c+h_c)}$ | × | Minimise |
| $\mathrm{OSAGE}_{(t_c+h_c)}$ | × | Minimise |
| $\max(R(E^{\mathrm{T}}))_{h_c}$ | Minimise | Minimise |
| $\max(R(E^{\mathrm{T}}))_{t_c}$ | Minimise | Minimise |
| $\|u\|_{h_c}$ | $\leq 7$ | $\leq 7$ |
| $\|u\|_{t_c}$ | $\leq 7$ | $\leq 7$ |

### 2.7.4.4  Greenhouse climate models: $f_{h_c}$ and $f_{t_c}$

As mentioned above, greenhouse temperature and humidity are strongly coupled variables. Identifying long-term predictive models of coupled variables can be cast as the problem of finding a set of coupled models satisfying some pre-specified criteria. As long as autoregressive (AR) model structures are to be selected based on long-term prediction errors, both models must be determined prior to evaluation. Having these considerations in mind, the problem can be stated as the search for a pair of coupled humidity–temperature models which is best in some sense.

Given this problem statement, each individual chromosome is constructed by concatenating two strings of integers as described in Section 2.7.3.3. The number of neurons for each network is required to be between 2 and 15. $f_{h_c}$ and $f_{t_c}$ may have 16 and 32 inputs, respectively. The selective pressure, cross-over rate and mutation survival rate have the same values as for the weather models.

Table 2.2 shows how the two case studies considered were set up in terms of objectives. The OSATE and OSAGE objectives correspond to the sum of the values obtained by both models, as indicated with subscripts. For the remaining objectives the subscript identifies the corresponding model. The × symbol means the objective was ignored. The measures $\|u\|_{h_c}$ and $\|u\|_{t_c}$ were set up as restrictions with equal priority. For each case, five MOGA runs were executed.

## 2.7.5    Results and discussion

### 2.7.5.1    Weather models: $f_{sr_w}$ and $f_{t_w}$

Table 2.3 presents a summary of the number of individuals obtained in the preferable set (PS) [46], for both weather models. The three values correspond to the minimum, average and maximum values obtained over the four MOGA runs, where a total of 45 and 33 models were selected for the solar radiation and temperature models, respectively. The distributions of the number of neurons and number of inputs are shown in Figure 2.14. The dashed line and $y$-axis scale mark the maximum numbers of temperature and solar radiation models, respectively. Greater dispersion is observed in the number of inputs. Networks with more inputs are clearly favoured in both models. Regarding the number of neurons, distinct results are obtained for each model. For temperature networks a minimal number of neurons is selected whereas for solar radiation a larger number is obtained. In summary, these results indicate that better temperature models are obtained with a small number of neurons with more inputs, and that solar radiation is best predicted by networks with a larger number of neurons and inputs.

Table 2.4 presents results for the MOGA objectives, the number of parameters of the networks, the OSA validation error (OSAVE) and $\max(R(E^T))_{DS_v}$ values obtained. OSAVE and $\max(R(E^T))_{DS_v}$ values are computed using the $DS_v$ data set. NP stands for the number of neural network parameters.

These results show that in general the selected networks have good generalisation capabilities, as it can be seen by comparing the OSATE and OSAGE, and observing the small degradation obtained for the OSAVE. Especially evident for the solar radiation model, is the fact that minimising OSATE and $\max(R(E^T))$ yields networks whose OSATE values are very close to the restriction imposed (see Table 2.1). An analysis of the models obtained in the non-dominated set reveals that many individuals have smaller OSATE, but, in general, worse generalisation and long-term prediction abilities. These facts point out that minimising OSA prediction errors does not lead to the best prediction results. A maximum degradation of 41.13 $Wm^{-2}$ and 0.82°C is obtained when comparing $\max(R(E^T))$ with $\max(R(E^T))_{DS_v}$, again revealing good generalisation properties of both weather models. Recall that during the MOGA optimisation, prediction is computed starting at 1 h intervals to reduce computational effort. The last row in Table 2.4 presents $\max(R(E^T))_{DS_v}$ results with prediction starting at 5 min intervals, that is, the whole data set.

*Table 2.3    Weather models MOGA results*

| Solar radiation | Temperature |
| --- | --- |
| 7 | 4 |
| 11.5 | 9 |
| 16 | 13 |

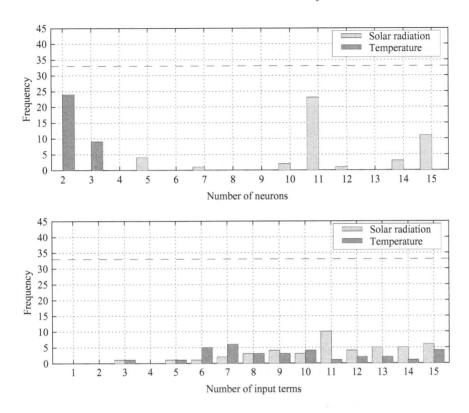

*Figure 2.14    Frequencies of the number of neurons and number of inputs in weather models*

*Table 2.4    MOGA objectives summary for the weather models obtained*

| Objectives | Solar radiation | | | Temperature | | |
|---|---|---|---|---|---|---|
| | Min. | Average | Max. | Min. | Average | Max. |
| $\|u\|$ | 1383.5 | 2439.5 | 4527.1 | 32.0405 | 61.1362 | 79.8225 |
| OSATE | 44.5397 | 47.5724 | 49.9504 | 0.1694 | 0.2378 | 0.4890 |
| OSAGE | 53.0043 | 62.6532 | 98.0064 | 0.1790 | 0.2677 | 0.5869 |
| $\max(R(E^T))$ | 94.0175 | 130.7545 | 304.6906 | 1.1420 | 1.2859 | 2.3949 |
| $\max(R(E^T))_{DS_v}$ | 133.2116 | 168.3182 | 345.8190 | 1.9472 | 2.4567 | 3.2138 |
| NP | 44 | 142.8 | 240 | 12 | 23 | 39 |
| OSAVE | 51.7991 | 65.7983 | 106.4085 | 0.2236 | 0.3630 | 0.8867 |
| $\max(R(E^T))_{DS_v}$ | 116.5187 | 146.8342 | 308.9832 | 1.9094 | 2.4076 | 3.1397 |

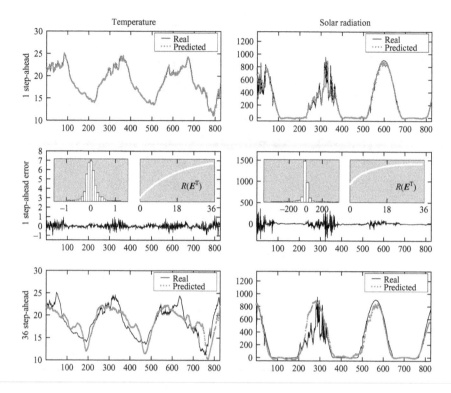

*Figure 2.15    Weather models, fittings*

The number of model parameters obtained indicate that solar radiation prediction is a harder task when compared to temperature prediction, as all $sr_w$ models have larger complexity than $t_w$ models. Looking at the results presented in Figure 2.14 it can be concluded that the number of neurons motivates this difference in model complexity. This increase in the number of neurons is probably due to stronger and uncertain dynamics associated with cloudiness.

Figure 2.15 presents graphics regarding one temperature model (left column of plots) and one solar radiation model (right column of plots). The one-step-ahead fitting is shown by the top plots. The plots in the middle row present the corresponding error and two subgraphs: a histogram of the plotted error and the evolution of $R(E^T)$ with the prediction horizon. The bottom plots depict the 3 h ahead prediction. All the fittings correspond to the $DS_v$ data set.

The histograms of input terms employed in the weather models are shown in Figure 2.16. The smaller histograms embedded in each model lags histogram, help to evaluate the input selectivity achieved. They show that $y$ lags ($y$-axis) are used $x$ times ($x$-axis) in the models. For example, regarding solar radiation, there is 1 lag which is employed in 43 models and 26 lags which are used only in 1 model. The main histograms' $y$-scale corresponds to the number of models obtained over the four MOGA runs.

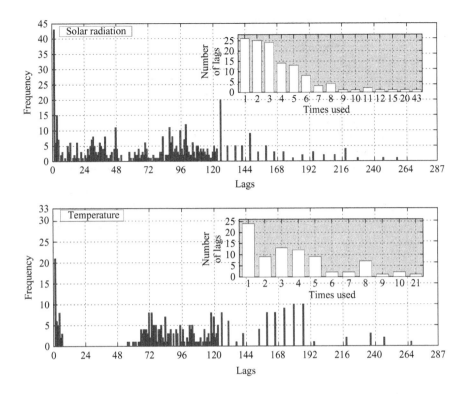

*Figure 2.16    Input terms frequency for weather models*

For solar radiation, 125 out of 144 possible terms are present in the 45 models obtained. The corresponding smaller histogram shows that a large number of lags are used very few times. As an example, 110 lags (the sum of the $y$-values for the first 6 bars in the small histogram) are present a maximum of 5 times, and 15 are used in 7 or more models. If the lags used 6 or fewer times are not considered, than there are only 15 lags left. Regarding the temperature models, 82 out of 144 available terms are used in the 33 models selected. Discarding lags employed 5 or fewer times leaves only 15 lags. Although the lags histograms seem to indicate a better input selection for the temperature model, the numbers are not so different when filtering less used lags. The most relevant difference relies on the fact that lag 0 is relatively more frequent in the solar radiation models, being absent only in two models.

### 2.7.5.2    Greenhouse climate models: $f_{h_c}$ and $f_{t_c}$

Table 2.5 presents some general results regarding the two case studies described in Section 2.7.4.4, Table 2.2. Where the notation $x$, $y$, $z$ is found, $x$, $y$ and $z$ correspond to the minimum, mean and maximum values obtained over the five MOGA runs.

Case 2 has a larger objective space, originating a PS with more individuals than in case 1. The numbers of distinct models indicate greater humidity model similarity

*Table 2.5    MOGA runs results*

|  | Case 1 | Case 2 |
|---|---|---|
| Generations | 350 | 500 |
| PS(#) | 19, 24.4, 33 | 123, 134.8, 157 |
| Distinct pairs of models (PS) | 16, 20.8, 25 | 113, 125, 148 |
| Distinct humidity models (PS) | 15, 19.8, 25 | 67, 84.2, 110 |
| Distinct temperature models (PS) | 12, 17.4, 25 | 101, 116, 140 |

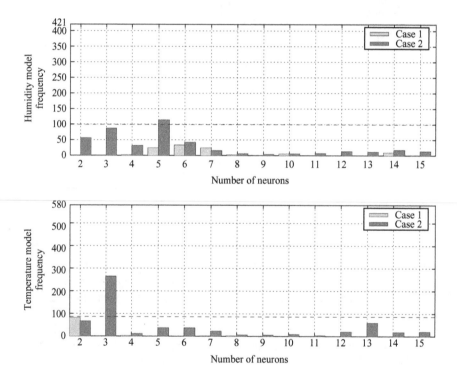

*Figure 2.17    Frequencies of the number of neurons in climate models*

in case 2, whereas for case 1 this is verified by the temperature models (see also Figure 2.19). Figure 2.17 shows the distribution of the number of neurons in the preferred humidity and temperature models found in the five MOGA runs. The dashed line and the $y$-axis scale correspond to the number of models obtained in cases 1 and 2, respectively. The numbers show a greater dispersion in case 2, probably due to the presence of the OSA prediction error objectives, which favour a larger number of neurons. Clearly, networks with fewer neurons are favoured, and better long-term prediction results were observed for such networks in general.

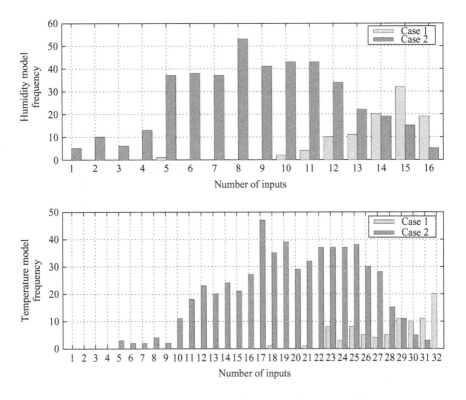

*Figure 2.18    Frequencies of the number of inputs in climate models*

Figure 2.18 presents the distribution of the number of inputs in the humidity and temperature models found in the five runs. Again the results for case 2 show much greater diversity. As opposed to the number of neurons, networks with more inputs are favoured and generally presented better long-term prediction results. In summary, the best predictors are obtained by means of networks with few neurons and many inputs. Table 2.6 compares both cases in terms of the MOGA objectives, the OSA validation error ($\text{OSAVE}_{(t_c+h_c)}$) and the values of $\max(R(E^T))$ computed over the $\text{DS}_v$ data set for both models.

It is clear that case 2 models are generally better OSA predictors, although some exhibit worse $\text{OSATE}_{(t_c+h_c)}$ and $\text{OSAGE}_{(t_c+h_c)}$ values. This is caused by ill-conditioned models as can be seen by the results in $\|u\|_{h_c}$ and $\|u\|_{t_c}$. The fact that case 1 models are not as good OSA predictors is justified by the absence of small lag terms, namely $t-1$ appearing only in about 30 per cent of the models obtained. Considering this, it is not surprising that the relation between $\text{OSAVE}_{(t_c+h_c)}$ and the other OSA measures favours case 2 models. On the other hand, case 1 models present the best results regarding long-term prediction, both in the MOGA objectives and when evaluated with the $\text{DS}_v$ data set (2 last rows). Figure 2.19 compares the attainment surfaces (ASs) [130] obtained in both case studies, regarding the long-term prediction objectives only.

*Table 2.6    MOGA objectives in the two runs, regarding distinct pairs of models in the PS*

| Measure | Minimum | | Average | | Maximum | |
|---|---|---|---|---|---|---|
| | Case 1 | Case 2 | Case 1 | Case 2 | Case 1 | Case 2 |
| $OSATE_{(t_c+h_c)}$ | 0.0238 | **0.0085** | 0.0360 | **0.0197** | **0.0552** | 0.2281 |
| $OSAGE_{(t_c+h_c)}$ | 0.0338 | **0.0123** | 0.0622 | **0.0304** | **0.1030** | 0.3191 |
| $\max(R(E^T))_{h_c}$ | **4.19%** | 5.46% | **6.71%** | 8.82% | **24.37%** | 32.11% |
| $\max(R(E^T))_{t_c}$ | 1.18°C | **1.17°C** | **1.41°C** | 2.00°C | **2.70°C** | 9.87°C |
| $\|u\|_{h_c}$ | 1.0492 | **0.6025** | **1.8663** | 2.9753 | **5.0475** | 6.9696 |
| $\|u\|_{t_c}$ | 1.0723 | **0.9138** | **2.5793** | 3.0127 | **4.2869** | 6.8532 |
| $OSAVE_{(t_c+h_c)}$ | 0.0521 | **0.0129** | 0.1122 | **0.0445** | **0.1577** | 0.3502 |
| $\max(R(E^T))_{h_c[DS_v]}$ | 8.88% | **8.61%** | **12.16%** | 13.65% | **44.61%** | 49.38% |
| $\max(R(E^T))_{t_c[DS_v]}$ | 2.71°C | **2.37°C** | **3.10°C** | 3.67°C | **5.82°C** | 26.65°C |

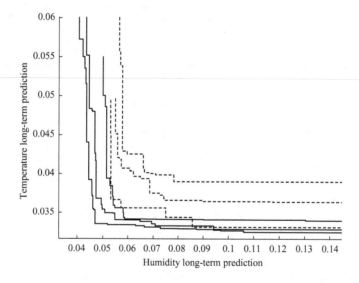

*Figure 2.19    Attainment surfaces for the two case studies (Solid line: case 1. Dashed line: case 2)*

For each case study, the three lines (from the lower left to the upper right corner) depict the estimates of the 0, 50 and 100 per cent ASs. The region attained in case 2 is almost entirely covered by that of case 1, with the exception of a small region of very good temperature models but poor humidity predictors. It is important to note that the models obtained in case 1 are generally non-dominated with respect to those obtained in case 2. This is an indication that in case 2 the MOGA was unable to cover the trade-off surface to its full extent, possibly due to the increased dimensionality of the objective space. However, goal levels for each objective could be provided to the MOGA, which would reduce the size of the preferred region and focus the search, eventually providing better performance [47].

Regarding input term selection, good agreement could be found for a considerable number of either specific terms or terms with very similar lags. Figures 2.20 and 2.21 show input lags histograms for each input variable, regarding humidity and temperature models selected in case 1, respectively. The $y$-axis is scaled to the number of models obtained over the five MOGA runs. For the humidity models, humidity and temperature input terms have percentages of use of 42.9 and 57.1 per cent, respectively. These values account for the relevance of the two variables in humidity prediction; 109 out of 144 lags are present for humidity and 113 out of 144 for temperature. Considering humidity inputs, 18 lags are employed 11 or more times and 8 lags are present 14 or more times. The most frequent lag is employed in 40 models. For temperature inputs, 16 lags appear 13 or more times and 9 are present in 19 or more models. The most frequent lag appears in 53 models. Considering the temperature models, the input variables $t_c$, $h_c$, $t_w$ and $sr_w$, have percentages of use of 27.9, 22.3, 14.9 and 34.9, respectively.

The first row in Table 2.7 shows how many (out of 144 possible) input terms appear in the models, for each variable. The two subsequent rows show the minimum

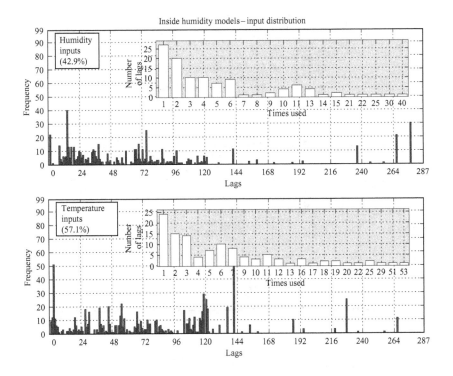

*Figure 2.20    Humidity model input selection*

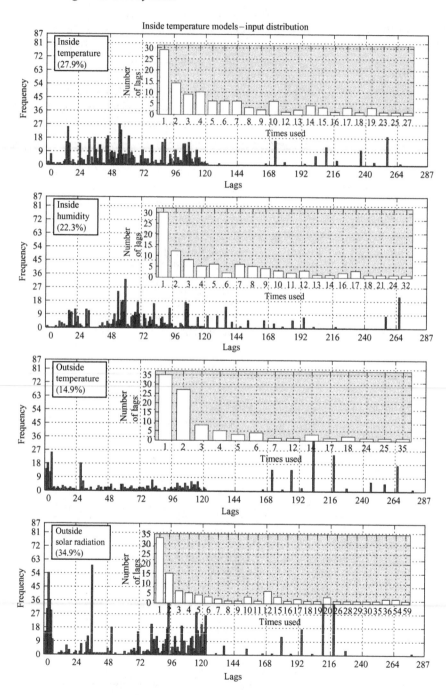

*Figure 2.21    Temperature model input selection*

*Table 2.7* Input selection details for the $t_c$ models

|  | $t_c$ | $h_c$ | $t_w$ | $sr_w$ |
|---|---|---|---|---|
| Out of 144 | 112 | 97 | 93 | 101 |
| Most frequent 8 lags frequency | 17+ | 16+ | 14+ | 29+ |
| Most frequent 16 lags frequency | 14+ | 11+ | 5+ | 17+ |
| Most frequent | 27 | 32 | 35 | 59 |

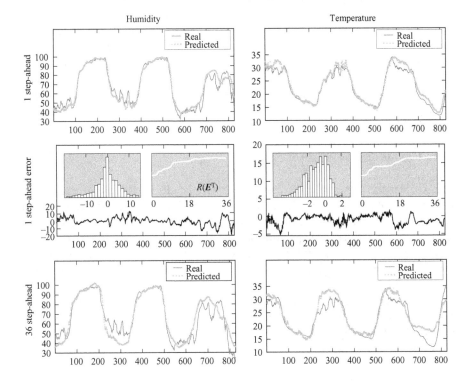

*Figure 2.22* Climate models, fittings

number of models where the 8th and 16th more frequent lags appear. Finally, the last row shows the number of models where the most frequent lag is employed.

Outside temperature is clearly the variable with less relevance to the inside temperature models, but that does not mean that the models would benefit from its removal. For example, its six more frequent input lags have comparable frequencies with those from $t_c$ and $h_c$. These results suggest that a smaller number of outside temperature lags should be considered when compared to the other input variables. Another possibility, motivated by the percentage of use of this variable, would be to remove it from the design problem, rerun the MOGA, compute the results and evaluate them, in order to decide if the removal would be beneficial or not. In fact, this iterative method was already used to perform input space reduction in a previous work [121], resulting in the four variables considered here. The most frequent variable in the models is the outside solar radiation, considering both the percentage of terms and their frequency. This result has some physical meaning, given the known effect of light integration in the inside temperature, caused by the greenhouse cover. Certainly, input term selection would gain from a reduction in search space dimension by further restricting the number of neurons and inputs. Such restrictions can be guided by the results obtained regarding those parameters.

Long-term prediction capabilities over the $DS_v$ data set, for a model pair with a good compromise between humidity and temperature prediction, selected from a run in case 1, is shown in Figure 2.22. The networks have 6 neurons and 12 inputs for the humidity model, and 2 neurons and 14 inputs for the temperature model. Good fittings are obtained for the prediction in 1 and 36 steps ahead. Both figures also show the evolution of $\max(R(E^T))$ over the prediction horizon. The mean and maximum absolute error values obtained for humidity and temperature are 5.77 and 26.15 per cent, and 1.78°C and 7.76°C.

## 2.8    Concluding remarks

This chapter has introduced the basic concepts related with the use of neural networks for nonlinear systems identification, and has briefly reviewed neuro-control approaches. Several important issues for the design of data-driven models, as neural networks are, such as data acquisition and the design of excitation signals could not be covered here, but they will be discussed in Chapter 5. On purpose, important topics such as neuro-fuzzy and local linear models were not discussed, as they will be treated in other chapters of this book.

Neural network modelling is an iterative process, requiring, at the present stage, substantial skills and knowledge from the designer. It is our view that the methodology presented in the example, employing multi-objective evolutionary algorithms for the design of neural models, is a suitable tool to aid the designer in this task. It incorporates input model order and structure selection, as well as parameter estimation, providing the designer with a good number of well performing models with varying degrees of complexity. It also allows the incorporation of objectives which are specific for the ultimate use of the model. Through the analysis of the results obtained in one iteration, the search space can be reduced for future iterations, therefore allowing a more refined search in promising model regions.

### Acknowledgements

The authors wish to acknowledge the support of Fundação para a Ciência e Tecnologia (project POCTI/33906/MGS/2000 and grant SFRH/BD/1236/2000) and the project Inovalgarve 04-02.

# References

1  HAYKIN, S.: 'Neural networks: a comprehensive foundation' (Prentice-Hall, New Jersey, 1999, 2nd edn.)

2  PRINCIPE, J. C., EULIANO, N. R., and LEFEBVRE, W. C.: 'Neural and adaptive systems: fundamentals through simulations' (John Wiley & Sons, New York, 1999)

3  NELLES, O.: 'Nonlinear systems identification: from classical approaches to neural networks and fuzzy models' (Springer-Verlag, Berlin, 2000)

4  BROWN, M., and HARRIS, C. J.: 'Neurofuzzy adaptive modelling and control' (Prentice-Hall, Hemel Hempstead, 1994)

5  HARRIS, C., HONG, X., and GAN, Q.: 'Adaptive modelling, estimation and fusion from data: a neurofuzzy approach' (Springer-Verlag, Berlin, 2002)

6  RUMELHART, D., McCLELLAND, J., and the PDP Research Group: 'Parallel distributed processing, vol. 1' (MIT Press, Cambridge, MA, 1986)

7  FUNAHASHI, K.: 'On the approximate realization of continuous mappings by neural networks', *Neural Networks*, 1989, **2**, pp. 183–92

8  HORNIK, K., STINCHCOMBE, M., and WHITE, H.: 'Multilayer feed-forward networks are universal approximators', *Neural Networks*, 1989, **2**, pp. 359–66

9  BROOMHEAD, D., and LOWE, D.: 'Multivariable functional interpolation and adaptive networks', *Complex Systems*, 1988, **2**, pp. 321–55

10  GIROSI, F., and POGGIO, T.: 'Networks and the best approximation property', *Biological Cybernetics*, 1990, **63**, pp. 169–76

11  WERBOS, P.: 'Beyond regression: new tools for prediction and analysis in the behavioral sciences'. Doctoral dissertation, Applied Mathematics, Harvard University, USA, 1974

12  WATROUS, R.: 'Learning algorithms for connectionist networks: applied gradient methods of nonlinear optimization'. Proceedings of first IEEE international conference on *Neural networks*, San Diego, USA, 1987, vol. 2, pp. 619–27

13  JACOBS, R.: 'Increased rates of convergence through learning rate adaptation', *Neural Networks*, 1988, **1**, pp. 295–307

14  LEONARD, J., and KRAMER, M. A.: 'Improvement of the backpropagation algorithm for training neural networks', *Computers and Chemical Engineering*, 1990, **14** (3), pp. 337–41

15  GILL, P., MURRAY, W., and WRIGHT, M.: 'Practical optimization' (Academic Press Limited, London, 1981)

16  LEVENBERG, K.: 'A method for the solution of certain problems in least squares', *Quarterly Applied Mathematics*, 1944, **2**, pp. 164–8

17  MARQUARDT, D.: 'An algorithm for least-squares estimation of non-linear parameters', *SIAM Journal of Applied Mathematics*, 1963, **11**, pp. 431–41

18  RUANO, A. E., FLEMING, P. J., and JONES, D.: 'A connectionist approach to PID autotuning'. Proceedings of IEE international conference on *CONTROL 91*, Edinburgh, UK, 1991, vol. 2, pp. 762–7

19  FERREIRA, P. M., and RUANO, A. E.: 'Exploiting the separability of linear and nonlinear parameters in radial basis function networks'. Proceedings symposium 2000 on *Adaptive systems for signal processing, communications and control* (*ASSPCC*), Lake Louise, Canada, 2000, pp. 321–6

20  RUANO, A. E., CABRITA, C., OLIVEIRA, J. V., and KÓCZY, L. T.: 'Supervised training algorithms for B-spline neural networks and neuro-fuzzy systems', *International Journal of Systems Science*, 2002, **33** (8), pp. 689–711

21  RUANO, A. E., JONES, D., and FLEMING, P. J.: 'A new formulation of the learning problem for a neural network controller'. Proceedings of the thirtieth IEEE conference on *Decision and control*, Brighton, England, 1991, vol. 1, pp. 865–6

22  RUANO, A. E.: 'Applications of neural networks to control systems'. Ph.D. thesis, University College of North Wales, UK, 1992

23  YAO, X.: 'Evolving artificial neural networks', *Proceedings of the IEEE*, 1999, **87** (9), pp. 1423–47

24  FREAN, M.: 'The upstar algorithm: a method for constructing and training feedforward neural networks', *Neural Computation*, 1990, **2** (2), pp. 198–209

25  WIDROW, B., and LEHR, M. A.: '30 years of adaptive neural networks: perceptron, madaline and backpropagation', *Proceedings of the IEEE*, 1990, **78** (9), pp. 1415–42

26  MOODY, J., and DARKEN, C. J.: 'Fast learning in networks of locally-tuned processing units', *Neural Computation*, 1989, **1**, pp. 281–94

27  NELLES, O., and ISERMANN, R.: 'A new technique for determination of hidden layer parameters in RBF networks'. Proceedings of IFAC World Congress, San Francisco, CA, USA, 1996, pp. 453–7

28  CHINRUNGRUENG, C., and SQUIN, C. H.: 'Optimal adaptive $k$-means algorithm with dynamic adjustment of learning rate', *IEEE Transactions on Neural Networks*, 1995, **6** (1), pp. 157–69

29  LIMA, J. M., and RUANO, A. E.: 'Comparison of off-line and on-line performance of alternative neural network models'. Proceedings on *Information processing and management of uncertainity in knowledge based systems* (IPMU 2000), Madrid, Spain, 2000, vol. 2, pp. 1219–25

30  AN, P. E., BROWN, M., and HARRIS, C. J.: 'Aspects of on-line learning rules'. Proceedings of IEE international conference on *CONTROL 94*, 1994, vol. 1, pp. 646–51

31  ASIRVADAM, V. J.: 'On-line learning and construction of neural networks'. Ph.D. thesis, University of Belfast, UK, 2002

32  NAGUNO, J. I., and NODA, A.: 'A learning method for system identification', *IEEE Transactions on Automatic Control*, 1967, **12** (3), pp. 282–8

33  PARKS, P. C., and MILITZER, J.: 'A comparison of five algorithms for the training of CMAC memories for learning control systems', *Automatica*, 1992, **28** (5), pp. 1027–35

34  LJUNG, L.: 'System identification: theory for the user' (Prentice-Hall, NJ, 1999, 2nd edn.)

35 YEE, P., and HAYKIN, S.: 'A dynamic regularized radial basis function network for nonlinear, nonstationary time series prediction', *IEEE Transactions on Signal Processing*, 1999, **47** (9), pp. 2503–21

36 DARKEN, C., CHANG, J., and MOODY, J.: 'Learning rate schedules for faster stochastic gradient search', in KUNG, S. Y., FALLSIDE, F., SØRENSEN, J. A., and KRAMM, C. A. (Eds): 'Neural networks for signal processing, vol. 2' (IEEE Press, Piscataway, NJ, 1992) pp. 3–13

37 CHAN, L. W., and FALLSIDE, F.: 'An adaptive training algorithm for back-propagation networks', *Computer Speech and Language*, 1987, **2**, pp. 205–18

38 BATITI, R.: 'Accelerated backpropagation learning: two optimization methods', *Complex Systems*, 1989, **3**, pp. 331–42

39 CHEN, S., and BILLINGS, S. A.: 'Recursive prediction error parameter estimator for non-linear models', *International Journal of Control*, 1989, **49** (2), pp. 569–94

40 NGIA, L. S. H., and SJÖBERG, J.: 'Efficient training of neural nets for nonlinear adaptive filtering using a recursive Levenberg–Marquardt algorithm', *IEEE Transactions on Signal Processing*, 2000, **48** (7), pp. 1915–26

41 CHEN, S., COWAN, C. F. N., BILLINGS, S. A., and GRANT, P. M.: 'Parallel recursive prediction error algorithm for training layered networks', *International Journal of Control*, 1990, **51** (6), pp. 1215–28

42 NERRAND, O., ROUSSEL-RAGOT, P., PERSONNAZ, L., DREFYS, G., and MARCOS, S.: 'Neural networks and nonlinear adaptive filtering: unifying concepts and new algorithms', *Neural Computation*, 1993, **5** (2), pp. 165–99

43 FERREIRA, P. M., FARIA, E. A., and RUANO, A. E.: 'Neural network models in greenhouse air temperature prediction', *Neurocomputing*, 2002, **43** (1), pp. 51–75

44 HABER, R., and UNBEHAUEN, H.: 'Structure identification of nonlinear dynamic systems: a survey on input/output approaches', *Automatica*, 1990, **26** (4), pp. 651–77

45 VAPNIK, V. N.: 'Statistical learning theory' (John Wiley & Sons, New York, 1998)

46 FONSECA, C. M., and FLEMING, P. J.: 'Multiobjective optimization and multiple constraint handling with evolutionary algorithms – part I: a unified formulation', *IEEE Transactions on Systems, Man, and Cybernetics – Part A*, 1998, **28** (1), pp. 26–37

47 FONSECA, C. M., and FLEMING, P. J.: 'Multiobjective optimization and multiple constraint handling with evolutionary algorithms – part II: application example', *IEEE Transactions on Systems, Man, and Cybernetics – Part A*, 1998, **28** (1), pp. 38–47

48 CHERKASSKY, V. S., and MULIER, F. M.: 'Learning from data: concepts, theory, and methods' (John Wiley & Sons, New York, 1998)

49 HWANG, J., LAY, S., MAECHLER, M., DOUGLAS, R., and SCHIMET, J.: 'Regression modelling in back-propagation and project pursuit learning', *IEEE Transactions on Neural Networks*, 1994, **5** (3), pp. 342–52

50  INTRATOR, N.: 'Combining explanatory projection pursuit and project pursuit regression with application to neural networks', *Neural Computation*, 1993, **5** (3), pp. 443–57

51  FAHLMAN, S. E., and LEBIERE, C.: 'The cascade-correlation learning architecture', in TOURETZKY, D. S. (Ed.): 'Advances in neural information processing systems, vol. 2' (Morgan Kaufman, San Francisco, CA, USA, 1990), pp. 524–32

52  REED, R.: 'Pruning algorithms: a survey', *IEEE Transactions on Neural Networks*, 1993, **4** (5), pp. 740–7

53  WEIGEND, A. S., RUMELHART, D. E., and HUBERMAN, B. A.: 'Generalization by weight-elimination with application to forecasting', in LIPPMANN, R., MOODY, J., and TOURETZKY, D. *et al.* (Eds): 'Advances in natural information processing, vol. 3' (Morgan Kaufman, San Mateo, CA, 1991), pp. 875–82

54  MOODY, J. E., and ROGNVALDSSON, T.: 'Smoothing regularizers for projective basis function networks'. Proceedings on *Neural information processing systems* (NIPS 1996), 1997, vol. 9, pp. 558–91

55  LeCUN, Y., DENKER, J., and SOLLA, S. A.: 'Optimal brain damage', in TOURETZKY, D. S. (Ed.): 'Advances in neural information processing systems, vol. 2' (Morgan Kaufman, San Francisco, CA, USA, 1990), pp. 918–24

56  HASSIBI, B., STORK, D. G., and WOLFF, G. J.: 'Optimal brain surgeon and general network pruning'. Proceedings of IEEE international conference on *Neural networks*, vol. 1, 1992, pp. 293–9

57  CHEN, S., COWAN, C. F. N., and GRANT, P. M.: 'Orthogonal least-squares learning algorithm for radial basis function networks', *IEEE Transactions on Neural Networks*, 1991, **2** (2), pp. 302–9

58  CHEN, S., BILLINGS, S. A., and LUO, W.: 'Orthogonal least-squares and their application to nonlinear system identification', *International Journal of Control*, 1989, **50** (5), pp. 1873–96

59  ORR, M. J. L.: 'Regularization in the selection of radial basis function centers', *Neural Computation*, 1995, **7** (3), pp. 606–23

60  PLATT, J.: 'A resource allocating network for function interpolation', *Neural Computation*, 1991, **3**, pp. 213–25

61  KADIRKAMANATHAN, V., and NIRANJAN, M.: 'A function estimation approach to sequential learning with neural networks', *Neural Computation*, 1993, **5**, pp. 954–75

62  YINGWEI, L., SUNDARARAJAN, N., and SARATCHANDRAN, P.: 'A sequential learning scheme for function approximation using minimal radial basis function neural networks', *Neural Computation*, 1997, **9**, pp. 461–78

63  YINGWEI, L., SUNDARARAJAN, N., and SARATCHANDRAN, P.: 'Performance evaluation of a sequential minimal radial basis function (RBF) neural network learning algorithm', *IEEE Transactions on Neural Networks*, 1998, **9** (2), pp. 308–18

64  KAVLI, T.: 'ASMOD. An algorithm for adaptive spline modelling of observation data', *International Journal of Control*, 1993, **58** (4), pp. 947–67

65 CABRITA, C., RUANO, A. E., and FONSECA, C. M.: 'Single and multi-objective genetic programming design for B-spline neural networks and neuro-fuzzy systems'. Proceedings of IFAC workshop on *Advanced fuzzy/neural control*, Valência, Spain, 2001, pp. 93–8

66 CABRITA, C., BOTZHEIM, J., RUANO, A. E., and KÓCZY, L. T.: 'Genetic programming and bacterial algorithm for neural networks and fuzzy systems design: a comparison'. Proceedings of IFAC international conference on *Intelligent control and signal processing* (ICONS 2003), Faro, Portugal, 2003, pp. 443–8

67 PSALTIS, D., SIDERIS, A., and YAMAMURA, A.: 'Neural controllers'. Proceedings of IEEE first conference on *Neural networks*, 1987, vol. 4, pp. 551–8

68 SJOBERG, J., ZHANG, Q., LJUNG, L. *et al.*: 'Nonlinear black-box modelling in system identification. A unified overview', *Automatica*, 1995, **31** (12), pp. 1691–724

69 WILLIAMS, R. J., and ZIPSER, D.: 'A learning algorithm for continually running fully recurrent neural networks', *Neural Computation*, 1989, **1** (2), pp. 270–80

70 ELMAN, L. J.: 'Finding structure in time', *Cognitive Science*, 1990, **14**, pp. 179–211

71 JORDAN, M. I., and JACOBS, R. A.: 'Attractor dynamics and parallelism in a connectionist sequential machine'. Proceedings of the eighth annual conference of the cognitive science society, 1986, pp. 531–46

72 SCHENKER, B. G. E.: 'Prediction and control using feedback neural networks and partial models'. Ph.D. thesis, Swiss Federal Institute of Technology, Zurich, Switzerland, 1996

73 TSOI, A., and BACK, A. D.: 'Locally recurrent globally feedforward networks: a critical review of architectures', *IEEE Transactions on Neural Networks*, 1994, **5** (2), pp. 229–39

74 HABER, R., and UNBEHAUEN, H.: 'Structure identification of nonlinear dynamical systems: a survey on input/output methods', *Automatica*, 1985, **26** (4), pp. 651–77

75 HE, X., and ASADA, H.: 'A new method for identifying orders in input–output models for nonlinear dynamical systems'. Proceedings of American control conference, San Francisco, CA, USA, 1993, pp. 2520–3

76 SÖDERSTRÖM, T., and STOICA, P.: 'On covariance function tests used in system identification', *Automatica*, 1990, **26** (1), pp. 834–49

77 BILLINGS, S., and VOON, W. S. F.: 'Correlation based validity tests for nonlinear models', *International Journal of Control*, 1986, **44** (1), pp. 235–44

78 BILLINGS, S. A., and ZHU, Q. M.: 'Nonlinear model validation using correlation tests', *International Journal of Control*, 1994, **60** (6), pp. 1107–20

79 HUNT, K. J., SBARBARO, D., ZBIKOWSKI, R., and GAWTHROP, P. J.: 'Neural networks for control system – a survey', *Automatica*, 1992, **28** (6), pp. 1083–112

80  LIGHTBODY, G., and IRWIN, G. W.: 'Nonlinear control structures based on embedded neural system models', *IEEE Transactions on Neural Networks*, 1997, **8** (3), pp. 553–67

81  NARENDRA, K. S., and LEWIS, F. L. (Eds): 'Special issue on neural network feedback control', *Automatica*, 2001, **37** (8), 1147–301

82  JONCKHEERE, E. A., YU, G. R., and CHIEN, C. C.: 'Gain scheduling for lateral motion of propulsion controlled aircraft using neural networks'. Proceedings of 1997 American control conference, Albuquerque, USA, 1997, pp. 3321–5

83  JENG, J. T., and LEE, T. T.: 'A neural gain scheduling network controller for nonholonomic systems', *IEEE Transactions on Systems, Man and Cybernetics – Part A*, 1999, **29** (6), pp. 654–61

84  NARENDRA, K. S., and PARTHASARATHY, K.: 'Identification and control of dynamical systems using neural networks', *IEEE Transactions on Neural Networks*, 1990, **1** (1), pp. 4–27

85  PHAM, D. T., and OH, S. J.: 'Identification of plant inverse dynamics using neural networks', *Artificial Intelligence in Engineering*, 1999, **13** (3), pp. 309–20

86  CABRERA, J. B. D., and NARENDRA, K. S.: 'Issues in the application of neural networks for tracking based on inverse control', *IEEE Transactions on Automatic Control*, 1999, **44** (11), pp. 2007–27

87  PLETT, G. L.: 'Adaptive inverse control of linear and nonlinear systems using dynamic neural networks', *IEEE Transactions on Neural Networks*, 2003, **14** (2), pp. 360–76

88  HUNT, K., and SBARBARO, D.: 'Neural networks for nonlinear internal model control', *IEE Proceedings on Control Theory and Applications*, 1991, **138** (5), pp. 431–8

89  AOYAMA, A., and VENKATASUBRAMANIAN, V.: 'Internal model control framework for the modelling and control of a bioreactor', *Engineering Applications of Artificial Intelligence*, 1995, **8** (6), pp. 689–701

90  BROWN, M., LIGHTBODY, G., and IRWIN, G. W.: 'Nonlinear internal model control using local model networks', *IEE Proceedings on Control Theory and Applications*, 1997, **144** (4), pp. 505–14

91  DIAZ, G., SEN, M., YANG, K. T., and McCLAIN, R. L.: 'Dynamic prediction and control of heat exchangers using artificial neural networks', *International Journal of Heat and Mass Transfer*, 2001, **44** (9), pp. 1671–9

92  RIVALS, I., and PERSONNAZ, L.: 'Nonlinear internal model control using neural networks; application to processes with delay and design issues', *IEEE Transactions on Neural Networks*, 2000, **11** (1), pp. 80–90

93  RUANO, A. E., FLEMING, P. J., and JONES, D. I.: 'A connectionist approach to PID autotuning', *IEE Proceedings on Control Theory and Applications*, 1992, **139** (3), pp. 279–85

94  OMATU, S., HALID, M., and YUSOF, R.: 'Neuro-control and its applications' (Springer-Verlag, Berlin, 1996)

95 LIMA, J. M., and RUANO, A. E.: 'Neuro-genetic PID autotuning: time invariant case', *IMACS Journal of Mathematics and Computers in Simulation*, 2000, **51**, pp. 287–300

96 DUARTE, N. M., RUANO, A. E., FONSECA, C. M., and FLEMING, P. J.: 'Accelerating multi-objective control system design using a neuro-genetic approach'. Proceedings of eighth IFAC symposium on *Computer aided control systems design*, Salford, UK, 2000

97 WILLIS, M. J., DIMASSIMO, C., MONTAGUE, G. A., THAM, M. T., and MORRIS, A. J.: 'Artificial neural networks in process engineering', *IEE Proceedings on Control Theory and Applications*, 1991, **138** (3), pp. 256–66

98 MILLS, P. M., ZOMAYA, A. Y., and TADE, M. O.: 'Adaptive control using neural networks', *International Journal of Control*, 1994, **60** (6), pp. 1163–92

99 TURNER, P., MONTAGUE, G., and MORRIS, J.: 'Dynamic neural networks in nonlinear predictive control (an industrial application)', *Computers and Chemical Engineering*, 1996, **20**, pp. 937–42

100 ISERMANN, R., and HAFNER, M.: 'Mechatronic combustion engines – from modeling to control', *European Journal of Control*, 2001, **7** (2–3), pp. 220–47

101 FINK, A., TOPFER, S., and ISERMANN, R.: 'Nonlinear model-based predictive control with local linear neuro-fuzzy', *Archive of Applied Mechanics*, 2003, **72** (11–12), pp. 911–22

102 ISIDORI, A.: 'Non-linear control systems' (Springer-Verlag, 1995, 3rd edn.)

103 HANCOCK, E., and FALLSIDE, F.: 'A direct control method for a class of nonlinear systems using neural networks'. Technical report 65, Engineering Department, Cambridge University, UK, 1991

104 HANCOCK, E., and FALLSIDE, F.: 'Stable control of nonlinear systems using neural networks'. Technical report 81, Engineering Department, Cambridge University, UK, 1991

105 POLYCARPOU, M., and IOANNOU, P.: 'Modelling, identification and stable adaptive control of continuous time nonlinear dynamical systems using neural networks'. Proceedings of American control conference, 1992, pp. 36–40

106 SANNER, M., and SLOTINE, J.: 'Gaussian networks for direct adaptive control', *IEEE Transactions on Neural Networks*, 1992, **3** (6), pp. 55–62

107 YESILDIREK, A., and LEWIS, F. L.: 'Feedback linearization using neural networks', *Automatica*, 1995, **31** (11), pp. 1659–64

108 LI, Q., BABUSKA, R., and VERHAEGEN, M.: 'Adaptive output tracking of nonlinear systems using neural networks'. Proceedings of the fourteenth IFAC World Congress, Beijing, China, 1999, pp. 339–44

109 ROVITHAKIS, G., and CHRISTODOULOU, M.: 'Adaptive control of unknown plants using dynamical neural networks', *IEEE Transactions on Systems, Man and Cybernetics*, 1994, **24** (3), pp. 400–12

110 PONZYAK, A., YU, W., SANCHEZ, E., and PEREZ, P.: 'Nonlinear adaptive tracking using dynamical neural networks', *IEEE Transactions on Neural Networks*, 1999, **10** (6), pp. 1402–11

111  DELGADO, A., KAMBAHMPATI, C., and WARWICK, K.: 'Dynamic recurrent neural networks for systems identification and control', *IEE Proceedings on Control Theory and Applications*, 1995, **142** (4), pp. 307–14

112  BRDYS, M., KULAWSKI, G., and QUEVEDO, J.: 'Recurrent networks for nonlinear adaptive control', *IEE Proceedings on Control Theory and Applications*, 1998, **145** (2), pp. 177–88

113  KULAWSKI, G., and BRDYS, M.: 'Stable adaptive control with recurrent networks', *Automatica*, 2000, **36** (1), pp. 5–22

114  LEVIN, A., and NARENDRA, K.: 'Control of non-linear dynamical systems using neural networks: part I, controlability and stabilization', *IEEE Transactions on Neural Networks*, 1993, **4** (2), pp. 192–206

115  LEVIN, A., and NARENDRA, K.: 'Control of non-linear dynamical systems using neural networks: part II, observability, identification and control', *IEEE Transactions on Neural Networks*, 1996, **7** (1), pp. 30–42

116  CHEN, F., and KHALIL, H.: 'Adaptive control of a class of non-linear discrete time systems using neural networks', *IEEE Transactions on Automatic Control*, 1995, **40** (5), pp. 791–801

117  JAGANNATHAN, S., LEWIS, F., and PASTRAVANU, O.: 'Discrete time model reference adaptive control of non-linear dynamical systems using neural networks', *International Journal of Control*, 1996, **64** (2), pp. 107–30

118  TeBRAAKE, H., BOTTO, M., CAN, J., Sá da COSTA, J., and VERBRUGGEN, H.: 'Linear predictive control based on approximate input–output feedback linearization', *IEE Proceedings on Control Theory and Applications*, 1999, **146** (4), pp. 295–300

119  SUYKENS, K., DeMOOR, B., and VANDEWALLW, J.: 'NLq theory: a neural control framework with global asymptotic stability criteria', *Neural Networks*, 1997, **10**, pp. 615–37

120  JIN, L., NIKIFORUK, P., and GUPTA, M.: 'Dynamic recurrent neural network for control of unknown nonlinear systems', *Journal of Dynamical Systems, Measurement and Control*, 1994, **116**, pp. 567–76

121  FERREIRA, P. M., RUANO, A. E., and FONSECA, C. M.: 'Genetic assisted selection of RBF model structures for greenhouse inside air temperature prediction'. IEEE conference on *Control applications*, Istanbul, Turkey, June 2003

122  FERREIRA, P. M., FARIA, E. A., and RUANO, A. E.: 'Design and implementation of a real-time data acquisition system for the identification of dynamic temperature models in a hydroponic greenhouse', *Acta Horticulturae*, 2000, **519**, pp. 191–7

123  FERREIRA, P. M., FARIA, E. A., and RUANO, A. E.: 'Application of radial basis function neural networks to a greenhouse inside air temperature model'. Proceedings of IFAC international conference on *Modelling and control in agriculture, horticulture and post-harvested processing* (Agricontrol 2000), Wageningen, The Netherlands, 2000, vol. 2, pp. 172–7

124 FERREIRA, P. M., FARIA, E. A., and RUANO, A. E.: 'Comparison of on-line learning algorithms for radial basis function models in greenhouse environmental control'. Proceedings of fourth Portuguese conference on *Automatic control* (Controlo 2000), Guimarães, Portugal, 2000

125 FERREIRA, P. M., and RUANO, A. E.: 'Predicting the greenhouse inside air temperature with rbf neural networks'. Preprints of the second IFAC-CIGR workshop on *Intelligent control for agricultural applications*, Bali, Indonesia, 2001, pp. 67–72

126 FERREIRA, P. M., and RUANO, A. E.: 'Choice of rbf model structure for predicting greenhouse inside air temperature'. Proceedings of fifteenth IFAC World Congress, Barcelona, Spain, 2002

127 FERREIRA, P. M., and RUANO, A. E.: 'Predicting solar radiation with rbf neural networks'. Proceedings of the sixth Portuguese conference on *Automatic control* (Controlo 2004), Faro, Portugal, 2004, pp. 31–6

128 FONSECA, C. M., and FLEMING, P. J.: 'Non-linear system identification with multiobjective genetic algorithms'. Proceedings of the thirteenth IFAC World Congress, San Francisco, CA, USA, 1996, vol. C, pp. 187–92

129 RODRÍGUEZ-VÁSQUEZ, K., FONSECA, C. M., and FLEMING, P. J.: 'Identifying the structure of nonlinear dynamic systems using multiobjective genetic programming', *IEEE Transactions on Systems, Man, and Cybernetics – Part A: Systems and Humans*, 2004, **34** (4), pp. 531–45

130 FONSECA, C. M., and FLEMING, P. J.: 'On the performance assessment and comparison of stochastic multiobjective optimizers', *Parallel Problem Solving from Nature – PPSN IV*, Lecture Notes in Computer Science, 1141, Springer-Verlag, Berlin, Germany, 1996, pp. 584–93

*Chapter 3*

# Multi-objective evolutionary computing solutions for control and system identification

*Ian Griffin and Peter John Fleming*

## 3.1 Introduction

Systems engineers are increasingly faced with problems for which traditional tools and techniques are ill suited. Such problems are posed by systems that are complex, uncertain and not conducive to deterministic analysis. Modern tools and techniques are being applied in these subject areas in order to address such problems. This chapter introduces the multi-objective genetic algorithm (MOGA) of Fonseca and Fleming [1] as one of the tools for addressing these problems. Background to genetic algorithms (GAs) is given in terms of their operators and abstraction of the problem domain. Multi-objective optimisation is introduced using the concept of Pareto dominance and trade-off between competing objectives. Visualisation techniques are illustrated in which 'many-objective' problems may be handled and preference articulation may be implemented. The motivation for using GAs for multi-objective problems such as control systems design and systems identification is given. The chapter concludes with case studies illustrating the use of MOGA for control system design and multi-objective genetic programming (MOGP) for system identification.

## 3.2 What is a genetic algorithm?

Genetic algorithms (GAs) are stochastic global search algorithms that form a subset of evolutionary algorithms (EAs). The search method is drawn from the principles of natural selection [2] and population genetics [3]. GAs were first proposed by Holland [4] and popularised by Goldberg [5]. GAs work with a population of potential solutions to a problem (see Figure 3.1). This population-based approach results in a parallel search strategy in which a multiple number of points in the search space

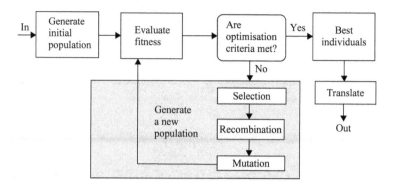

*Figure 3.1    Flowchart of a genetic algorithm*

are investigated simultaneously. The investigation of multiple points allows the GA to search for the global solution to a problem rather than a locally optimal solution. The GA is also iterative, selection and random variation techniques being used to modify the composition of the current set of potential solutions, thereby creating a subsequent generation. Thus, the population evolves with each generation in an effort to identify an optimal solution.

Each individual within the population is usually expressed using some form of genetic code. The encoding may be of any type that allows variation operators to be used to evolve the population. The encoded representation is referred to as the geno-type, while the real decision variable values are known as the phenotype. Common choices are binary representations and direct floating-point representations [6]. The algorithm is initiated by randomly generating a set of individuals that lie within the user-defined bounded search space of each decision variable.

The GA requires a scalar value, or measure of fitness, that represents the quality of each individual in the population. An individual's raw performance measures are assessed and then converted to a fitness value. Fitness values may be assigned in a variety of ways, ranking being a popular approach for multi-objective problems.

Assuming that the termination criteria are not met, the genetic operators are then applied to selected individuals. The selection process is stochastic, with the prob-ability of a given individual being selected determined by its fitness value; the fitter an individual, the more likely it is to be chosen. A number of selection strategies can be found in the literature. These strategies, such as roulette wheel selection and stochastic universal sampling [7], influence the search in different ways [8].

The individuals selected from the population are then subject to variation oper-ators. These operators fall into two basic categories: recombination and mutation. Recombination, also known as crossover, aims to direct the search towards superior areas of the search space that have already been identified within the population. This is referred to as exploitation.[1] Figure 3.2 shows how recombination may be applied to pairs of binary decision variable representations.

In recombination, one or more points along the strings of the selected parent chromosomes are identified. At these points, the strings are broken and subsequent

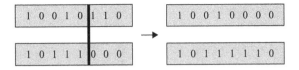

Figure 3.2   Binary recombination (crossover)

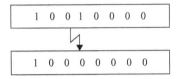

Figure 3.3   Binary mutation

genetic material exchanged between the parents. The objective is to encourage the formation of a fitter individual. If this proves unsuccessful, less fit individuals are unlikely to survive subsequent selection processes.

The purpose of mutation is to explore new areas of the search space and ensure that genetic material cannot be lost irretrievably. This is known as exploration. For mutation, a bit in the binary string is selected and then flipped as shown in Figure 3.3. Recombination and mutation are stochastic operators with parameters determining the probability and nature of each application. The choice of recombination and mutation parameters is critical to the effectiveness of the search. These parameters may also be varied during the search process and subjected to evolution [9].

Once the genetic operators have been applied, the new individuals are reinserted into the population. This may be done such that the new individuals constitute all or part of the population according to the generation gap specified by the user. The number of individuals in the population usually remains static from generation to generation and is commonly between 20 and 100. A typical population size is $10n$, where $n$ is the number of decision variables. At each new generation the population may contain randomly generated individuals. These perform a task similar to mutation in that they ensure that there is a non-zero probability of searching any area of the search space. High quality individuals from previous generations may also be inserted into the population in a strategy known as elitism. The GA continues to iterate until the termination criteria are met. The termination criteria of a GA are user-defined and frequently subjective. A software implementation of a GA is illustrated in pseudo-code in Figure 3.4. First the population is randomly initialised with the decision variables lying within user-defined ranges. This initial population is then evaluated in the problem domain such that objective function values are found. Having created and evaluated the initial population, the algorithm then iterates through subsequent generations by selecting and breeding pairs of individuals and then re-evaluating the modified population until a termination criterion is met.

```
procedure GA
begin
        t = 0;
        initialize P(t);
        evaluate P(t);
        while not finished do
        begin
                t = t + 1;
                select P(t) from P(t - 1);
                reproduce pairs in P(t);
                evaluate P(t);
        end
end
```

*Figure 3.4    Genetic algorithm in pseudo-code*

## 3.3    Why use GAs for control?

Technological advances are increasingly producing new challenges for which current control technology is unprepared. Assumptions required for classical control systems' design methods may be inappropriate. The systems being addressed may be nonlinear, stochastic or have a number of different decision variable data types. Systems can often be poorly understood, difficult to formulate mathematically and therefore not conducive to analysis and conventional search methods. The GA's lack of reliance on domain-specific heuristics makes it attractive for such problems as very little *a priori* information is required [7].

Genetic algorithms constitute a flexible, non-specific tool, being suitable for a wide range of applications without major modification. Unlike conventional optimisers, a number of different decision variable formats can be represented and manipulated due to the use of an encoded representation of the parameter set. This allows the GA to optimise parameters that are of different units or qualitatively different, e.g. choice of actuator and gain value. GAs are also able to assess the quality of potential solutions to a problem using a wide variety of cost function types. Whereas conventional optimisers may rely on using conveniently well-structured functions such as the rms error between the desired and actual values of a chosen metric, there is no such reliance for GAs.

Unlike conventional optimisers, GAs are robust to different characteristics in the cost landscape. The GA can search globally across landscapes that are multimodal, discontinuous, time varying, random and noisy. Moreover, GAs are a directed search tool, making them potentially more efficient than random or enumerative search methods.

Engineering problems are usually subject to constraints. These often represent physical limitations of the plant or the limits of acceptable system behaviour. GAs are able to deal with constraints in a number of ways. The most efficient approach is to enforce the constraint via the bounds on the decision variable, thereby making it impossible to evolve an unfeasible solution. This does, however, require prior knowledge of the decision variable and cost landscape and so may prove impossible for many practical applications. Alternatively, a penalty may be applied to the fitness value of an individual that does not conform to the constraints of the problem. This approach is crude but has often proved effective in practice. Other approaches, such as repair algorithms that convert unfeasible solutions to feasible ones and modification of the genotype to phenotype conversion procedure are also possible. A further means of addressing constraints is through the use of a multi-objective framework. This is discussed in Section 3.4.

Genetic algorithms are well suited to new problem areas in which conventional techniques would require more *a priori* information than is available. GAs are how-ever less well suited to linear, well-understood problems to which trusted solutions exist. Conventional solutions are best where available. Due to the stochastic nature of the algorithm, the GA does not lend itself to mission-critical or safety-critical real-time applications as the quality of the result provided by the GA cannot be guaranteed. The computationally intensive nature of the algorithm also makes it unsuited to most real-time applications.

## 3.4   Multi-objective optimisation

Most engineering problems require the designer to satisfy a number of criteria. These objectives may be in conflict with each other, an improvement in one leading to deterioration in another. Multi-objective problems in which there is competi-tion between objectives may have no single, unique optimal solution. In such cases a set of solutions exists, each of which has a unique mapping to the objective space of the problem. A solution is said to be non-dominated or Pareto-optimal if there exists no other solution within the current set that gives improved performance with regard to all the objectives. Consider, e.g. the two-objective problem:

minimise $[f_1, f_2]$ w.r.t. $x$.

Figure 3.5 shows two non-dominated candidate solutions, A and B, to a two-objective problem. A multi-objective optimisation process attempts to find solutions that move the front of non-dominated solutions as close as possible to the utopian point (assuming minimisation and non-negative objectives) at the origin of the objective space. Figure 3.6 shows how the principle can be extended to the three-dimensional case where the competing objectives are performance, reliability and cost.

An alternative visualisation framework is required to visualise problems with $n$ objectives. Figure 3.7 shows a parallel coordinates graph in which the performance levels of a number of non-dominated solutions are shown with regard to two objectives to provide a comparison to the previous approach. The two-objective function values are shown on each of the vertical axes. Each line on the graph represents the perfor-mance of a single candidate solution. In this configuration, lines that run concurrently between adjacent objectives show that candidate solutions do not compete with regard

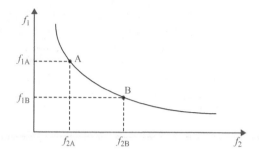

*Figure 3.5    Multi-objective optimisation – two-dimensional representation*

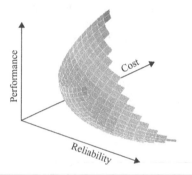

*Figure 3.6    Multi-objective optimisation – three-dimensional representation*

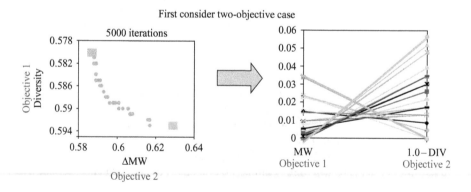

*Figure 3.7    Parallel coordinates representation of a multi-objective problem domain*

to these objectives while crossing lines indicate competition and therefore a trade-off between the objectives. In Figure 3.7, the crossing lines between objectives 1 and 2 show competing objectives with those highlighted being from opposite ends of the Pareto front.

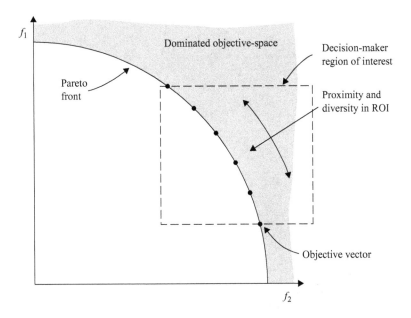

*Figure 3.8    Proximity, diversity and pertinency in the Pareto framework*

## 3.5    Multi-objective search with GAs

The task of a multi-objective optimiser is to provide the designer with good information about the trade-off surface. This involves generating a set of non-dominated solutions that have good proximity, diversity and pertinency to the true Pareto front. Figure 3.8 illustrates these properties.

Proximity is the closeness of the solution set to the true Pareto front. Diversity is a measure of the spread of individuals along the Pareto front. For most multi-objective optimisations, the designer defines an area of objective space within which solutions are considered acceptable. This area is referred to as the region of interest (ROI). Pertinency is a measure of the extent to which individuals are concentrated in the region of interest.

Genetic algorithms lend themselves well to addressing multi-objective problems. The population-based approach of GAs results in a parallel search strategy. This parallelism provides results for a number of points within the objective space from each optimiser run. Therefore, GAs can provide good proximity, diversity and pertinency from a single optimiser run assuming the parameters of the genetic operators are well specified for the given problem. This is in contrast to conventional optimisers that result in a single solution from an optimisation and therefore no information regarding the trade-offs within the problem can be obtained without running multiple optimisations. GAs can be augmented from the standard genetic algorithm (SGA) proposed by Goldberg [5] in order to have properties ideally suited for multi-objective search. Ranking as a form of fitness assignment can be used to identify non-dominated

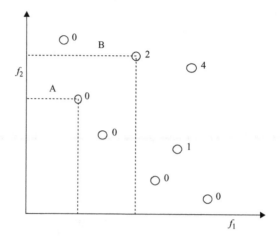

*Figure 3.9    Multi-objective ranking*

individuals within a multi-objective space. Figure 3.9 shows the ranking technique used in the MOGA of Fonseca and Fleming [1].

In Figure 3.9, individuals are ranked according to the number of other potential solutions that lie in the rectangle formed between the individual and the origin (e.g. see rectangles A and B). Any individuals that lie within that rectangle dominate the individual being assessed. Those individuals that are non-dominated are therefore ranked zero. The fittest individuals within a population can be identified as having lower ranking values.

Fitness sharing may also be employed in order to improve the GA's suitability for multi-objective optimisation. Fitness sharing acts to reduce the fitness value of individuals that lie in densely populated areas of the search space. If a large number of fit individuals from a small area of the search space were allowed to go forward for selection, there could be bunching of solutions. Fitness sharing promotes diversity of solutions. Fitness sharing strategies such as Epanechnikov [1] act to reduce the fitness of individuals that lie in densely populated areas and prevent a high population density from distorting the search.

## 3.6    Multi-objective optimisation for control

Control system design problems almost always involve simultaneously satisfying multiple requirements, viz:

$$\underset{\mathbf{x} \in \Omega}{\text{Min}}\, \mathbf{f}(\mathbf{x}), \tag{3.1}$$

where $\mathbf{x} = [x_1, x_2, \dots, x_q]$ and $\Omega$ define the set of free variables, $\mathbf{x}$, subject to any constraints and $\mathbf{f}(\mathbf{x}) = [f_1(\mathbf{x}), f_2(\mathbf{x}), \dots, f_n(\mathbf{x})]$ contains the design objectives to be minimised.

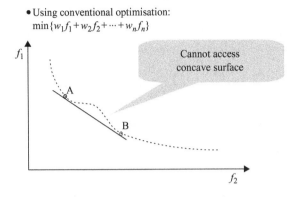

*Figure 3.10    Weighted sum approach to multi-objective problems*

Control systems may have requirements specified in both time and frequency domain metrics for both steady-state and transient behaviour. Other characteristics may also be stipulated such as complexity, power requirements and cost. Traditional, single objective search methods can be adapted in order to address multi-objective problems. This is generally done using the weighted sum approach. A weighting value is assigned to each objective function value prior to the search process. The weighting effectively indicates the relative importance of each objective. The values of the weightings determine the location of the solution on the Pareto front that results from such an optimisation. Inappropriate weightings may therefore produce solutions that do not lie within the region of interest.

The use of the weighted sum approach can be unsatisfactory for a number of reasons. In most practical applications, there will be no means of determining the most suitable values of the weightings in advance of the search. The designer therefore has to pre-select weighting values that may be unsuitable combinations in hindsight. Multiple optimisation runs may have to be performed using different weighting function settings in order to find a suitable result. Also, small changes in the weighting function values may result in large changes in the objective vectors; conversely, large changes in weights may result in similar objective vectors.

A further deficiency of the weighted sum approach is that it cannot access non-convex areas of the search space as shown in Figure 3.10. Use of a truly multi-objective approach, such as MOGA, yields the full trade-off surface in a single run.

## 3.7    Case study: direct controller tuning

The following example shows how MOGA can be used to directly tune the parameters of a three-term controller. The system to be controlled is a three-state, single-input–single-output (SISO) industrial system.[2] The open loop system is stable with one real and two complex conjugate poles and two left-half-plane, complex conjugate zeros. A schematic of the closed loop system is as shown in Figure 3.11.

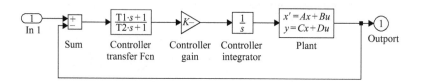

*Figure 3.11    Closed-loop control system of industrial problem*

*Table 3.1    The decision variables for the three-term controller*

| Decision variable | Lower bound | Upper bound | Precision | Coding protocol |
|---|---|---|---|---|
| Proportional gain | 0 | 100 | 16 | Linear with bounds |
| Integral gain | 0 | 3 | 16 | Linear with bounds |
| Differential gain | 0 | 2 | 16 | Linear with bounds |

The decision variables to be manipulated during the search are the proportional gain, $K$, lead term, T1, and lag term, T2, of the controller. Table 3.1 shows the parameters chosen for the decision variable encoding. A Grey code binary representation of the decision variables is used with a population size of 50. Grey code is used in preference to standard binary as it has a reduced Hamming distance giving improved variation characteristics. The search space for each parameter was established from a brief initial 'hand-tuning' exercise. This involved making a small number of iterative adjustments to the controller parameters in the Simulink representation shown in Figure 3.11 in order to establish a time-domain response of moderate quality. Frequency-domain measures were not considered in this initial exercise in order to save time. The step response achieved in this manner is shown in Figure 3.14 where it is contrasted to the optimised response. These preliminary controller settings were then used as guidelines for defining the region over which MOGA would search for each parameter. The precision value denotes the number of binary bits to be used to represent each decision variable.

The objective functions listed in Table 3.2 are direct measures of the time-domain performance of the closed-loop system and the frequency-domain performance of the loop transfer function. The time-domain metrics are obtained from a step response of the Simulink representation of the closed-loop system shown in Figure 3.11 while the frequency domain measures are calculated within MATLAB. The choice of objectives is typical of industrial applications of this type.

Note that MOGA performs search and optimisation through minimisation of cost functions. Minimisation can be applied directly to objective functions 1–5 where it is desirable to reduce the values returned from the Simulink representation. For gain and phase margin however, it is desirable to maximise the values to optimise stability. This is done by negating both the goal values and the raw measures obtained for each individual.

*Table 3.2* *The objectives used for the three-term controller optimisation*

| Objective number | Objective | Goal value |
|---|---|---|
| 1 | Overshoot | 3% |
| 2 | Overshoot time | 3 s |
| 3 | Undershoot | 1% |
| 4 | Settling time | 5 s |
| 5 | Integral of absolute error | 100 |
| 6 | Gain margin | 12 dB |
| 7 | Phase margin | 60° |
| 8 | Bandwidth | 1, 2 or 3 rad/s |
| 9 | Noise assessment | 0.01 |

*Table 3.3* *The genetic operator values used for the three-term controller optimisation*

| MOGA parameter | Value specified |
|---|---|
| Number of individuals in a generation | 50 |
| Maximum number of generations | 25 |
| Number of random immigrants | 5 |
| Probability of recombination | 0.7 |
| Probability of mutation | 0.1 |

For this problem, the required bandwidth was a function of other system parameters. In order to make the technique generic, an objective function was constructed such that the required bandwith was calculated and a delta, $\Delta$, representing the bandwidth of each individual relative to the required value was used as the objective function where,

$\Delta$ = Required bandwidth − bandwidth of individual.

The noise reduction metric was specified in the industrial literature. This metric is a function of controller structure and was ignored during this optimisation as the controller structure was fixed.

The parameters of the genetic operators need to be defined before the search process is undertaken. These are shown in Table 3.3, their choice stemming from previous experience of similar problems.

Following the definition of all the required parameters, the search process is initiated and a number of non-dominated candidates appear on the parallel coordinates graph. The pattern of these individuals provides the designer with information about the problem domain. It can be seen in Figure 3.12 that all the individuals featured

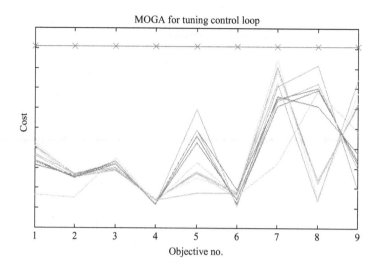

*Figure 3.12  Parallel coordinates graph resulting from the optimisation*

on the parallel coordinates graph achieve the required performance set by the goal values as indicated by the crosses.

Given that all the goals are satisfied, the designer can then select an individual from the graph to be the chosen solution. The choice of individual is made by tightening up the goal values on the parallel coordinates graph. This progressively eliminates individuals from the graph as they no longer meet the new tighter goals. The designer continues to express preference by iteratively tightening the goals in order to isolate a single individual that conforms most closely to the designer's perception of the ideal solution. The adjusted parallel coordinates graph is shown in Figure 3.13, where the new goals are depicted and a single solution is isolated.

Note that in cases where no individual can be found that satisfies the initial goal values, the designer may adjust the preferences during the search process. This may help to identify optimal areas of the search space that would best conform to the goals. Figure 3.14 shows the improvement obtained in step response performance from the initial values taken from the hand-tuning exercise and the MOGA-optimised controller.

## 3.8  Case study: gasifier control design

### 3.8.1  Introduction

In contrast to the previous example, the following case study illustrates how MOGA may be used to design a controller indirectly through the optimisation of weighting functions. MOGA is used to select suitable weighting coefficients for the $H_\infty$ loop-shaping design approach, where weights are manipulated to realise additional design specifications. Modern awareness of environmental issues has led to the desire for low

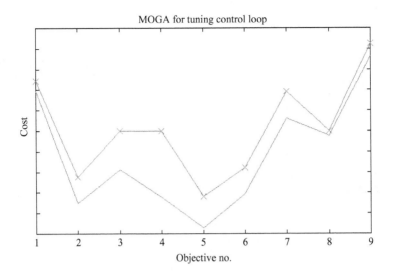

*Figure 3.13    Adjusted parallel coordinates graph*

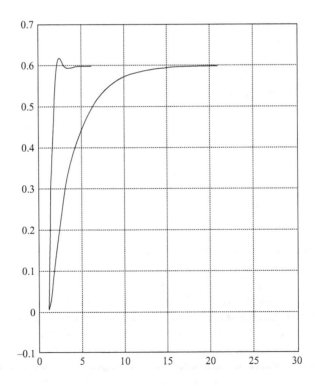

*Figure 3.14    Initial and optimised closed-loop step responses*

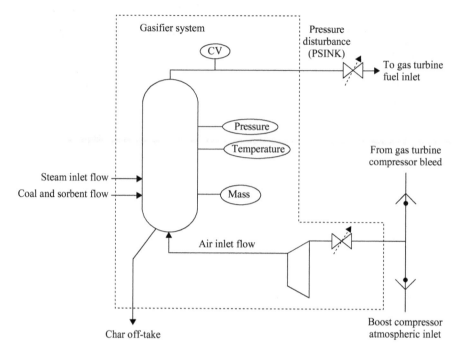

*Figure 3.15   Gasification plant*

emissions power generation techniques. A member of the clean coal power generation group (CCPGG) has developed an advanced coal-based demonstrator system aimed at satisfying the requirements of greater efficiency and lower harmful emissions levels. This prototype integrated plant (PIP) is based on the air blown gasification cycle (ABGC) and contains a novel component known as the gasification plant or gasifier. The gasifier uses the spouted fluidised bed concept developed by British Coal in order to generate a low calorific value fuel gas for gas turbine-driven electricity generation.

A mixture of ten parts coal to one part limestone is pulverised and introduced into the gasifier chamber in a stream of air and steam. Once in the gasifier, the air and steam react with the carbon and other volatiles in the coal. This reaction results in a low calorific fuel gas and residual ash and limestone derivatives. This residue falls to the bottom of the gasifier and is removed at a controlled rate. The fuel gas escapes through the top of the gasifier and is cleaned before being used to power a gas turbine. The gasifier is shown in Figure 3.15.

The design of a control system for the gasifier was the subject of a Benchmark Challenge issued to UK Universities in 1997 [10]. The Challenge required the design of a controller for a single linear operating point that represented the gasifier at the 100 per cent load case. Input and output constraints were specified in terms of actuator saturation and rate limits and maximum allowed deviation of controlled variables from the operating point. The gasifier was modelled as a six-input, four-output system as described in Table 3.4.

*Table 3.4   Gasifier inputs and outputs*

| Inputs | Outputs |
|---|---|
| 1. WCHR – char extraction flow (kg/s) | 1. CVGAS – fuel gas calorific value |
| 2. WAIR – air mass flow (kg/s) | 2. MASS – bed mass (kg) |
| 3. WCOL – coal flow (kg/s) | 3. PGAS – fuel gas pressure ($N/m^2$) |
| 4. WSTM – steam mass flow (kg/s) | 4. TGAS – fuel gas temperature (K) |
| 5. WLS – limestone mass flow (kg/s) | |
| 6. PSINK – sink pressure ($N/m^2$) | |

It was further specified that the controller developed should be tested at two other linear operating points. These represented the 50 and 0 per cent load cases. This testing was to be performed offline using the MATLAB/Simulink analysis and simulation package in order to assess robustness. The state–space representations of the system were linear, continuous, time-invariant models and had the inputs and outputs ordered as shown in Table 3.4. They represent the system in open loop and are of twenty-fifth order.

The purpose of introducing limestone into the gasifier is to lower the level of harmful emissions. Limestone absorbs sulphur in the coal and therefore needs to be introduced at a rate proportional to that of the coal. It was specified in the Challenge that the limestone flow rate should be set to a constant ratio of 1:10, limestone to coal, making limestone a dependent input. Given that PSINK is a disturbance input, this leaves four degrees of freedom for the controller design. The controller design can therefore be approached as a square, 4 × 4 problem.

The underlying objective of the controller design is to regulate the outputs during disturbance events using the controllable inputs. The outputs of the linear plant models are required to remain within a certain range of the operating point being assessed. These limits are expressed in units relative to the operating point and are therefore the same for all three operating points. They are as follows:

- The fuel gas calorific value (CVGAS) fluctuation should be minimised, but must always be less than ±10 kJ/kg.
- The bed mass (MASS) fluctuation should be minimised, but must always be less than 5 per cent of the nominal for that operating point.
- The fuel gas pressure (PGAS) fluctuation should be minimised, but must always be less than ±0.1 bar.
- The fuel gas temperature (TGAS) fluctuation should be minimised, but must always be less than ±1°C.

The disturbance events stipulated by the Challenge were chosen in order to provide a difficult control design problem. The sink pressure (PSINK), represents the pressure upstream of the gas turbine which the gasifier is ultimately powering. Fluctuations (in PSINK) represent adjustments to the position of the gas turbine fuel valve. The Challenge specified that the value of PSINK should be adjusted in the following ways to generate disturbance events.

- Apply a step change to PSINK of $-0.2$ bar 30 s into the simulation. The simulation should be run for a total of 300 s.
- Apply a sine wave to PSINK of amplitude 0.2 bar and frequency 0.0 4 Hz for the duration of the simulation. The simulation should be run for a total of 300 s.

The controller design technique with which MOGA will interact for this problem is the $H_\infty$ loop shaping design procedure (LSDP).

### 3.8.2   Loop shaping design

The $H_\infty$ loop shaping design procedure is essentially a two-stage controller design technique [11]. First, performance requirements are addressed by shaping the frequency response of the open-loop plant in a manner analogous to that of classical loop shaping. Second, robustness requirements are addressed using $H_\infty$ optimisation [12] to stabilise the shaped plant, given a range of possible model uncertainty. The result is a single degree of freedom controller, a configuration adequate for the disturbance rejection task of the Challenge. The design procedure assumes positive feedback in the closed-loop system.

### 3.8.3   Loop shaping

In classical linear SISO loop shaping, the magnitude of the open-loop transfer function is a function of frequency and is manipulated in order to meet system requirements. The gain of a multi-input–multi-output (MIMO) plant, however, varies at any given frequency with the direction of the input vector. No unique gain value can be given for a multivariable system as a function of frequency. A measure analogous to that of SISO plant gain is required for multivariable systems if loop shaping is to be employed. Eigenvalues are unsuitable for this task as they only provide a measure of gain for the specific case of a square system whose input and output vectors are in the direction of an eigenvector. However, an accurate representation of the gain of a multivariable system can be found using the singular value decomposition. The singular value decomposition of any $l \times m$ matrix $\mathbf{G}$ can be written as

$$\mathbf{G} = \mathbf{U}\mathbf{\Sigma}\mathbf{V}^{\mathrm{H}}, \tag{3.2}$$

where $\mathbf{V}^{\mathrm{H}}$ is the complex conjugate transpose of $\mathbf{V}$. Each column vector of matrix $\mathbf{U}$ represents the direction of the vector output signal produced by the plant $\mathbf{G}$ subject to an input in the direction of the corresponding column vector of matrix $\mathbf{V}$. These column vectors are each of unit length. Matrix $\mathbf{\Sigma}$ is a diagonal matrix of $\min\{l, m\}$ non-negative singular values in descending order of magnitude, the remaining diagonal elements being zero. These singular values represent the gain of $\mathbf{G}$ for the corresponding input and output directions in

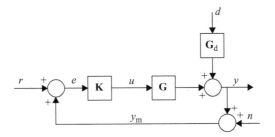

*Figure 3.16    One degree of freedom feedback control system*

$\mathbf{V}$ and $\mathbf{U}$ and can be computed as the positive square roots of the eigenvalues of $\mathbf{G}^H\mathbf{G}$.

$$\sigma_i(\mathbf{G}) = \sqrt{\lambda_t(\mathbf{G}^H\mathbf{G})}, \tag{3.3}$$

where $\sigma(\cdot)$ denotes a singular value and $\lambda(\cdot)$ denotes an eigenvalue.

Hence, the maximum and minimum singular values, $\sigma_{max}(\mathbf{G})$ and $\sigma_{min}(\mathbf{G})$, constitute the upper and lower bounds on the range of system gains in response to all possible input directions at a given frequency. In order to determine what constitutes a desirable shape for the plant singular values, the closed-loop configuration in Figure 3.16 can be analysed.

From this configuration the output $y$ can be derived as being

$$y = (\mathbf{I} - \mathbf{G_sK})^{-1}\mathbf{G_sK}r + (\mathbf{I} - \mathbf{G_sK})^{-1}\mathbf{G_d}d + (\mathbf{I} - \mathbf{G_sK})^{-1}\mathbf{G_sK}n, \tag{3.4}$$

where $r$ is the reference signal, $d$ is the disturbance, $n$ is the measurement noise, $u$ is the plant input, $y$ is the actual output and $y_m$ is the measured output.

From equation (3.4) it can be seen that when $|\mathbf{G_sK}|$ is large, reference signals are propagated while disturbances are attenuated. However, a large value of $|\mathbf{G_sK}|$ fails to subdue measurement noise and a trade-off situation arises. A compromise can be found because reference signals and disturbances are usually low-frequency events while measurement noise is prevalent over a much wider bandwidth. Acceptable performance can therefore be attained by shaping the singular values of $\mathbf{G_sK}$ to give high gain at low frequency for disturbance rejection and reduced gain at higher frequency for noise suppression [13].

For this particular design procedure, the purpose of $\mathbf{K}$ is to robustly stabilise the shaped plant as described in the next section. The shaping of the plant cannot be accomplished through the manipulation of $\mathbf{K}$. Hence we define $\mathbf{G_s}$ to be the augmented plant, $\mathbf{G_s} = \mathbf{W_2GW_1}$, where $\mathbf{G}$ represents the fixed plant. This structure allows the designer to shape the singular values of the augmented plant $\mathbf{G_s}$, through the selection of appropriate weighting matrices $\mathbf{W_1}$ and $\mathbf{W_2}$. The selection of these matrices is therefore the key element in attaining the performance requirements of the system and is the focal point of this design technique. This design task will be performed using a MOGA as outlined in Section 3.8.6.

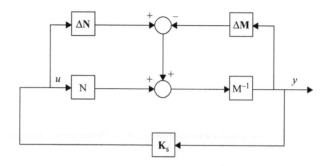

*Figure 3.17    Robust stabilisation with respect to coprime factor uncertainty*

### 3.8.4    Robust stabilisation

The normalised left coprime factorisation (NLCF) of a plant $\mathbf{G}$ is given by $\mathbf{G} = \mathbf{M}^{-1}\mathbf{N}$. A perturbed plant model $\mathbf{G_p}$ is then given by,

$$\mathbf{G_p} = (\mathbf{M} + \Delta\mathbf{M})^{-1}(\mathbf{N} + \Delta\mathbf{N}). \tag{3.5}$$

To maximise this class of perturbed models such that the configuration shown in Figure 3.17 is stable, a controller $\mathbf{K_s}$ that stabilises the nominal closed-loop system and minimises $\gamma$ must be found, where

$$\gamma = \left\| \begin{bmatrix} \mathbf{K_s} \\ \mathbf{I} \end{bmatrix} (\mathbf{I} - \mathbf{GK_s})^{-1}\mathbf{M}^{-1} \right\|_\infty. \tag{3.6}$$

This is the problem of robust stabilisation of normalised coprime factor plant descriptions [14]. From the small gain theorem [11], the closed-loop plant will remain stable if,

$$\left\| \begin{bmatrix} \Delta\mathbf{N} \\ \Delta\mathbf{M} \end{bmatrix} \right\|_\infty < \gamma^{-1}. \tag{3.7}$$

The lowest possible value of $\gamma$ and hence the highest achievable stability margin is given by $\gamma_{\min} = (1 + \rho(ZX))^{1/2}$, where $\rho$ is the spectral radius and $Z$ and $X$ are the solutions to the following algebraic Riccati equations,

$$(\mathbf{A} - \mathbf{B}S^{-1}\mathbf{D}^T\mathbf{C})Z + Z(\mathbf{A} - \mathbf{B}S^{-1}\mathbf{D}^T\mathbf{C})^T - Z\mathbf{C}^T R^{-1}\mathbf{C}Z + \mathbf{B}S^{-1}\mathbf{B}^T = 0,$$
$$\tag{3.8}$$

$$(\mathbf{A} - \mathbf{B}S^{-1}\mathbf{D}^T\mathbf{C})^T X + X(\mathbf{A} - \mathbf{B}S^{-1}\mathbf{D}^T\mathbf{C}) - X\mathbf{B}S^{-1}\mathbf{B}^T X + \mathbf{C}^T R^{-1}\mathbf{C} = 0,$$
$$\tag{3.9}$$

where $\mathbf{A}$, $\mathbf{B}$, $\mathbf{C}$ and $\mathbf{D}$ are the state–space matrices of $\mathbf{G}$ and,

$$R = \mathbf{I} + \mathbf{DD}^T, \qquad S = \mathbf{I} + \mathbf{D}^T\mathbf{D}. \tag{3.10}$$

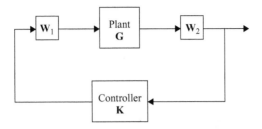

*Figure 3.18    Loop-shaping controller structure*

By solving these equations, the state–space controller, $\mathbf{K}_s$, can be generated explicitly [11]. This controller gives no guarantee of the system's performance, only that it is robustly stable. It is therefore necessary to shape the system's response with both pre- and post-plant weighting function matrices $\mathbf{W}_1$ and $\mathbf{W}_2$ as shown in Figure 3.18. This will ensure that the closed-loop performance meets the specifications required.

### 3.8.5    Augmenting the $H_\infty$ approach with additional objectives

A multi-objective optimisation strategy is compatible with the $H_\infty$ controller design method. The inherent compromise between performance and robustness, which is prevalent in all control system design approaches, lends itself to formulation as a multi-objective $H_\infty$ optimisation. Within an $H_\infty$ controller design framework, MOGA exercises control over the selection of suitable LSDP weighting matrices to satisfy the performance objectives. These performance objectives are the multiple objectives considered by MOGA. Operating in this way, each controller is assured of being robust, as it has been obtained via the $H_\infty$ design process. The task of the designer will be to select a suitable controller from a range of non-dominated options. The mapping of the system specifications to the multi-objective problem formulation is shown in Figure 3.19.

### 3.8.6    Application of design technique to gasifier problem

The weighting function structures used were those of a diagonal matrix of first-order lags for $\mathbf{W}_1$ and a diagonal matrix of gains for $\mathbf{W}_2$. The first-order lag structure of the diagonal elements of $\mathbf{W}_1$ was considered sufficient to break any algebraic loops that may appear in simulation due to the non-zero $\mathbf{D}$ matrix in the linear model. As the linear model of the gasifier contains 25 states, this design technique produces controllers that are at least of the order of the plant. The terms in $\mathbf{W}_2$ were specified as stateless in order to minimise the order of the resulting controllers. The controller was designed for the 100 per cent load linear model as specified in the Challenge.

Each controller's performance was then evaluated by running simulations at all three operating points using both the step and sine-wave disturbance signals. As the optimisation philosophy of MOGA is to minimise objective function values, the plant model was not offset, relative values about the operating point being preferred

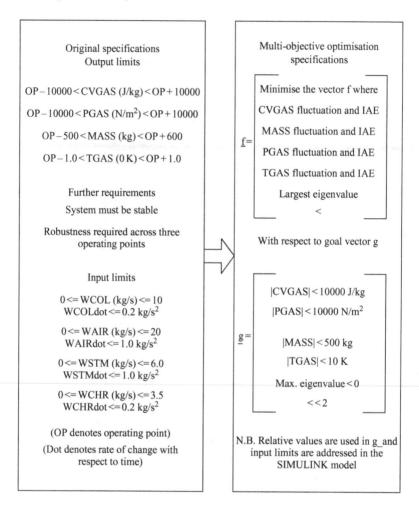

*Figure 3.19    Mapping of system specifications into multi-objective formulation*

to absolute input–output values. This allowed the objective function to assess the peak deviations in gasifier outputs produced by each candidate controller by taking the maximum absolute value of each output vector. Input constraints were observed by placing saturation and rate-limit blocks on the inputs of the simulated system representation containing relative values appropriate to the operating point. Stability of the closed-loop system was guaranteed by minimising the real part of the closed-loop continuous eigenvalue having the largest real part, and discarding any individual in the population that did not result in stable closed-loop eigenvalues. One further objective attempted to minimise the $H_\infty$ norm, $\gamma$, in order to maximise the robustness of the closed-loop control system. The objectives for the gasifier problem are shown in Table 3.5.

Figure 3.20 shows the parallel coordinates graph that resulted from the above formulation of the gasifier problem. The final choice of controller was made with

*Table 3.5    Gasifier design objectives*

| Objective number | Objective description |
| --- | --- |
| 1 | Peak fluctuation of CVGAS from 100% operating point |
| 2 | Peak fluctuation of MASS from 100% operating point |
| 3 | Peak fluctuation of PGAS from 100% operating point |
| 4 | Peak fluctuation of TGAS from 100% operating point |
| 5 | Peak fluctuation of CVGAS from 50% operating point |
| 6 | Peak fluctuation of MASS from 50% operating point |
| 7 | Peak fluctuation of PGAS from 50% operating point |
| 8 | Peak fluctuation of TGAS from 50% operating point |
| 9 | Peak fluctuation of CVGAS from 0% operating point |
| 10 | Peak fluctuation of MASS from 0% operating point |
| 11 | Peak fluctuation of PGAS from 0% operating point |
| 12 | Peak fluctuation of TGAS from 0% operating point |
| 13 | Maximum continuous eigenvalue of closed-loop system |
| 14 | $H_\infty$ robustness norm, $\gamma$ |

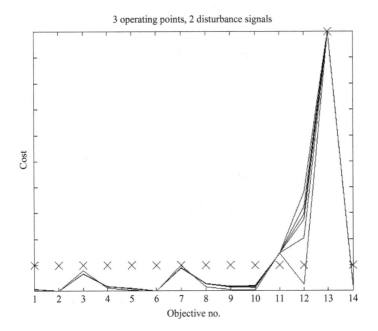

*Figure 3.20    Gasifier trade-off graph*

reference to the performance requirements specified in the Challenge. As can be seen from Figure 3.20, one potential solution satisfied more performance requirements than the others, that being the only one to meet objective 12, the peak temperature fluctuation for the 0 per cent operating point. This individual was therefore selected

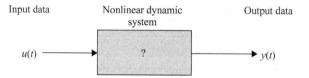

*Figure 3.21   Generic nonlinear model*

as the final choice of controller. A full set of results and data may be found in Griffin [15].

## 3.9   System identification

### 3.9.1   *Use of GAs and special operators*

This section presents a multi-objective optimisation approach to the use of a Nonlinear Auto-Regressive Moving Average with eXtra inputs (NARMAX) model-based approach to nonlinear system identification. The generic problem of system identification is shown in Figure 3.21 where the input–output relationship for a given system is to be represented mathematically.

The approach is demonstrated for a SISO system and applied to discrete input and output time history data. The generic form of a NARMAX model is shown in equation (3.11).

$$y(t) = F(y(t-1), \ldots, y(t-n_y), u(t-1), \ldots, u(t-n_u),$$
$$e(t-1), \ldots, e(t-n_e)) + e(t). \tag{3.11}$$

The nonlinear function $F(\cdot)$ may contain a range of nonlinear combinations of $y$, $u$ and $e$. Here, this generic form is truncated such that the error terms, $e(t)$, $e(t-1), \ldots, e(t-n_e)$, are omitted so giving a NARX model.

For a given set of terms, linear regression is then applied in order to identify the optimum coefficient values, $\theta_i$, for each term, $p_i$, resulting in the model shown in equation (3.12).

$$y(t) = \sum_{i=1}^{M} q_i p_i(t) + x(t), \quad t = 1, \ldots, N. \tag{3.12}$$

These terms may be raised to a given maximum power. In conjunction with the maximum lags, $n_y, n_u$ and $n_e$, this gives a set of possible terms that may be used to construct a model of the system that best conforms to the time history data. The choice of a maximum degree and maximum time lag define a finite set of possible terms, $p_i$, which could be used to construct a model to fit the time history data. From this set, a subset of terms can then be chosen to form the model. The number of terms in this subset, $k$, is pre-defined. MOGA is applied in order to choose $k$ terms from the set of all possible terms in order to form the optimal model.

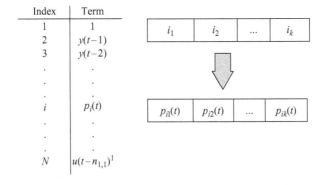

| Index | Term |
|-------|------|
| 1 | 1 |
| 2 | $y(t-1)$ |
| 3 | $y(t-2)$ |
| . | . |
| . | . |
| . | . |
| $i$ | $p_i(t)$ |
| . | . |
| . | . |
| . | . |
| $N$ | $u(t-n_{1,1})^1$ |

*Figure 3.22    Chromosome for system identification*

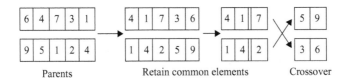

| 6 | 4 | 7 | 3 | 1 |    | 4 | 1 | 7 | 3 | 6 |    | 4 | 1 | 7 |    | 5 | 9 |
| 9 | 5 | 1 | 2 | 4 |    | 1 | 4 | 2 | 5 | 9 |    | 1 | 4 | 2 |    | 3 | 6 |

Parents            Retain common elements            Crossover

*Figure 3.23    Identity preserving crossover*

Each of the possible nonlinear terms, $p_i$, is indexed. The chromosome of each individual within the MOGA population contains the indices of the $k$ terms to be chosen from the set of all possible terms as shown in Figure 3.22.

Given that the model requires $k$ distinct terms, it is necessary to prevent the duplication of indices within the chromosome. The genetic operators of recombination and mutation are therefore applied in a modified manner. Full identity preserving crossover ensures that the parents with common terms in their chromosome do not produce offspring with duplicated terms. This is done by reordering the chromosomes of the parents such that duplicated terms are not included in that part of the chromosome available for crossover. Those elements to the right of the crossover point are then exchanged producing offspring with no duplicated terms as shown in Figure 3.23.

Trade mutation ensures that no area of the search space has a zero probability of being explored while also ensuring that there are no duplicated terms in the chromosome. A term in the chromosome that is to be mutated is exchanged with a term from those remaining outside the chromosome of that individual. This is shown in Figure 3.24.

In order to reduce the computational burden of assessing the individuals, it is normal to train the model on a short section of data points in the system time history. This approach is designed to minimise the residual variance of the model where residual variance measures the fidelity of the model to the data set over which it was fitted. However, since the method produces a population of solutions, it is interesting

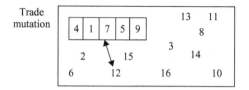

*Figure 3.24    Trade mutation*

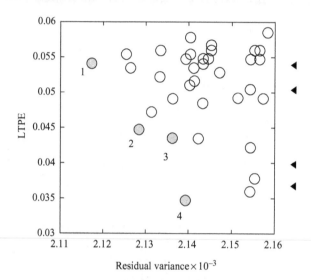

*Figure 3.25    Objective space for system identification*

to see how members of the population perform with respect to long-term prediction error (LTPE), where LTPE is a measure of the model's ability to predict the output from previously unseen input data. A Pareto front is visualised as a two-objective problem, from which the designer can assess both characteristics of each model. This is shown in Figure 3.25.

### 3.9.2    Use of MOGP

It should be noted that the MOGA-based approach for the identification of NARX models requires the number of terms to be used to be specified in advance. A more flexible approach in which the number of terms to be used could also be selected by the evolutionary procedure would be desirable.

Genetic programming (GP) [16] is a subclass of GAs in which the potential solutions are expressed as programmes instead of individuals represented by genotypic strings. The fact that many problems can be expressed as computer programmes makes the tree-based encoded GP a more powerful tool than a GA, which uses linear

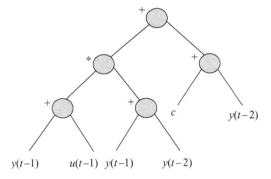

*Figure 3.26   GP tree structure*

encoding. In GP, the trees are composed of functions and terminals that are appropriate to the problem domain. The programme is then encoded as a hierarchical tree
structure that provides a dynamic and variable representation. Figure 3.26 shows a
tree structure that may be used to represent the 'programme' given in equation (3.13).

$$y(t) = c + a_1 y(t-2) + a_2 y(t-1)^2 + a_3 y(t-1)y(t-2)$$
$$+ a_4 y(t-1)u(t-1) + a_5 y(t-2)u(t-1). \tag{3.13}$$

The internal nodes of the tree structure are elements from the function set (operations),
$G$, and the leaf nodes are the input data from the terminal set, $T$. For the above example
$T = \{c, u(t-1), y(t-1)\}$ and $G = \{`*`, `+`\}$.

As for the standard GA, the genetic operators of recombination and mutation are
applied in order to evolve the population. Recombination produces a pair of offspring
that inherits characteristics from both parent programmes. A random node in each of
the parent tree structures is selected and the subsequent expressions are exchanged
between the parents to form the offspring. Due to the dynamic representation used in
GP, the parents are typically of different size, shape and content and the offspring are
also generally different. Equations (3.13) and (3.14) describe the parent trees shown
in Figure 3.27 prior to crossover. Equations (3.15) and (3.16) describe the offspring
trees resulting from crossover shown in Figure 3.28.

$$y(t) = a_1 y(t-3) + a_2 y(t-1)u(t-1) + a_3 u(t-1)^2. \tag{3.14}$$

$$y(t) = c + a_1 y(t-2) + a_2 y(t-1)y(t-3) + a_3 y(t-3)u(t-1). \tag{3.15}$$

$$y(t) = a_1 y(t-1) + a_2 y(t-2) + a_3 y(t-1)u(t-1) + a_4 u(t-1)^2. \tag{3.16}$$

Mutation is performed by randomly selecting a node that can be an internal or
terminal node and replacing the associated sub-expression with a randomly generated
sub-programme. Figures 3.29 and 3.30 show a tree structure before and after mutation.

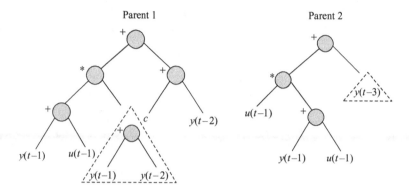

*Figure 3.27    Parental nodes selected for GP crossover*

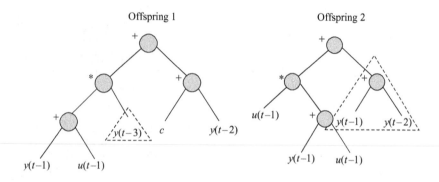

*Figure 3.28    Offspring resulting from GP crossover*

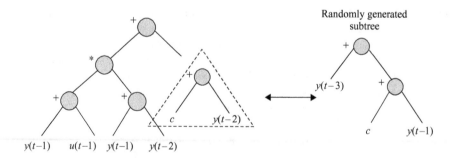

*Figure 3.29    GP mutation before operation*

In order to generate programs that represent not only valid models of the system but also parsimonious models, a set of objectives is specified which addresses the two main themes of (1) model structure (complexity) and (2) model performance.

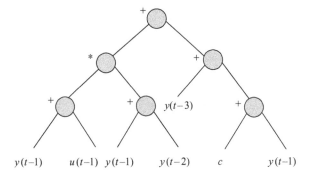

*Figure 3.30   GP mutation after operation*

### 3.9.2.1   Case study: simple Wiener process

Based on the attributes of complexity and performance, the MOGP method described above is demonstrated on the simple Wiener model and compared with two conventional identification techniques, stepwise regression and orthogonal regression [17,18].

The simple Wiener process embodies a linear dynamic part defined by the differential equation,

$$10\dot{v}(t) + v(t) = u(t), \tag{3.17}$$

and a static nonlinear part expressed by

$$y(k) = 2 + v(k) + v^2(k). \tag{3.18}$$

The input–output data used here are defined in Haber and Unbehauen [17], as described in Figure 3.31.

The MOGP approach was run considering five objectives representing the structure and the performance of the models. These were: the number of terms, NT, degree of nonlinearity, DEG, maximum lag, LAG, residual variance, VAR, and LTPE. Crossover and mutation probabilities were 0.9 and 0.1, respectively. The MOGP method evolved for 100 generations using a population of 200 tree expressions. The method was run several times and produced similar families of solutions each time.

For the purpose of analysis, results of one run are presented in Table 3.6. In terms of performance (VAR and LTPE) all models emerging from the MOGP approach dominate those obtained by the stepwise and orthogonal regression methods. For models $MOGP_3$ and $MOGP_4$, an improvement in the VAR criterion is also achieved – that which the stepwise and orthogonal regression methods explicitly targetted.

In terms of model complexity, Table 3.7 shows the structures of the polynomial NARX models which are similar and have some terms in common. This identification example is a simple one based upon simulated data; however it serves to illustrate the potential of the MOGP identification.

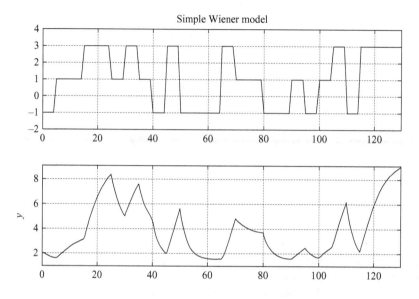

*Figure 3.31    Simple Wiener process of input–output data*

*Table 3.6    Comparative performance of the identification methods*

| Model | NT | DEG | LAG | VAR×10⁻³ | LTPE×10⁻³ |
|---|---|---|---|---|---|
| $MOGP_1$ | 6 | 2 | 1 | 2.3839 | 6.0221 |
| $MOGP_2$ | 6 | 2 | 2 | 2.1978 | 6.7967 |
| $MOGP_3$ | 7 | 2 | 2 | 1.6484 | 6.4279 |
| $MOGP_4$ | 7 | 2 | 2 | 1.6474 | 7.8151 |
| Stepwise | 7 | 2 | 2 | 1.6808 | 7.8526 |
| Orthogonal | 7 | 2 | 2 | 5.2243 | 26.8080 |

## 3.10    Conclusions

This chapter has introduced GA and its usage in a multi-objective framework. The benefits of using GA for multi-objective optimisation have been described by contrasting the capabilities of evolutionary methods to traditional optimisers when addressing multi-objective problems. The ability to identify the Pareto-optimal surface of a problem domain in a single run is a significant advantage in the use of MOGA. The modifications that can be applied to the standard GA in order to provide suitable characteristics for multi-objective optimisation are given. The case studies given in this chapter highlight the use of preference articulation and illustrate the mapping of

*Table 3.7    Simple Wiener model structures*

| Term | $MOGP_1$ | $MOGP_2$ | $MOGP_3$ | $MOGP_4$ | Stepwise | Orthogonal |
|---|---|---|---|---|---|---|
| $c$ | $\sigma$ | $\sigma$ | $\sigma$ | $\sigma$ | $\sigma$ | $\sigma$ |
| $y(k-1)$ | $\sigma$ | $\sigma$ | $\sigma$ | $\sigma$ | $\sigma$ | $\sigma$ |
| $y(k-2)$ | | | | $\sigma$ | $\sigma$ | $\sigma$ |
| $u(k-1)$ | $\sigma$ | | $\sigma$ | | | $\sigma$ |
| $u(k-2)$ | | $\sigma$ | | | | $\sigma$ |
| $y(k-1)^2$ | $\sigma$ | $\sigma$ | $\sigma$ | $\sigma$ | $\sigma$ | |
| $y(k-1)y(k-2)$ | | | | $\sigma$ | | |
| $y(k-2)^2$ | | | | | $\sigma$ | |
| $y(k-1)u(k-1)$ | $\sigma$ | $\sigma$ | | $\sigma$ | $\sigma$ | $\sigma$ |
| $u(k-1)^2$ | $\sigma$ | $\sigma$ | $\sigma$ | $\sigma$ | $\sigma$ | $\sigma$ |
| $y(k-2)u(k-1)$ | | | $\sigma$ | | | |
| $u(k-1)u(k-2)$ | | | $\sigma$ | | | |

complex control design problems into the MOGA framework. The use of MOGA for system identification has also been shown along with the development of GP.

## Notes

1. As an introduction to GAs, it is convenient to think of recombination as an exploitation operator. However, upon further consideration, it will be recognised that it has both exploitation and exploration roles.
2. Commercial confidentiality prevents us from full disclosure at this point. However, our MOGA tool is currently being used in industry and producing significant savings in design time.

## References

1 FONSECA, C. M., and FLEMING, P. J.: 'Genetic algorithms for multi objective optimisation: formulation, discussion and generalization'. Proceedings of the fifth international conference on *Genetic algorithms*, San Mateo, CA, 1993, pp. 416–23
2 DARWIN, C.: 'The origin of species' (John Murray, London, 1859)
3 FISHER, R. A.: 'The genetical theory of natural selection' (Clarendon Press, Oxford, 1930)
4 HOLLAND, J. H.: 'Adaptation in natural and artificial systems' (The University of Michigan Press, Ann Arbor, MI, 1975)
5 GOLDBERG, D. E.: 'Genetic algorithms in search, optimisation and machine learning' (Addison-Wesley, Reading, MA, 1989)
6 MICHALEWICZ, Z.: 'Genetic algorithms + data structures = evolution programmes' (Springer, Berlin, 1996)

7 FLEMING, P. J., and PURSHOUSE, R. C.: 'Evolutionary algorithms in control systems engineering: a survey', *Control Engineering Practice*, 2002, **10**, pp. 1223–41

8 FONSECA, C. M., and FLEMING, P. J.: 'An overview of evolutionary algorithms in multi-objective optimisation', *Evolutionary Computation*, 1995, **3** (1), pp. 1–16

9 RECHENBERG, I.: 'Evolutionsstrategie: optimierung technischer systeme nach prinzipien der biologoschen evolution' (Frommann-Holzboog, Stuttgart, 1973)

10 DIXON, R., PIKE, A. W., and DONNE, M. S.: 'The ALSTOM benchmark challenge on gasifier control', *Proceedings of the Institution of Mechanical Engineers, Part I: Journal of Systems and Control Engineering*, 2000, **214**, pp. 389–94

11 SKOGESTAD, S., and POSTLETHWAITE, I.: 'Multivariable feedback: control, analysis and design' (John Wiley & Sons Ltd, Chichester, England, 1996)

12 WILLIAMS, S. J.: '$H_\infty$ for the layman', *Measurement and Control*, 1991, **24** (2), pp. 18–21

13 MACIEJOWSKI, J. M.: 'Multivariable feedback control' (Addison-Wesley, Wokingham, UK, 1989)

14 GLOVER, K., and MACFARLANE, D. C.: 'Robust stabilisation of normalised coprime factor plant descriptions with $H_\infty$ bounded uncertainties', *IEEE Transactions of Automatic Control*, 1989, **34** (8), pp. 821–30

15 GRIFFIN, I. A.: 'Multivariable control methods for gas turbine engines'. Ph.D. dissertation, University of Sheffield, UK, 2002

16 KOZA, J. R.: 'Genetic programming: on the programming of computers by means of natural selection' (MIT Press, Boston, MA, 1992)

17 HABER, R., and UNBEHAUEN, H.: 'Structure identification of non-linear dynamic systems: a survey of input/output approaches', *Automatica*, 1990, **26** (4), pp. 651–77

18 RODRIGUEZ-VAZQUEZ, K., FONSECA, C.-M., and FLEMING, P.-J.: 'Identifying the structure of nonlinear dynamic systems using multi objective genetic programming', *IEEE Transactions on Systems, Man & Cybernetics – Part A Systems & Humans*, 2004, **34** (4), pp. 531–45

*Chapter 4*

# Adaptive local linear modelling and control of nonlinear dynamical systems

*Deniz Erdogmus, Jeongho Cho, Jing Lan,*
*Mark Motter and Jose C. Principe*

## 4.1  Introduction

Systems theory is a well-established and mature area of engineering research, where many strong general mathematical results are available. Especially the analysis of linear identification and control systems has been pursued by many researchers leading to a complete understanding of various mechanisms that are effective in the stability, controllability and observability of these. Due to the availability of such an extensive knowledge base about linear systems, modern industrial control applications are still typically designed utilising the results from linear control systems theory. Nevertheless, academic research has been concentrating around problems involving the stability, identification and control of nonlinear dynamical systems in the last few decades. These efforts have now also matured into a broad theory of nonlinear systems, their identification and control. Initial efforts in this area pursued parametric approaches, inspired by the established linear systems theory, where the system dynamical equations are generally assumed to be known from physical principles, possibly with some uncertainty in the values of certain parameters. In this framework, the system identification and system control problems are decoupled, therefore can be solved sequentially. More recently, adaptive system identification and control methodologies have also been investigated, once again leading to a very good understanding of the adaptation in linear systems and a satisfactorily general insight to adaptation in nonlinear control systems. The latter problem, however, is implicitly extremely difficult to tackle and although nice mathematical results are obtained, practicality of these nonlinear techniques is yet difficult to achieve.

Control theory deals with the problem of manipulating the behaviour of dynamical systems to satisfy certain desired outputs from the system. Classically, as mentioned above, the design procedure will follow the system identification and controller selection stages in the parametric approach to system modelling and control. In the case of traditional identification based on models derived from physical principles, the data are used to estimate the unknown parameters [1,2], whereas modern approaches stemming from the advances in neural network theory introduce black-box function approximation schemes in parts of the models [3–7]. The neural network modelling capabilities may be further enhanced using multiple such sub-models in the context of switching between adaptive models, as we will present in more detail in this chapter, to obtain closed-loop control systems that enhance transient behaviour and cope better with modelling uncertainties and sudden model changes [8,9]. Following the system identification stage, depending on the modelling approach taken, the controller is designed typically using classical techniques based on linear systems theory, such as gain scheduling [10], switching between multiple fixed or adaptive controllers [11,12], as well as classical or neural-network-based nonlinear techniques [13–15].

A large class of real-world systems can be reasonably approximated by nonlinear, time-invariant mathematical models. Therefore, our discussions here will focus on this class of systems, although we will briefly describe how to extend the presented approaches to the more general nonlinear time-varying system scenarios. Note that, however, the latter is an extremely difficult problem to solve. Even the global modelling of time-invariant nonlinear systems and designing corresponding controllers is itself a daunting task, let alone dealing successfully with time-variability in cases except where the variations are slow so that available adaptation tools can cope with the task of tracking the changing models.

A principle that is adopted with great enthusiasm in the statistical function approximation literature is the divide-and-conquer approach that dictates solving complicated problems by breaking them up into smaller and easier pieces that can be managed by simpler topologies. The method presented here follows along the lines of this principle. Therefore, conceptually, this modelling technique can be regarded as a piecewise modelling approach, where the pieces are then patched together to form an approximate but successful global model. Specifically, when each of the model pieces are selected to be linear, the resulting model is a piecewise linear dynamical approximation to the globally nonlinear dynamical system. The advantages of such a partitioning approach are threefold: system identification complexity is reduced significantly due to the scaling down of the optimisation problem from one large task to multiple small and simple tasks, the piecewise model easily scales up to encompass more volume in the state–space of the dynamical system by the addition of new patches as data from previously unseen portions of the state–space is acquired, and the design of a control system for the nonlinear system can be reduced to the design of multiple simple and local controllers among which switching or cooperation is possible to generate a single control command to the actual plant. Especially with the selection of local linear dynamical models, the global nonlinear controller design reduces to the much simpler problem of designing multiple linear controllers for linear systems, a problem for which there are many extremely strong tools available in the linear control systems literature [16–18].

In the local modelling approach, there are two possibilities for utilising the individual local models to generate a single global value: select one model at a given time (winner-take-all), or take a weighted combination of the models (mixture-of-experts). Both approaches will be discussed in detail in the following sections. In particular, the winner-take-all approach will make explicit use of the self-organising maps (SOM) [19] in order to select which model–controller pair to switch to at every time instant (as opposed to the output-tracking-error-based switching criterion proposed by Narendra and co-workers [8,11]), and the mixture-of-experts approach will utilise the finite Gaussian mixture models (GMM) for a statistical interpretation of the local model contributions through the components of the mixture density model. Although the output-error approach is also commonly utilised in switching expert systems, it requires adjusting switching criterion parameters in noisy situations or for different systems, whereas in the SOM-based switching modality, these considerations are automatically taken care of in the SOM-training phase through the statistical interpretation of the data by the self-organisation algorithm. The trade-off in this is the requirement that the multidimensional state–space of the system is sufficiently covered by the SOM, whereas the output error approach operates in the lower dimensional output space. In addition, the SOM can be trained to classify the current state of the system directly from an input vector that is representative of this state, rather than the indirect measure of output-error. Consequently, the model selection sequence obtained using an SOM is expected to be more in tune with the actual state–space transition than the dynamical system experiences. The GMM-based models will also be partitioned based on the same representative state vector as the SOM, consisting of past values of the plant's input and output, leading to the questions of validity and accuracy of such state representations.

It is well known that linear dynamical systems expressed by an observable state–space equation set can be equivalently described by autoregressive moving average (ARMA) difference equations (or differential equations in continuous time), which are essentially recursive expressions for the current output of the system in terms of its past inputs and outputs. The existence of such input–output recursive representations for nonlinear systems is also dependent on the extended definitions of observability for nonlinear systems. Results demonstrate that a wide class of nonlinear systems, called generically observable systems, also possess such nonlinear ARMA (NARMA) models that are valid at least locally, and sometimes even globally, in the state–space [2,20–23]. In summary, we can conclude that the behaviour of nonlinear dynamical systems can at least locally be described well by NARMA equations, which is of crucial importance in the case of state–space reconstruction for dynamical systems where the internal state variables are not accessible. In such cases, the time-delay embedding method [24] has to be used in order to create local NARMA or ARMA models that are representative of the system dynamics.

## 4.2 Motivation for local linear modelling and time-delay embedding

Consider, without loss of generality, a single-input single-output (SISO) nonlinear time-invariant dynamical system with state vector $\mathbf{x} \in \mathfrak{R}^n$, input $u \in \mathfrak{R}$ and output

$y \in \Re$ with the following set of state equations and output mapping:

$$\mathbf{x}_{k+1} = \mathbf{f}(\mathbf{x}_k, u_k),$$
$$y_k = h(\mathbf{x}_k).$$
(4.1)

Notice that the consecutive outputs are (following the reasoning in Reference 3 and with o denoting composite functions):

$$y_k = h(\mathbf{x}_k) = \phi_1(\mathbf{x}_k)$$
$$y_{k+1} = h \circ f(\mathbf{x}_k, u_k) = \phi_2(\mathbf{x}_k, u_k)$$
$$\vdots$$
$$y_{k+n-1} = h \circ f \circ \cdots \circ f(\mathbf{x}_k, u_k) = \phi_n(\mathbf{x}_k, u_k, u_{k+1}, \ldots, u_{k+n-2}).$$
(4.2)

Defining $\mathbf{y}_k^n = [y_k \ \cdots \ y_{k+n-1}]^{\mathrm{T}}$ and $\mathbf{u}_k^{n-1} = [u_k \ \cdots \ u_{k+n-2}]^{\mathrm{T}}$, (4.2) can be collected in a vector-valued function form as $\mathbf{y}_k^n = \mathbf{\Phi}(\mathbf{x}_k, \mathbf{u}_k^{n-1})$.

**Implicit Function Theorem [24]**   *Let* $\mathbf{f}$ *be a* $C^1$ *mapping of an open set* $E \subset \Re^{n+m}$ *into* $\Re^n$ *such that* $\mathbf{f}(\mathbf{a}, \mathbf{b}) = \mathbf{0}$ *for some point* $(\mathbf{a}, \mathbf{b})$ *in* $E$. *Let* $\mathbf{f}_\mathbf{x}(\mathbf{x}, \mathbf{y})$ *denote the Jacobian of* $\mathbf{f}$ *with respect to* $\mathbf{x}$ *at the point* $(\mathbf{x}, \mathbf{y})$ *in* $E$. *If* $\mathbf{f}_\mathbf{x}(\mathbf{a}, \mathbf{b})$ *is invertible, then there exists an open set* $U \subset \Re^{n+m}$ *and* $W \subset \Re^m$ *with* $(\mathbf{a}, \mathbf{b}) \in U$ *and* $\mathbf{b} \in W$ *such that to every* $\mathbf{y} \in W$ *there corresponds a unique* $\mathbf{x}$ *satisfying* $\mathbf{f}(\mathbf{x}, \mathbf{y}) = \mathbf{0}$, $(\mathbf{x}, \mathbf{y}) \in U$. *If this* $\mathbf{x}$ *is defined to be* $\mathbf{g}(\mathbf{y})$, *then* $\mathbf{g} \in C^1$ *and is a mapping of* $W$ *into* $\Re^n$, $\mathbf{g}(\mathbf{b}) = \mathbf{a}$, *and* $\mathbf{f}(\mathbf{g}(\mathbf{y}), \mathbf{y}) = \mathbf{0}$ *for all* $\mathbf{y} \in W$.

The Implicit Function Theorem basically states that the condition of local invertibility for a nonlinear function is that its Jacobian is locally nonsingular. Employing this theorem on the vector-valued function representation of (4.2), we conclude that if the Jacobian $\partial \mathbf{\Phi} / \partial \mathbf{x}$ is nonsingular at a stationary point in the state–space of the unforced system, then $\mathbf{x}_k$ can be expressed locally in terms of $\mathbf{y}_k^n$ and $\mathbf{u}_k^{n-1}$. However, since by definition $\mathbf{x}_{k+n}$ depends on the inputs $u_k, \ldots, u_{k+n-1}$ and the initial state $\mathbf{x}_k$, there exists a unique local nonlinear input–output mapping of the form

$$y_{k+1} = F(y_k, y_{k-1}, \ldots, y_{k-n+1}, u_k, u_{k-1}, \ldots, u_{k-n+1})$$
(4.3)

valid in an open set in the state–space encompassing the stationary point of linearisation. The same conclusion result could also be obtained with the brute force method of linearising the nonlinear dynamics around a stationary point in the state–space and defining a state transformation from the actual state vector to incremental changes in the states, such that the system is locally represented by an ARMA process with state-dependent coefficients, which essentially becomes a NARMA equation as in (4.3). Conversely, it is possible to express the local NARMA process of (4.3) by a set of switching local ARMA processes, where each linearisation is carried out at the current operating point. Effectively, at a given operating point

$(\mathbf{y}_k^{n*}, \mathbf{u}_k^{n*}) = (y_k^*, \ldots, y_{k-n+1}^*, u_k^*, \ldots, u_{k-m+1}^*)$, the approximate ARMA process is

$$\hat{y}_{k+1} = F(\mathbf{y}_k^{n*}, \mathbf{u}_k^{n*}) + \left[\nabla_{\mathbf{y}} F(\mathbf{y}_k^{n*}, \mathbf{u}_k^{n*}) \quad \nabla_{\mathbf{u}} F(\mathbf{y}_k^{n*}, \mathbf{u}_k^{n*})\right] \cdot \begin{bmatrix} \hat{\mathbf{y}}_k^n - \mathbf{y}_k^{n*} \\ \hat{\mathbf{u}}_k^n - \mathbf{u}_k^{n*} \end{bmatrix}$$

$$= \left[\nabla_{\mathbf{y}} F(\mathbf{y}_k^{n*}, \mathbf{u}_k^{n*}) \quad \nabla_{\mathbf{u}} F(\mathbf{y}_k^{n*}, \mathbf{u}_k^{n*})\right] \cdot \begin{bmatrix} \hat{\mathbf{y}}_k^n \\ \hat{\mathbf{u}}_k^n \end{bmatrix}$$

$$+ \left(F(\mathbf{y}_k^{n*}, \mathbf{u}_k^{n*}) - \left[\nabla_{\mathbf{y}} F(\mathbf{y}_k^{n*}, \mathbf{u}_k^{n*}) \quad \nabla_{\mathbf{u}} F(\mathbf{y}_k^{n*}, \mathbf{u}_k^{n*})\right] \cdot \begin{bmatrix} \mathbf{y}_k^{n*} \\ \mathbf{u}_k^{n*} \end{bmatrix}\right), \quad (4.4)$$

where the model output is a linear combination of the reconstructed state variables plus a bias term. For smoothly varying nonlinear dynamical systems, this ARMA model can further be accurately approximated by a purely linear combination of the reconstructed state variables as $\hat{y}_{k+1} = \mathbf{a}^{\mathsf{T}}\hat{\mathbf{y}}_k^n + \mathbf{b}^{\mathsf{T}}\hat{\mathbf{u}}_k^n$, where the bias term is implicitly embedded in the local model coefficient vectors $\mathbf{a}$ and $\mathbf{b}$ through the least squares type consideration of the approximation error and the mean state vector value in the neighbourhood of approximation. This latter approximation is necessary when it is desired to design a local linear controller for the local linear ARMA model. The elimination of the bias term makes the local model truly linear in terms of its inputs and outputs, not just linear in its coefficients.

In this piecewise linear approximation of the original nonlinear system, the coefficients of the locally effective ARMA model are determined by the current state of the nonlinear system, which is expressed in terms of the past values of the system input and output. This is an important observation, since in general, if the mathematical model of the plant is not known, its physically meaningful internal state variables are not accessible either. Under such conditions, the past values of the input and the output signals can be utilised to generate a representative state vector to identify the local behaviour of the system.

In chaos theory and nonlinear time-series analysis, this method of reconstructing a state vector is referred to as time-delay embedding, and there are strong theoretical results that demonstrate the mathematical validity of this approach for the case of autonomous nonlinear systems [22,23]. In particular, Takens' embedding theorem states, in plain words, that there exists an invertible (i.e. a one-to-one and onto) between the original state dynamics and the reconstructed state dynamics provided that the embedding dimension (the number of lags in the reconstructed state vector) is sufficiently large (specifically greater than two times the original state dimension). A similar result was also demonstrated by Aeyels that stated almost any autonomous system of the form $\dot{\mathbf{x}} = \mathbf{f}(\mathbf{x}), y = h(\mathbf{x})$ is generically observable if $2n+1$ samples of the output is taken in a manner similar to (4.2) [20].

These theoretical results on the local NARMA representations combined with the observability of nonlinear systems and the utility of time-delay embedding reconstructions of the state vector allow the construction of a piecewise linear dynamical model approximation for a nonlinear system, which can be determined and optimised completely from input–output data collected from the original system. The literature is rich in multiple model approaches for nonlinear modelling [1,25], where the general consensus is that local modelling typically outperforms global modelling with a single highly complicated neural network in input–output modelling scenarios [2,26–30], despite the intrinsic simplicity, and the input–output delay embedding approach has been adopted commonly based on results from nonlinear time-series analysis, such as Takens' theorem and its extensions [28]. A question of practical importance in this black-box input–output modelling approach is how to choose the number of embedding lags for both the input and the output. The next section deals with this question.

The motivation presented above mainly dealt with noise-free deterministic dynamical systems, whereas in practice, the available data is certainly noisy. From a mathematical perspective the suitability of the local modelling approach is justified by the above discussion. The practical aspects when noisy data is utilised is going to be investigated in the following sections whenever necessary.

## 4.3   Selecting the embedding dimension

If the number of physical dynamical states of the actual nonlinear system is known *a priori*, one can select the length of the embedding tap-delay lines for the input and output in accordance with the theoretical results by Takens and Aeyels. For complete practicality of the proposed local linear modelling approach for unknown systems, however, a truly data-driven methodology for determining the embedding dimensions for the input and output signals is required.

The problem of determining accurate input–output models from training data generated by the system has been addressed by many researchers [1–3,7,9,30,31], where the selection of the number of lags for the input and output signals (which is essentially a question of model-order selection) has always been an issue of practical importance. A useful solution to determine the embedding dimensions for input–output models is outlined by He and Asada [32], where the model order is determined based on the Lipschitz index calculated using the training data and the corresponding optimal model outputs for various embedding dimensions.

1.  Select candidate output and input embedding dimensions $n$ and $m$.
2.  From now on consider all past values of input and output as input variables $x_1, \ldots, x_n, x_{n+1}, \ldots, x_{n+m}$. Let the model output be denoted by $y$. Denote the $i$th input vector sample by $\mathbf{x}_i$ and the output by $y_i$, $i = 1, \ldots, N$.
3.  For every pair of samples evaluate the Lipschitz quotient: $q_{ij} = |y_i - y_j|/ \|\mathbf{x}_i - \mathbf{x}_j\|$, $i \neq j$, $i, j = 1, \ldots, N$.
4.  Let $q^{(n+m)}(k)$ denote the $k$th largest quotient $q_{ij}$.
5.  Evaluate the Lipschitz index: $q^{(n+m)} = (\prod_{k=1}^{p} \sqrt{n} q^{(n+m)}(k))^{1/p}$, where $p$ is an integer in the range $[0.01N, 0.02N]$.
6.  Go to step 1 and evaluate the Lipschitz index for a different set of embedding dimensions. The appropriate values of embedding dimensions will be indicated by the convergence index of the decreasing Lipschitz index as the embedding dimensions are increased one by one.

According to the theory, the appropriate embedding dimension pair for the input and the output is indicated by the convergence of the index. In other words, the embedding dimension, where the index stops decreasing (significantly), is to be selected as the model order.

## 4.4 Determining the local linear models

The local linear modelling approach can be broken into two consecutive parts: clustering/quantising the reconstructed state vector $\mathbf{x}_k^r = [\mathbf{y}_k^{nT} \ \mathbf{u}_k^{nT}]^T$ adaptively using a statistically sound approach and optimising the local linear models corresponding to each cluster of samples with least squares (or some other criterion) using data from only that cluster.[1] There are many possible techniques for tackling each of these individual problems available in the literature. For example, the first step (data clustering) can be achieved by using a standard clustering algorithm (such as $k$-means clustering [23]) or vector quantisation methods [33], numerous variants of self-organising maps (referred to as SOM) [19], or probability density mixture models (specifically the Gaussian mixture models – GMM) [34]; the second step (model optimisation) can be achieved using the analytical least squares solution (also referred to as the Wiener solution) [35], the least-mean-squares (LMS) algorithm [36], the recursive least squares algorithm [35] or the Kalman filter [37] if the mean-squared-error (MSE) is the optimality criterion of choice.[2]

In this section, we will focus on two of the clustering methods listed above: SOM and GMM. Based on the approach selected in modelling, the principle behind the local linear models will be either competitive (for the SOM) or cooperative (for the GMM). Later on, the controller designs will also be slightly different, due to this difference in the nature of the two approaches. In any case, the local model representation regimes will be selected optimally according to the criteria that these clustering methods utilise marking the main difference of the proposed adaptive local linear modelling approach from the standard gain-scheduling-like traditional approaches where the operating points of these local linear models are typically selected to be the stationary points of the state dynamics. In the case of completely unknown dynamics, this option is out of the question any way. Therefore, the methods presented here can be successfully applied both to cases where the actual state vector is accessible and where it is not available (so that input–output modelling is required).

### 4.4.1 Clustering the reconstructed state vector using self-organising maps

#### 4.4.1.1 Offline training phase

Suppose that input–output training data pairs of the form $\{(u_1, y_1), \ldots, (u_N, y_N)\}$, where $u$ is the input signal and $y$ is the output signal, are available from a SISO system for system identification. Under the conditions stated earlier, a nonlinear time-invariant dynamical system can be approximated locally by a linear ARMA system of the form[3]

$$y_k = a_1 y_{k-1} + \cdots + a_n y_{k-n} + b_1 u_{k-1} + \cdots + b_m u_{k-m} + c$$
$$= \mathbf{a}^T \mathbf{y}_{k-1}^n + \mathbf{b}^T \mathbf{u}_{k-1}^m + c. \tag{4.5}$$

The reconstructed state vector $\mathbf{x}_k^{n,m} = [y_{k-1} \cdots y_{k-n} \, u_{k-1} \cdots u_{k-m}]^T$ can be adaptively achieved using an SOM with any topology of choice (triangular or rectangular grids are possible). The SOM consists of an array of neurons with weight vectors $\mathbf{w}_i$ that are trained competitively on its input vectors $\mathbf{x}_k^{n,m}$.[4] In training, these input vector samples are presented to the SOM one at a time in multiple epochs, and preferably in each epoch, the presentation order of the samples is randomly shuffled to prevent memorising and/or oscillatory learning behaviour.

At every iteration, the winner neuron is selected as the one that minimises instantaneously the Mahalonobis distance $d(\mathbf{w}, \mathbf{x}_k) = (\mathbf{w}-\mathbf{x}_k)^T \mathbf{\Sigma}^{-1}(\mathbf{w}-\mathbf{x}_k)$ between the weight vector and the current training sample. Then the winner weight and its topological neighbours are updated using the following stochastic incremental learning rules [19], where $\mathbf{w}_w$ and $\mathbf{w}_n$ denote the winner and neighbour neuron weights, respectively:

$$
\begin{aligned}
\mathbf{w}_w(t+1) &\leftarrow \mathbf{w}_w(t) + \eta(t)\mathbf{\Sigma}^{-1}(\mathbf{x}_k - \mathbf{w}_w(t)), \\
\mathbf{w}_n(t+1) &\leftarrow \mathbf{w}_n(t) + \eta(t)h(\|\mathbf{w}_n(t) - \mathbf{w}_w(t)\|, \sigma(t))\mathbf{\Sigma}^{-1}(\mathbf{x}_k - \mathbf{w}_n(t)).
\end{aligned}
$$

$$(4.6)$$

In the Mahalonobis distance, the scaling matrix $\mathbf{\Sigma}$ can be selected as the input covariance to emphasise various directions in the updates in accordance with the data structure, or it can be set to identity. The neighbourhood function $h(\cdot, \sigma)$ is a monotonically decreasing function in its first argument and it is unity when evaluated at zero and zero when evaluated at infinity. This allows the neighbouring neurons to be updated proportionally to their distance to the instantaneous winner. The neighbourhood radius $\sigma$ is slowly annealed as well as the learning rate $\eta$. The neighbourhood radius is initially set such that most of the network is included in this region, but in time, as the neurons start specialising in their distinct regions of quantisation, this radius decreases to small enough values to cover only the winner neuron effectively. Typically, the neighbourhood function is selected as a one-sided Gaussian function with the standard deviation parameter controlling the neighbourhood radius. Both the radius and the learning rate can be annealed linearly or exponentially in terms of iterations or epochs.

The trained SOM can be regarded as a vector quantiser with the special topology preserving property. In particular, the SOM quantises its input space while preserving the topological structure of the manifold that the samples come from, resulting in strong neighbourhood relationships between the neurons; neighbouring input vectors are mapped to neighbouring neurons, thus in the next step of local linear modelling, neighbouring models will be structurally similar. The input space of the SOM, which is equivalently the reconstructed state–space of the system under consideration, is partitioned into smaller non-overlapping sets that are typically illustrated by a Voronoi diagram. The input samples $\{(\mathbf{x}_{i1}, y_{i1}), \ldots, (\mathbf{x}_{iN_i}, y_{iN_i})\}$, which are in the $i$th Voronoi region, are associated with the corresponding neuron with weight vector $\mathbf{w}_i$, where $N_i$ is the number of training samples in this region. In this local modelling scheme, besides the SOM weight vectors, each neuron also has a vector of local linear model coefficients associated with it, denoted by $\mathbf{a}_i$ and $\mathbf{b}_i$ separately for the output and input portions of the reconstructed state vector, respectively. These models can be optimised using the training data clustered to the $i$th Voronoi region and the MSE criterion. This results in the following least squares optimal local linear model coefficients [35, 36]:[5]

$$\mathbf{c}_i = \begin{bmatrix} \mathbf{a}_i \\ \mathbf{b}_i \end{bmatrix} = \begin{bmatrix} \mathbf{R}_i^{yy} & \mathbf{R}_i^{yu} \\ \mathbf{R}_i^{uy} & \mathbf{R}_i^{uu} \end{bmatrix}^{-1} \begin{bmatrix} \mathbf{P}_i^{dy} \\ \mathbf{P}_i^{du} \end{bmatrix} = \mathbf{R}_i^{-1} \mathbf{P}_i, \tag{4.7}$$

where the blocks of the input autocorrelation matrix are obtained from the training samples using

$$\mathbf{R}_i^{yy} = \frac{1}{N_i} \sum_k \mathbf{y}_k^n \mathbf{y}_k^{nT} \qquad \mathbf{R}_i^{uu} = \frac{1}{N_i} \sum_k \mathbf{u}_k^n \mathbf{u}_k^{nT}$$

$$\mathbf{R}_i^{yu} = \frac{1}{N_i} \sum_k \mathbf{y}_k^n \mathbf{u}_k^{nT} \qquad \mathbf{R}_i^{uy} = \frac{1}{N_i} \sum_k \mathbf{u}_k^n \mathbf{y}_k^{nT} \tag{4.8}$$

and the input-desired cross-correlation vector blocks are estimated from samples using

$$\mathbf{P}_i^{dy} = \frac{1}{N_i} \sum_k y_k \mathbf{y}_k^n \qquad \mathbf{P}_i^{du} = \frac{1}{N_i} \sum_k y_k \mathbf{u}_k^n. \tag{4.9}$$

Finally, the output of the $i$th local linear model with the optimised coefficients is given by $\hat{y}_k = [\mathbf{a}_i^T \ \mathbf{b}_i^T] \mathbf{x}_k$. More generally, introducing the model output weighting term $p_{ik}$, the overall local linear model system output is expressed as a switching (weighted) combination of the $M$ individual model outputs:

$$\hat{y}_k = \sum_{i=1}^M p_{ik} \mathbf{c}_i^T \mathbf{x}_k. \tag{4.10}$$

For hard competition, the weighting coefficients $p_{ik}$ take only the values 0 or 1 at every time instant $k$. The selection completely depends on the $i$th neuron winning for input vector $\mathbf{x}_k$. A simple modification that one can introduce to the SOM-based local linear models to make the overall model cooperative rather than competitive is to allow other weighting values for the models. For example, a weighted average type combination based on the distances of the current sample to the neuron weights would have

$$p_{ik} = \frac{f(d(\mathbf{w}_i, \mathbf{x}_k), \sigma)}{\sum_{j=1}^M f(d(\mathbf{w}_j, \mathbf{x}_k), \sigma)}, \tag{4.11}$$

where a monotonically increasing emphasis function $f(\cdot, \sigma)$ is combined with the Mahalonobis distance $d(\cdot, \cdot)$. The choice of a linear emphasis function would exactly be weighted averaging based on Mahalonobis distances. For clarity, the overall SOM-based local modelling topology is illustrated in Figure 4.1.

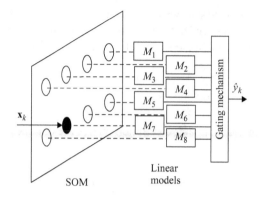

*Figure 4.1    Schematic diagram of the SOM-based local linear modelling approach for system identification. The instantaneous input vector determines the winner neuron, and thus the winning linear model that is activated*

### 4.4.1.2    Online training phase

In most cases, a batch-training phase as described earlier is beneficial for control system performance on the actual system. The online training procedure, although it could be employed immediately to the unknown system with random initialisation of all the weights and coefficients to be optimised, could require a large number of samples and/or time to become sufficiently accurate, while in the mean time the system operates under an improper controller system. Nevertheless, the online training algorithm presented here could be especially useful in fine-tuning the existing models or introducing additional local models to the archive whenever modelling performance of the existing system drops below acceptable levels. In addition, for identifying time-varying systems, the local models can be continuously adapted using online data. For only fine-tuning of the existing models, one only needs to continue updating the SOM weights as well as the local linear model coefficients on a sample-by-sample basis in real time. The SOM weights can be continued to be updated using the original update rules given in (4.6). The local linear model coefficients, however, must be updated by one of the many existing online linear model weight update rules from the literature. These online training rules for linear models include LMS and RLS (for recursive least squares) [35].[6]

When using LMS or RLS, only the coefficients of the linear model associated with the instantaneous winner neuron (whose weight vector is updated recently using the SOM learning algorithm) are updated. The least mean square (LMS) update rule for the coefficients of the winner model (assuming neuron $i$ is the winner) is given by

$$\mathbf{c}_i \leftarrow \mathbf{c}_i + \mu(y_k - \mathbf{c}_i^T \mathbf{x}_k)\mathbf{x}_k, \tag{4.12}$$

where $\mu$ is the LMS step size. Since LMS uses stochastic gradient updates for the model coefficients, it exhibits a misadjustment associated with the power of the inputs and the step size. However, its computational complexity is very low, suitable for fast real-time applications. On the other hand, RLS is a fixed-point algorithm that can track the analytical Wiener solution via sample-by-sample updates. The drawback is its increased complexity compared to LMS. The coefficient updates for the winner model according to RLS are given by the following iterations:

$$\mathbf{k}_i \leftarrow (\lambda^{-1}\mathbf{R}_i\mathbf{x}_k)/(1+\lambda^{-1}\mathbf{x}_k^T\mathbf{R}_i\mathbf{x}_k),$$

$$\mathbf{c}_i \leftarrow \mathbf{c}_i + \mathbf{k}_i(y_k - \mathbf{c}_i^T\mathbf{x}_k), \qquad\qquad (4.13)$$

$$\mathbf{R}_i \leftarrow \lambda^{-1}\mathbf{R}_i - \lambda^{-1}\mathbf{k}_i\mathbf{x}_k^T\mathbf{R}_i,$$

where the input autocorrelation matrix and weight vectors are initialised to $\mathbf{R}_i = \delta^{-1}\mathbf{I}$, $\mathbf{c}_i = \mathbf{0}$, $\delta$ being a small positive value.

Besides MSE, alternative model optimisation criteria such as alternative lags of error correlation [40], higher order error moments [36], or error entropy [38], can be utilised. Similar online update algorithms can be derived for these alternative criteria.

In some situations, simply fine-tuning of existing local models might not be sufficient to meet performance requirements in a sustained manner. Especially if, in actual operation, situations that are not encompassed in the training data set are encountered then a new local model might need to be introduced to the system of models. This could be achieved by utilising a growing SOM (GSOM) [41]. The most suitable grid structure for the GSOM is triangular. The neuron weights are still updated using (4.6). Contrary to a static SOM, in the GSOM, once in a while (e.g. at the end of every epoch), a new neuron is inserted (generated) in the weight space to the midpoint of the line segment connecting the neuron with the highest winning frequency and the neuron farthest from it. A similar neuron-killing criterion can be developed to eliminate infrequently activated neurons. This procedure is repeated until a balanced distribution of input samples per neuron is obtained in the Voronoi diagram. In the process of generating and killing neurons, the triangular topology of the SOM must be preserved, so the new neighbourhood connections must be selected accordingly.

### 4.4.2   Clustering the reconstructed state vector using Gaussian mixture models

#### 4.4.2.1   Offline training phase

Suppose that input–output training data pairs of the form $\{(u_1, y_1), \ldots, (u_N, y_N)\}$, where $u$ is the input signal and $y$ is the output signal, are available from a SISO system for system identification. The linear models described by (4.5) are still valid locally. In contrast to the SOM clustering of the reconstructed state vector $\mathbf{x}_k^{n,m} = [y_{k-1} \cdots y_{k-n} \quad u_{k-1} \cdots u_{k-m}]^T$, which trains the cluster centres (neurons) competitively, the Gaussian mixture model considers the possibility of multiple modes generating the same state. In particular, it is assumed that the probability distribution of the state vector is given by

$$p(\mathbf{x}_k) = \sum_{i=1}^{M} \alpha_i G(\mathbf{x}_k; \boldsymbol{\mu}_i, \boldsymbol{\Sigma}_i), \qquad\qquad (4.14)$$

where $G(\mathbf{x}; \boldsymbol{\mu}, \boldsymbol{\Sigma})$ is a multivariate Gaussian density with mean $\boldsymbol{\mu}$ and covariance $\boldsymbol{\Sigma}$. The coefficient $\alpha_i$ denotes the probability of occurrence of the $i$th mode in the GMM, which in turn reflects the probability of the corresponding local model being effective. Given the training data and once the state vectors are reconstructed using embedding, the maximum likelihood solution for the parameters $\alpha_i$, $\boldsymbol{\mu}_i$ and $\boldsymbol{\Sigma}_i$ can be determined using the expectation maximisation (EM) algorithm [34]. The EM algorithm can be outlined as follows:[7]

1. E-Step: Compute the expectation of the log-likelihood of the complete data conditioned by the observed samples assuming the most recent solution for the mixture parameters, which is given by $Q(\vartheta | \vartheta_t) = \sum_k E[\log p(\mathbf{v} | \vartheta) | \mathbf{x}_k, \vartheta_t]$, where the parameter vector is defined to include all means, covariances and weights in the mixture model: $\vartheta = [\alpha_1 \cdots \alpha_M \ \boldsymbol{\mu}_1 \cdots \boldsymbol{\mu}_M \ \text{vec}(\boldsymbol{\Sigma}_1) \cdots \text{vec}(\boldsymbol{\Sigma}_M)]^T$.

2. M-Step: Update parameter estimates to $\vartheta_{t+1}$, which is the maximiser of $Q(\vartheta | \vartheta_t)$.

Similar to the SOM-based modelling, in GMM-based local linear models, each Gaussian mode has a vector of local linear model coefficients associated with it, denoted by $\mathbf{a}_i$ and $\mathbf{b}_i$, again, for output and input portions of the reconstructed state vector, respectively. The output of the $i$th local linear model is given by $\hat{y}_k = [\mathbf{a}_i^T \ \mathbf{b}_i^T]\mathbf{x}_k$. The overall model output is a weighted combination of the $M$ individual outputs as in (4.10), $\hat{y}_k = \sum_{i=1}^M p_{ik}\mathbf{c}_i^T\mathbf{x}_k$, where $p_{ik} = \alpha_i G(\mathbf{x}_k; \boldsymbol{\mu}_i, \boldsymbol{\Sigma}_i)$. The linear model coefficients can be collectively optimised using a modified Wiener solution similar to that in (4.7). The modification involves the model activation probabilities, $p_{ik}$, and is explicitly given as $\boldsymbol{\theta} = \mathbf{R}^{-1}\mathbf{P}$, where $\boldsymbol{\theta} = [\mathbf{a}_1^T \ \mathbf{b}_1^T \ \cdots \ \mathbf{a}_M^T \ \mathbf{b}_M^T]$. The input autocorrelation matrix and the input–output cross-correlation vector are defined using the modified input vector $\mathbf{z}_k = [p_{1k}\mathbf{x}_k^T \ \cdots \ p_{Mk}\mathbf{x}_k^T]$.[8] Specifically

$$\mathbf{R} = \frac{1}{N}\sum_k \mathbf{z}_k \mathbf{z}_k^T \qquad \mathbf{P} = \frac{1}{N}\sum_k y_k \mathbf{z}_k. \qquad (4.15)$$

For completeness, the GMM-based local modelling topology, which is similar to the SOM-based topology in many aspects, is shown in Figure 4.2.

### 4.4.2.2 Online training phase

After the offline training procedure described earlier, the GMM-based model can be put to practice, while small adjustments to the existing parameters and model coefficients could be carried out in operation on a sample-by-sample basis, although this would be computationally extremely expensive. The EM algorithm could still be iterated by including one more sample to the probability density evaluations at every time instant. Alternatively, the EM algorithm could be replaced by a gradient-based maximum likelihood algorithm that can be operated in a stochastic manner (similar to LMS) to update the GMM parameters. Similarly, the linear model coefficients can be updated online using LMS or RLS with the modified input vectors $\mathbf{z}_k$ [35]. Particularly, the LMS update for the linear model coefficients is given by

$$\boldsymbol{\theta} \leftarrow \boldsymbol{\theta} + \mu(y_k - \boldsymbol{\theta}^T\mathbf{z}_k)\mathbf{z}_k \qquad (4.16)$$

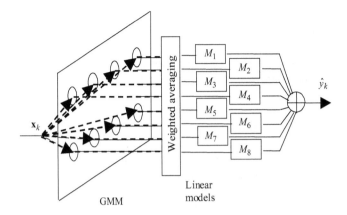

*Figure 4.2*   *Schematic diagram of the GMM-based local linear modelling approach for system identification. The instantaneous input vector activates all Gaussian modes determining the weights of the individual linear models*

and the corresponding RLS update is similar to (4.13), but the input and weight vectors are modified:

$$\mathbf{k} \leftarrow (\lambda^{-1}\mathbf{R}\mathbf{z}_k)/(1 + \lambda^{-1}\mathbf{z}_k^T\mathbf{R}\mathbf{z}_k),$$
$$\boldsymbol{\theta} \leftarrow \boldsymbol{\theta} + \mathbf{k}(y_k - \boldsymbol{\theta}^T\mathbf{z}_k), \tag{4.17}$$
$$\mathbf{R} \leftarrow \lambda^{-1}\mathbf{R} - \lambda^{-1}\mathbf{k}\mathbf{z}_k^T\mathbf{R}.$$

Alternative model optimisation criteria such as alternative lags of error correlation [40], higher order error moments [36], or error entropy [38], can also be utilised in this case with appropriate modifications.

### 4.4.3   Extending the methodology to local nonlinear modelling

It was made clear that the general nonlinear system of (4.1) is, in general, approximated locally by a NARMA process and we went one step further in the approximation to replace the local NARMA approximation by piecewise linear dynamics. One obvious modification would be to allow the local models to be nonlinear input–output dynamical recursive systems, such as time-delay neural networks (TDNN) [42]. The TDNN, being an extension of multilayer perceptrons (MLPs) to time-series processing, still possesses the universal approximation capabilities of MLPs, however, for the restricted class of dynamical systems with myopic memories (i.e. systems where a finite number of past values of the input affect the output, in a manner similar to the observability conditions discussed earlier in this chapter) [43,44]. A TDNN basically consists of a number of FIR filters in parallel whose outputs are modified by sigmoid nonlinearities and then linearly combined by the output layer. More generally, multiple layers of nonlinear FIR filter banks can be employed, but it is known that a sufficiently large single hidden layer TDNN has enough approximation capability. Training is typically performed via backpropagation of MSE [42].

Another feasible alternative that has a smaller approximation capability, but significantly simple to optimise is a Hammerstein dynamical system [45,46]. A Hammerstein structure consists of a static nonlinearity that transforms the input followed by a linear dynamical system. Once the input nonlinearities are set, the training of the linear dynamical portions is similar to the linear models discussed earlier (using the properly modified input autocorrelation matrix in the case of MSE optimality criterion).

Another possibility in nonlinear modelling is to use Volterra series approximation [47]. Volterra series expansion is an extension of Taylor series expansion to dynamical systems. It is based on multidimensional convolution integrals and the first-order Volterra approximation is simply a linear convolutive system. Typically, Volterra series approximations are truncated at most at the third-order convolution and separability of the multidimensional impulse responses is assumed for simplicity. This, of course, limits the approximation capability of the model in addition to the fact that the least-squares optimisation of the model coefficients is not necessarily simplified; local minima problems still exist.

Finally, as a direct consequence of Taylor series expansion, the local linear models can be extended to include higher order polynomial factors of delayed values of the input and the output. This is the Kolmogorov–Gabor polynomial modelling approach [48]. The number of coefficients to be optimised grows combinatorially with the order of the polynomial model, creating the main drawback of this approach.

## 4.5   Designing the local linear controllers

An added advantage of the proposed local linear modelling approach is that it greatly simplifies the design of control systems for nonlinear plants. In general, this is a daunting task and typically practical solutions involve linearisation of the dynamics and then employing well-established controller design techniques from linear control systems theory. While designing globally stable nonlinear controllers with satisfactory performance at every point in the state–space of the closed loop control system is extremely difficult, and perhaps impossible to achieve especially in the case of unknown plant dynamics, using the local linear modelling technique presented above, coupled with strong controller design techniques from linear control theory [18,49] and recent theoretical results on switching control systems [8,11,18], it becomes possible to achieve this goal through the use of this much simpler approach of local modelling. The topology of local linear controllers that naturally arise from the local linear modelling approach is illustrated in Figure 4.3.

### 4.5.1   Inverse error dynamics controller design

Once the optimal local linear models have been identified from the training input–output data available for system identification, one can use any standard linear controller design techniques to meet predefined regulation, stabilisation or tracking performance goals. Possibilities include stabilisation with linear state feedback (the

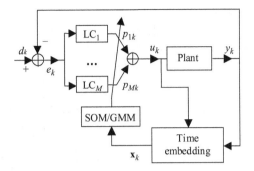

*Figure 4.3    Schematic diagram of the switching/weighted linear controller based on the SOM/GMM assessment of contributions*

individual ARMA systems can be expressed in controllable canonical form to design their corresponding state–space controllers), regulation or tracking a time-varying desired output response signal by a PID controller or more generally an inverse error dynamics controller. In this section, we will focus on the latter, inverse error dynamics controller scheme as it includes the PID controllers [50–53] and the exact tracking control [3,54], commonly utilised in practice.

The principle behind inverse error dynamics controller design is pole placement. Simply, it can be described as selecting a set of stable poles for the tracking error signal dynamics. If we denote the desired plant output at time $k$ by $d_k$ and the actual plant output by $y_k$, then the instantaneous tracking error is simply given by $e_k = d_k - y_k$.[9] The goal of this controller design technique is to guarantee that the error signal obeys the following dynamical equation:

$$e_{k+1} + \lambda_1 e_k + \lambda_2 e_{k-1} + \cdots + \lambda_l e_{k-l+1} = 0. \tag{4.18}$$

The parameters $\boldsymbol{\lambda} = [\lambda_1, \ldots, \lambda_l]^{\mathrm{T}}$ are selected such that the roots of the polynomial $1 + \lambda_1 x + \cdots + \lambda_l x^l$ are inside the unit circle. This guarantees the global stability of the closed loop control system provided that there is no noise, disturbance or modelling error. We will address how to handle these unwanted effects by designing the parameter vector appropriately later in this section. Of particular interest are inverse error dynamic equations of order 0 to 3. These are explicitly

Order 0:   $e_{k+1} = 0$                                          Exact tracking controller
Order 1:   $e_{k+1} + \lambda e_k = 0$                            First-order linear decaying error dynamics
Order 2:   $e_{k+1} + \lambda_1 e_k + \lambda_2 e_{k-1} = 0$      Second-order linear decaying error dynamics
Order 3:   $e_{k+1} + \lambda_1 e_k + \lambda_2 e_{k-1} + \lambda_3 e_{k-2} = 0$   PID controller

$$\tag{4.19}$$

The exact tracking controller simply solves for the error equation to determine the control input necessary for the next error value to be identically zero. This strategy, obviously, is not robust to modelling errors or external disturbances such as measurement noise. The first- and second-order error dynamic controllers are easy to design, since analytical expressions of their corresponding poles and the associated dynamical behaviours are well understood in linear system theory. Especially the second-order controller allows the designer to choose both overshoot and settling time simultaneously. The third-order error dynamics controller is effectively a PID controller, since PID controllers, when discretised, allow the designer to place the closed loop system poles to any location on a three-dimensional manifold in the $n$-dimensional space of the plant dynamics.

In the most general case, defined by (4.18), the control input is solved for as follows:

$$e_{k+1} = d_{k+1} - \hat{y}_{k+1} = -(\lambda_1 e_k + \lambda_2 e_{k-1} + \cdots + \lambda_l e_{k-l+1}) = -\boldsymbol{\lambda}^T \mathbf{e}_k^l.$$

(4.20)

Recall that explicitly the model output is given by $\hat{y}_{k+1} = \sum_{i=1}^{M} p_{ik} \mathbf{c}_i^T \mathbf{x}_k^{n,m} = \sum_{i=1}^{M} p_{ik} \mathbf{a}_i^T \mathbf{y}_k^n + \sum_{i=1}^{M} p_{ik} \mathbf{b}_i^T \mathbf{u}_k^m$. We are particularly interested in the terms involving $u_k$. Separating these terms from the others in the expression, the model output is $\hat{y}_{k+1} = \left( \sum_{i=1}^{M} p_{ik} \mathbf{a}_i^T \mathbf{y}_k^n + \sum_{i=1}^{M} p_{ik} \tilde{\mathbf{b}}_i^T \mathbf{u}_{k-1}^{m-1} \right) + \left( \sum_{i=1}^{M} p_{ik} b_{1i} \right) u_k = v_k + \tilde{b}_1 u_k$, where all new variables are defined in accordance with the equalities here. Introducing this new expression in (4.20) and solving for the input $u_k$, we obtain

$$u_k = \frac{1}{\tilde{b}_1} \left( d_{k+1} + \boldsymbol{\lambda}^T \mathbf{e}_k^l - v_k \right).$$

(4.21)

Notice that we can define $u_{ik} = (d_{k+1} - (\mathbf{a}_i^T \mathbf{y}_k^n + \tilde{\mathbf{b}}_i^T \mathbf{u}_{k-1}^{m-1}) + \boldsymbol{\lambda}^T \mathbf{e})/b_{1i}$ as the control input suggested by the $i$th model. With this definition, the control input of (4.21) can be expressed as

$$u_k = \frac{\sum_{i=1}^{M} p_{ik} b_{1i} u_{ik}}{\sum_{i=1}^{M} p_{ik} b_{1i}} = \sum_{i=1}^{M} \alpha_{ik} u_{ik},$$

(4.22)

where $\sum_{i=1}^{M} \alpha_{ik} = 1$, thus (4.21) is equivalent to a weighted average of the individual control input suggestions by the $M$ local linear models, which motivates the parallel controller topology shown in Figure 4.3.

For various choices of the error dynamics order and pole locations, the control law will drive the tracking closed-loop system in various different ways, while the error signal will always satisfy the linear difference equation given in (4.18), which is selected by the designer. An alternative controller design technique for both local linear models and local nonlinear models (should this be the designer's choice) is sliding mode control [55], which is essentially a nonlinear extension of the linear inverse error dynamics controller design methodology presented here. Sliding mode control is also well understood in the nonlinear control literature and it is known to be more robust to modelling errors and external disturbances compared to its linear counterpart [56]. In the linear case, which is being considered in this chapter, while any stable choice of stable poles will work satisfactorily and as expected in no-error/disturbance scenarios, in practice, due to the piecewise nature of the local models and sensor noise, these undesirable effects will always corrupt the calculated control input in (4.21). This can be countered by carefully selecting the error dynamic pole locations such that the parameter vector $\lambda$ represents a discrete-time filter that eliminates most of the power that is contained in these disturbances. Clearly, this requires the *a priori* knowledge of the statistical characteristics of these disturbances. Specifically, one needs the power spectral density information so that the filter can be designed to suppress high-energy frequency bands.

## 4.5.2 Selecting the error dynamics pole locations to suppress disturbance effects

Recall the tracking error dynamical equation in (4.18), $\hat{e}_{k+1} = d_{k+1} - \hat{y}_{k+1} = -\lambda^T \hat{e}_k^l$, obtained using the local linear models with noisy input–output measurements from the system according to the prediction equation $\hat{y}_{k+1} = \theta^T \hat{z}_k = y_{k+1} + n_k$, where $\hat{z}_k$ is the vector of noisy input–output measurements, $y_{k+1}$ is the true plant output corresponding to the planned control input and $n_k$ is the overall error term that includes all external disturbances and modelling errors. With this notation, also including the measurement noise present in $-\lambda^T \hat{e}_k^l$ in $n_k$, the dynamical equation that the true tracking error obeys becomes stochastic: $e_{k+1} = -\lambda^T e_k^l + n_k$. In effect, this represents an all-poles filter from the noise term $n_k$ to $e_k$, whose pole locations are exactly determined by $\lambda$. Hence, besides the stability constraint, the poles should be located inside the unit circle such that the all-pole filter from noise to tracking error has the transfer function $H(z) = 1/(1 + \lambda_1 z^{-1} + \cdots + \lambda_l z^{-l})$.

This observation inspires an alternative linear error dynamics approach to controller design in the case of measurement noise, external disturbances and modelling errors. If the equation in (4.18) is properly modified, it might be possible to achieve arbitrary transfer functions $H(z)$ from noise to tracking error (i.e. transfer functions with zeros and poles simultaneously). To achieve this, the error dynamic equation must be extended to further error predictions into the future using the model. Specifically, if the following error dynamic equation is used:

$$\eta_q \hat{e}_{k+q} + \cdots + \eta_1 \hat{e}_{k+1} + \lambda_1 e_k + \lambda_2 e_{k-1} + \cdots + \lambda_l e_{k-l+1} = 0, \tag{4.23}$$

then the noise and error terms, again collected in a single $n_k$, drive the following ARMA system from the disturbance to the tracking error:

$$[\eta^T \quad \lambda^T] e_{k+q}^{q+l} = \eta^T n_{k+q}^q. \tag{4.24}$$

Effectively, this corresponds to the following transfer function from $n_k$ to $e_k$:

$$H(z) = \frac{\eta_1 + \eta_2 z^{-1} + \cdots + \eta_q z^{1-q}}{1 + \eta_1 z^{-1} + \cdots + \eta_q z^{1-q} + \lambda_1 z^{-q} + \cdots + \lambda_l z^{1-l-q}}. \tag{4.25}$$

Consequently, the parameter vectors $\eta$ and $\lambda$ must be selected to place the zeros and poles of the transfer function in (4.25) to maximally filter out noise, using the spectral density information. The noise PSD can be estimated from training data, which is collected from the original system under noisy measurement conditions. These procedures are out of the scope of this chapter; therefore, we will not go into the details. However, spectral estimation is a mature and well-established research area [57].

## 4.6   Simulations

The local linear modelling and control technique that is presented above has been tested on a variety of nonlinear system identification and control problems including chaotic time-series prediction, synthetic nonlinear system identification, simplified missile lateral dynamics identification and control, NASA Langley transonic wind tunnel and NASA LoFlyte waverider aircraft. In this section, we will provide a compilation of simulation and experimental results obtained from the application of these principles to the listed problems.

### 4.6.1   Chaotic time-series prediction

In this example, we will demonstrate the use of the Lipschitz index for determining the model order for an SOM-based local linear modelling system using data generated by the Mackey–Glass attractor and the Lorenz attractor. The Mackey–Glass chaotic system is governed by the following differential equation [58]:

$$\dot{x}(t) = bx(t) + \frac{\alpha x(t - \gamma)}{1 + x^\rho(t - \gamma)}. \tag{4.26}$$

The delay $\gamma$ controls the depth of the underlying chaotic motion. The parameters are set to $\alpha = 0.2, \beta = -0.1$ and $\rho = 10$. The system in (4.26) is iterated using the Runge–Kutta-4 integration method with a time step of 1 and the signal is downsampled by a factor of 6. An embedding dimension of 6 and an embedding delay of 2 were determined based on the Lipschitz index and auto-mutual-information[10] of the signal, shown in Figure 4.4. An SOM trained on 1000 samples of input vector generated as described exhibits the Mackey–Glass attractor's topology, as shown in Figure 4.5. In addition, evaluation of the single step prediction performance using local linear models (LLM) attached to the neurons of the SOM demonstrates the high accuracy: prediction signal-to-error ratios (SER) of 33.80, 33.39 and 31.05 dB are obtained using a $15 \times 15$ rectangular SOM–LLM, a 40-neuron single layer TDNN and a 100-basis radial basis function (RBF) network, respectively.

The second chaotic system that will be considered here is the Lorenz attractor. The dynamics of this system are governed by the following system of differential equations [58]:

$$\begin{aligned} \dot{x} &= \sigma(y - x), \\ \dot{y} &= -y - xz + Rx, \\ \dot{z} &= xy - bz, \end{aligned} \tag{4.27}$$

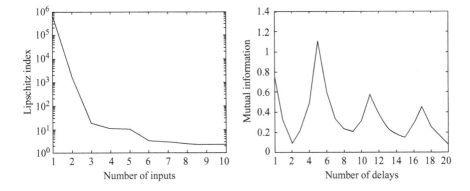

*Figure 4.4*    *Determining the embedding dimension for the chaotic Mackey–Glass attractor using the Lipschitz index (left) and the embedding delay using mutual information (right)*

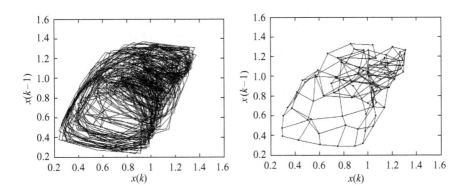

*Figure 4.5*    *The Mackey–Glass attractor is illustrated on the left using two consecutive samples from the time series. The trained SOM on the right captures the essence of this attractor in its topology*

where we selected $R = 28$, $\sigma = 10$ and $b = 8/3$ to obtain chaotic dynamics and the first state as the output. Using Runge–Kutta-4 integration with time step 0.01, 9000 samples were generated for analysis. Using the Lipschitz index and mutual information analysis (as shown in Figure 4.6) the embedding dimension is determined to be 3 and the embedding delay is 2. The first 8000 samples from the output time-series are used to create the reconstructed state samples, according to which the SOM and the local linear models are trained for single step prediction. As seen in Figure 4.7, the SOM represents the Lorenz attractor accurately. In addition, the single step prediction performance, again evaluated in terms of SER is found to be 53.74, 49.23 and 50.46 dB for the 20 × 20 SOM–LLM, 100-neuron TDNN and 150-basis RBF network, respectively, on the remaining 1000-sample test set.

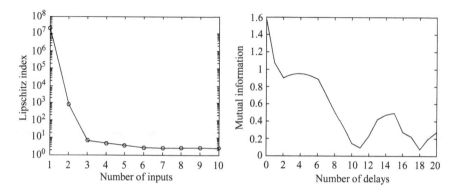

*Figure 4.6    Determining the embedding dimension for the chaotic Lorenz attractor using the Lipschitz index (left) and the embedding delay using mutual information (right)*

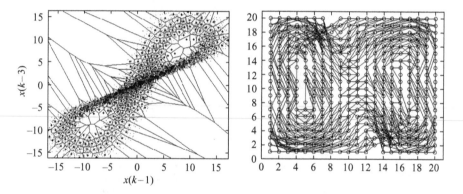

*Figure 4.7    The Lorenz attractor is captured in the topology of the trained SOM on the right. The trajectory of the winning neurons is illustrated in the rectangular SOM grid on the left showing the topological neighbourhood preserving property of the SOM*

These examples illustrate the modelling capability of the LLM technique on chaotic signal prediction, which is a benchmark problem in system identification. In addition, the validity of the Lipschitz index for selecting the embedding dimension (i.e. model order) is demonstrated by the successful reconstruction of the two chaotic attractors with the dimensions indicated by this index.

### 4.6.2    Synthetic nonlinear system identification

The first nonlinear system that is considered here is defined by the following state dynamics and output mapping [13]. Since the input is cubed, a finite number of linear models will not have good global approximation performance. The approximation will only be valid in the range of inputs available in the training data.

$$x_{k+1} = \frac{x_k}{1 + x_k^2} + u_k^3,$$

(4.28)

$$y_k = x_k.$$

Input–output training data is generated using white input signal distributed uniformly in $[-2, 2]$. Since the input is independent from itself at any delay, the embedding delay is taken to be unity. The embedding dimensions for the input and the output are both 2. Consequently, an SOM is trained using the input vector $[y_k, y_{k-1}, u_k, u_{k-1}]^T$ and corresponding local linear models are optimised using least squares. For a comparison, a local quadratic polynomial Hammerstein modelling (LPM) approach is also implemented [46]. The reconstructed state vector for this model is $[y_k, u_k, u_k^2]^T$. The size of the rectangular grid structure is selected based on the generalisation MSE of both models on a validation set. A size of $15 \times 15$ is selected for both SOMs. The system identification capabilities of the SOM–LLM, the SOM–LPM, a 150-basis RBF network and a 13-neuron focused gamma neural network (FGNN) [60] are compared.[11] Their performances on the test set are respectively 12.73, 47.95, 43.56 and 44.59 dB. Clearly, the cubic input term is not sufficiently approximated by the selected number of linear models and increasing the size of the network results in poor generalisation. As expected, the quadratic polynomial is certainly much more effective in approximating the cubic nonlinearity locally with the same number of local models.

The second nonlinear system that is considered is an FIR filter followed by a static nonlinearity defined by:

$$y_{k+1} = \arctan(u_k - 0.5u_{k-1}).$$

(4.29)

The system is excited by white input uniform on $[-2, 2]$ and system identification was carried out using the SOM–LPM topology explained above. The system identification testing results yielded SER values of 47.96, 49.99 and 52.30 dB for 8-neuron FGNN, 41-basis RBF and $15 \times 15$ SOM–LPM, respectively.

### 4.6.3   Simplified lateral missile dynamics identification and control

Under the assumptions of constant mass, zero roll and pitch angles and angular speed, the yaw dynamics of a missile can be expressed by [5]:

$$\dot{x}_1 = x_2 - 0.1\cos(x_1)(5x_1 - 4x_1^3 + x_1^5) - 0.5\cos(x_1)u,$$
$$\dot{x}_2 = -65x_1 + 50x_1^3 - 15x_1^5 - x_2 100u,$$
$$y = x_1.$$

(4.30)

The input is the rudder deflection and it is limited by $\pm 0.5$ rad. Two local linear models, SOM–LLM and GMM–LMM were identified using 6000 samples of input–output data, where the system is excited by white input uniform on $[-0.5, 0.5]$. A discretisation time step of 0.05 s was used (making the training set correspond to 300 s of flight time). The embedding delays for both the input and the output were found to be 2 using the Lipschitz index. The SOM consisted of a $15 \times 15$ rectangular grid, while the GMM had 5 Gaussian modes. Both models were tested on 1000 independently generated test samples generated using a random input sequence. The SOM–LLM and GMM–LLM performances were 31.7 and 31.0 dB in terms of SER, respectively. In other modelling problems, it was observed that the GMM–LLM approach required a smaller number of linear models, in general, compared to the SOM–LLM approach.

In addition to system identification, local linear PID controllers were designed for both models placing the poles of the closed-loop response at 0, 0, $0.05 + i0.3$ and $0.05 - i0.3$. For each linear model, the PID coefficients are set to these locations by adjusting their corresponding $\lambda$ vectors. The closed-loop regulation and tracking performances are tested by forcing the system output to track step changes and smooth sinusoidal changes in the desired output. The tracking performances of the SOM–LLM and GMM–LLM based PID controllers are shown in Figure 4.8. As a comparison, the tracking performance of a TDNN-based global adaptive model-controller pair is also presented in Figure 4.9. Clearly, the local PID controllers outperform the adaptive globally nonlinear technique in terms of both overshoot and convergence time.

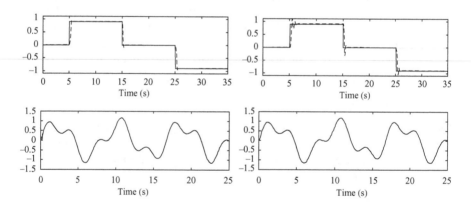

*Figure 4.8*   *Simplified missile model tracking step and smoothly changing desired outputs under SOM–LLM (left) and GMM–LLM (right) control rules*

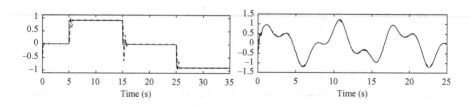

*Figure 4.9*   *Simplified missile model tracking step and smoothly changing desired outputs under TDNN controller*

### 4.6.4   NASA LoFlyte waverider aircraft identification and control

The NASA LoFlyte is an unmanned aerial vehicle (UAV) designed by Accurate Automation Corporation (AAC) and an illustrative picture is shown in Figure 4.10. The LoFlyte program is an active test program at the Air Force Flight Test Center of the Edwards Air Force Base with the objective of developing the technologies necessary to design, fabricate and flight test a Mach 5 waverider aircraft [59,61]. The LoFlyte UAV is also used to understand the low speed characteristics of a hypersonic shape and to demonstrate several innovative flight control technologies. The task of CNEL is to develop modelling and control strategies for LoFlyte based solely on input–output data.

According to classical aerodynamics, the flight dynamics of any rigid body are determined by movements along or around three body axes: roll, pitch (longitudinal motion) and yaw (lateral motion). The elevator $\delta_e$ is mainly responsible for controlling the longitudinal motion state variables (pitch angle, $\theta$ and pitch rate, $q$), the rudder $\delta_r$ primarily controls the lateral motion state variables (yaw angle, $\psi$ and yaw rate, $r$), the aileron $\delta_a$ mainly controls the roll motion state variables (roll angle, $\phi$ and roll rate, $p$). Finally, the throttle $\delta_t$ largely controls the aircraft's longitudinal speed, while in some aircraft, deflectable thrust vectors might allow yaw and roll contributions from the engine power. Typically, under certain symmetry assumptions for the aircraft body, the state dynamics of the rigid-body aircraft are represented around its centre of gravity as follows (see Reference 62 or any standard text book on flight dynamics and control):

$$\dot{u} = -(wq - vr) - g\sin\theta + F_x/m,$$
$$\dot{v} = -(ur - wp) + g\cos\theta\sin\phi + F_y/m,$$
$$\dot{w} = -(vp - uq) + g\cos\theta\cos\phi + F_z/m,$$
$$\dot{p} = ((I_{yy} - I_{zz})qr + I_{xz}(qr - pq) + L)/I_{xx},$$
$$\dot{q} = ((I_{zz} - I_{xx})rp + I_{xz}(r^2 - p^2) + M)/I_{yy}, \tag{4.31}$$

*Figure 4.10   The NASA LoFlyte UAV*

$$\dot{r} = ((I_{xx} - I_{yy})pq + I_{xz}(pq - qr) + N)/I_{zz},$$
$$\dot{\phi} = p + q \sin\phi \tan\theta + r \cos\phi \tan\theta,$$
$$\dot{\theta} = q \cos\phi - r \sin\theta,$$
$$\dot{\psi} = q \sin\phi \sec\theta + r \cos\phi \sec\theta.$$

In (4.31), $u$, $v$, $w$ are the speed components of the aircraft along its body $x$, $y$, $z$ axes, respectively. Similarly, $p$, $q$, $r$ are angular speeds around these axes, and $\phi$, $\theta$, $\psi$ are the Euler angles that define the rotation matrix between the body coordinate frame and the inertial coordinate frame (e.g. the north-east-down system in the case of short-duration, short distance flights within the atmosphere, under the flat-earth assumption). The gravity $g$ is along the *down* direction of the inertial frame. The engine power and aerodynamic effects generate the forces $F_x$, $F_y$ and $F_z$ as well as the moments $L$, $M$, $N$. The coefficients $m$, $I_{xx}$, $I_{yy}$, $I_{zz}$ and $I_{xz}$ are the aircraft mass and moments of inertia determined by its geometry.

The LoFlyte aircraft is simulated using a software by AAC and is assumed to be the true plant. It is assumed that the throttle is constant and state variables $p$, $q$, $r$, $u$, $v$, $w$ are available for external measurement. The goal of the identification and control problem is to determine local linear models from the three inputs (aileron, elevator and rudder) to these six state variables (outputs) and to control them in order to track a desired trajectory of flight.

Input–output training data is generated using the ACC flight simulator by manually flying the model aircraft (with a joystick) to imitate a test flight. The embedding dimensions for both input and the output are selected to be 3. SOM-based local linear models are trained from all three inputs to all six outputs, quantising the reconstructed state–space formed by the vector of delayed past output values. In these local models, in order to reduce model complexity, the coupling between the state variables is ignored, while all three inputs are still assumed to affect all six outputs. In essence, in matrix-vector form, the models are of the form

$$\hat{\mathbf{y}}_{k+1} = \sum_{i=0}^{3} \mathbf{A}_i^{\mathrm{T}} \mathbf{y}_{k-i} + \sum_{i=0}^{3} \mathbf{B}_i^{\mathrm{T}} \mathbf{u}_{k-i}, \tag{4.32}$$

where $\hat{\mathbf{y}}$ denotes the vector of outputs (the six measured states) and $\mathbf{u}$ denotes the vector of inputs (the three deflector angles). Since the output coupling is ignored, $\mathbf{A}_i$ are diagonal matrices, while $\mathbf{B}_i$ are full matrices that allow coupling effects from all inputs to all outputs.

Using 5000 training samples and a $10 \times 10$ rectangular SOM grid, whose size was determined according to the validation set performance (see Figure 4.11), the system identification performance was found to be 28.12, 20.21, 28.56, 74.73, 27.83 and 37.98 dB for the six outputs, respectively, in terms of SER on a 1000-sample test set. The same data was used in training a global 12-neuron TDNN model, which achieved 27.63, 19.57, 27.34, 47.56, 27.24 and 36.57 dB, respectively, for the outputs in the test set.

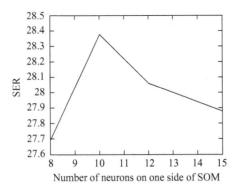

*Figure 4.11    Validation set SER of system identification*

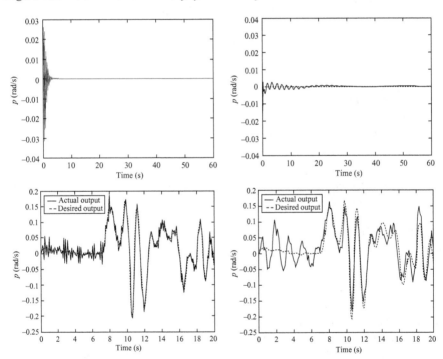

*Figure 4.12    Regulation and tracking performance of the roll rate controllers: SOM–LLM (left) and TDNN (right)*

An order-0 inverse error dynamic controller was designed for $p$, $q$ and $r$ according to equations (4.19) and (4.21). For comparison, an adaptive TDNN inverse controller based on the TDNN was also designed for the aircraft. The performances of the controllers are tested in set-point regulation and arbitrary output tracking. The tracking results of both controller systems are presented in Figures 4.12 and 4.13 [54].

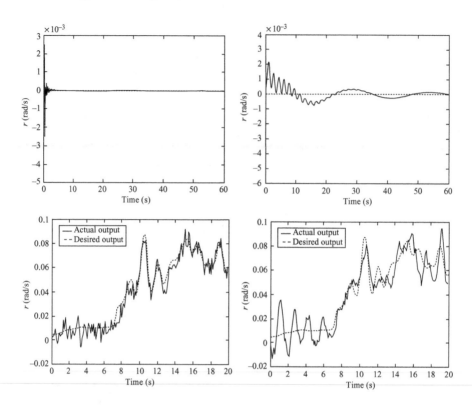

*Figure 4.13    Regulation and tracking performance of the yaw rate controllers: SOM–LLM (left) and TDNN (right)*

Clearly the local linear controllers exhibit better overshoot and convergence speed characteristics compared to the nonlinear counterpart.

### 4.6.5    NASA Langley transonic wind tunnel identification and control

In this final application example, the performance of the proposed local linear modelling and control approach in the identification and control of the NASA Langley 16-foot transonic tunnel will be presented [12]. This wind tunnel, whose picture is shown in Figure 4.14 to provide some perspective, is driven by a simple bang-zero-bang type control input with three possible values: $-1$, $0$, $+1$. The plant output is the Mach number (in the range 0.20–1.30) achieved around the experimental aircraft model whose dynamics are being studied as shown in Figure 4.15. The possible input value sequences of length $p$ were considered, resulting in a total of $3^p$ possible sequences. These sequences were partitioned to 9 sets, which were experimentally determined to meet performance needs with low computational requirements. Seven of these prototype input sequences were 50-samples long, while the remaining two were only 10-samples long [12]. For each of these input partitions, the tunnel's Mach number responses were clustered using a 20-node linear SOM. Finally, each neuron

*Figure 4.14    NASA Langley 16-foot transonic tunnel*

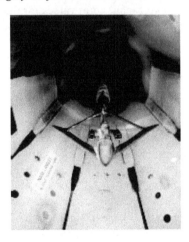

*Figure 4.15    The model aircraft under investigation*

of each SOM has a linear predictor of Mach number associated with it, as in the SOM–LLM framework, that evaluates the suitability of each input sequence by comparing the predicted Mach number with the desired value in the following $p$ time steps (either 50 or 10 depending on the input sequence being evaluated). The control input that produces the best Mach number match to the desired is selected and employed.

The identified local linear models and associated controllers are tested on the actual tunnel and the performance is compared to that of an expert human operator and the existing controller scheme. Typically, acceptable performance is maintaining a Mach number regulation error within $\pm 0.003$ of the set point while completing an angle-of-attack ($\alpha$) sweep in as small time as possible (due to power considerations).

In the first experiment, each controller is required to maintain 0.85 Mach within specifications for 15 min while an $\alpha$-sweep is completed. Minimum control activity is a plus. The performance of the three controllers is shown in Figure 4.16.

The average Mach number of the existing controller, the expert operator, and the SOM–LMM controller (denoted by NNCPC in figures 4.16 and 4.17) are 0.8497, 0.8500 and 0.8497, respectively, with standard deviations 0.001527, 0.001358 and 0.001226. The amount of time these controllers were out of tolerance was 46.5, 34.52 and 33.2 s. The $L_1$ norms of the control inputs were 10.6, 12.33 and 6.33, respectively. Clearly, the local linear controllers outperformed both competitors in terms of meeting tolerance bounds with minimum controller activity.

As a second test, all controllers were required to track a step-wise changing Mach number set point, again with as small as possible control effort, in a 28-min experiment. The Mach number tracking performances of the controllers are shown in Figure 4.17.

The existing controller was out of the tolerance bounds for 329 s, the expert operator was out of tolerance for 310 s and the SOM–LLM was out of bounds for 266 s. Their respective $L_1$ control input norm values for the duration of this experiment were 424.2, 466.2 and 374.3, again indicating the superior performance of the proposed local linear control approach.

## 4.7   Conclusions

In this chapter, the problem of nonlinear system identification and control system design was addressed under the divide-and-conquer principle. This principle motivated the use of multiple local models for system identification in order to simplify the modelling task. Especially in the case of unknown dynamics, where only input–output data from the plant is available, the proposed method is able to approximate the nonlinear dynamics of the plant using a piecewise linear dynamical model that is optimised solely from the available data. Especially when local linear models are used as described, it also became possible to design a piecewise linear controller for the plant, whose design is based on the identified model.

The questions of the existence and validity of input–output models as described and utilised was addressed theoretically using the implicit function inversion theorem that points out the observability conditions under which such models are possible to build from input–output data alone. The performance of the proposed local linear modelling scheme and the associated local linear controllers was tested on a variety of nonlinear dynamical systems including chaotic systems, a NASA aircraft and the NASA Langley transonic wind tunnel.

It was seen that the designed closed-loop control systems are extremely successful; in fact, in the experimental comparisons of performance at the NASA Langley tunnel, the proposed local linear controllers outperformed the existing computer controller and a human expert operator. These are encouraging results that motivate the use of this modelling and control technique for various other control applications. The capabilities of the local linear modelling approach are not limited to system

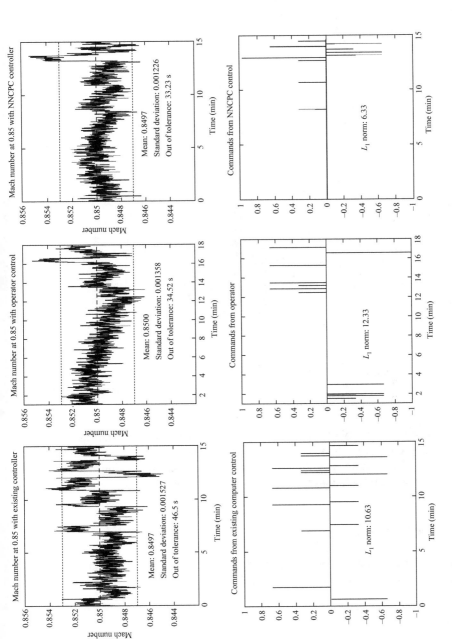

*Figure 4.16   Comparison of the existing controller (left), expert human operator (middle) and the SOM–LLM controller (right). The achieved Mach numbers for 15 min superimposed on the set point and the tolerance bounds (top), and control inputs produced by the controllers (bottom)*

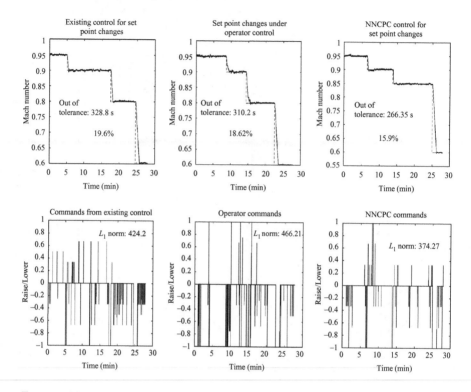

*Figure 4.17    Comparison of the existing controller (left), expert human operator (middle) and the SOM–LLM controller (right). Mach number tracking step changes in set point for 28 min (top), and control inputs produced by the controllers (bottom)*

identification and control applications. There are a wide range of nonlinear signal processing problems, such as magnetic resonance imaging, speech processing and computer vision, where the local linear signal processing can be employed to obtain simple but successful solutions to difficult nonlinear problems.

## Acknowledgements

This work was partially supported by the Accurate Automation Corporation under grant #463 and by NASA under grant NAG-1-02068.

## Notes

1.  Although we focus on local linear modelling in this chapter, the principles and methodologies outlined here can immediately be extended to local nonlinear modelling. In fact, we will briefly investigate this latter choice later.

2. Alternative optimality criteria include other moments [38] and entropy [39] of the modelling error.
3. The bias term $c$ is required for mathematical consistency, however, in practice it is optional and can be removed due to reasons discussed before.
4. The superscript $n, m$ will be dropped from now on whenever unnecessary.
5. If the bias term is included in the linear model, then the Wiener solution in (4.7) must be modified accordingly.
6. These update rules can be extended to the updating of nonlinear model weights [40]. The extension of LMS is trivial. The RLS algorithm is, in principle, an implementation of the Kalman filter considering the adaptive weights as states. Hence, extensions to nonlinear systems (such as neural networks) are achieved through the formulation of the learning problem as an extended Kalman filtering problem [41].
7. The EM algorithm is essentially a fixed-point update rule for the mixture density parameters to maximise the likelihood of the data.
8. Notice that the GMM-based model output is equivalently expressed as $\hat{y}_k = \boldsymbol{\theta}^T \mathbf{z}_k$.
9. Although we are considering discrete-time controller design here, the idea is easily applied to continuous-time controller design as well. In addition, generalisation to MIMO systems is also achieved simply by defining a vector-valued error signal, whose entries may or may not interact in the desired error dynamics equation through the use of non-diagonal or diagonal coefficient matrices.
10. Information dimension is a standard method for determining the delay amount in the embedding. The first zero or the minimum of the auto-mutual-information [59] is used as the embedding delay.
11. The FGNN is a generalisation of the TDNN where the input tap-delay line is replaced by a tap-gamma line.

# References

1 NELLES, O.: 'Nonlinear system identification' (Springer, New York, 2001)
2 LEONTARITIS, I. J., and BILLINGS, S. A.: 'Input–output parametric models for nonlinear systems part I: deterministic nonlinear systems', *International Journal of Control*, 1985, **41** (2), pp. 303–28
3 NARENDRA, K. S.: 'Neural networks for control: theory and practice', *Proceedings of IEEE*, 1996, **84** (10), pp. 1385–406
4 JOHANSEN, T. A., and FOSS, B. A.: 'Constructing NARMAX models using ARMAX models', *International Journal of Control*, 1993, **58** (5), pp. 1125–53
5 NI, X., VERHAEGEN, M., KRIJGSMAN, A. J., and VERBRUGGEN, H. B.: 'A new method for identification and control of nonlinear dynamic systems', *Engineering Applications of Artificial Intelligence*, 1996, **9** (3), pp. 231–43
6 WALKER, D. M., TUFILLARO, N. B., and GROSS, P.: 'Radial-basis models for feedback systems with fading memory', *IEEE Transactions on Circuits and Systems*, 2001, **48** (9), pp. 1147–51

7  KIM, B. S., and CALISE, A. J.: 'Nonlinear flight control using neural networks', *Journal of Guidance, Control, and Dynamics*, 1997, **20** (1), pp. 26–33

8  NARENDRA, K. S., BALAKRISHNAN, J., and CILIZ, M. K.: 'Adaptation and learning using multiple models, switching, and tuning', *IEEE Control Systems Magazine*, 1995, **15** (3), pp. 37–51

9  PRINCIPE, J. C., WANG, L., and MOTTER, M. A.: 'Local dynamic modeling with self-organizing maps and applications to nonlinear system identification and control', *Proceedings of IEEE*, 1998, **86** (11), pp. 2240–58

10  LEE, C. H., and CHUNG, M. J.: 'Gain-scheduled state feedback control design technique for flight vehicles', *IEEE Transactions on Aerospace and Electronic Systems*, 2001, **37** (1), pp. 173–82

11  NARENDRA, K. S., and XIANG, C.: 'Adaptive control of discrete-time systems using multiple models', *IEEE Transactions on Automatic Control*, 2000, **45** (9), pp. 1669–86

12  MOTTER, M. A.: 'Control of the NASA Langley 16-foot transonic tunnel with the self-organizing feature map'. Ph.D. dissertation, University of Florida, Gainesville, FL, 1997

13  NARENDRA, K. S., and PARTHASARATHY, K.: 'Identification and control of dynamical systems using neural networks', *IEEE Transactions on Neural Networks*, 1990, **1** (1), pp. 4–27

14  MILLER, W. T.: 'Real-time neural network control of a biped walking robot', *IEEE Control Systems Magazine*, 1994, **14** (1), pp. 41–8

15  BOSKOVIC, J. D., and NARENDRA, K. S.: 'Comparison of linear, nonlinear, and neural network based adaptive controllers for a class of fed-batch fermentation process', *Automatica*, 1995, **31** (6), pp. 817–40

16  CHEN, C. T.: 'Introduction to linear system theory' (Holt, Rinehart, and Winston, New York, 1970)

17  DORF, R. C., and BISHOP, R. H.: 'Modern control systems' (Addison-Wesley, New York, 1998, 8th edn.)

18  OGATA, K.: 'Modern control engineering' (Prentice Hall, London, 2001, 4th edn.)

19  KOHONEN, T.: 'Self-organizing maps' (Springer, New York, 1995)

20  AEYELS, D.: 'Generic observability of differentiable systems', *SIAM Journal of Control and Optimization*, 1979, **19**, pp. 139–51

21  SONTAG, E. D.: 'On the observability of polynomial systems', *SIAM Journal of Control and Optimization*, 1979, **17**, pp. 139–51

22  TAKENS, F.: 'On numerical determination of the dimension of an attractor', in RAND, D., and YOUNG, L. S. (Eds): 'Dynamical systems and turbulence' (Warwick 1980, *Lecture Notes in Mathematics*, vol. 898) (Springer-Verlag, Berlin, 1981) pp. 366–81

23  STARK, J., BROOMHEAD, D. S., DAVIES, M. E., and HUKE, J.: 'Takens embedding theorems for forced and stochastic systems', *Nonlinear Analysis: Theory Methods, and Applications*, 1997, **30** (8), pp. 5303–14

24  RUDIN, W.: 'Principles of mathematical analysis' (McGraw-Hill, New York, 1976)

25  MURRAY-SMITH, R., and JOHANSEN, T. A.: 'Multiple model approaches to modeling and control' (Taylor & Francis, New York, 1997)

26 SIDOROWICH, J. J.: 'Modeling of chaotic time series for prediction, interpolation, and smoothing'. Proceedings of ICASSP'92, 1992, pp. 121–4

27 SINGER, A. C., WORNELL, G. W., and OPPENHEIM, A. V.: 'Codebook prediction: a nonlinear signal modeling paradigm'. Proceedings of ICASSP'92, 1992, pp. 325–8

28 STARK, J., BROOMHEAD, D. S., DAVIES, M. E., and HUKE, J.: 'Takens embedding theorems for forced and stochastic systems', *Nonlinear Analysis: Theory, Methods, and Applications*, 1997, **30** (8), pp. 5303–14

29 WALTER, J., RITTER, H., and SCHULTEN, K.: 'Nonlinear prediction with self-organizing maps'. Proceedings of IJCNN'90, 1990, pp. 589–94

30 CASDAGLI, M.: 'Nonlinear prediction of chaotic time series', *Physica D*, 1989, **35** (3), pp. 335–56

31 WALKER, D. M., TUFILLARO, N. B., and GROSS, P.: 'Radial-basis models for feedback systems with fading memory', *IEEE Transactions on Circuits and Systems*, 2001, **48** (9), pp. 1147–51

32 HE, X., and ASADA, H.: 'A new method for identifying orders of input–output models for nonlinear dynamic systems'. Proceedings of ACC'93, 1993 pp. 2520–3

33 MARTINETZ, T., RITTER, H., and SCHULTEN, K.: 'Neural-gas network for vector quantization and its application to time-series prediction', *IEEE Transactions on Neural Networks*, 1993, **4** (4), pp. 558–68

34 McLACHLAN, G. J., and PEEL, D.: 'Finite mixture models' (Wiley, New York, 2001)

35 HAYKIN, S.: 'Adaptive filter theory' (Prentice-Hall, New York, 2001, 4th edn.)

36 WIDROW, B., and STEARNS, S.: 'Adaptive signal processing' (Prentice-Hall, New York, 1985)

37 SCHONER, B.: 'Probabilistic characterization and synthesis of complex driven systems'. Ph.D. dissertation, MIT, Cambridge, MA, 1996

38 ERDOGMUS, D., and PRINCIPE, J. C.: 'An error-entropy minimization algorithm for supervised training of nonlinear adaptive systems', *IEEE Transactions on Signal Processing*, 2002, **50** (7), pp. 1780–6

39 SINGHAL, S., and WU, L.: 'Training multilayer perceptrons with the extended Kalman algorithm'. Proceedings of NIPS '91, Denver, CO, 1991, pp. 133–40

40 PRINCIPE, J. C., RAO, Y. N., and ERDOGMUS, D.: 'Error whitening Wiener filters: theory and algorithms', in HAYKIN, S., and WIDROW, B. (Eds): 'Least-mean-square adaptive filters' (Wiley, New York, 2003)

41 FRITZKE, B.: 'Growing cell structures – a self-organizing network for supervised and unsupervised learning', *IEEE Transactions on Neural Networks*, 1994, **7** (9), pp. 1441–60

42 HAYKIN, S.: 'Neural networks: a comprehensive foundation' (Prentice-Hall, Englewood Cliffs, NJ, 1998, 2nd edn.)

43 HORNIK, K.: 'Approximation capabilities of multilayer feedforward networks', *Neural Networks*, 1991, **4**, pp. 251–7

44 CYBENKO, G.: 'Approximation by superposition of a sigmoidal function', *Mathematics of Control, Signals, Systems*, 1989, **2**, pp. 303–14

45  ESKINAT, E., JOHNSON, S., and LUYBEN, W. L.: 'Use of Hammerstein models in identification of nonlinear systems', *AIChE Journal*, 1991, **37**, pp. 255–68

46  CHO, J., PRINCIPE, J. C., and MOTTER, M. A.: 'Local Hammerstein modeling based on self-organizing map'. Proceedings of NNSP'03, Toulouse, France, 2003, pp. 809–18

47  WRAY, J., and GREEN, G. G. R.: 'Calculation of the Volterra kernels of nonlinear dynamic systems using an artificial neural network', *Biological Cybernetics*, 1994, **71** (3), pp. 187–95

48  MADALA, H. R., and IVAKHNENKO, A. G.: 'Inductive learning algorithms for complex systems modeling' (CRC Press, Boca Raton, FL, 1994)

49  NARENDRA, K. S., and BALAKRISHNAN, J.: 'Adaptive control using multiple models', *IEEE Transactions on Automatic Control*, 1997, **42** (2), pp. 171–87

50  BROWN, R. E., MALIOTIS, G. N., and GIBBY, J. A.: 'PID self-tuning controller for aluminum rolling mill', *IEEE Transactions on Industry Applications*, 1993, **29** (3), pp. 578–83

51  VU, K. M.: 'Optimal setting for discrete PID controllers', *IEE Proceedings D*, 1992, **139** (1), pp. 31–40

52  BAO, J., FORBES, J. F., and McLELLAN, P. J.: 'Robust multiloop PID controller design: a successive semidefinite programming approach', *Industrial and Engineering Chemistry Research*, 1999, **38**, pp. 3407–19

53  LAN, J., CHO, J., ERDOGMUS, D., PRINCIPE, J. C., MOTTER, M., and XU, J.: 'Local linear PID controllers for nonlinear control', *International Journal of Control and Intelligent Systems*, 2005, **33**(1), pp. 26–35

54  CHO, J., PRINCIPE, J. C., ERDOGMUS, D., and MOTTER, M. A.: 'Modeling and inverse controller design for an unmanned aerial vehicle based on the self organizing map', submitted to *IEEE Transactions on Neural Networks*, 2003

55  HUNG, J. Y., GAO, W., and HUNG, J. C.: 'Variable structure control: a survey', *IEEE Transactions on Industrial Electronics*, 1993, **40** (1), pp. 2–22

56  ERDOGMUS, D.: 'Optimal trajectory tracking guidance of an aircraft with final velocity constraint'. MS thesis, Middle East Technical University, Ankara, Turkey, 1999

57  KAY, S. M.: 'Modern spectral estimation: theory and application' (Prentice-Hall, Englewood Cliffs, NJ, 1988)

58  KAPLAN, D., and GLASS, L.: 'Understanding nonlinear dynamics' (Springer-Verlag, New York, 1995)

59  COX, C., NEIDHOEFER, J., SAEKS, R., and LENDARIS, G.: 'Neural adaptive control of LoFLYTE', *Proceedings of ACC'01*, 2001, **4**, pp. 2913–7

60  PRINCIPE, J. C., EULIANO, N., and LEFEBVRE, C.: 'Neural and adaptive systems: fundamentals through simulations' (Wiley, New York, 1999)

61  COX, C., MATHIA, K., and SAEKS, R.: 'Learning flight control and LoFLYTE'. Proceedings of WESCON'95, 1995, pp. 720–3

62  SCHMIDT, L. V.: 'Introduction to aircraft flight dynamics' (American Institute of Aeronautics and Astronautics, Reston, VA, 1998)

*Chapter 5*

# Nonlinear system identification with local linear neuro-fuzzy models

*Alexander Fink, Ralf Zimmerschied, Michael Vogt and Rolf Isermann*

## 5.1 Introduction

Nonlinear models are required in a variety of different tasks. They are needed for system analysis, simulation, optimisation and fault diagnosis of nonlinear processes. Furthermore, they are the basis for nonlinear model-based control approaches. These models can be obtained by theoretical or experimental modelling [1]. Theoretical models are based on first principles and are also called white-box models because of their transparency. In experimental modelling (or identification), the model is solely built from measurement data. Since these models usually lack physical inter-pretability, they are called black-box models. To make such models more transparent, grey-box models [2] try to incorporate prior physical knowledge in some way and therefore combine black-box and white-box modelling.

Contrary to linear systems [3,4], general methods for nonlinear identification are not available because a large number of different nonlinearities and a variety of nonlinear model architectures exist, e.g. Hammerstein and Wiener models, Volterra series and Kolmogorov–Garbor polynomials. Generally, a nonlinear model has a flexible structure, and its parameters are adjusted by an optimisation algorithm to capture the desired nonlinear behaviour. In recent years, also neural network and fuzzy logic approaches have been successfully applied.

Artificial neural networks (ANNs) are motivated by biological neural structures regarding learning and information processing. They consist of a large number of simple but strongly interconnected units: the neurons. ANNs belong to the class of black-box models and usually require nonlinear optimisation methods to adapt their parameters. Three basic architectures can be found among ANNs, namely basis function approaches, local model approaches and memory-based approaches [5].

Fuzzy systems are based on fuzzy logic [6], which is an extension of the classical boolean logic allowing variables to take gradual values between 0 and 1. Fuzzy logic offers a way to qualitatively describe a process by linguistic IF–THEN rules resulting in an interpretable grey-box model. On the other hand, fuzzy systems are not as precise as other model types and may suffer from the 'curse of dimensionality' [7] because of their grid-based structure. Neuro-fuzzy networks combine the advantages of both approaches [8,9], because their fuzzy rule base is at least partly built from measurement data instead of expert knowledge. For that, neuro-fuzzy models allow the use of sophisticated neural network learning methods while keeping their transparent structure. Local model approaches, which will be considered throughout this chapter, often belong to this class of models. Their idea to decompose a complex task into smaller problems has also been shown to be reasonable in many applications, e.g. in control design.

## 5.2   Dynamic local linear neuro-fuzzy models

Physical processes often have dynamic behaviour, i.e. the process output is not only dependent on the current state (as for static processes) but also on the previous ones. This section shows how such processes can be modelled by local linear structures.

### 5.2.1   Nonlinear dynamic models

A large class of nonlinear processes with $p$ physical inputs $u_1, \ldots, u_p$ and output $y$ can be described by the discrete-time input-output equation

$$y(k) = f(\boldsymbol{\varphi}(k)), \tag{5.1}$$

for the time sample $k$ [10]. The function $f(\cdot)$ is a nonlinear mapping of the input or regression vector

$$\begin{aligned}
\boldsymbol{\varphi}(k) = [&u_1(k - d_{u_1} - 1)\, u_1(k - d_{u_1} - 2) \, \cdots \, u_1(k - d_{u_1} - n_{u_1}) \\
&u_2(k - d_{u_2} - 1)\, u_2(k - d_{u_2} - 2) \, \cdots \, u_2(k - d_{u_2} - n_{u_2}) \\
&\vdots \\
&u_p(k - d_{u_p} - 1)\, u_p(k - d_{u_p} - 2) \, \cdots \, u_p(k - d_{u_p} - n_{u_p}) \\
&y(k - 1)\, y(k - 2) \, \cdots \, y(k - n_y)]^{\mathrm{T}}
\end{aligned} \tag{5.2}$$

composed out of previous values of the process inputs $u_i$, $i = 1, \ldots, p$, with according dead times $d_{u_i}$ and the output $y$. The variables $n_{u_i}$ and $n_y$ denote the dynamic orders of the inputs $u_i$ and the output $y$, respectively. The restriction to MISO processes (multiple input – single output) is made for the sake of simplicity. However, MIMO processes (multiple input – multiple output) can easily be realised by including more outputs in (5.1) and adding previous values of these outputs to the regression vector $\boldsymbol{\varphi}(k)$ in (5.2).

For modelling nonlinear *dynamic* processes based on the extension of nonlinear static models, there are two different approaches [11]: while the internal dynamics approach incorporates dynamic elements into the model itself, the external dynamics approach separates the nonlinear dynamic model into an external dynamic filter bank and a nonlinear static approximator. According to the model representation by its input-output behaviour (5.1) with (5.2), the external filters are realised as simple time delays $q^{-1}$ resulting in so-called tapped delay lines. Note, however, that the external dynamics approach usually yields a high dimensionality of the input space for the nonlinear function $f(\cdot)$. So the approximator should be able to handle high dimensional problems.

## 5.2.2 Local linear model structure

Local model architectures are based on the decomposition of the overall operating range of a complex system into a number of smaller operating regimes [12]. Within these local regimes, simpler subsystems are valid which are then combined to yield a global model. In general, any type of local submodel can be implemented within the local model framework. Typically, however, linear models are chosen for the submodels. They have a lot of advantages compared to other model types in terms of model identification, model interpretation as well as the later controller design based on the model [9,12–14]. In the following, solely linear submodels will be utilised.

According to the local model approach, the nonlinear function $f(\cdot)$ in (5.1) is approximated in the following form

$$\hat{y} = \sum_{j=1}^{M} l_j(\mathbf{x}) \cdot \Phi_j(\mathbf{z}). \tag{5.3}$$

The model output $y$ consists of an interpolation of $M$ local linear submodels $l_j$, $j = 1, \ldots, M$, weighted with the corresponding validity or activation functions $\Phi_j$. These weighting functions represent the partitioning of the input space and form a partition of unity, i.e.

$$\sum_{j=1}^{M} \Phi_j(\mathbf{z}) = 1 \tag{5.4}$$

is true for any input vector $\mathbf{z}$. Weighting functions satisfying this property are called normalised. Only normalised weighting functions allow the interpretation as validity functions since they ensure the individual contributions of the local submodels are summing up to 100 per cent. For weighting functions $\mu_j(\mathbf{z})$ that are not normalised, normalised validity functions $\Phi_j(\mathbf{z})$ can be generated by the following equation:

$$\Phi_j(\mathbf{z}) = \frac{\mu_j(\mathbf{z})}{\sum_{i=1}^{M} \mu_i(\mathbf{z})}. \tag{5.5}$$

In the context of neural networks, (5.3) can be seen as a local model network. But this local linear model description can also be interpreted as a fuzzy system. In this case, the model is represented by a Takagi–Sugeno (TS) fuzzy system [15].

Here, (5.3) is realised by a fuzzy rule base consisting of $M$ rules. The $j$th fuzzy rule, $j = 1, \ldots, M$,

$$\mathcal{R}_j : \text{ IF } z_1 \text{ is } \widetilde{Z}_{j,1} \text{ AND } z_2 \text{ is } \widetilde{Z}_{j,2} \text{ AND } \cdots \text{ AND } z_{n_z} \text{ is } \widetilde{Z}_{j,n_z}$$
$$\text{THEN } \hat{y} = l_j(\mathbf{x}) \tag{5.6}$$

corresponds to the $j$th submodel of the local model network. The fuzzy sets $\widetilde{Z}_{j,n}$ in the rule premise are defined on the universe of discourse of the according input $z_n$ with $n = 1, \ldots, n_z$ and the rule consequent is described by the linear function $l_j$ of the input $\mathbf{x}$. In both representations, (5.3) as well as (5.6), the $n_z$-dimensional vector $\mathbf{z} = [z_1 \; z_2 \cdots z_{n_z}]^T$ and the $n_x$-dimensional vector $\mathbf{x} = [x_1 \; x_2 \cdots x_{n_x}]^T$ are subsets of the input vector $\boldsymbol{\varphi}$ in (5.2):

$$\mathbf{x} \subseteq \boldsymbol{\varphi}, \qquad \mathbf{z} \subseteq \boldsymbol{\varphi}. \tag{5.7}$$

Consequently, the local linear submodels $l_j$ can have different input variables than the validity functions $\Phi_j$. This distinction has a significant advantage for the model identification and will be discussed later. However, only for the special case $\mathbf{z} = \boldsymbol{\varphi}$, it can be shown that the local linear model architecture presented in (5.3) is a universal approximator [16].

The conditions that the local model is equivalent to a TS fuzzy model are the following [17]: the membership functions of the fuzzy system have to be Gaussian, the $t$-norm (conjunction) must be the product operator, and the validity functions have to be axis-orthogonal. These conditions are not very restrictive, they rather describe a common choice. Local linear models are also a straightforward extension of normalised radial basis function (NRBF) networks [18]: they weight each validity function with a linear model instead of a scalar.

### 5.2.3   Dynamic local linear models

In the framework of local linear model approaches, the ARX (autoregressive with exogenous input) model is by far the most often utilised local linear submodel. The ARX model is an architecture with output feedback, meaning that the output signal is fed back and past values of the output act again as model inputs. In order to keep equations simple, only SISO-systems, i.e. systems with only one input, will be considered, and dead times are neglected.

Each local ARX model LLM$_j$ with $j = 1, \ldots, M$ and $M$ being the total number of local submodels, has the same structure. Its output $\hat{y}_j$ can be evaluated by the following difference equation

$$\hat{y}_j(k) = b_{j,1}u(k-1) + \cdots + b_{j,n_u}u(k-n_u) - a_{j,1}y(k-1) - \cdots - a_{j,n_y}y(k-n_y). \tag{5.8}$$

By arranging the past input and output values in the regression vector $\boldsymbol{\varphi}$, the input vector $\mathbf{x} = \boldsymbol{\varphi}$ becomes

$$\mathbf{x} = [u(k-1) \; u(k-2) \; \cdots \; u(k-n_u) \; y(k-1) \; y(k-2) \; \cdots \; y(k-n_y)]^T. \tag{5.9}$$

From this, the local linear submodel $l_j(\mathbf{x}) = \hat{y}_j$ can be defined with the help of the local submodel parameters $w_{j,n}$ as a linear combination of the input $\mathbf{x}$

$$l_j(\mathbf{x}) = w_{j,0} + w_{j,1}x_1 + w_{j,2}x_2 + \cdots + w_{j,n_x}x_{n_x}. \tag{5.10}$$

Note that (5.10) has been extended by the term $w_{j,0}$ compared to (5.8). This offset parameter $w_{j,0}$ defines the operating point of the $j$th local linear submodel.

The local linear model output $\hat{y}(k)$ can be described by substituting $l_j(\mathbf{x}(k))$ in (5.3) by (5.10) as

$$\hat{y}(k) = \sum_{j=1}^{M}(w_{j,0} + w_{j,1}x_1(k) + \cdots + w_{j,n_x}x_{n_x}(k)) \cdot \Phi_j(\mathbf{z}(k)), \tag{5.11}$$

which can be rewritten as

$$\hat{y}(k) = \left(\sum_{j=1}^{M} w_{j,0} \cdot \Phi_j(\mathbf{z}(k))\right) + \left(\sum_{j=1}^{M} w_{j,1} \cdot \Phi_j(\mathbf{z}(k))\right) x_1(k)$$

$$+ \cdots + \left(\sum_{j=1}^{M} w_{j,n_x} \cdot \Phi_j(\mathbf{z}(k))\right) x_{n_x}(k)$$

$$= \tilde{w}_0(k) + \tilde{w}_1(k)x_1(k) + \cdots + \tilde{w}_{n_x}(k)x_{n_x}(k), \tag{5.12}$$

which describes a linear system with operating point dependent on time-varying parameters $\tilde{w}_i(k)$. These parameters $\tilde{w}_i(k)$ are a superposition of the local linear submodel parameters $w_{j,i}$ weighted with the validity function values $\Phi_j(\mathbf{z}(k))$. Since the relationship between the model output $\hat{y}(k)$ is pseudo-linear in the input $\mathbf{x}(k)$, (5.12) is called a linear parameter varying (LPV) system. Based on the description of the local linear submodel $l_j(\mathbf{x})$ in (5.10), the local output $\hat{y}_j$ of a system with one input can be written as

$$\hat{y}_j(k) = \sum_{n=1}^{n_u} b_{j,n}u(k-n) - \sum_{n=1}^{n_y} a_{j,n}y(k-n) + o_j, \tag{5.13}$$

where $o_j$ is the offset parameter $w_{j,0}$ defining the operating point of the local submodel. In principle, this difference equation corresponds to the linear transfer function

$$G_j(z^{-1}) = \frac{\hat{y}_j(z^{-1})}{u(z^{-1})} = \frac{b_{j,1}z^{-1} + b_{j,2}z^{-2} + \cdots + b_{j,n_u}z^{-n_u}}{1 + a_{j,1}z^{-1} + a_{j,2}z^{-2} + \cdots + a_{j,n_y}z^{-n_y}}. \tag{5.14}$$

Based on these local transfer functions, it is possible to describe the overall nonlinear system by means of local gains or poles describing the local process behaviour. Following the idea of linear parameter varying systems, a linear transfer function $\tilde{G}(z^{-1})$ with operating point dependent on time-varying parameters

$$\tilde{G}(k, z^{-1}) = \frac{\hat{y}(z^{-1})}{u(z^{-1})} = \frac{\tilde{b}_1(k)z^{-1} + \tilde{b}_2(k)z^{-2} + \cdots + \tilde{b}_{n_u}(k)z^{-n_u}}{1 + \tilde{a}_1(k)z^{-1} + \tilde{a}_2(k)z^{-2} + \cdots + \tilde{a}_{n_y}(k)z^{-n_y}} \tag{5.15}$$

can be obtained.

## 5.3   Identification of local linear neuro-fuzzy models

The identification of local linear models is twofold: on the one hand, the local operating regimes have to be determined, on the other hand, the parameters describing the local linear submodels have to be estimated [19]. Both tasks can be solved simultaneously; however, this leads to complex nonlinear optimisation problems. Therefore, a variety of techniques, especially heuristic methods, solve both tasks in a nested or staggered approach which can significantly reduce the optimisation effort. Both problems will be considered separately in the following sections.

### 5.3.1   Parameter estimation

When the nonlinear model structure is known, i.e. the validity functions $\Phi_j(\mathbf{z})$, $j = 1, \ldots, M$ of the local model network are defined, the estimation of the submodel parameters is a fairly simple task. If the local submodels $l_j$ in (5.3) are linear in the parameters (as they are in (5.10)) the estimation problem can be solved by linear regression techniques.

In this study, the *local* estimation of the submodel parameters is employed. It is based on the minimisation of the local criteria, $j = 1, \ldots, M$,

$$I_j = \sum_{k=1}^{N} \Phi_j(\mathbf{z}(k))e_j^2(k). \tag{5.16}$$

The criterion $I_j$ leads to a locally weighted estimation of the parameters $w_{j,n}$ of the $j$th submodel (assuming a partition of unity). The error $e_j(k) = y(k) - \hat{y}_j(k)$ is the error between the measured output $y(k)$, $k = 1, \ldots, N$, and the local submodel output $\hat{\mathbf{y}}_j = [\hat{y}_j(1)\ \hat{y}_j(2) \cdots \hat{y}_j(N)]^T$ obtained by the local regression model

$$\hat{\mathbf{y}}_j = \mathbf{X}_j \mathbf{w}_j \tag{5.17}$$

with the corresponding local regression matrix[1]

$$\mathbf{X}_j = \begin{bmatrix} 1 & x_1(1) & x_2(1) & \cdots & x_{n_x}(1) \\ 1 & x_1(2) & x_2(2) & \cdots & x_{n_x}(2) \\ \vdots & \vdots & \vdots & & \vdots \\ 1 & x_1(N) & x_2(N) & \cdots & x_{n_x}(N) \end{bmatrix}. \tag{5.18}$$

The optimal parameter vector

$$\mathbf{w}_j = [w_{j,0}\ w_{j,1} \cdots w_{j,n_x}]^T \tag{5.19}$$

can be evaluated by the *weighted linear least squares* solution

$$\hat{\mathbf{w}}_j = (\mathbf{X}_j^T \mathbf{Q}_j \mathbf{X}_j)^{-1} \mathbf{X}_j^T \mathbf{Q}_j \mathbf{y} \tag{5.20}$$

with the $N \times N$ diagonal weighting matrix

$$\mathbf{Q}_j = \text{diag}(\Phi_j(\mathbf{z}(1)), \Phi_j(\mathbf{z}(2)), \ldots, \Phi_j(\mathbf{z}(N))). \tag{5.21}$$

In the local estimation approach, $M$ independent estimations (5.20) of the submodels are performed by neglecting their overlap. In principle, the data of the whole input space is used to estimate the parameters of a local submodel. However, by weighting the data with the validity functions $\Phi_j$, the data points closer to the centre of the $j$th submodel (corresponding to larger values of $\Phi_j(\mathbf{z}(k))$) are more relevant than the data far away from the centre (with values $\Phi_j(\mathbf{z}(k)) \rightarrow 0$). This yields locally valid submodels that capture the local behaviour of the process in contrast to the global estimation where this cannot be ensured. The local description is especially important if the local model network should be interpretable or a subsequent local controller design is based on the model. The computational complexity of the local estimation is considerably lower than for the global estimation approach. However, since the interaction of the submodels is ignored for the estimation, the model has a systematic error. This bias becomes bigger for a larger overlap of the validity functions. On the other hand, the limited flexibility of the local estimation reduces the variance error due to the bias/variance dilemma [20,21]. Consequently, the local estimation approach has a regularisation effect on the estimation [9, 22].

### 5.3.2 Structure identification

The structure optimisation determines the parameters describing the membership functions $\mu_j(\mathbf{z})$ or the validity functions $\Phi_j(\mathbf{z})$, respectively, $j = 1, \ldots, M$ and generally also the number of local submodels $M$. In other words, it determines the partitioning of the input space. Generally, the structure identification problem is much harder to solve than the parameter estimation. The most efficient way is to partition the input space based on process knowledge. If this knowledge is not available, data-driven concepts have to be employed. Several methods exist while the choice of the partitioning strategy has a significant influence on the overall model behaviour. The most common approaches will be described in the sequel; for a more detailed discussion refer to References 9 and 12.

- Clustering methods combine data points with similar properties to classes that result in local submodels. The most commonly used approach is the supervised *product space* clustering in the input/output space utilising the Gustafson–Kessel clustering algorithm to find hyperplanes in the product space [23].
- Neuro-fuzzy approaches exploit the similarities between neural networks and fuzzy systems and optimise the parameters defining the model structure utilising nonlinear optimisation techniques. These approaches usually yield a good modelling performance but are computationally expensive. Well-known model types are the adaptive-networks-based fuzzy inference systems (ANFIS) [24] and adaptive spline modelling of observation data (ASMOD) [8,25].
- Genetic algorithms and evolutionary strategies belong to the class of zeroth order optimisation methods and can be employed for structure search. However, their convergence is usually slow.
- Construction algorithms cover a large variety of different partitioning strategies starting from grid partitioning approaches leading to hierarchical construction techniques. Since the grid-based methods strongly underlie the 'curse of

dimensionality', the latter approaches are usually favoured. They construct the local model structure iteratively by refining the input space into smaller segments. In the context of neural networks and fuzzy logic, several local model approaches exist. They vary from flexible structures with arbitrary partitions to more restrictive structures with axis-orthogonal partitions [15,22,26–28]. The simpler the partitioning strategy, the lower the computational effort will usually be. The local linear model tree (LOLIMOT) algorithm [9] described later belongs to this category.

Figure 5.1 depicts four different strategies for the decomposition of the input space. It can be seen that the grid-based structure in Figure 5.1(a) does not exploit the local complexity of the process compared to the other approaches. The axis-orthogonal partition in Figure 5.1(b) is very transparent and can easily be interpreted in terms of fuzzy logic by projecting the multivariate validity functions onto the input axes where univariate fuzzy membership functions can be defined. The more flexible approaches in Figure 5.1(c) and (d) can yield better approximation results, but their interpretation as fuzzy rules is not as simple. Furthermore, if the multivariate functions have to be projected on to the input axes, this will introduce a loss of accuracy. The LOLIMOT algorithm is based on the partition shown in Figure 5.1(b).

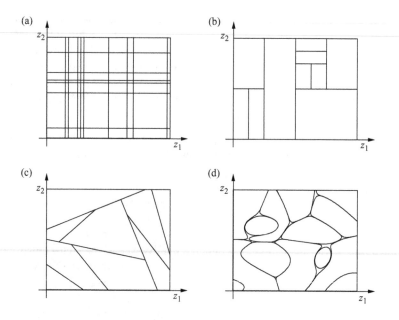

*Figure 5.1*   *Different partitioning strategies for the input space* **z** *for local model approaches: (a) grid partitioning (non-equidistant), (b) hierarchical partitioning (axis-orthogonal), (c) hierarchical partitioning (axis-oblique) and (d) arbitrary partitioning*

### 5.3.3   The LOLIMOT algorithm

In the previous section, different approaches for the parameter estimation and the structure identification of local linear models have been presented. This section will describe the LOLIMOT (Local Linear Model Tree) algorithm, which has originally been developed by Nelles [9]. It is an incremental input space partition algorithm combined with a local parameter estimation approach.

The LOLIMOT algorithm performs the parameter estimation and the structure identification of the local linear network in a nested way. In the outer loop, the input space is decomposed into hyper-rectangles by axis-orthogonal cuts based on a tree-structure. Each hyper-rectangle describes one local linear submodel. At each iteration of the algorithm, the $j$th local linear submodel with the worst error measure

$$I_j = \sum_{k=1}^{N} \Phi_j(\mathbf{z}(k))(y(k) - \hat{y}(k))^2 \tag{5.22}$$

is split into two halves. For the local loss function $I_j$, the $N$ measured data points are weighted with the degree of validity of the corresponding local submodel $j$. In the inner loop, the linear submodel parameters are estimated using a local weighted linear least squares algorithm as described in Section 5.3.1.

The LOLIMOT algorithm utilises multivariable Gaussian membership functions

$$\mu_j = \exp\left(-\frac{1}{2}\left(\frac{(z_1 - c_{j,1})^2}{\sigma_{j,1}^2} + \frac{(z_2 - c_{j,2})^2}{\sigma_{j,2}^2} + \cdots + \frac{(z_{n_z} - c_{j,n_z})^2}{\sigma_{j,n_z}^2}\right)\right)$$

$$= \exp\left(-\frac{1}{2}\frac{(z_1 - c_{j,1})^2}{\sigma_{j,1}^2}\right) \cdot \exp\left(-\frac{1}{2}\frac{(z_2 - c_{j,2})^2}{\sigma_{j,2}^2}\right)$$

$$\times \cdots \times \exp\left(-\frac{1}{2}\frac{(z_{n_z} - c_{j,n_z})^2}{\sigma_{j,n_z}^2}\right), \tag{5.23}$$

where the $c_{j,n}$ denote the centres and $\sigma_{j,n}$ the standard deviations in dimension $n$ for the membership function associated with the $j$th local submodel. The membership functions are placed in the centres of the hyper-rectangles and the standard deviations are chosen proportional to the extension of the hyper-rectangles. Consequently, the size of the validity region of the local linear submodel reflects the size of the according hyper-rectangle.

Figure 5.2 depicts the construction of the local linear network by the LOLIMOT algorithm. The first four iterations of the algorithm for a two-dimensional input space $\mathbf{z}$ are shown. The algorithm can be summarised as follows [9].

- *Initial (linear) model.* The algorithm starts with a single local linear submodel[2] ($M = 1$). This is actually a globally linear model since its valid-ity function $\Phi_1(\mathbf{z}) = 1$ covers the whole input space. The submodel parameters are estimated by the linear least-squares algorithm.

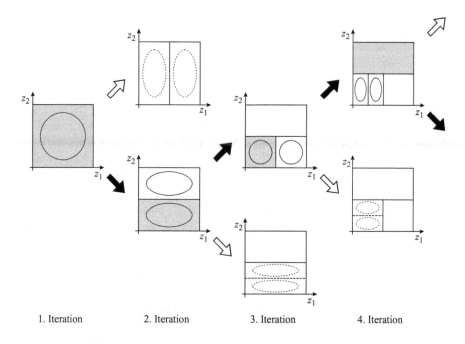

1. Iteration        2. Iteration        3. Iteration        4. Iteration

*Figure 5.2    Tree construction and input space decomposition by LOLIMOT. The*
*ellipses in the rectangles denote the contour lines of the membership*
*functions of the corresponding local submodels. The shaded rectangles*
*depict the currently worst performing LLM subject to division and the*
*filled black arrows indicate the split chosen from the alternatives*

- *Find worst submodel.*   The local loss functions $I_j$ are evaluated for all $j = 1, \ldots, M$ submodels according to (5.22). The worst performing submodel with $\xi = \max_j I_j$ is subject for further refinement.
- *Check all possible divisions.*   The hyper-rectangle corresponding to the $\xi$th submodel is split into two halves by an axis-orthogonal cut.[3] Divisions in all $n_z$ dimensions are tested. For each tried split, the membership functions are constructed and the parameters of the two newly generated local submodels are estimated using the local weighted linear least-squares algorithm. Finally, the global loss function is evaluated for the current overall model.
- *Find best division.*   The division yielding the best overall model error is selected from the tested alternatives and the validity functions and the submodel parameters determined in the previous step are adopted. The number of local submodels is increased to $M + 1$.
- *Convergence test.*   If the termination criterion is met, the algorithm terminates. Otherwise, the algorithm proceeds with the next iteration at the second step.

Various methods can be included in the termination criterion, e.g. a desired model error, a maximum model complexity meaning a maximum number of local submodels or statistical validation tests or information criteria. LOLIMOT utilises the Akaike's information criterion (AIC) [29].

For the local loss function (5.22), the local sum of squared errors and not their mean is used to compare the performance of the local submodels. As a consequence, the algorithm tends to divide these rectangles where more measurement data are available. This behaviour is desired since more data points also allow the estimation of more parameters and, typically, more data are acquired where the process nonlinearity is stronger.

Additionally, the LOLIMOT algorithm includes an orthogonal least-squares (OLS) algorithm to select the regressors in the linear submodels. This feature can be used for an automatic structure selection as model order determination or dead time evaluation [30]. For additional aspects of the LOLIMOT algorithm refer to Reference 9.

The LOLIMOT algorithm is a construction method resulting in an axis-orthogonal partition which can be conveniently interpreted. Its heuristic search approach for the partition, only the worst submodel is subject for division and is split in the middle, avoids the use of time-consuming nonlinear optimisation which yields a very fast training. Certainly, this simple decomposition algorithm leads to suboptimal results; however, it is more robust against overfitting. Furthermore, only a few parameters are estimated in each iteration. This local parameter estimation also yields an additional regularisation effect, compare Section 5.3.1. Finally, the LOLIMOT algorithm uses the one-step-ahead-prediction configuration (for the parameter estimation) and the simulation configuration (for the input space partitioning strategy) combining a fast linear estimation with the desired simulation behaviour of the model in a simple and efficient way.

## 5.4 Incorporation of prior process knowledge

The identification approaches discussed so far are data-driven methods and the modelling result is solely based on the available measurement data. In practice, however, the input space cannot always evenly be covered with enough measurement data and the model quality will deteriorate. In this case, it is desirable to incorporate all available sources of information into the modelling procedure. This section discusses the integration of prior process knowledge into the local linear modelling approach. Generally, local linear models have the advantage that they are transparent and interpretable which is the basic requirement for the incorporation of prior knowledge and, hence, are an adequate framework for grey-box modelling [2].

Process knowledge can be distinguished into the following three information sources: basic technical knowledge or expert knowledge, theoretical modelling based on first principles, preliminary tests as impulse or step responses. The knowledge from these sources is often qualitative rather than quantitative, e.g. the operator has only some rough knowledge about the process behaviour or theoretical modelling yields the general relationships between the inputs and outputs but exact parameter values are not available or too costly to determine. Then, the qualitative knowledge can be

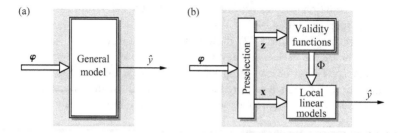

*Figure 5.3* *(a) Global and (b) distinct input spaces for local linear submodels and validity functions*

exploited to initialise the model structure and data-driven methods can be utilised to determine the missing model parameters.

For linear systems, it is comparably simple to tailor a model to match a given linear process structure. The task mainly reduces to identifying the inputs that influence the process output, determining the process order and finding the dead times for the individual inputs based on prior knowledge. The determination of the structure for nonlinear processes, on the other hand, is much more complex. There are various ways to incorporate prior knowledge into the nonlinear modelling procedure. A thorough discussion can be found in References 9 and 31. The most common methods particularly for local linear architectures will be discussed in the following.

- Many local linear model architectures allow the distinction between the input space of the local submodels $l_j$ given by $\mathbf{x}$ and of the validity functions $\Phi_j$ given by $\mathbf{z}$. The input vector $\mathbf{z}$ merely comprises those inputs of the vector $\varphi$ having a significant nonlinear behaviour which cannot be explained by the local submodels. Here, process knowledge can usually be exploited to define the input spaces or nonlinearity tests can be performed. Figure 5.3 shows a general nonlinear model with the input space $\varphi$ and a model structure with different input spaces, respectively. Distinct input spaces have the advantage that the dimensionality of the structure identification problem can be considerably reduced.
- If knowledge about the nonlinear process structure is available, this can be included in the form of membership functions formulated by linguistic fuzzy rules. This generally results in an axis-orthogonal partitioning of the input space.
- In principle, any type of model can be chosen for the local submodels. Linear models have proven to be a good compromise between submodel complexity and number of necessary submodels. However, it is also possible to predefine more complex structures based on prior knowledge. In the extreme case, fully physically determined models or even black-box models can be included. Here, local model architectures combined with local submodel estimation offer the advantage that different optimisation methods can be applied to the individual submodels.
- For dynamic systems, the user can specify the dynamic model order and dead times based, e.g. on prior experiments. Furthermore, it is possible to define gains, time constants and damping factors for the submodels if knowledge is accessible.

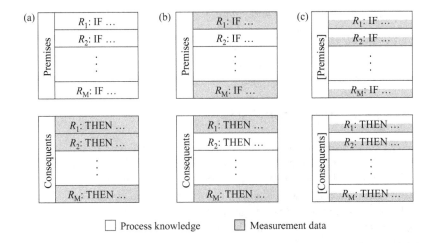

*Figure 5.4*    *Integration of information based on process knowledge and measurement data for local models: (a) parameter dependent, (b) operating point dependent and (c) successive approach*

Figure 5.4 illustrates different ways to combine knowledge-based and data-driven modelling for local model networks. The most common approach is shown in Figure 5.4(a) where the nonlinear structure of the model is defined based on the available process knowledge and the submodels are then estimated from data. An alternative is depicted in Figure 5.4(b). Here, some local models are defined solely based on prior knowledge and others are fully optimised based on measurement data. This is done, e.g. if the process can exactly be described by first principles in some operating points while process knowledge is not available for other regimes. Furthermore, prior knowledge can be exploited in regimes where no or only little measurement data is available due to process or productivity reasons or safety restrictions. Another approach is depicted in Figure 5.4(c). The model is initialised based on process knowledge followed by a data-driven fine-tuning of the model parameters. In the optimisation stage, care must be taken to preserve the prior knowledge [32]. Finally, all three individual methods in Figure 5.4 can be combined to fit the specific modelling needs of the process under investigation.

The local model structure obtained with the LOLIMOT algorithm is well suited for the incorporation of prior knowledge because of several reasons. Although the axis-orthogonal partitioning is somewhat restrictive, the model architecture is highly transparent and can directly be interpreted in terms of fuzzy logic. Contrary to this, the multivariate fuzzy sets of local model structures with arbitrary input space partitions have to be projected with accuracy losses onto univariate membership functions. The fuzzy logic interpretation is often the basis for the exploitation of prior knowledge. Second, the LOLIMOT algorithm distinguishes between the structure identification and submodel parameter estimation in a nested fashion. Consequently, different model structures can be predefined for different submodels. Product space clustering, for instance, finds the partitioning and submodel parameters simultaneously; all submodels are linear and different submodel structures are generally not realizable. The identification of the cross-flow heat exchanger in Section 5.6 will exemplarily show how to incorporate process knowledge into the local model design.

## 5.5    Design of identification signals

Although a lot of prior process knowledge can be incorporated into local linear model structures, the model accuracy strongly depends on the information present in the identification data. The design of an appropriate excitation signal for gathering identification data is therefore a crucial task in system identification. This step is even more decisive for nonlinear than linear models because nonlinear dynamic models are significantly more complex and thus the data must contain considerably more information.

For linear systems, guidelines for the design of excitation signals have been presented, which often result in the use of the so-called pseudo-random binary signal (PRBS) [3]. The PRBS switches between the minimum and maximum amplitudes in order to maximise the power of the input signal. The parameters of this signal which define the frequency spectrum are chosen according to the dynamics of the process.

For nonlinear systems, however, besides the frequency properties of the excitation signal, the amplitudes have to be chosen properly to cover all operating conditions of interest. Further the dynamic behaviour of the process often depends on the operating point. Therefore, the design of the excitation signal requires an individual design for each process where the following aspects should always be considered because they are more or less general [9,31].

- *Purpose of modelling.* The application of the model (e.g. model-based control, fault diagnosis, etc.) should be specified before the identification. Thereby, the required model precision for the different operating conditions and frequency ranges is determined.
- *Maximum length of the training data set.* The more training data can be measured the more precise the model will be if a reasonable data distribution is assumed. However, in industrial applications the measurement time is limited especially for configuration experiments.
- *Characteristics of different input signals.* For each input of the system, it must be checked whether dynamic excitation is necessary (e.g. for the manipulated variable in control systems) or if a static signal is sufficient (e.g. slowly changing measurable disturbances in control systems).
- *Range of input signals.* The process should be driven through all operating regimes that might occur in real operation. It is important that the data covers the limits of the input range because model extrapolation is more critical than interpolation.
- *Equal data distribution.* In particular for control purposes, the data at the process output should be equally distributed in order to contain the same amount of information about each setpoint.

- *Dynamic properties*. Because the dynamic properties of nonlinear processes often depend on the operating point, the frequency spectrum of the excitation signals must be designed with respect to operating point dependent time constants.

Finally, it is certainly a good idea to gather more information in operating regimes that are assumed to be more complex and/or to be more relevant than others.

From this list of general ideas, it follows that prior process knowledge is required for the design of an excitation signal. For this reason prior knowledge can be utilised to improve the model accuracy not only by finding an appropriate model structure but also by designing an excitation signal which improves the estimation of the parameters of the local submodels.

A basic signal for the identification of nonlinear dynamic processes is the so-called amplitude modulated PRBS (APRBS) which is an extension of the PRBS where each step is given a different amplitude [9,31]. An important design parameter of this signal is the minimum hold time $T_h$ which strongly influences its frequency characteristics. It should be chosen about equal to the dominant (largest) time constant of the process:

$$T_h \approx T_{dom}.$$

Because of changing dynamic properties of many nonlinear processes the minimum hold time also has to vary depending on the operating point.

## 5.6 Local linear modelling of a cross-flow heat exchanger

In this section, the local linear modelling technique is applied to an industrial-scale cross-flow heat exchanger whereas the aim of modelling is controller design. The static as well as dynamic behaviour of this water/air heat exchanger displays a significant nonlinearity. Its process gains, time constants as well as dead times strongly depend on the actual operating point described by the flow rates of both media.

### 5.6.1 Process description

The pilot plant depicted in Figure 5.5 comprises two heat exchangers. $E_1$ is a tubular steam/water heat exchanger, and $E_2$ is the cross-flow water/air heat exchanger under investigation. In the primary circuit, the electric steam generator G ($P = 54\,kW$) produces saturated steam at a pressure of 6 bar. The steam flow rate $Q_S$ is manipulated by a motor-driven valve whose position is regulated by an inner control loop. The steam condensates in the tubular heat exchanger $E_1$ and the liquid condensate is pumped back to the steam generator. By means of $E_1$ the water in the secondary circuit is heated. A predictive functional controller based on semi-physical modelling regulates the water temperature $T_{WI}$ by sending reference values to the inner control loop of the steam valve.

On the left side of the secondary circuit, the water is cooled down in the cross-flow water/air heat exchanger $E_2$. Cold air from the environment (temperature $T_{AI}$) is sucked in by the fan. After passing the heat exchanger and the fan, the air is blown out back to the environment. The water temperature $T_{WO}$ is to be controlled

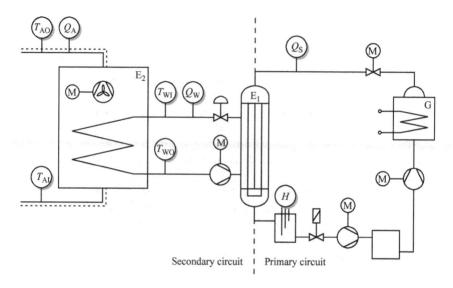

*Figure 5.5    Scheme of the thermal test plant*

by manipulating the air flow $Q_A$ by changing the fan speed. Besides the manipulated variable $Q_A$, the control variable $T_{WO}$ also depends on the disturbances: inlet temperature $T_{WI}$, air temperature $T_{AI}$ and water flow rate $Q_W$. The latter highly influences the static behaviour of the heat exchanger $E_2$. This behaviour has to be captured by the local linear model to be developed. The signal processing is performed on a PC under MATLAB/Simulink. A self-made Simulink block addresses an AD/DA converter board. The sample time is chosen as $T_0 = 1$ s, which was chosen based on initial experiments.

### 5.6.2    Modelling of the cooling process

Theoretical modelling of the dynamic behaviour of heat exchangers turns out to be extremely cumbersome. First, physical modelling is quite complex. Second, the modelling requires the knowledge of several physical parameters as, e.g. heat transfer coefficients, which often cannot exactly be determined. Here, the heat exchanger is modelled by a local linear model. This neuro-fuzzy architecture is a grey-box approach [2] where different sources of process information can be integrated for the modelling. Furthermore, the resulting model can conveniently be interpreted. In the following, it will be demonstrated how prior process knowledge and experimental data can be combined for the local modelling approach.

At first, some prior experiments can be conducted. Simple step responses provide basic information about the general process behaviour in different operating points. The experiments reveal that the process can satisfactorily be modelled by a first-order time-lag system with a dead time where the model parameters are dependent on the actual operating point. Figure 5.6 provides some characteristics of the cooling process. The water flow rate $Q_W$ highly influences the static behaviour as can be seen from the

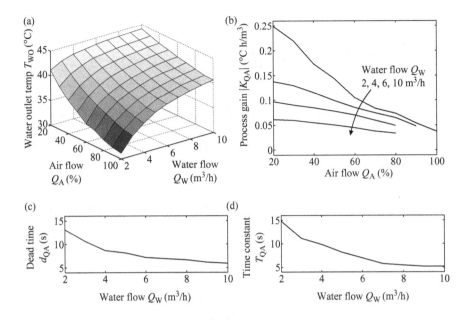

*Figure 5.6* *Characteristics of the cooling process* $T_{WO}/Q_A$ ($T_{WI} = 45°C$, $T_{AI} = 10°C$): *(a) static mapping, (b) process gain* $K_{Q_A} = \Delta T_{WO}/\Delta Q_A$, *(c) major dead time* $d_{Q_A}$ *and (d) dominant time constant* $T_{Q_A}$

static mapping in Figure 5.6(a) ($T_{WI} = 45°C$, $T_{AI} = 10°C$). Figure 5.6(b) depicts the corresponding static gain $K_{Q_A} = \Delta T_{WO}/\Delta Q_A$ versus the air flow rate $Q_A$. The gain changes by a factor of about five for different water flow rates $Q_W$. Moreover, different water flow rates $Q_W$ induce varying dead times $d_{Q_A}$ and time constants $T_{Q_A}$. Note that the time constant and dead time are also slightly influenced by the air flow $Q_A$. The plots in Figure 5.6(c) and (d) show the values averaged for different air flows.

Based on the gathered process knowledge, an identification signal can be designed based on the guidelines given in Section 5.5. The excitation signal together with the measured output $T_{WO}$ is shown in Figure 5.7. The allowable range for the air flow rate $Q_A$ is 20–100 per cent, smaller flow rates are not admissible because of security reasons. The valve controlling the steam flow $Q_S$ in the primary circuit does not shut completely and the secondary circuit is permanently heated by the heat exchanger $E_1$. Thus, the water circuit has to be continuously cooled to prevent overheating of the system. The air flow rate is dynamically excited over the whole operating range. The water inlet temperature $T_{WI}$ performs steps between 45°C and 47°C because it constantly undergoes fluctuations during the control experiment. Whenever the energy demand in the secondary circuit changes strongly, the predictive functional controller for the inlet temperature $T_{WI}$ cannot react quickly enough and temperature shows some undershoots or overshoots. This can also be noticed during the identification run when steps in the air or water flow rate occur. Since the water flow rate $Q_W$ changes only slowly, it is sufficient to excite the water flow only quasi-static. It is

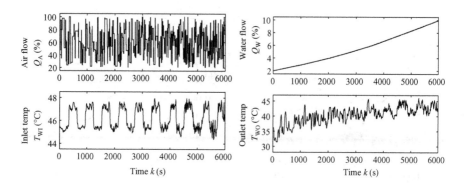

*Figure 5.7    Identification data of the heat exchanger. The air temperature is $T_{AI} = 10°C$*

varied in operating regimes from 2 to $10\,\text{m}^3/\text{h}$. The air inlet temperature $T_{AI}$ cannot be influenced and, hence, is not included in the model. It is widely constant during the experiment at about $10°C$.

The air flow rate $Q_A$ is excited with an APRBS with evenly distributed amplitudes over the whole operating range. Based on the varying time constant, compare Figure 5.6(d), the minimum hold time of the APRBS is to be reduced for larger water flows. The water flow rate $Q_W$ is excited by a ramp which is slightly hyperbolic because simulations have shown that model accuracy is improved if each operating regime is excited with the same number of steps. The hyperbolic behaviour compensates the effect of longer minimum hold time for smaller water flows.

Based on the prior knowledge and the acquired measurement data, the local linear model for the cross-flow heat exchanger can be generated. The available information is parameter dependent. While the parameters describing the nonlinearity of the cooling process are obtained from process knowledge, the local submodel parameters are estimated from measurement data, compare Figure 5.4(a).

The plots in Figure 5.6(a) show that the process nonlinearity depends on both media flows. Consequently, regressors of the air flow rate $Q_A$ and the water flow rate $Q_W$ should span the input space for the activity functions. Based on prior experiments, the input vector

$$\mathbf{z}(k) = [Q_A(k - 7)\ Q_W(k - 1)] \tag{5.24}$$

is chosen. In order to account for all possible dead times of the air flow $Q_A$, more regressors with different delays should be included in $\mathbf{z}$. However, the limitation to a two-dimensional input space results in a much more transparent and consequently interpretable description of the model. Additionally, the input space can be covered more evenly by the measurement data. Here, the smallest possible dead time of 6 s (compare Figure 5.6(d)) plus 1 s unit delay is selected for the input vector. Prior identification runs with the LOLIMOT algorithm, showed that the input space partition shown in Figure 5.8(a) captures the process nonlinearity well. A model structure with more local submodels does only yield a slight approximation improvement

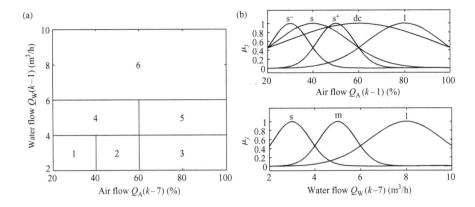

*Figure 5.8*    *Partition of the input space and corresponding membership functions:*
*(a) input space partition and (b) membership functions*

which is not required for this application. The membership functions describing the
partitioning are given in Figure 5.8(b) for both flow rates. The partition corresponds
to the following rule base:

$\mathcal{R}_1$: IF $Q_A(k-7)$ is small$^-$ AND $Q_W(k-1)$ is small

$\mathcal{R}_2$: IF $Q_A(k-7)$ is small$^+$ AND $Q_W(k-1)$ is small

$\mathcal{R}_3$: IF $Q_A(k-7)$ is large     AND $Q_W(k-1)$ is small

$\mathcal{R}_4$: IF $Q_A(k-7)$ is small    AND $Q_W(k-1)$ is medium      (5.25)

$\mathcal{R}_5$: IF $Q_A(k-7)$ is large     AND $Q_W(k-1)$ is medium

$\mathcal{R}_6$: IF $Q_A(k-7)$ is dc       AND $Q_W(k-1)$ is large.

By introducing the fuzzy attribute 'dc' (don't care), multidimensional membership
functions can directly be generated from the one-dimensional membership functions.
This results in the equivalence of the fuzzy description to local networks based on
multidimensional validity functions. The static mapping in Figure 5.6 also confirms
this partition. The nonlinearity is stronger for small water flow rates $Q_W$ requiring more local submodels inducing a stronger segmentation in this operating range.
Furthermore, the process gain is significantly dependent on the air flow rate $Q_A$ for
smaller values of $Q_W$ and widely independent of $Q_A$ for large water flow rates. For
each operating regime described by the fuzzy rule base, a local submodel is defined
whose parameters are estimated from the experimental data.

For the local submodels, the following regressors are used:

$$\mathbf{x}(k) = [Q_A(k-7)\, Q_A(k-8) \cdots Q_A(k-15)\, Q_W(k-1)$$
$$T_{WI}(k-4)T_{WI}(k-5) \cdots T_{WI}(k-16)\, T_{WO}(k-1)]. \quad (5.26)$$

The delays allow variable dead times between 6 s and 14 s for the air flow rate $Q_A$,
compare Figure 5.6(c), and delays between 4 s and 16 s for the water inlet temperature

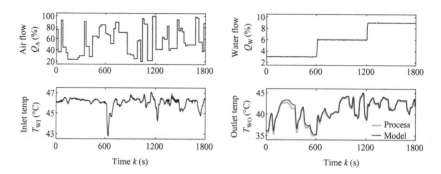

*Figure 5.9    Modelling performance on a validation data set*

$T_{WI}$ depending on the current water flow rate $Q_W$ which influences the process output directly without dead time. The incorporation of $T_{WO}(k - 1)$ yields first-order time-lag dynamics. For the individual first-order submodels, it is sufficient to select only one regressor from $\mathbf{x}$ for $Q_A$ and $T_{WI}$, respectively. Since the delay for the inlet temperature $T_{WI}$ is strictly based on the transportation process between the temperature sensors of the water inlet and outlet of the cooler, the delay can be obtained based on the transport time

$$T_{\text{Transport}} = \frac{l_{\text{Pipe}} A_{\text{Pipe}}}{Q_W}, \tag{5.27}$$

where $l_{\text{Pipe}}$ and $A_{\text{Pipe}}$ are the pipe length between the sensors and the cross-sectional area of the pipe in the cooler, respectively. The averaged delay of the individual submodels is chosen as the closest regressor in $\mathbf{x}$ to $T_{\text{Transport}}$ where the transport time is evaluated for the water flow rate $Q_W$ at the centre of each submodel. For the identification, only this submodel parameter is estimated that corresponds to the chosen delay while the remaining parameters corresponding to other delays are set to zero. This is a further example of how process knowledge can advantageously be incorporated into the local modelling approach. The description of the dead time for the air flow $Q_A$ cannot be described by a simple relationship and, thus, is evaluated based on numerical optimisation. For each local submodel, the dead time $d_{Q_W}$ minimising the local error is selected.

The performance of the model can also be seen on a validation data set in Figure 5.9. The air flow rate $Q_A$ is excited with an APRBS with a minimum hold time $T_h = 20$ s whereas the water flow rate $Q_W$ just consists of three steps. The inlet temperature $T_{WI}$ should be kept constant at 46°C, which was not possible especially after changes in the water flow rate $Q_W$. The measured water outlet temperature $T_{WO}$ together with the simulated model output show that the identified local linear model is able to capture the process behaviour well. The maximum model error is about 1°C. During the first 600 s there is an additional offset error because of a lower air inlet temperature $T_{AI}$ in comparison to the identification data set. The influence of the air inlet temperature which is an unmodelled disturbance can be compensated by an online adaptation of the model [33].

*Table 5.1    Characteristic values of the local linear submodels*

| Submodel $j$ | Local times (in s) | | Local gains | | |
|---|---|---|---|---|---|
| | $d_{QA_j}$ | $T_{QA_j}$ | $K_{QA_j}$ | $K_{QW_j}$ | $K_{TWI_j}$ |
| 1 | 9 | 14.0 | −0.17 | 1.89 | 0.89 |
| 2 | 8 | 14.9 | −0.14 | 2.21 | 0.75 |
| 3 | 8 | 17.4 | −0.11 | 2.57 | 0.78 |
| 4 | 7 | 9.1 | −0.10 | 0.95 | 0.92 |
| 5 | 6 | 10.9 | −0.07 | 1.28 | 0.78 |
| 6 | 6 | 6.2 | −0.06 | 0.54 | 0.93 |

Table 5.1 provides some important characteristics of the submodels for the identified model. The model parameters match the corresponding values of the process very well as can be seen by comparing them with the plots in Figure 5.6. They also comply with the values obtained by theoretical modelling of this cross-flow heat exchanger [14].

## 5.7   Conclusions

The local linear modelling approach is capable of modelling complex nonlinear processes. The transparency of the model architecture permits the incorporation of prior knowledge in different ways and the integration of process knowledge and experimental data has been demonstrated in detail for the modelling of the heat exchanger. Utilising an arbitrary local modelling strategy does not automatically guarantee a transparent model. The employed LOLIMOT algorithm has proven to yield transparent and well interpretable models. The appropriate choice of an excitation signal is vital for a good model. Here, the transparency of the local structure again aids the engineer in the design and the assessment of the excitation signal. The interpretability of the local structure offers simple ways for the validation of the identified model based on local characteristic values as process gains or time constants. The obtained local linear models can further be utilised in model-based control approaches as well as process monitoring or fault detection.

## Notes

1. The local regression matrix $\mathbf{X}_j$ of each submodel can be built up with an individual set of regressors in order to use different model structures for each operating regime.
2. It is also possible to pre-define an initial input space partition consisting of multiple submodels with $M > 1$. Then, the validity functions for all submodels

have to be evaluated and their parameters have to be estimated using the local weighted least-squares algorithm.

3. In Figure 5.2, the rectangles are divided in the middle. It is also possible to divide at different ratios. Then, all possible ratios in all dimensions have to be tested.

## References

1 ISERMANN, R.: 'Mechatronic systems' (Springer Verlag, Berlin, 2003)
2 TULLEKEN, H. J. A. F.: 'Grey-box modelling and identification using prior knowledge and Bayesian techniques', *Automatica*, 1993, **29** (2), pp. 285–308
3 ISERMANN, R.: 'Identifikation dynamischer Systeme – Band 1 (Identification of dynamic systems) (Springer Verlag, Berlin, 1992, Vol. 1) (German)
4 LJUNG, L.: 'System identification – theory for the user' (Prentice-Hall, Englewood Cliffs, NJ, 1987)
5 FINK, A., TÖPFER, S., and ISERMANN, R.: 'Nonlinear model-based control with local linear neuro-fuzzy models', *Archive of Applied Mechanics*, 2003, **72** (11–12), pp. 911–22
6 ZADEH, L. A.: 'Fuzzy sets', *Information and Control*, 1965, **8**, pp. 338–53
7 BELLMAN, R. E.: 'Adaptive control processes' (Princeton University Press, Princeton, NJ, 1961)
8 BROWN, M., and HARRIS, C. J.: 'Neurofuzzy adaptive modelling and control' (Prentice-Hall, New York, 1994)
9 NELLES, O.: 'Nonlinear system identification' (Springer Verlag, Heidelberg, 2000)
10 LEONTARITIS, I. J., and BILLINGS, S. A.: 'Input–output parametric models for nonlinear systems, part 1: Deterministic nonlinear systems', *International Journal of Control*, 1985, **41** (2), pp. 303–28
11 SJÖBERG, J., ZHANG, Q., LJUNG, L., *et al.*: 'Nonlinear black-box modeling in system identification: a unified overview', *Automatica*, 1995, **31** (12), pp. 1691–724
12 MURRAY-SMITH, R., and JOHANSON, T. A.: 'Multiple model approaches to modelling and control' (Taylor & Francis, London, 1997)
13 BABUŠKA, R., and VERBRUGGEN, H. B.: 'An overview of fuzzy modeling for control', *Control Engineering Practice*, 1996, **4** (11), pp. 1593–606
14 FISCHER, M.: 'Fuzzy-modellbasierte Regelung nichtlinearer Prozesse (Fuzzy model-based control of nonlinear processes)'. Ph.D. thesis, Darmstadt University of Technology, Fortschritt-Berichte VDI, Reihe 8, Nr. 750. VDI Verlag, Düsseldorf, Germany, 1999 (German)
15 TAKAGI, T., and SUGENO, M.: 'Fuzzy identification of systems and its application to modelling and control', *IEEE Transactions on Systems, Man, and Cybernetics*, 1985, **15** (1), pp. 116–32
16 CAO, S. G., REES, N. W., and FENG, G.: 'Analysis and design for a class of complex control systems – part i: Fuzzy modelling and identification', *Automatica*, 1997, **33** (6), pp. 1017–28

17 HUNT, K. J., HAAS, R., and MURRAY-SMITH, R.: 'Extending the functional equivalence of radial basis functions networks and fuzzy inference systems', *IEEE Transactions on Neural Networks*, 1996, **7** (3), pp. 776–81

18 MOODY, J., and DARKEN, C. J.: 'Fast learning in networks of locally-tuned processing units', *Neural Computation*, 1989, **1** (2), pp. 281–94

19 JOHANSEN, T. A., and FOSS, B. A.: 'Constructing NARMAX models using ARMAX models', *International Journal of Control*, 1993, **58** (5), pp. 1125–53

20 GEMAN, S., BIENENSTOCK, E., and DOURSAT, R.: 'Neural networks and the bias/variance dilemma', *Neural Computation*, 1992, **4** (1), pp. 1–58

21 HASTIE, T. J., and TIBSHIRANI, R. J.: 'Generalized additive models'. Monographs on Statistics and Applied Probability 43 (Chapman and Hall, London, 1990)

22 MURRAY-SMITH, R.: 'A local model network approach to nonlinear modeling'. Ph.D. thesis, University of Strathclyde, Department of Computer Science, Glasgow, Scotland, 1994

23 BABUŠKA, R.: 'Fuzzy modeling and identification'. Ph.D. thesis, Department of Electrical Engineering, Delft University of Technology, The Netherlands, 1996

24 JANG, J.-S. R., SUN, C.-T., and MIZUTANI, E.: 'Neuro-fuzzy and soft computing: a computational approach to learning and machine intelligence' (Prentice-Hall, Englewood Cliffs, NJ, 1997)

25 KAVLI, T.: 'ASMOD: an algorithm for adaptive spline modeling of observation data', *International Journal of Control*, 1993, **58** (4), pp. 947–67

26 ERNST, S.: 'Hinging hyperplane trees for approximation and identification'. IEEE conference on *Decision and control* (CDC), Tampa, FL, 1998, pp. 1261–77

27 SUGENO, M., and KANG, G. T.: 'Structure identification of fuzzy model', *Fuzzy Sets and Systems*, 1988, **28** (1), pp. 15–33

28 JOHANSEN, T. A.: 'Operating regime based process modeling and identification'. Ph.D. thesis, Department of Engineering Cybernetics, The Norwegian Institute of Technology – University of Trondheim, Trondheim, Norway, 1994

29 AKAIKE, H.: 'A new look at the statistical model identification', *IEEE Transactions on Automatic Control*, 1974, **19** (6), pp. 716–23

30 NELLES, O., HECKER, O., and ISERMANN, R.: 'Automatic model selection in local linear model trees for nonlinear system identification of a transport delay process'. IFAC symposium on *System identification* (SYSID), Fukuoka, Japan, July 1997, pp. 727–32

31 FISCHER, M., NELLES, O., and ISERMANN, R.: 'Exploiting prior knowledge in fuzzy model identification of a heat exchanger'. IFAC symposium on *Artificial intelligence in real-time control* (AIRTC), Kuala Lumpur, Malaysia, September 1997, pp. 445–50

32 LINDSKOG, P.: 'Methods, algorithms and tools for system identification based on prior knowledge'. Ph.D. thesis, Department of Electrical Engineering, Linköping University, Linköping, Sweden, 1996

33 FINK, A., FISCHER, M., NELLES, O., and ISERMANN, R.: 'Supervision of nonlinear adaptive controllers based on fuzzy models', *Control Engineering Practice*, 2000, **8** (10), pp. 1093–105

*Chapter 6*

# Gaussian process approaches to nonlinear modelling for control

*Gregor Gregorčič and Gordon Lightbody*

## 6.1  Introduction

In the past years many approaches to modelling of nonlinear systems using neural networks and fuzzy models have been proposed [1–3]. The difficulties associated with these black-box modelling techniques are mainly related to the curse of dimensionality and lack of transparency of the model. The local modelling approach has been proposed to increase transparency as well as reduce the curse of dimensionality [4]. Difficulties related to partitioning of the operating space, structure determination, local model identification and off-equilibrium dynamics are the main drawbacks of such local modelling techniques. To improve the off-equilibrium behaviour, the use of non-parametric probabilistic models, such as Gaussian process priors was proposed [5]. The Gaussian process prior approach was first introduced in Reference 6 and revised in References 7–9. The ability to make a robust estimation in the transient region, where only a limited number of data points is available, is one of the advantages of the Gaussian process in comparison to the local model network.

The number of tunable parameters for a Gaussian process model is dramatically reduced in comparison to typical neural networks. These parameters need to be trained from training data or provided from prior knowledge. In common with neural networks, a Gaussian process model is a black-box model, which will not provide any physical knowledge about the modelled system. However, a Gaussian process model will provide an estimate of the variance of its predicted output, which can be interpreted as a level of confidence of the model. This is a major advantage of this approach in comparison to neural networks and fuzzy models, as it indicates when the model can be trusted, or needs retraining.

An overview of the probabilistic Bayesian background is given in Section 6.2 of this chapter. A Gaussian process model and how this technique is related to the radial basis function neural network is shown in Section 6.3. Also, the advantage of the predicted variance as a measure of the accuracy of the model prediction in comparison with the uncertainty in the parameters of the parametric model is explained.

Section 6.4 demonstrates the use of the Gaussian process prior approach for control. The model confidence is utilised to extend the nonlinear internal model control structure [10]. In this approach, proposed in Reference 11, the predicted variance is used to constrain the control effort in such a way that the internal model and hence the controller (which is chosen to be an inverse of the internal model) do not leave the operating space they were designed for. Since the Gaussian process model is not analytically invertible, a numerical search for the inverse has to be utilised each sampling time, which unfortunately increases the already large computational load.

The degree of transparency, the importance of selection of the dimension of the input space, invertibility and the problem of the large computation load are highlighted in Section 6.5.

To increase the transparency of the model and reduce computational load, a divide and conquer approach using Gaussian processes is presented in Section 6.6. If the operating space is partitioned into operating regimes and local Gaussian process models are found for each operating regime, then these local models can be combined to form a local Gaussian process model network. Each cluster of data consists of much less data points than the whole operating space, and hence the computational load for each local Gaussian process model will therefore be decreased, and hence the overall computational cost will also be dramatically reduced.

Assuming a linear approximation of the underlying nonlinear system, at each operating regime, a local Gaussian process model based on a linear covariance function can be utilised. In comparison with traditional linear regression, linear Gaussian process models are more robust, less prone to bias, and they provide a measure of uncertainty in the prediction. In order to model rapidly changing surfaces, across the wider operating space, a set of linear Gaussian process models can be used to form a global network [12]. Uncertainty available for each local linear Gaussian Process model can be used for the optimisation of the network structure. This network is less computationally expensive, more transparent, analytically invertible and still provides an estimate of the variance.

A possible use of the linear local Gaussian process network for control applications is given in Section 6.6.4.

## 6.2   Bayesian background

### 6.2.1   Bayesian regression

Given a set of data, a set of probabilistic models can be found. Let a member model of this set be denoted by $\mathcal{M}_a$. The prior belief about which model is most suitable to model the data, can be expressed as a probability distribution $P(\mathcal{M}_a)$, over all the possible models. If the model $\mathcal{M}_a$ has a set of parameters $\mathcal{W}_a$, another prior distribution, about the parameters $P(\mathcal{W}_a|\mathcal{M}_a)$ ('the probability of $\mathcal{W}_a$ given $\mathcal{M}_a$')

can be expressed. Since the parameters depend on the function which relates the parameters with the model, the distribution $P(\mathcal{W}_a|\mathcal{M}_a)$ is conditional on $\mathcal{M}_a$.

For given data set $\mathcal{D} = \{D_1, \ldots, D_i, \ldots, D_N\}$, each of the models makes its own predictions about how likely the observed data $D_i$ is generated by that particular model with its specific set of parameters. Given a single data point $D_i$, these predictions are described by a probability distribution $P(D_i|\mathcal{W}_a, \mathcal{M}_a)$. Taking into account the whole set of data $\mathcal{D}$, for each model, the predictions are combined in the distribution:

$$P(\mathcal{D}|\mathcal{W}_a, \mathcal{M}_a) = \prod_{i=1}^{N} P(D_i|\mathcal{W}_a, \mathcal{M}_a). \tag{6.1}$$

The Bayesian approach to data modelling utilises Bayes' theorem at two levels of inference. At the first level, the Bayes' theorem combines the prior knowledge about the choice of parameters of the particular model $P(\mathcal{W}_a|\mathcal{M}_a)$ with the knowledge gained from the data $P(\mathcal{D}|\mathcal{W}_a, \mathcal{M}_a)$:

$$P(\mathcal{W}_a|\mathcal{D}, \mathcal{M}_a) = \frac{P(\mathcal{W}_a|\mathcal{M}_a)P(\mathcal{D}|\mathcal{W}_a, \mathcal{M}_a)}{P(\mathcal{D}|\mathcal{M}_a)}. \tag{6.2}$$

The probability $P(\mathcal{W}_a|\mathcal{D}, \mathcal{M}_a)$ is called the posterior which carries the information about the value of the parameters. $P(\mathcal{D}|\mathcal{W}_a, \mathcal{M}_a)$ is known as the likelihood and $P(\mathcal{D}|\mathcal{M}_a)$ is called the evidence. In general the Bayes' theorem can be written as:

$$\text{posterior} = \frac{\text{Prior} \cdot \text{likelihood}}{\text{evidence}}. \tag{6.3}$$

The evidence $P(\mathcal{D}|\mathcal{M}_a)$, of equation (6.2) carries information about the credibility of each of the models according to given data. If $P(\mathcal{D}|\mathcal{M}_a)$ can be calculated, the information contributed by $P(\mathcal{D}|\mathcal{M}_a)$ can be incorporated with the prior belief $P(\mathcal{M}_a)$, by applying the Bayes' theorem again:

$$P(\mathcal{M}_a|\mathcal{D}) = \frac{P(\mathcal{M}_a)P(\mathcal{D}|\mathcal{M}_a)}{P(\mathcal{D})}. \tag{6.4}$$

This is the second level of inference. At this level the particular model $\mathcal{M}_a$ is compared with the others. The denominator of equation (6.4) does not depend on the model parameters. $P(\mathcal{D})$ is a normalising constant and equation (6.4) can be written as:

$$P(\mathcal{M}_a|\mathcal{D}) = \text{constant} \cdot P(\mathcal{M}_a)P(\mathcal{D}|\mathcal{M}_a),$$

$$\text{posterior} = \text{constant} \cdot \text{Prior} \cdot \text{likelihood}, \tag{6.5}$$

$$\text{posterior} \propto \text{prior} \cdot \text{likelihood}.$$

At the second level of inference, the posterior distribution $P(\mathcal{M}_a|\mathcal{D})$ provides information about the ability of different models to model the data, which allows for the ranking of the different models.

Consider as an example, four models, $\mathcal{M}_1, \mathcal{M}_2, \mathcal{M}_3$ and $\mathcal{M}_4$, which generate the following curves: $f_1(u), f_2(u), f_3(u)$ and $f_4(u)$. A prior belief about how accurately each model can model the data can be quantified by a list of prior probabilities,

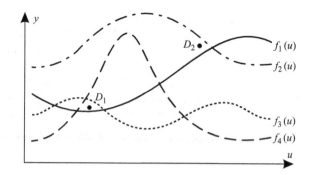

*Figure 6.1    Bayesian model*

$P(\mathcal{M}_a), a \in \{1, 2, 3, 4\}$, which sum to 1. If there is no prior reason to prefer one curve over another, then the prior probability of each individual model is $\frac{1}{4}$.

Given a set of data as input–output pairs $D_k = (u(k), y(k)), k \in \{1, \ldots, N\}$ and assuming that the output $y(k)$ is generated by the unknown function and corrupted with zero-mean Gaussian, a noise of variance $\sigma_\eta^2$, the likelihood of an individual model, when the particular data point is observed can be calculated as:

$$P(y(k)|u(k), \mathcal{M}_a) = \frac{1}{(2\pi\sigma_\eta^2)^{1/2}} \exp\left(-\frac{(y(k) - f_a(u(k)))^2}{2\sigma_\eta^2}\right). \tag{6.6}$$

The likelihood of each model given all $N$ data points is $\prod_k P(y(k)|u(k), \mathcal{M}_a)$. It is assumed that the noise variance is small, much smaller than the overall $y$ scale on Figure 6.1.

Observing a single data point, the Bayes' theorem describes how the beliefs in the models should be updated based on the information that the data has provided. The posterior distribution is an information about plausibility of the model, that is, how possible it is, that the data point was generated by an individual model. It is obtained by multiplying two quantities: how plausible the particular model was believed to be, before the data contributed its information, and how much the particular model predicted the observed data. To make the final beliefs add up to 1, the product has to be normalised and it can be written as equation (6.4).

It can be seen in Figure 6.1, when the data point $D_1$ is observed, the likelihood of equation (6.6) is much higher for curves $f_1(u)$, $f_3(u)$, $f_4(u)$ than for curve $f_2(u)$. The posterior distribution will have more weight on models $\mathcal{M}_1$, $\mathcal{M}_3$ and $\mathcal{M}_4$ and less weight on model $\mathcal{M}_2$. When the second data point $D_2$ is observed, it can be seen that only curve $f_1(u)$ fits well to both data points. The posterior distribution will then have most of its weight concentrated on the model $\mathcal{M}_1$. For prediction at the new input point, the model $\mathcal{M}_1$ will have the larger contribution.

For this simple example, calculating the posterior distribution for each model and then combining them to make a prediction is elementary. For more complex problems, the Bayesian framework might be difficult to implement. Most of the difficulties are related to calculating the evidence at the first level of inference. The evidence is one of the most important quantities in the Bayesian approach; however, to be able to evaluate this evidence, approximations and simplifications are often necessary.

In general, placing a prior can also be problematic. The prior depends on the knowledge of the modelled system and can be subjective. A philosophical argument about the meaning of probability related to the subjectivity of the Bayesian framework is pointed out in Reference 13. It is clearly explained in Reference 8, that the subjectivity of the prior belief does not make the Bayesian theory questionable. For the prior to be objective, the only requirement to be satisfied is that people having the same knowledge about the modelled system use the same prior. If different people have a different prior knowledge, then it is expected to use different priors. However, the main difficulty related to prior beliefs is not subjectivity of the prior, but assigning the probability to the prior knowledge. Bayesian theory automatically embodies Occam's Razor[1] principle [14]. Occam's Razor principle states that unnecessarily complex models should not be preferred to the simpler models, when extra complexity is not required to explain the data. Thus applications of Bayesian techniques will prevent overfitting the data, without the introduction of the especially set penalty terms [13], unless higher complexity is specifically defined through a prior over models [8]. As shown in References 13 and 14, the Bayesian Occam's Razor directly penalises complex models. In the neural network and system identification community the Occam's Razor principle applied to non-probabilistic models is known as a bias-variance trade-off [4,15] and is closely related to regularisation [16,17].

The Bayesian model comparison is based on the assumption that the true model is included in the set of available models. If the true model is not included in the set of possible candidates, then the particular model will never be compared with the true model and a Bayesian comparison cannot be used successfully. There exists no Bayesian criterion to determine if the set of available models is correct.

Even after assuming that the set of the possible models is correct, making a prediction based on a model comparison is still not easy. The likelihood and the prior are usually some nonlinear function of parameters and placing a prior over the parameters is based on prior beliefs, which often result in analytically unsolvable integrals. In this case a Monte Carlo approximation of the methods based on the most possible parameters can be used.

Gaussian processes are a way to set the prior directly on the model functions, which is often simpler than placing a prior on the model parameters. To make a prediction, the Gaussian process approach does not involve unsolvable integrals. The prediction can be made using linear algebra.

## 6.3 Gaussian process model

A stochastic process is a collection of random variables. Any finite set of random variables $\underline{y}$, which have a joint Gaussian distribution:

$$P(\underline{y}|\mathbf{C}, \Phi_N) = \frac{1}{Q} \exp\left(-\frac{1}{2}(\underline{y} - \mu)^\mathrm{T} \mathbf{C}^{-1}(\underline{y} - \mu)\right), \tag{6.7}$$

is a Gaussian process (GP). Here $\mathbf{C}$ is the covariance matrix of the data, defined by the covariance function, $Q$ is the normalising constant, $\underline{u}$ is any collection of inputs and $\mu$ is the mean of the distribution. The Gaussian process is fully represented by its mean and covariance function $C(\cdot)$, which defines a covariance matrix $\mathbf{C}$. In this chapter, a zero-mean distribution is assumed:

$$P(\underline{y}|\mathbf{C}, \Phi_N) = \mathcal{N}(0, \mathbf{C}). \tag{6.8}$$

Obviously not all data can be modelled as a zero-mean process. If the data is properly scaled and detrended, then the assumption about the zero-mean distribution is correct.

### 6.3.1   Constructing the Gaussian process model

Consider a noisy input–output set of data $\mathcal{D}$. The full matrix $\Phi_N$ of $N$ $d$-dimensional input vectors is constructed as follows:

$$\Phi_N = \begin{bmatrix} u_1(1) & u_2(1) & \cdots & u_d(1) \\ u_1(2) & u_2(2) & \cdots & u_d(2) \\ \vdots & \vdots & \vdots & \vdots \\ u_1(k) & u_2(k) & \cdots & u_d(k) \\ \vdots & \vdots & \vdots & \vdots \\ u_1(N) & u_2(N) & \cdots & u_d(N) \end{bmatrix}. \tag{6.9}$$

Scalar outputs are arranged in the output vector $\underline{y}_N$:

$$\underline{y}_N = [y(1) \quad y(2) \quad \cdots \quad y(k) \quad \cdots \quad y(N)]^T. \tag{6.10}$$

The aim is to construct the model, and then at some new input vector:

$$\underline{u}^T(N+1) = [u_1(N+1) \quad u_2(N+1) \quad \cdots \quad u_d(N+1)] \notin \mathcal{D}, \tag{6.11}$$

find the distribution of the corresponding output $y(N+1)$. A general model for the $k$th output can be written as:

$$y(k) = y_f(\underline{u}(k)) + \eta(k), \tag{6.12}$$

where $y(k)$ is a noisy output, $y_f$ is the modelling function which produces noise-free output from the input vector $\underline{u}(k)$ and $\eta(k)$ is additive noise. The prior over the space of possible functions to model the data can be defined as $P(y_f|\underline{\alpha})$, where $\underline{\alpha}$ is some set of hyperparameters.[2] A prior over the noise $P(\underline{\eta}|\underline{\beta})$ can also be defined, where $\underline{\eta}$ is the vector of noise values $\underline{\eta} = [\eta(1) \quad \eta(2) \quad \cdots \quad \eta(k) \quad \cdots \quad \eta(N)]$ and $\underline{\beta}$ is the set of hyperparameters. The probability of the data given hyperparameters $\underline{\alpha}$ and $\underline{\beta}$ can be written as:

$$P(\underline{y}_N|\underline{\alpha}, \underline{\beta}, \Phi_N) = \int P(\underline{y}_N|\Phi_N, y_f, \underline{\eta}) P(y_f|\underline{\alpha}) P(\underline{\eta}|\underline{\beta}) dy_f \, d\underline{\eta}. \tag{6.13}$$

Defining an extension of the vector $\underline{y}_N$:

$$\underline{y}_{N+1} = \left[\underline{y}_N^T \quad y(N+1)\right]^T, \tag{6.14}$$

the conditional distribution of $y(N + 1)$ can then be written as:

$$P(y(N + 1)|\mathcal{D}, \underline{\alpha}, \beta, \underline{u}(N + 1)) = \frac{P(\underline{y}_{N+1}|\Phi_N, \underline{\alpha}, \beta, \underline{u}(N + 1))}{P(\underline{y}_N|\Phi_N, \underline{\alpha}, \beta)}. \tag{6.15}$$

This conditional distribution can be used to make a prediction about $y(N + 1)$. The nominator of equation (6.15), that is the integral in equation (6.13) is complicated. Some of the standard approaches to solving such problems are reviewed in Reference 8.

Assuming that the additive noise is Gaussian and the output vector $\underline{y}_N$ is generated by a Gaussian process of equation (6.7), the conditional distribution of equation (6.15) is also Gaussian [18]. If the probability distribution of the data is defined to be Gaussian, then it is not necessary to place an explicit prior on the noise or on the modelling function. Both of the priors now have been fused in the covariance matrix $\mathbf{C}$, generated from data by the covariance function $C(\cdot)$. The conditional distribution over $y(N + 1)$ of equation (6.15), can then be written as:

$$P(y(N + 1)|\mathcal{D}, C(\cdot), \underline{u}(N + 1))$$

$$= \frac{P(\underline{y}_{N+1}|C(\cdot), \Phi_N, \underline{u}(N + 1))}{P(\underline{y}_N|C(\cdot), \Phi_N)}$$

$$= \frac{Q_N}{Q_{N+1}} \exp\left(-\frac{1}{2}\left(\underline{y}_{N+1}^T \mathbf{C}_{N+1}^{-1}\underline{y}_{N+1} - \underline{y}_N^T \mathbf{C}_N^{-1}\underline{y}_N\right)\right)$$

$$= \frac{1}{Q} \exp\left(-\frac{1}{2}\left(\underline{y}_{N+1}^T \mathbf{C}_{N+1}^{-1}\underline{y}_{N+1}\right)\right), \tag{6.16}$$

where $Q_N$, $Q_{N+1}$ and $Q$ are appropriate normalising constants. The mean and standard deviation of the distribution of $y(N + 1)$ given in equation (6.16), can be evaluated by the brute force inversion of $\mathbf{C}_{N+1}$. If the matrix is partitioned as shown in Figure 6.2(a), a more elegant way of inversion of $\mathbf{C}_{N+1}$ can be used by utilising the partitioned inverse equation [19].

The $\mathbf{C}_{N+1}^{-1}$ can then be constructed as shown in Figure 6.2(b), where:

$$\tilde{v} = \left(v - \underline{v}_{N+1}^T \mathbf{C}_N^{-1}\underline{v}_{N+1}\right)^{-1},$$

$$\underline{\tilde{v}}_{N+1} = -\tilde{v}\mathbf{C}_N^{-1}\underline{v}_{N+1}, \tag{6.17}$$

$$\tilde{\mathbf{C}}_N = \mathbf{C}_N^{-1} + \frac{1}{\tilde{v}}\underline{\tilde{v}}_{N+1}\underline{\tilde{v}}_{N+1}^T.$$

Calculating the inverse using the partitioned inverse equation is useful when predictions are to be made at a number of new points on the basis of the data set of size $N$.

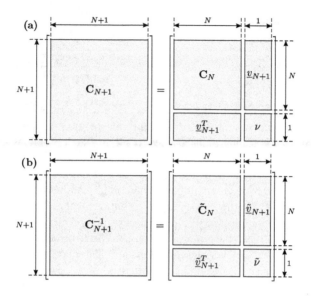

*Figure 6.2   Construction of the covariance matrix* $\mathbf{C}_{N+1}$

By substitution of matrix $\mathbf{C}_{N+1}^{-1}$ into equation (6.16), the Gaussian conditional distribution of $y(N+1)$ at the new input vector $\underline{u}(N+1)$ can be written as:

$$P(y(N+1)|\mathcal{D}, C(\cdot), \underline{u}(N+1)) = \frac{1}{Q}\exp\left(-\frac{1}{2}\frac{(y(N+1)-\mu_{\hat{y}(N+1)})^2}{\sigma_{\hat{y}(N+1)}^2}\right),$$

(6.18)

where

$$\mu_{\hat{y}(N+1)} = \underline{v}_{N+1}^{\mathrm{T}}\mathbf{C}_N^{-1}\underline{y}_N$$

(6.19)

and

$$\sigma_{\hat{y}(N+1)}^2 = v - \underline{v}_{N+1}^{\mathrm{T}}\mathbf{C}_N^{-1}\underline{v}_{N+1}.$$

(6.20)

Here $\mu_{\hat{y}(N+1)}$ is the mean prediction for the new input vector and $\sigma_{\hat{y}(N+1)}^2$ is the variance of this prediction. From the system identification point of view, the Gaussian Process model can be interpreted as follows.

Given model input vector $u$, the predicted model output $\hat{y}$ is the mean of the Gaussian distribution: $\hat{y} = \mu_{\hat{y}(N+1)}$. The model uncertainty about this prediction is then defined by $\sigma_{\hat{y}(N+1)}$. This uncertainty or level of confidence in the model can be also represented as the error bars on model prediction. The estimate of the uncertainty in model prediction is a main advantage of a Gaussian process model in comparison to neural networks and parametrical models.

It is important to highlight here, that in order to make a prediction for input vector $\underline{u}(N+1)$ utilising equations (6.19) and (6.20), only $\mathbf{C}_N^{-1}$ needs to be inverted and not $\mathbf{C}_{N+1}^{-1}$. This means that the prediction, assuming fixed hyperparameters and a fixed data set requires the inversion of an $N \times N$ matrix. A more detailed derivation of equations (6.19) and (6.20) from the conditional distribution given in equation (6.16), is given in References 8 and 20.

The predictions produced by a Gaussian process depend entirely on the covariance matrix $\mathbf{C}$. The covariance matrix $\mathbf{C}$ is produced from the inputs by the covariance function $C(\cdot)$. In the next section, the most widely used form of the covariance function is discussed.

### 6.3.2   The covariance function

In the prediction equations (6.19) and (6.20), the $\mathbf{C}_N \in \mathfrak{R}^{N \times N}$, $\underline{v}_N \in \mathfrak{R}^N$ and the scalar $v$ can be constructed as follows:

$$
\mathbf{C}_N = \begin{bmatrix} C_{1,1} & \cdots & C_{1,n} & \cdots & C_{1,N} \\ \vdots & \ddots & \vdots & \ddots & \vdots \\ C_{m,1} & \cdots & C_{m,n} & \cdots & C_{m,N} \\ \vdots & \ddots & \vdots & \ddots & \vdots \\ C_{N,1} & \cdots & C_{N,n} & \cdots & C_{N,N} \end{bmatrix},
\tag{6.21}
$$

where

$$
C_{m,n} = C(\underline{u}(m), \underline{u}(n)),
\tag{6.22}
$$

$$
\underline{v}_{N+1} = [C(\underline{u}(1), \underline{u}(N+1)) \cdots C(\underline{u}(N), \underline{u}(N+1))]^{\mathsf{T}}
\tag{6.23}
$$

and

$$
v = C(\underline{u}(N+1), \underline{u}(N+1)).
\tag{6.24}
$$

Here $C(\cdot)$ is the covariance function (in the machine learning literature it can be called a kernel function), and is a function of inputs only. Any choice of the covariance function, which will generate a positive definite[3] covariance matrix for any set of input points, can be chosen [20]. If the covariance matrix $\mathbf{C}_N$ is not positive definite, the distribution of equation (6.7) cannot be normalised. The flexibility of choice of the covariance function offers the ability to include the prior knowledge in the model. A sum of positive definite functions is also a positive definite function; similarly a product of positive definite functions is also a positive definite function. These properties give the ability to construct complex covariance functions from a set of simple functions.

The covariance function of equation (6.25) has proven to work well and it has been widely used in practice:

$$
C(\underline{u}(m), \underline{u}(n)) = w_0 \exp\left(-\frac{1}{2} \sum_{l=1}^{d} w_l (\underline{u}_l(m) - \underline{u}_l(n))^2\right) + w_\eta \delta(m, n),
$$

$$
\tag{6.25}
$$

where $\underline{\theta} = [w_0 \ w_1 \ \cdots \ w_d \ w_\eta]^T$ is the vector of hyperparameters, $d$ is the dimension of the input space and $\delta(m, n)$ is a Kronecker delta defined as:

$$\delta(m, n) = \begin{cases} 1 & \text{for } m = n, \\ 0 & \text{for } m \neq n. \end{cases} \tag{6.26}$$

This covariance function assigns a similar prediction to the input points which are close together. The $w_l$ parameters allow a different distance measure in each input dimension. Parameter $w_0$ controls the overall scale of the local correlation. The hyperparameter $w_\eta$ is the estimate of the noise variance. The term $w_\eta \delta(m, n)$, takes into account the noise on the data. The additive noise is assumed to be white and zero-mean Gaussian. If noise is correlated then the noise model $w_\eta \delta(m, n)$ can be modified as shown in Reference 21.

To be able to make a prediction, using equations (6.19) and (6.20), the hyperparameters have to be provided either as prior knowledge, or trained from the training data.

### 6.3.3   Training a Gaussian process model

In cases, when the hyperparameters are not provided as prior knowledge, the Bayesian framework can be utilised to find an appropriate set of hyperparameters. The prior on the hyperparameters $P(\underline{\theta}|C(\cdot))$ can be placed and the integral of equation (6.27) over the resulting posterior has to be found:

$$P(y(N+1)|\underline{u}(N+1), \mathcal{D}, C(\cdot))$$

$$= \int P(y(N+1)|\underline{u}(N+1), \mathcal{D}, C(\cdot), \underline{\theta}) P(\underline{\theta}|C(\cdot)) d\underline{\theta}. \tag{6.27}$$

For most covariance functions $C(\cdot)$, the likelihood $P(y(N+1)|\underline{u}(N+1), \mathcal{D}, C(\cdot), \underline{\theta})$ has a complex form and the integral of equation (6.27) is analytically unsolvable. To solve the integral, two approaches have been proposed in literature:

- The integral can be solved numerically using the Monte Carlo approach.
- A maximum likelihood[4] approach can be used to adjust the hyperparameters to maximise the likelihood of the training data.

#### 6.3.3.1   Monte Carlo approach

In the Monte Carlo approach sampling methods are used to calculate the predictive distribution. By approximating the integral in equation (6.27) by a Markov chain,[5] the average of the series of samples is used to approximate the predictive distribution. This predictive distribution is a mixture of Gaussians and as the number of samples increases, the approximation approaches to the true predictive distribution. The Monte Carlo approach is very flexible; however, a large memory storage for large matrices is required. For small data sets, where matrix storage is not an important issue, the Monte Carlo approach will give better results for fixed CPU time. For larger data sets, the maximum likelihood approach is faster and the trained model performs better [7,8].

In this chapter, the maximum likelihood approach for Gaussian process training will be utilised.

### 6.3.3.2 Maximum likelihood

The integral of equation (6.27) can be approximated by an average prediction over all possible values of the hyperparameters using the most probable set of hyperparameters $\theta_{MP}$. Assuming that the distribution $P(\theta|\mathcal{D}, C(\cdot))$ over $\theta$ is sharply peaked around the most probable set of hyperparameters $\theta_{MP}$ relative to the predictive distribution $P(y(N+1)|\underline{u}(N+1), \mathcal{D}, C(\cdot))$, the integral of equation (6.27) can be approximated as:

$$P(y(N+1)|\underline{u}(N+1), \mathcal{D}, C(\cdot))$$
$$\approx P(y(N+1)|\underline{u}(N+1), \mathcal{D}, C(\cdot), \theta_{MP}) P(\theta_{MP}|C(\cdot)) \Delta\theta, \qquad (6.28)$$

where $\Delta\theta$ is the width of the distribution $P(\theta|\mathcal{D}, C(\cdot))$ at $\theta_{MP}$, as shown in Figure 6.3.

Assuming $P(y(N+1)|\underline{u}(N+1), \mathcal{D}, C(\cdot), \theta_{MP})$ is a Gaussian with mean and variance given by equations (6.19) and (6.20), then at $\theta = \theta_{MP}$, the $P(y(N+1)|\underline{u}(N+1), \mathcal{D}, C(\cdot))$ is also Gaussian with the same mean and variance. It has been shown in Reference 23 that this approximation is good and the maximum likelihood framework usually provides the same prediction as using the true predictive distribution.

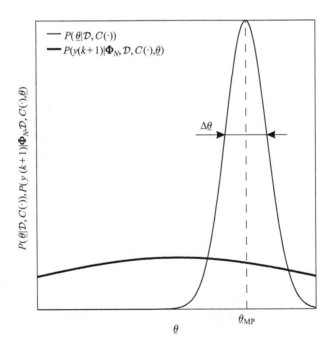

*Figure 6.3*    *Distribution $P(\theta|\mathcal{D}, C(\cdot))$ over $\theta$, is assumed to be sharply peaked (small $\Delta\theta$) around the most probable set of hyperparameters $\theta_{MP}$ in comparison to the predictive distribution $P(y(N+1)|\underline{u}(N+1), \mathcal{D}, C(\cdot))$*

The idea of the maximum likelihood framework is to find the most probable set of hyperparameters by utilising standard gradient-based optimisation routines such as the conjugate gradient technique. This type of optimisation optimises the function using only the derivatives of the function with respect to the parameters and evaluation of the function itself is not needed.

The posterior distribution of $\underline{\theta}$ can be written as:

$$P(\underline{\theta}|\mathcal{D}, C(\cdot)) = \frac{P(\underline{y}_N|\mathbf{\Phi}_N, C(\cdot), \underline{\theta})P(\underline{\theta})}{P(\underline{y}_N|\mathbf{\Phi}_N, C(\cdot))}. \tag{6.29}$$

Since the denominator of the posterior is independent of $\underline{\theta}$, it does not play a role in the derivation of derivatives and it can be ignored. The nominator, likelihood of $\underline{\theta}$ multiplied by a prior of $\underline{\theta}$ is a Gaussian process. The negative log-likelihood $\mathcal{L}$ for a Gaussian process is given as:

$$\mathcal{L} = -\frac{1}{2}\left(\log(\det(\mathbf{C}_N)) + \underline{y}_N^T\mathbf{C}_N^{-1}\underline{y}_N + N\log(2\pi)\right). \tag{6.30}$$

Jacobi's formula for the derivative of a determinant is given by

$$\frac{\partial \det(\mathbf{A})}{\partial x} = \mathrm{Tr}\left(\mathrm{adj}(\mathbf{A})\frac{\partial \mathbf{A}}{\partial x}\right), \tag{6.31}$$

and the partial derivatives of $\mathcal{L}$ with respect to individual hyperparameter $\theta_i \in \underline{\theta}$ can be expressed as:

$$\frac{\partial \mathcal{L}}{\partial \theta_i} = -\frac{1}{2}\left(\mathrm{Tr}\left(\mathbf{C}_N^{-1}\frac{\partial \mathbf{C}_N}{\partial \theta_i}\right) - \underline{y}_N^T\mathbf{C}_N^{-1}\frac{\partial \mathbf{C}_N}{\partial \theta_i}\mathbf{C}_N^{-1}\underline{y}_N\right). \tag{6.32}$$

The likelihood function of equation (6.30) is in general multimodal with respect to $\underline{\theta}$. Some of the modes often correspond to non-sensible solutions. By utilising a conjugate gradient optimisation technique, the algorithm can get stuck in local minima. This means that the set of hyperparameters $\underline{\theta}_{MP}$ that is found by the optimisation routine is dependent on the initial values of hyperparameters $\underline{\theta}_{ini}$. Multiple random restarts of the optimisation algorithm is one possible approach to overcome this problem. This technique is heavily time consuming. Rasmussen in Reference 7 suggests a single run, allowing a fixed number (about 150) of evaluations, by which time the likelihood is changing very slowly.

Inappropriate choice of initial values makes the partial derivatives of the likelihood small, which creates a problem for the optimisation algorithm. Choice of initial values: $w_0 = 1$, $w_i = 1/d$ and $w_\eta = \exp(-2)$, suggested in Reference 7, has proven to work well. Since the hyperparameters themselves must be always positive, the initial values $\underline{\theta}_{ini}$ can be chosen as $\underline{\theta}_{ini} = \log(\underline{\theta})$, where the log is applied elementwise $\underline{\theta}_{ini} = [\log(w_0)\ \log(w_1)\ \cdots\ \log(w_d)\ \log(w_\eta)]^T$. In this case the unconstrained optimisation can be used.

Taking into account the recommendation about the choice of the initial condition, experience shows that multimodality does not create a major problem. The biggest drawback of using the maximum likelihood algorithm is that each evaluation of the gradient of the log-likelihood requires the inversion of the $N \times N$ covariance matrix $\mathbf{C}_N$. The exact inversion of the $N \times N$ matrix has an associated computation cost of $\mathcal{O}(N^3)$, which makes training of the Gaussian process model extensively time consuming for large training data sets.

### 6.3.4 Relation of the Gaussian process model to the radial basis function network

From the neural networks point of view, a Gaussian process model can be characterised as a form of radial basis function network (RBFN) [24]. In general, the prediction equation of the RBFN is given as:

$$y(k+1) = \sum_{i=1}^{M} g_i(\underline{u}(k))W_i = \underline{g}(k)^T \underline{W}, \tag{6.33}$$

where $g_i(\cdot)$ is chosen to be a Gaussian basis function[6] and $W_i$ is the corresponding weight. To be able to make a prediction, based on equation (6.33), the structure of the network has to be defined and the vector of tunable parameters $\underline{W}$ has to be found. The structure definition involves choice of a number of basis functions $M$, choice of their widths $\sigma_g$ and placing their centres on the input space. The weights are tuned to achieve a desirable performance of the network. The problem of the structure selection and especially the difficulty involved in tuning a large number of parameters in a large scale network lead to the idea of investigating the network parameters as a set of random variables [7]. In this framework, the distribution of the parameters defines the network performance, rather than the parameters themselves. Assuming that the weights of an RBFN of equation (6.33) are Gaussian distributed, then for any given finite set of inputs, the network outputs will also be Gaussian distributed [25]. This is a defining property of Gaussian process. A specific choice of structure for the RBFN simplifies the model even more. By placing the centres of basis functions on each input point of the input space and assuming that the number of data points and hence number of basis functions goes to infinity, the covariance function of equation (6.25) for Gaussian process model has been derived [8,25].

In the parametric approach to RBFN parameter tuning, the vector of weights $\underline{W}$ is usually calculated using least-squares regression:

$$\underline{W} = (\mathbf{G}^T\mathbf{G})^{-1}\mathbf{G}^T\underline{y}_N, \tag{6.34}$$

where $\mathbf{G} \in \Re^{N \times d}$ is defined as a matrix of basis functions at the input data points $\mathbf{G} = [\underline{g}^T(1) \cdots \underline{g}(N)^T]^T$ and $\underline{y}_N$ is a vector of measured outputs (or targets). In general, instead of the exact calculation of $(\mathbf{G}^T\mathbf{G})^{-1}\mathbf{G}^T$, the pseudo-inverse $\mathbf{G}^+$ is used. The pseudo-inverse utilises the singular value decomposition [26], to find the minimum norm solution. This is very useful when the problem is ill-conditioned.[7] By replacement of $(\mathbf{G}^T\mathbf{G})^{-1}\mathbf{G}^T$ with $\mathbf{G}^+$ in equation (6.34), the prediction equation (6.33), can be written as:

$$\hat{y}(k+1) = \underline{g}(k)^T \underline{W} = \underline{g}(k)^T \mathbf{G}^+ \underline{y}_N. \tag{6.35}$$

A comparison of prediction equations (6.35) and (6.19) clearly shows that the Gaussian process model using covariance function of equation (6.25) and the RBFN are the same, if the centres of the basis function of RBFN network are placed on each training data point, the widths are chosen to be the same for all basis functions and the weights are assumed to be zero-mean Gaussian distributed, $P(\underline{W}) = \mathcal{N}(0, \sigma_W^2)$. In this representation, the RBFN can be characterised by the same set of hyperparameters as the parametrisation of the Gaussian process covariance function of equation (6.25). The hyperparameter $w_0$, in this interpretation, controls the variance of the distribution of weights: $w_0 \propto \sigma_W^2$. Hyperparameters $w_l$, where $l = 1, \ldots, d$, are inversely proportional to the widths $\sigma_g$ of the basis functions $w_l \propto 1/\sigma_{g_l}$. Subscript $l$ emphasises that the scale of widths can be different for each input dimension. By including a noise model in the covariance function, the total number of tunable parameters of the Gaussian process model is $d + 2$. This is a significant reduction of the number of tunable parameters in comparison to RBFN, where for a given network structure, which is comparable with a Gaussian process model, a much larger number of parameters needs to be found.

### 6.3.5    *Linear regression versus Gaussian process models*

To illustrate the benefits of the Gaussian process approach in comparison with linear regression, consider a single-input, single-output system:

$$y(k) = Ku(k) + \eta(k), \tag{6.36}$$

where $K$ is an unknown gain and $\eta(k)$ is a zero-mean Gaussian white noise. For each pair of data $\{u(k), y(k)\}$ for $k = 1, \ldots, N$, the parameter $K$ can be calculated. Since $y(k)$ data is corrupted with noise, every calculation gives a different value of $K$ (Figure 6.4(a)).

Parameter $\hat{K}$ can be estimated using the least-squares algorithm. This approach is equivalent to fitting the best line through the set of data $\mathcal{D}$. The prediction at point $u(N + 1)$ can be written as:

$$\hat{y}(N+1) = \hat{K}u(N+1). \tag{6.37}$$

A different approach can be taken for the prediction of $y(N + 1)$ at the point $u(N + 1)$ using the same set of data. Assuming that outputs $y(k)$ have a Gaussian distribution, a Gaussian process model can be utilised. Given the training data and the hyperparameters, the prediction at the point $u(N + 1)$ can be made using equations (6.19) and (6.20).

The difference between equation (6.37) and equations (6.19) and (6.20) is obvious. Given $\hat{K}$ at the new input $u(N+1)$, equation (6.37) gives a point prediction $\hat{y}(N + 1)$. The accuracy of this point prediction depends only on the estimated parameter $\hat{K}$. The prediction based on Gaussian process model is more powerful. Given $\mathbf{C}_N$, $\underline{v}_{N+1}$ and $v$, the prediction at the new input $u(N + 1)$ is not a point prediction, but is a Gaussian

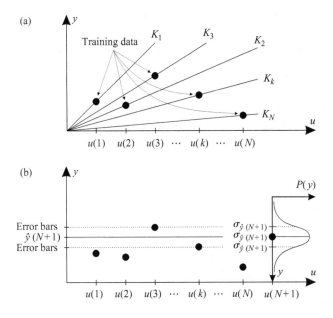

*Figure 6.4* (a) *Training set of data. (b) Mean* $\mu_{\hat{y}(N+1)}$ *of the prediction distribution is the predicted value of* $\hat{y}(N+1)$ *at* $u(N+1)$. *The standard deviation* $\sigma_{\hat{y}(N+1)}$ *defines error bars* $\hat{y}(N+1) \pm \sigma_{\hat{y}(N+1)}$ *on predicted point*

distribution. This means that at the point $u(N+1)$ the output $y(N+1)$ is most likely going to be the mean of the prediction, $\mu_{\hat{y}(N+1)}$. The certainty of this prediction is given by the standard deviation $\sigma_{\hat{y}(N+1)}$ of the predictive distribution. This is demonstrated in Figure 6.4(b). The standard deviation $\sigma_{\hat{y}(N+1)}$ of the predictive distribution as a measure of confidence in the model prediction can be used to define error bars for this prediction.

### 6.3.6 Uncertainty of the model, confidence and the error bars

For parametric models, the uncertainty is usually expressed as an uncertainty in parameters, and does not take into account uncertainty about the model structure and the distance of the current point prediction from the training data used to estimate model parameters [27,28]. In an attempt to estimate the model uncertainty an extension to the RBFN, known as the VI net was proposed in Reference 29. Based on the training data density, the VI net has the ability to indicate when the network is extrapolating and to calculate confidence limits for its prediction. In Bayesian regression models, the uncertainty in the model prediction is a combination of uncertainties in parameters, model structure, distance of prediction from the training data set and the measurement noise [30,31].

There is a significant difference in the uncertainty information between parametrical models and Bayesian models. Uncertainty for parametrical models provides information about how well the parameters of the model are estimated and level of confidence does not depend on the input point for which the prediction is made [32]. This effectively means that the model is incapable of detecting if its prediction is valid for the given input point. The uncertainty of the Gaussian process model carries more information. Here the model prediction and hence its estimate of uncertainty are based on the training set of data. This allows the model to measure the distance, in the input space, between the input point for which the prediction is made and the training data set. Increasing the distance will increase the model uncertainty. This is important when a nonlinear function is modelled, based on locally collected data. When the model is asked to predict the output on the subspace, where the training data is available, the model will produce a relatively accurate prediction, which will also result in a low uncertainty. If the model is asked to predict the output based on the input point, which is further away from the well-modelled subspace, the model will produce a prediction, but the increase of uncertainty will indicate that this prediction is less accurate. The model uncertainty also depends on the model structure. A more complex model will result in a lower uncertainty. Finally, the ability to incorporate noise information in the model will significantly improve the accuracy of the uncertainty estimate.

To demonstrate the accuracy of the uncertainty estimate of the Gaussian process model in comparison with least-squares regression, consider the following example: a noisy training set of ten data points, shown in Figure 6.5(a), was generated by a static nonlinear function

$$y = f(u) + \eta, \tag{6.38}$$

where $\eta$ is a white zero-mean Gaussian noise. First, a second-order polynomial was fitted through the data. The least-squares algorithm was used to find the polynomial parameters. Figure 6.5(b) shows the parametric least-squares model prediction $\hat{y}_{LS}$ and $\hat{y}_{LS} \pm 2\sigma_{LS}$ error bars. $\sigma_{LS}^2$ is given as [33]:

$$\sigma_{LS}^2 = \frac{1}{N} \sum_{k=1}^{N} (e(k) - \bar{e})^2, \tag{6.39}$$

where $e(k)$ is an error of the $k$th prediction point, $\bar{e}$ is an average of the prediction error over the $N$ prediction points.

It can be seen that the prediction fits well to a true function in the region where data is available. Further away the prediction is not accurate and yet error bars on prediction do not change across the input space. The uncertainty in this case provides an information on how well the parameters are estimated and does not depend on the distance of the input point from training data. This clearly shows, that the model is not capable of providing an accurate estimate of uncertainty.

In Figure 6.5(c) the Gaussian process model prediction $\hat{y}_{GP}$ is shown. Conjugate gradient optimisation was used to find suitable hyperparameters. Twenty runs were performed with different, randomly chosen sets of initial conditions to avoid local minima. The prediction fits better to the true function in the region where training data are available. In the region where data are not available, the prediction does not fit to the true function which is clearly indicated by the wider error bars. In comparison

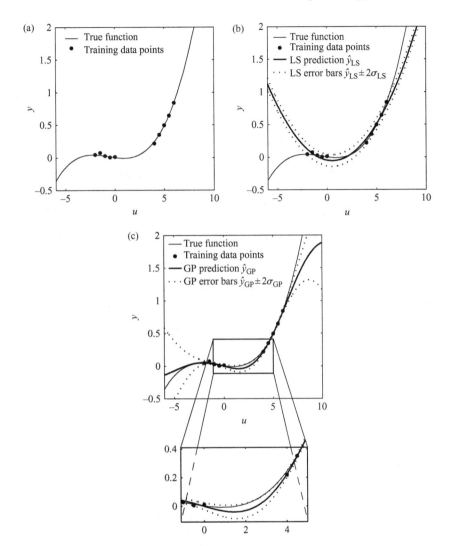

*Figure 6.5* *(a) Training set of data generated by the nonlinear function. (b) Prediction of the parametric model $\hat{y}_{LS} = 0.0277u^2 - 0.0285u - 0.0455$ and its error bars. The error bars do not depend on the distance of the input point from the training data. (c) The Gaussian process prediction and its error bars. The error bars get wider away from the training data. Note the lack of data in the region $0 \leq u \leq 4$. The prediction there is less accurate, which is indicated with wider error bars*

to the parametric model, where error bars do not change across $u$, the Gaussian process model is able to provide better uncertainty estimate. For inputs which are closer to training data, the error bars are tighter and for inputs farther away from training data the error bars are wider.

### 6.3.6.1   Gaussian process model of a dynamic system

To illustrate the construction of a Gaussian process model of a dynamical system, consider the example given below.

A first-order discrete dynamic system is described by the nonlinear function:

$$y(k + 1) = 0.95 \tanh(y(k)) + \sin(u(k)) + \eta(k), \tag{6.40}$$

where $u(k)$ and $y(k)$ are the input and output of the system, $y(k + 1)$ is the one-step ahead predicted output and $\eta(k)$ is a zero-mean white Gaussian noise. To model the system using a Gaussian process model, the dimension of the input space should be first selected. The problem is similar to the structure selection of an ARX model [34]. It has been shown in Reference 35 that in the nonlinear case, the choice of the dimension of the input space is not trivial. Selection of the dimension of the input space is discussed in Section 6.5. For clarity, the input space in this example is chosen to be two-dimensional. One delayed input and output are required to make a one-step ahead prediction.

The system was simulated to generate representative data. A Gaussian process model was trained as a one-step ahead prediction model. The training data points were arranged as follows:

$$\Phi_N = \begin{bmatrix} u(1) & u(2) & \cdots & u(k) & \cdots & u(N-1) \\ y(1) & y(2) & \cdots & y(k) & \cdots & y(N-1) \end{bmatrix}^{\mathrm{T}}$$

$$\underline{y}_N = \begin{bmatrix} y(2) & y(3) & \cdots & y(k+1) & \cdots & y(N) \end{bmatrix}^{\mathrm{T}}. \tag{6.41}$$

The noise variance was 0.01 and the sampling time was 0.1 s. The training data is shown in Figure 6.6(a).

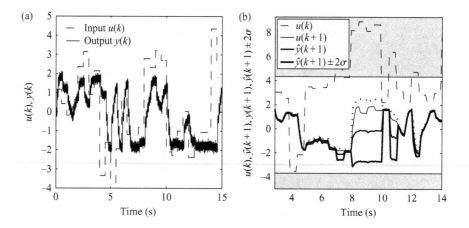

*Figure 6.6*    *(a) Training data set. (b) Validation of the model. In the range away from operating space spanned by the training data, the prediction does not fit well to true output. The less certain prediction in this region is confirmed by the wider error bars*

The maximum likelihood framework was used to determine the hyperparameters and the following set was found to be optimal: $\underline{\theta} = [w_1\ w_2\ w_0\ w_\eta]^T = [0.1971\ 0.1796\ 1.6931\ 0.0104]^T$. The Gaussian process model was then tested on a validation data set. Figure 6.6(b) shows the input $u(k)$, true noise-free output $y(k+1)$, the prediction $\hat{y}(k+1)$ and the error bars of $\pm 2$ standard deviation from the predicted mean. In the range of the operating space, where the model was trained (white region), the predicted output fits well to the true output. As soon as the input moves away from the well-modelled region, the prediction does not fit to the true output. In the time interval between 8 and 11 s, the system was driven well into an untrained region of the operating space. The wide error bars in this region clearly indicate that the model is less certain about its prediction.

The noise-free version of the system of equation (6.40) can be presented as a three-dimensional surface $y(k+1) = f(u(k), y(k))$ as shown in Figure 6.7(a). Figure 6.7(b) shows the approximation of the surface using a Gaussian process model. The dots represent the training set of data.

A better view of the model accuracy is provided by the contour and gradient plots of the surfaces, shown in Figure 6.8. It can be seen on the magnified portion of the plot that the model accurately represents the true function on the region where the model has been trained. Away from that region, however, the model is no longer a good representation of the true function.

Figure 6.9 shows the error surface $y(k+1) - \hat{y}(k+1)$ and the uncertainty of the model $2\sigma_{\hat{y}(N+1)}$. The lower surface represents the error plot. In the region of the operating space where training data (represented with dots) are available, the error is close to zero. The uncertainty plot, given above the error surface, shows that in the region where the error is small, the uncertainty is low and it rapidly increases away from the well-modelled region.

This uncertainty information can be used to improve model-based control algorithms. Some model-based control algorithms, using Gaussian process models are reviewed in the next section.

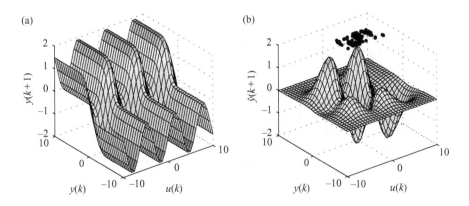

*Figure 6.7* (a) True process surface $y(k+1) = f(u(k), y(k))$. (b) Gaussian process approximation of the process surface $\hat{y}(k+1)$

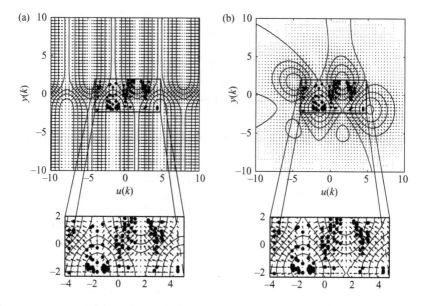

*Figure 6.8*    *(a) Contour and gradient plot of the true process function. (b) Gaussian process approximation of the process function. The magnified portion of the plot shows the close-up of the operating space where the model was trained*

## 6.4    Gaussian process model-based control

Linear control algorithms, such as pole placement, have been successfully applied to nonlinear control since they can adapt their parameters to cope with the nonlinear characteristics of the real system. However, the performance of such adaptive control techniques degrades if the controlled system rapidly changes its characteristics, due, e.g. to disturbance or set-point change [36]. Model-based techniques, where a control law is determined using a full nonlinear model, such as a neural network [1], fuzzy model [2,3] or local model network [37], have been applied to help improve the transient response. In these approaches, it is assumed that the model behaves in the same way as the nonlinear system. It has been pointed out in Reference 27 that in order to improve the control law, the accuracy of the model prediction should be taken into account. The availability of an analytical expression of the model uncertainty has led to the use of Gaussian processes in minimum variance control [27]. In this approach the control law was analytically obtained by minimising the expected value of a quadratic cost function based on the variance of the model prediction. The extension of analytic multistep minimisation of the cost function with respect to controller action was proposed in Reference 28.

Gaussian process model and its estimate of the prediction uncertainty was used for model predictive control in References 38 and 39. Here the control effort was optimised over the prediction horizon. The model uncertainty is treated as a constraint

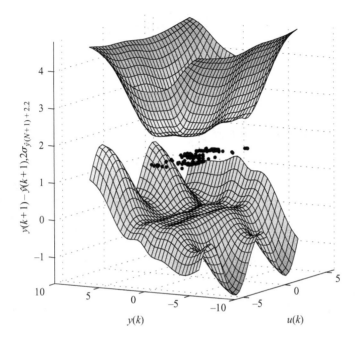

*Figure 6.9    Error and uncertainty plot. In the region where the error is small, the uncertainty is low. Away from the well-modelled region uncertainty increases rapidly. In order to increase clarity, the uncertainty surface has been raised by 2.2 in the 2 direction. The two dimensional u(k), y(k) data points are plotted on the 2=1.5 plane*

and together with the other physical constraints was integrated into the optimisation algorithm. For this control algorithm to work well, the propagation of variance over the prediction horizon is required [40], necessarily to utilise. An extension of this approach, where derivative observations are incorporated in a Gaussian process model was proposed [41]. This approach dramatically improves the efficiency, as well as robustness of the control loop.

The next section introduces the Gaussian process model for nonlinear internal model control. The use of the estimate of the model uncertainty is proposed to extend the nonlinear internal model control structure. A simulated example was used to illustrate the benefits of the proposed technique.

### 6.4.1    Gaussian process model for the internal model control

Internal model control is one of the most commonly used model-based techniques for the control of nonlinear systems. In this strategy the controller is chosen to be an inverse of the plant model. A general structure is shown in Figure 6.10. It was shown that with a perfect model, stability of closed-loop system is assured if the plant and the controller are stable. If the controller is chosen as the inverse of the model, perfect control can be achieved [10]. The filter in the loop defines the desired closed-loop behaviour of the closed-loop system and balances robustness to the model mismatch.

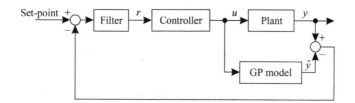

*Figure 6.10    General internal model control structure*

The main difference between the various internal model control approaches is in the choice of the internal model and its inverse. An analytical inverse of the nonlinear model, based on the physical understanding of the plant, was proposed in Reference 10. Also an analytical inverse of the local model network was shown in Reference 37. A numerical inverse, of the neural network models was utilised in Reference 42.

It was shown in Reference 11 that the Gaussian process model based on the covariance function of equation (6.25) is not analytically invertible. Instead of calculating the exact inverse, a numerical approach such as successive approximation or Newton–Raphson [42], can be used to find the control effort to solve the following equation:

$$f(\underline{u}(k), \underline{y}(k)) - r(k) = 0, \tag{6.42}$$

where

$$\underline{u}(k) = [u(k) \cdots u(k-m)]^{\mathrm{T}},$$
$$\underline{y}(k) = [y(k) \cdots y(k-n)]^{\mathrm{T}}, \tag{6.43}$$
$$f(\underline{u}(k), \ \underline{y}(k)) = \hat{y}(k+1),$$

and $r(k)$ is the controller input.

As an example of such an inverse, consider the first order nonlinear system:

$$y(t) = \frac{K(t)}{1 + p\tau(t)} u(t); \qquad p \triangleq \frac{\mathrm{d}}{\mathrm{d}t}; \begin{cases} K(t) = f(u(t), y(t)), \\ \tau(t) = g(u(t), y(t)). \end{cases} \tag{6.44}$$

The gain and time constant of the system change with the operating point. Figure 6.11 shows the open-loop system response at different operating points.

Simulation was used to generate 582 input–output training data points spanning the operating space. The sampling time was 0.5 s and the noise variance was set to $10^{-4}$. The Gaussian process model was trained as a one-step ahead prediction model. The internal model control strategy requires the use of the parallel model of equation $\hat{y}(k+1) = \hat{f}(u(k), \hat{y}(k))$. This Gaussian process model was then included in the internal model control structure and the numerical inverse of equation (6.42) was found at each sample.

As seen in Figure 6.12, the internal model control works well when the control input and the output of the system are in the region where the model was trained.

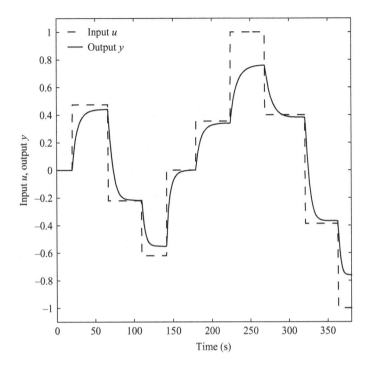

*Figure 6.11    Gain and time constant are nonlinear functions of the operating point*

As soon as the system moves away from the well-modelled region, equation (6.42) does not have a solution. This means that the control input $u$ on the left side of equation (6.42) cannot be found to force the right side of equation (6.42) to zero. The value of $u$ which drives equation (6.42) closest to zero is applied instead. This incorrect control input might drive the output of the model even further away from the well-modelled operating space. This behaviour can drive the closed-loop system to the point of instability, as can be seen in Figure 6.12, in the region after 60 s, where the model output does not follow the system output. The variance of the prediction, in this case, increases significantly.

### 6.4.1.1    Variance-constrained inverse of Gaussian process model

Since poor closed-loop performance is the result of the model being driven outside its trained region, the naive approach would be to constrain the control input. It could happen, however, that the operating space is not fully represented within the training data set. When the system is driven in this untrained portion of the operating space, the model will no longer represent the system well and the initial constraints of the control signal might not be relevant. However, the increase of the predicted variance will indicate a reduced confidence in the prediction. This increase of variance, can be used as a constraint in the optimisation algorithm utilised to solve equation (6.42). This concept is shown in Figure 6.13. The aim is to optimise the control effort, so that

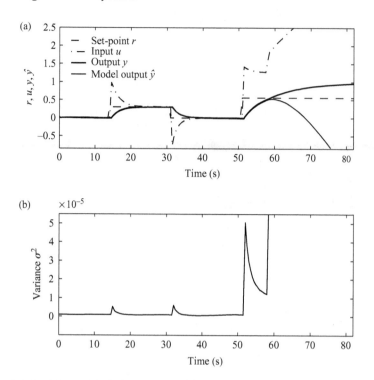

Figure 6.12   *Closed-loop response using internal model control strategy. Desired closed-loop behaviour was chosen as first-order response, with unity gain and time constant of 2 s*

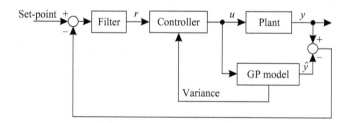

Figure 6.13   *Variance-constrained internal model control structure*

the variance does not increase above its predefined limit. A variance constraint should be defined by the designer and in general it can be a function of some scheduling variable. In this example, the variance constraint is set to be a constant value.

Constrained nonlinear programming with random restarts was used to optimise the control effort. In this example the variance was constrained to $4 \times 10^{-6}$. As seen in Figure 6.14, the closed-loop response has been improved. To achieve less aggressive,

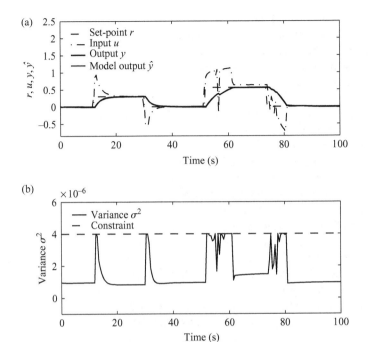

*Figure 6.14* *Improved closed-loop response using variance-constrained internal model control*

smoother response, it is possible to utilise the variance propagation [40] and use it as a constraint, instead of constraining the instantaneous variance.[8]

This example demonstrated a use of the prediction variance in internal model control, based on the Gaussian process model. It was shown that even for a simple, first-order nonlinear system, the model mismatch can produce an unstable closed-loop response. The predicted variance, as a measure of model mismatch, can be used as a constraint in the inversion algorithm, to improve the closed-loop response. The control effort can then be optimised, so that the variance does not increase above a predefined limit.

Since the Gaussian process model is not analytically invertible and numerical approaches have to be utilised to find the inverse of the model at each sample time, the associated computation load rises rapidly with the number of training data points. This is the main drawback of the Gaussian process modelling approach for internal model control.

## 6.5 Limitation of the Gaussian processes for control applications

The availability of a measure of uncertainty in the prediction, the small number of tuning parameters, model robustness and the ability to obtain a relatively good model when only a small training set is available are the main benefits of Gaussian process model in comparison to parametrical models and neural networks. Lack of transparency of the model and a large computation load are two main difficulties which limit the use of Gaussian process model for control applications. In this section these restrictions are discussed.

*Computation load.* To make a prediction using a Gaussian process model, an $N \times N$ covariance matrix needs to be inverted. Inversion of the $N \times N$ matrix has an associated computation cost of $\mathcal{O}(N^3)$. Training of the Gaussian process model also requires the inversion of the covariance matrix, which makes it time consuming for large training sets. The computation load associated with training of the model also increases with the number of hyperparameters. Since the number of hyperparameters is determined by the dimension of the input space, it is desirable, to choose the dimension of the input space as low as possible.

In many cases an identification task is a reconstruction of the dynamics of the modelled system from input–output samples. It was shown in Reference 34, that in a linear case, the order of the model $m_m$ should be selected as $m_m = m_o$, where $m_o$ is the order of an original system. The order of the model $m_m$ is the order of the difference equation that represents the physical system. The order of the modelled system $m_o$ is a length of the state vector of the continuous system. The state of a (linear or nonlinear) discrete-time system is the minimal amount of information about the past history of the system which is needed to predict its future behaviour [43]. Much work has been done on the estimation of model order of linear systems, if it is unknown [34,44].

It has been pointed out in Reference 35 that in order to accurately capture the dynamic of a nonlinear system a non-minimal realisation, with $m_m > m_o$ might be required. The Takens' embedding theorem for forced systems [45] provides a basis for the reconstruction of unknown dynamics by analysing the time-series data generated by a nonlinear system. In the noise-free case, a new state–space is constructed from time-series data. This new state–space can provide useful information about the selection of the order of the model. As shown in [55], on application of Takens' theorem to a model order selection problem, leads to the conclusion that for many practical applications $M_m$ should be searched over the interval $m_o \leq m_m \leq 2m_o + 1$. Theoretically, in the noise-free case, Takens' theorem provides sufficient conditions for the nonlinear, discrete-time model order selection if the order of the original system is known. However, no reference is made on how to estimate the order of the model in the case when the true order of the original nonlinear system is unknown [46].

From the Gaussian process point of view, the selection of the order of the model corresponds to the selection of the dimension of the input space. In the worst case, the dimension of the input space will be $2m_o + 1$. This increase of dimension of the input space from the minimal realisation will dramatically increase the computation load of the Gaussian process model.

*Transparency.* As with neural networks, Gaussian process models are also black-box in nature. Apart from the estimate of the variance of the noise, the hyperparameters themselves do not provide any interpretable information about the underlying system.

A level of interpretability of the original system from knowledge gained from the model is known as a degree of transparency. For many control applications, it is desirable to have a representation which is more transparent than the Gaussian process model presented so far.

*Invertibility.* The Gaussian process model based on covariance function (6.25) is not analytically invertible. As presented in Section 6.4.1, a numerical search for inverse needed for internal model control has to be utilised, which increases computation load. To be able to utilise an exact analytical inverse, the covariance function must be affine in the input. The simplest affine covariance function is a linear covariance function. Utilising a linear covariance function, the Gaussian process model would yield a linear approximation of the underlying system. To be able to model nonlinear systems using a linear covariance function, a 'divide and conquer' approach can be used.

## 6.6 Divide and conquer approach using Gaussian process models

The time $T_N$ required for training a Gaussian process model increases dramatically with the number of training data points $N$. The entire training data set can be divided into subsets, for each of which a simpler local model can be identified [35,47,48]. If each subset consists of less training points $N_{S_i} < N$, the associated computation time required for training of each local model would decrease dramatically: $T_{N_{S_i}} \ll T_N$. In order to cover an entire operating space, a number of local models for each subset have to be found. Taking into account the time required for clustering the training data $Tcl$ and assuming that each subset contains approximately the same number of data points, the time required for training all local models becomes $T_{N_{S_M}} = Tcl \sum_{i=1}^{M} T_{N_{S_i}}$, which significantly reduces the computation load. The overall time $T_{N_{S_M}}$ can be further reduced if the local models are trained simultaneously, using a parallel computer. These simpler local models are then blended to form a global network [4]. The drawback of this divide and conquer approach is that the operating space has to be partitioned by utilising an appropriate clustering algorithm, a number of local models have to be found and the network structure needs to be defined.

In general, any parametrisation of the covariance function can be used. In Reference 49 a nonlinear covariance function was used for local Gaussian process models and then these local models were combined to form a network. The advantage of this type of network is that it reduces the size of covariance matrices to the size of the training data for the individual subspace. This type of local model is very flexible, but since a nonlinear covariance function is used it is still a black-box model. If a linear covariance function is assumed, then the local Gaussian process models will each fit a plane to their data set. This linear Gaussian process (LGP) model,[9] can be seen as a linear local approximation of the nonlinear function for each subset of the input space. This allows a minimal realisation; $m_m = m_o$ for each submodel which provides a further significant reduction in the computation load.

In the following sections, an LGP model is derived and a local linear Gaussian process network (LGPN) model is introduced.

### 6.6.1   LGP model

If a linear relationship between input and output data is assumed, an ARX type of model can be written as:

$$\hat{y}(k+1) = a_1 y(k) + \cdots + a_p y(k-p+1)$$
$$+ b_1 u(k) + \cdots + b_q u(k-q+1) + d + \eta(k)$$
$$= \underline{\phi}(k)^{\mathrm{T}}\underline{\Theta} + \eta(k), \tag{6.45}$$

where:

$$\underline{\phi}(k) = [y(k) \cdots y(k-p+1)|u(k) \cdots u(k-q+1)|1]^{\mathrm{T}},$$
$$\underline{\Theta} = [a_1 \cdots a_p|b_1 \cdots b_q|d]^{\mathrm{T}}$$

and $\eta(k)$ is assumed to be a white zero-mean Gaussian noise of variance $\sigma_\eta^2$. The covariance function of the linear Gaussian process model can be written as follows:

$$C_{m,n} = E\{y(m), y(n)\} = \underline{\phi}(m)^{\mathrm{T}}\mathbf{C}_\Theta\underline{\phi}(n), \tag{6.46}$$

where

$$
\mathbf{C}_\Theta = \left[
\begin{array}{ccc|ccc|c}
a_1^2 & 0 & 0 & 0 & 0 & 0 & 0 \\
0 & \ddots & 0 & 0 & 0 & 0 & 0 \\
0 & 0 & a_p^2 & 0 & 0 & 0 & 0 \\
\hline
0 & 0 & 0 & b_1^2 & 0 & 0 & 0 \\
0 & 0 & 0 & 0 & \ddots & 0 & 0 \\
0 & 0 & 0 & 0 & 0 & b_q^2 & 0 \\
\hline
0 & 0 & 0 & 0 & 0 & 0 & d^2
\end{array}
\right]. \tag{6.47}
$$

Note, that $a_1^2, \ldots, a_p^2, b_1^2, \ldots, b_q^2, d^2$ can be seen as the hyperparameters and they have to be provided or trained from the data.

The covariance matrix $\mathbf{C}_N$ and vector $\underline{v}_{N+1}^{\mathrm{T}}$ can be written as follows:

$$\mathbf{C}_N = \mathbf{\Phi}_N \mathbf{C}_\Theta \mathbf{\Phi}_N^{\mathrm{T}} + \mathbf{I}\sigma_\eta^2, \tag{6.48}$$

$$\underline{v}_{N+1}^{\mathrm{T}} = \underline{\phi}(\kappa)^{\mathrm{T}}\mathbf{C}_\Theta \mathbf{\Phi}_N^{\mathrm{T}}. \tag{6.49}$$

The mean of the prediction equation can be written as:

$$\mu_{\mathrm{L}} = \mu_{\hat{y}(N+1)} = \underline{\phi}(\kappa)^{\mathrm{T}}\mathbf{C}_\Theta \mathbf{\Phi}_N^{\mathrm{T}}\mathbf{C}_N^{-1}\underline{y}_N = \underline{\phi}(\kappa)^{\mathrm{T}}\underline{\Theta}_{\mathrm{L}}, \tag{6.50}$$

where $\underline{\phi}(\kappa)^{\mathrm{T}}$ is a new test input and $\underline{\Theta}_{\mathrm{L}}$ is a constant vector, which is a function of the training data. The parameter vector $\underline{\Theta}_{\mathrm{L}}$ can be interpreted as parameters of a linear approximation of the modelled system. Unlike the hyperparameters, these parameters do provide a physical information about the system. This information therefore increases the degree of transparency.

The variance can be written as:

$$\sigma_{\mathrm{L}}^2 = \sigma_{\hat{y}(N+1)}^2 = \underline{\phi}(\kappa)^{\mathrm{T}}\mathbf{C}_\Theta\underline{\phi}(\kappa) - \underline{\phi}(\kappa)^{\mathrm{T}}\mathbf{C}_\Theta \mathbf{\Phi}_N^{\mathrm{T}}\mathbf{C}_N^{-1}\mathbf{\Phi}_N \mathbf{C}_\Theta^{\mathrm{T}}\underline{\phi}(\kappa)$$

$$= \underline{\phi}(\kappa)^{\mathrm{T}}\mathbf{V}_{\mathrm{L}}\underline{\phi}(\kappa), \tag{6.51}$$

where

$$\mathbf{V}_L = \mathbf{C}_\Theta - \mathbf{C}_\Theta \mathbf{\Phi}_N^T \mathbf{C}_N^{-1} \mathbf{\Phi}_N \mathbf{C}_\Theta^T. \tag{6.52}$$

The variance $\sigma_L^2$ depends on a new input vector $\underline{\phi}(\kappa)^T$ and the covariance matrix $\mathbf{V}_L$. $\mathbf{V}_L$ is a function of input data and does not depend on the target vector $\underline{y}_N$. The variance $\sigma_L^2$ and hence the standard deviation $\sigma_L$ is then a valid measure of locality of the model. This measure of locality of the local model is related to the uncertainty of the model prediction.

## 6.6.2  Local LGP network

### 6.6.2.1  Related work – mixtures of experts

Each LGP model represents a linear model of the process over a subset of the training data. In order to model a nonlinear process, a combination of a number of such local models to form a local model network is a well-established approach [4,35,47,48]. The Gaussian process model and consequently the LGP model are both probabilistic models. Many approaches for the mixture of probabilistic models (known as experts) have been proposed. An expert is referred to a probabilistic model which models a subset of the data and hence in this representation, an expert is equivalent to a local model. In Reference 50, e.g. the experts were each chosen to be a linear function of the inputs, and the outputs of each expert were combined to form the overall model output. The outputs of the experts proceed up the tree, being blended by the gating network outputs. This tree-structured mixture of experts, proposed in Reference 48 is called 'generalised linear' network. A similar approach was taken in Reference 49, where the experts were the Gaussian process models, each employing a nonlinear covariance function.

### 6.6.2.2  Blending the outputs of linear local Gaussian process models

The generalised linear network can be modified to give the multiple local models network structure shown in Figure 6.15. Here the local models are chosen to be LGP models. The output of the network is the weighted sum of outputs of each local model. Each of the weights $W_1, \ldots, W_M$ can be determined according to some scheduling vector $\underline{\psi}$. $W_i$ is then defined as:

$$W_i = \varphi_i(\underline{\psi}), \tag{6.53}$$

where $\varphi_i(\cdot)$, is some validity function.

In this representation, the output predictive distribution is a mixture of Gaussians and is in general multimodal, containing two or more peaks [51]. Multimodality of the distribution is often a strong indication that the distribution is not Gaussian. The mean and variance of a mixture of Gaussians is given by its first and second moments [49,50,52,53]:

$$\mu^* = \sum_{i=1}^{M} W_i \mu_{L_i} \tag{6.54}$$

$$\sigma^{*2} = \sum_{i=1}^{M} W_i(\sigma_{L_i}^2 + \mu_{L_i}^2) - \mu^{*2}. \tag{6.55}$$

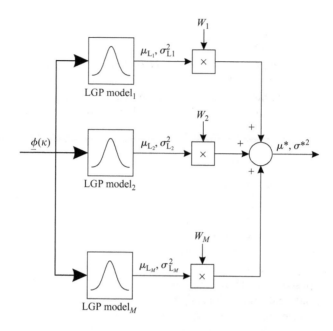

*Figure 6.15*   *Mixture of Gaussians can be modified to form a local model network, where the outputs of the local models are blended to form a predictive distribution*

Note, that the variance for a mixture of Gaussians as given in equation (6.55) is not the same as the sum of weighted variances:

$$\sigma^{*2} \neq \sum_{i=1}^{M} W_i \sigma_{L_i}^2. \tag{6.56}$$

Figure 6.16 shows a mixture of two Gaussian distributions of means $\mu_1 = -0.5$, $\mu_2 = 0.5$ and standard deviations $\sigma_1 = 0.25$ and $\sigma_2 = 0.5$. The weights are chosen to be equal $W_1 = W_2 = \frac{1}{2}$. It is clear, that the resulting mixture of distributions is not Gaussian.

Although the predictive distribution is multimodal, its mean and variance can still be calculated utilising equations (6.54) and (6.55). Therefore $\mu^* = 0$ and $\sigma_*^2 = 1.875$.

### 6.6.2.3   Blending the parameters of linear local Gaussian process models

A linear local Gaussian process model is fully defined by the vector of parameters $\underline{\Theta}_L$ and the covariance matrix $V_L$. Given a set of linear local Gaussian process models, instead of using the weighted sum of model outputs, the weighted sum of parameters can be used to form a local model network. This architecture is shown in Figure 6.17. Again the weights can be formed from blending functions, $W_i = \varphi_i(\underline{\psi})$, where $\underline{\psi}$ is a scheduling vector.

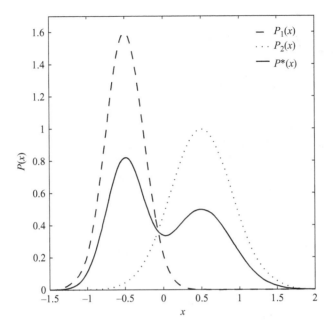

*Figure 6.16    Weighted sum of Gaussian distributions results in a non-Gaussian distribution*

In this case, the overall network will behave as a single LGP model for a particular choice of weights. Therefore, the predictive distribution will be Gaussian defined by its mean and variance:

$$\mu_L^* = \phi(\kappa)^T \sum_{i=1}^{M} W_i \Theta_{L_i} = \phi(\kappa)\Theta_L^* \tag{6.57}$$

$$\sigma_L^{*2} = \phi(\kappa)^T \left( \sum_{i=1}^{M} W_i \mathbf{V}_{L_i} \right) \phi(\kappa) = \phi(\kappa)^T \mathbf{V}_L^* \phi(\kappa). \tag{6.58}$$

Figure 6.18 shows a weighted mixture of two Gaussian distributions, where the parameters of local models are blended: $\mu_1 = -0.5$, $\mu_2 = 0.5$, $\sigma_1 = 0.25$, $\sigma_2 = 0.5$, $W_1 = W_2 = 0.5$. As it can be seen, the predictive distribution is Gaussian with the same mean as when the outputs are blended, $\mu_L^* = \mu^* = 0$, but different variance $\sigma_L^{*2} = 0.325 \neq \sigma^{*2}$.

For a particular choice of weights, the mixture of parameters of the network is defined as a linear combination of parameters of each local model. An observation of the parameters can provide useful information about the system at the particular operating point. This means that when the local LGP models are used the parameters of the network carry information of the linear approximation of the underlying system at each particular operating point. In comparison with output blending, a network which uses blending of the parameters has greater transparency.

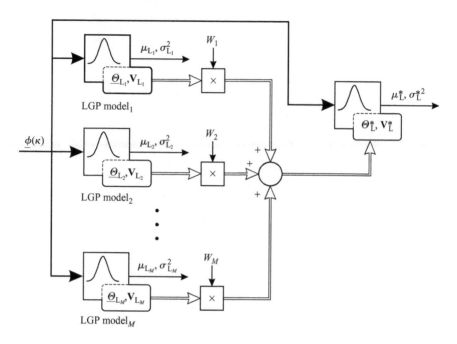

*Figure 6.17    Linear local Gaussian process network structure, where the parameters*
*of the local models are blended to form a predictive distribution*

### 6.6.3    Network structure optimisation

In order to form a global model, local models need to be blended using validity
functions. For this local model network architecture, the structure of the network
has to be well defined. This involves determination of the position of the validity
functions and the optimisation of their widths – this is a well-known problem related
to local model networks. The provided measure of locality of the local LGPs can be
used to improve the existing structure optimisation algorithms.

By combining the outputs of the local models, the predictive distribution is not
Gaussian and equation (6.55) clearly shows that the variance $\sigma^{*2}$ is not only dependent
on the training input points and a test point, but also on predictions of the local models.
The predictions of the local models can be far away from the training inputs. How far
away from the training inputs the predictions of the local models are depends on the
underlying function. This means that in the region between two local models, where
the uncertainty of the network prediction is expected to be higher, the variance does
not necessarily increase. This shows, that $\sigma^{*2}$ does not provide a valid measure of
network uncertainty in the prediction.

When the parameters of the local models are combined, the predictive distribution
will be Gaussian and the variance is now a valid measure of network uncertainty. This
can be used as a measure of the locality of each local model in the parameter blending
network architecture to help in the optimisation of the network structure. The validity

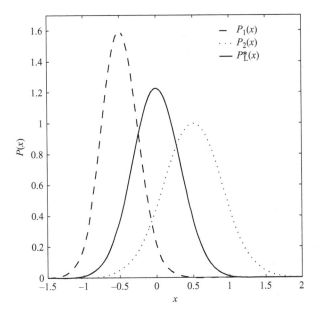

*Figure 6.18    Weighted sum of parameters of Gaussian distributions results in a Gaussian distribution*

functions $\varphi_i(\cdot)$, which generate weights $W_i$, can be optimised so that the uncertainty of the network is minimised.

When a parameter blending network architecture is assumed, then the uncertainty envelope of the linear local Gaussian process network can be defined as:

$$\varepsilon_L^* = \sum_{k=1}^{N} \sum_{i=1}^{M} (\mu_{L_i}(k) \pm \sigma_{L_i}(k))\varphi_i(\underline{\psi}(k)). \tag{6.59}$$

The upper envelope boundary can be defined as:

$$^+\varepsilon_L^* = \sum_{k=1}^{N} \sum_{i=1}^{M} (\mu_{L_i}(k) + \sigma_{L_i}(k))\varphi_i(\underline{\psi}(k)), \tag{6.60}$$

where $\varphi_i(\cdot)$ is the validity function, which defines the weighting of network parameters. The validity functions are frequently chosen to be Gaussian in shape:

$$\varphi_i(\underline{\psi}(k)) = \exp\left(-\frac{1}{2}\frac{(\psi(k) - \underline{c}_i)^{\mathrm{T}}(\psi(k) - \underline{c}_i)}{\sigma_i^2}\right), \tag{6.61}$$

where $\sigma_i$ is the width and $c_i$ is the centre of the validity function.

As can be seen from equation (6.60), the envelope is a function of the predictions, uncertainties and the validity functions of each local model. Therefore, the envelope is a combination of the orientation of the local plane and its uncertainty. If the position of each local model is fixed and if the parameters of local models are known in advance, minimising the cost of equation (6.62) by varying the widths of validity functions will result in a model, which will have the envelope as close as possible to its prediction. This 'envelope tightening' approach, first proposed in Reference 12, will also produce the best fit to the data. It is assumed here, that the positions of the local models in the input space are well defined with the centres of the validity functions set. The widths of the $\varphi_i(\cdot)$ are free to be optimised. The following modified cost function is proposed:

$$J = \sum_{k=1}^{N} \sum_{i=1}^{M} (\mu_{L_i}(k) + R\sigma_{L_i}(k))\varphi_i(\underline{\psi}(k)),$$
(6.62)

where $R$ is an appropriate regularisation constant. The choice of $R$ plays a major role in the performance of the algorithm, as it defines the impact of the variance on the curvature of the cost function. If $R$ is too small, the prediction part of the cost curvature will dominate. If $R$ is too large, the cost will be dominated by the variance, which will produce an averaging effect yielding validity regions for the local models, which will overlap too much.

### 6.6.3.1   Importance of normalisation of the validity functions

In order to optimise the structure of the network, using the proposed algorithm, normalised validity functions need to be used. If the validity functions are not normalised, the minimum envelope will be achieved when the widths are as small as possible. In this case the optimisation algorithm will force the widths to be zero.

To highlight this problem, consider an example, where three linear local Gaussian process models are used to approximate the nonlinear function $y = f(u)$, shown in Figure 6.19(a). An attempt to optimise the non-normalised validity functions was made. In order to examine the 'shrinking' effect, shown in Figure 6.19(b), the optimisation algorithm was stopped before the widths completely reduced to zero. The prediction is close to the modelled function only in the small regions where the validity functions are centred. Away from the centres, the output of the model is zero. At the same time, the envelope goes to zero, which does not indicate an inaccurate prediction. To prevent the widths shrinking to zero, and also to achieve a partition of unity, the validity function must be normalised. Normalisation effectively means that, across the operating space the sum of the contributions of the validity functions is one. This introduces an interaction between the validity functions which will prevent the widths shrinking to zero independently, as shown in Figure 6.19(c). At the edge of the training set a single (closest) local model is activated. It is obvious that this single local model is not an accurate approximation of the modelled function, which is indicated by significant increase of the envelope. Side effects of normalisation are discussed in Reference 4.

As expected, the network prediction is accurate in the region, where local models are available. As shown in Figure 6.19(c), in the region between the local models, the prediction is less accurate, which is indicated by a widening of the envelope.

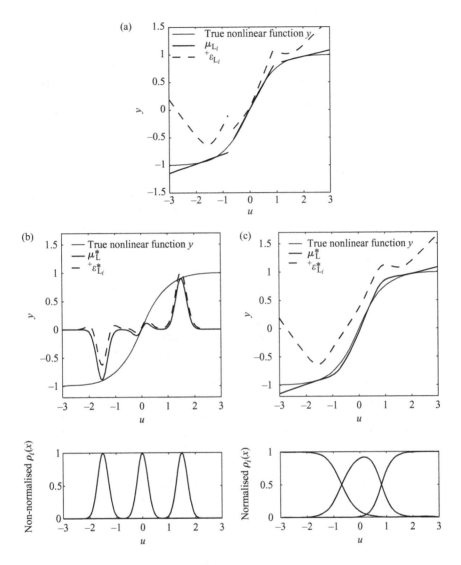

*Figure 6.19* (a) *Three linear local Gaussian process models and their uncertainty envelopes estimated at different points. The local models are a good linear approximation of the nonlinear function at the points where they were identified. The envelope is closer to the prediction at these points. Away from the well-modelled region, the envelope increases significantly. (b) If the validity functions are not normalised, the optimisation algorithm will force the widths of the validity functions to zero. (c) Normalisation of the validity functions assures good prediction across the well-modelled region. Outside the well-modelled region (note intervals* $(-\infty, -2]$ *and* $[2, \infty))$, *the increase of the envelope indicates inaccurate prediction*

### 6.6.4   *Possible use of the linear local Gaussian process network for control*

In many real industrial applications in the chemical, process and pharmaceutical industries the controlled system operates at a number of different operating points. Therefore, a large amount of data is available around the local equilibrium points. In this case it makes sense to utilise a divide and conquer approach. The linear local Gaussian process network is a possible application of the divide and conquer approach. In general this network is identical to the local model network architecture, with the added ability to provide a measure of model uncertainty.

The network optimisation using the envelope tightening algorithm, described above, works when only local data, collected at the equilibrium points is available. The interpolation between operating region is based on blending the local models to minimise the variance. In order to optimise a traditional local model network, off-equilibrium data is needed. Off-equilibrium data can be difficult to provide in real industrial processes.

A propagation of the variance, when the linear covariance function is utilised is in general simpler [54] than the variance propagation when a nonlinear covariance function is used. Simple variance propagation makes the linear local Gaussian Process network useful for model predictive control, where the Gaussian process model can be replaced with the linear local Gaussian process network. In comparison to the standard Gaussian process model, the linear local Gaussian process network is less computationally expensive and more transparent.

The biggest drawback of the Gaussian process model for internal model control is the increase of computation load for numerical search of the inverse. It has been shown in Reference 12 that if the local models are invertible, the linear local Gaussian process network is analytically invertible. Since the orientation of the inverted local models changes in comparison to original local models, the validity functions need to be re-optimised, using the envelope tightening algorithm. Utilisation of the inverse of the linear local Gaussian process network would dramatically reduce the computation load for the internal model control strategy.

## 6.7   Conclusions

Difficulties, such as lack of transparency and the limited ability to provide a measure of model uncertainty, related to parametrical models, such as neural networks, fuzzy models and local model networks, were pointed out in this chapter. Highlighting these problems should not be seen as a criticism of parametrical models and neural networks, but as an attempt to improve modelling techniques for control by utilising a Bayesian approach. An overview of Bayesian regression principles was given at the beginning of this chapter. The formulation of the Gaussian process model for modelling of nonlinear systems was presented and some training algorithms for Gaussian processes were explained. It was shown that as well as the prediction, the Gaussian process model also provides information about the confidence in its prediction.

This can be useful in model-based control applications, e.g. in model predictive control and internal model control to improve the closed-loop performance. The difficulty of a large computation load related to the Gaussian process model and the problem of selection of the input space dimension was explained. A linear local Gaussian process network, based on the divide and conquer approach was introduced, as an alternative, to reduce computation load and increase the transparency. Linear Gaussian process local models were used to form the network. The difference between combining the outputs and combining the parameters of the local models to form an overall network was shown. As with the standard Gaussian process model, this network provides uncertainty in prediction, which makes it very useful for control applications. The envelope tightening algorithm for the network structure optimisation, based on locally collected data, was explained and possible use of the linear local Gaussian process network for control applications was proposed.

## Acknowledgements

Gregor Gregorčič acknowledges the support of Ad futura, Science and Education Foundation of the Republic of Slovenia, grant 521-15/2002. The authors are grateful for comments and useful suggestions from Juš Kocijan and Cillian O'Driscoll.

## Notes

1. Occam's Razor is a logical principle attributed to the mediaeval philosopher William of Occam (also spelled Ockham). William of Occam was the most influential philosopher of the 14th century and a controversial theologian. His principle states: 'One should not increase, beyond what is necessary, the number of entities required to explain anything.'
2. Parameters which control the distribution are often referred to as hyperparameters.
3. A $p \times p$ real matrix $\mathbf{A}$ is called positive definite if $\underline{b}^T \mathbf{A} \underline{b} > 0$, for all non-zero vectors $\underline{b} \in \Re^p$.
4. Also known as evidence maximisation.
5. The integral is approximated by sum over the samples. See References 7, 8 and 22 for details.
6. It is important to emphasise here that the name 'Gaussian process model' does not refer to choice of the form of the basis function.
7. The matrix $\mathbf{G}^T \mathbf{G}$ is, or it is close to, a rank deficient.
8. Variance in this case can be seen as one-step ahead predicted variance, where the present information is not fed back to the input of the model and does not affect the variance prediction in the next step.
9. In this section, an LGP model is a Gaussian process model utilising a linear covariance function. This type of Gaussian process model is also known as a 'linear in parameters Gaussian process' model [27].

## References

1 NARENDRA, K. S., and PARTHASARATHY, K.: 'Identification and control of dynamical systems using neural networks', *IEEE Transactions on Neural Networks*, 1990, **1** (1), pp. 4–27

2 TAKAGI, T., and SUGENO, M.: 'Fuzzy identification of systems and its applications to modeling and control', *IEEE Transactions on Systems, Man and Cybernetics SMC*, 1985, **15** (1), pp. 116–32

3 ABONYI, J.: 'Fuzzy model identification for control' (Birkhäuser, Boston, MA, 2003)

4 MURRAY-SMITH, R., and JOHANSEN, T. A. (Eds): 'Multiple model approaches to modelling and control' (Taylor & Francis, London, 1997)

5 JOHANSEN, T. A., MURRAY-SMITH, R., and SHORTEN, R.: 'On transient dynamics, off-equilibrium behaviour and identification in blended multiple model structures', European Control Conference, Karlsruhe, Germany, 1999

6 O'HAGAN, A.: 'Curve fitting and optimal design for prediction (with discussion)', *Journal of the Royal Statistical Society B*, 1978, **40** (1), pp. 1–42

7 RASMUSSEN, C. E.: 'Evaluation of Gaussian processes and other methods for non-linear regression'. Ph.D. thesis, University of Toronto, 1996

8 GIBBS, M. N.: 'Bayesian Gaussian processes for regression and classification'. Ph.D. thesis, University of Cambridge, 1997

9 WILLIAMS, C. K. I.: 'Prediction with Gaussian processes: from linear regression to linear prediction and beyond', in JORDAN, M. I. (Ed.): 'Learning and inference in graphical models' (Kluwer Academic Press, Dordrecht, 1998)

10 ECONOMOU, C. G., MORARI, M., and PALSSON, B. O.: 'Internal model control: extension to non-linear systems', *Industrial and Engineering Chemical Process Design and Development*, 1986, **25**, pp. 403–11

11 GREGORČIČ, G., and LIGHTBODY, G.: 'Internal model control based on a Gaussian process prior model'. Proceedings of the *2003 American Control Conference*, Denver, CO, June 2003, pp. 4981–6

12 GREGORČIČ, G., and LIGHTBODY, G.: 'An affine local Gaussian process model network'. Proceedings of the IEE International Conference on *Systems Engineering*, Coventry, UK, September 2003, pp. 206–10

13 MACKAY, D. J. C.: 'Bayesian methods for adaptive models'. Ph.D. thesis, California Institute of Technology, 1991

14 RASMUSSEN, C. E., and GHAHRAMANI, Z.: 'Occam's razor', in LEEN, T., DIETTERICH, T. G., and TRESP, V. (Eds): 'Advances in neural information processing systems 13' (MIT Press, Cambridge, MA, 2001) pp. 294–300

15 JOHANSEN, T. A.: 'Identification of non-linear systems using empirical data and prior knowledge – an optimization approach', *Automatica*, 1996, **32**, pp. 337–56

16 JOHANSEN, T. A.: 'On tikhonov regularization, bias and variance in nonlinear system identification', *Automatica*, 1997, **33**, pp. 441–6

17 BISHOP, C. M., and TIPPING, M. E.: 'Bayesian regression and classification', in SUYKENS, J., HORVATH, G., BASU, S., MICCHELLI, C., and

VANDEWALLE, J. (Eds): 'Advances in learning theory: methods, models and applications', vol. 190 (IOS Press, NATO Science Series III: Computer and Systems Sciences, 2003) pp. 267–85

18 VON MISES, R.: 'Mathematical theory of probability and statistics' (Academic Press, New York, 1964)

19 BARNETT, S.: 'Matrix methods for engineers and scientists' (McGraw-Hill, London, 1979)

20 WILLIAMS, C. K. I., and RASMUSSEN, C. E.: 'Gaussian processes for regression', in HASSELMO TOURETZKY, M. E., and MOZER, M. C. (Eds): 'Advances in neural information processing systems 8' (MIT Press, Cambridge, MA, 1996) pp. 514–20

21 MURRAY-SMITH, R., and GIRARD, A.: 'Gaussian process priors with arma noise models'. Proceedings of *The Irish Signals and Systems Conference*, Maynooth, June 2001, pp. 147–52

22 MACKAY, D. J. C.: 'Introduction to Monte Carlo methods'. A review paper in the proceedings of an Erice summer school, 1999

23 MACKAY, D. J. C.: 'Comparison of approximate methods for handling hyper-parameters', *Neural Computation*, 1999, **11** (5), pp. 1035–68

24 GREGORČIČ, G., and LIGHTBODY, G.: 'From multiple model network to Gaussian processes prior models', in RUANO, A. E. (Ed.): Proceedings of the IFAC international conference on *Intelligent Control Systems and Signal Processing ICONS2003*, Centre for Intelligent Systems, University of Algarve, Portugal, April 2003, pp. 149–54

25 MACKAY, D. J. C.: 'Information theory, inference and learning algorithms' (Cambridge University Press, London, 2003)

26 STRANG, G.: 'Linear algebra and its applications' (Harcourt College Publishers, 1988, 3rd edn)

27 MURRAY-SMITH, R., and SBARBARO, D.: 'Nonlinear adaptive control using nonparametric Gaussian process models'. International federation of automatic control, 15th IFAC Triennial World Congress, 2002

28 MURRAY-SMITH, R., SBARBARO, D., RASMUSSEN, C. E., and GIRARD, A.: 'Adaptive, cautious, predictive control with Gaussian process priors', 13th IFAC symposium on *System Identification*, Rotterdam, 2003

29 LEONARD, J. A., KRAMER, M. A., and UNGAR, L. H.: 'Using radial basis function to approximate a function and its error bounds', *IEEE Transactions on Neural Networks*, 1992, **3** (4), pp. 624–7

30 QAZAZ, S. C., WILLIAMS, C. K. I., and BISHOP, C. M.: 'An upper bound on the Bayesian error bars for generalized linear regression'. Technical report NCRG/96/005, Neural Computing Research Group, Aston University, Birmingham, UK, 1995

31 WILLIAMS, C. K. I., QAZAZ, C., BISHOP, C. M., and ZHU, H.: 'On the relationship between Bayesian error bars and the input data density'. Proceedings of the fourth international conference on *Artificial Neural Networks*, 1995, pp. 165–9

32 KOCIJAN, J., GIRARD, A., BANKO, B., and MURRAY-SMITH, R.: 'Dynamic systems identification with Gaussian processes'. Proceedings of fourth Mathmod, Vienna, 2003, pp. 776–84

33 BRONSHTEIN, I. N., SEMENDYAYEV, K. A., MUSIOL, G., and MUHLIG, H.: 'Handbook of mathematics' (Springer, Berlin, 2004, 4th edn)

34 LJUNG, L.: 'System identification: theory for the user' (Prentice Hall PTR, Upper Saddle River, NJ, 1999, 2nd edn)

35 LEITH, D. J., LEITHEAD, W. E., SOLAK, E., and MURRAY-SMITH, R.: 'Divide & conquer identification using Gaussian process priors'. Conference on *Decision and Control*, Las Vegas, 2002

36 GREGORČIČ, G., and LIGHTBODY, G.: 'A comparison of multiple model and pole-placement self-tuning for the control of highly nonlinear processes', in FAGAN, A., and FEELY, O. (Eds): Proceedings of the *Irish Signals and Systems Conference*, Dublin, Ireland, 2000, pp. 303–11

37 BROWN, M. D., LIGHTBODY, G., and IRWIN, G. W.: 'Nonlinear internal model control using local model networks', *IEE Proceedings: Control Theory and Applications*, 1997, **144** (6), pp. 505–14

38 KOCIJAN, J., MURRAY-SMITH, R., RASMUSSEN, C. E., and LIKAR, B.: 'Predictive control with Gaussian process models'. The IEEE Region 8 EUROCON 2003: *Computer as a Tool*, University of Ljubljana, Slovenia, September 2003, pp. 352–6

39 KOCIJAN, J., MURRAY-SMITH, R., RASMUSSEN, C. E., and GIRARD, A.: 'Gaussian process model based predictive control'. Proceedings of the *2004 American Control Conference*, Boston, MA, USA, June 2004, pp. 2214–18

40 GIRARD, A., RASMUSSEN, C., QUINONERO CANDELA, J., and MURRAY-SMITH, R.: 'Multiple-step ahead prediction for non linear dynamic systems – a Gaussian process treatment wih propagation of the uncertainty', in BECKER, S., THRUN, S., and OBERMAYER, K. (Eds): 'Advances in neural information processing systems' (MIT Press, Cambridge, MA, 2002) pp. 545–52

41 KOCIJAN, J., and LEITH, D. J.: 'Derivative observations used in predictive control'. Proceedings of the 12th *IEEE Mediterranean Electrotechnical Conference, MELECON 2004*, Dubrovnik, Croatia, May 2004, pp. 379–82

42 NAHAS, E. P., HENSON, M. A., and SEBORG, D. E.: 'Non-linear internal model control strategy for neural network models', *Computers and Chemical Engineering*, 1992, **16**, pp. 1039–57

43 KALMAN, R. E.: 'A new approach to linear filtering and prediction problems', *Transactions of the ASME – Journal of Basic Engineering*, 1960, **82** (Series D), pp. 35–45

44 LARIMORE, W. E.: 'Optimal reduced rank modeling, prediction, monitoring, and control using canonical variate analysis'. IFAC 1997 international symposium on *Advanced Control of Chemical Processes*, Banff, Canada, 1997, pp. 61–6

45 STARK, J., BROOMHEAD, D. S., DAVIES, M. E., and HUKE, J.: 'Delay embeddings of forced systems: Ii stochastic forcing', *Journal of Nonlinear Science*, 2003, **13** (6), pp. 519–77

46  DODD, T. J., and HARRIS, C. J.: 'Identification of nonlinear time series via kernels', *International Journal of System Science*, 2002, **33** (9) pp. 737–50

47  LEITH, D. J., and LEITHEAD, W. E.: 'Analytic framework for blended multiple model systems using linear local models', *International Journal of Control*, 1999, **72** (7/8), pp. 605–19

48  JORDAN, M. I., and JACOBS, R. A.: 'Hierarchical mixtures of experts and the EM algorithm', *Neural Computation*, 1994, **6**, pp. 181–214

49  SHI, J. Q., MURRAY-SMITH, R., and TITTERINGTON, D. M.: 'Bayesian regression and classification using mixtures of Gaussian processes', *International Journal of Adaptive Control and Signal Processing*, 2003, **17** (2), pp. 149–61

50  WATERHOUSE, S., MACKAY, D., and ROBINSON, T.: 'Bayesian methods for mixtures of experts', in TOURETZKY, D. S., MOZER, M. C., and HASSELMO, M. E. (Eds): 'Advances in neural information processing systems', vol. 8 (The MIT Press, Cambridge, MA, 1996) pp. 351–7

51  UPTON, G., and COOK, I.: 'Introducing statistics' (Oxford University Press, Oxford, 2001)

52  BISHOP, C. M.: 'Mixture density networks'. Technical report NCRG/94/004, Neural Computing Research Group, Aston University, 1994

53  PENNY, W. D., HUSMEIER, D., and ROBERTS, S. J.: 'The Bayesian paradigm: second generation neural computing', in LISBOA, P. (Ed.): 'Artificial neural networks in biomedicine' (Springer-Verlag, Heidelberg, 1999)

54  KOCIJAN, J., GIRARD, A., and LEITH, D. J.: 'Incorporating linear local models in Gaussian process model'. Technical report DP-8895, Jožef Stefan Institute, Ljubljana, Slovenia, 2003

55  GREGORČIČ, G.: 'Data-based modelling of nonlinear systems for control'. Ph.d. thesis, University College Cork, 2004

*Chapter 7*

# Neuro-fuzzy model construction, design and estimation

*X. Hong and C. J. Harris*

## 7.1 Introduction

Over the past two decades there has been a strong resurgence in the field of data based nonlinear system modelling and identification involving researchers from diverse disciplines. Mathematical equations, or functional mappings, are constructed from observational data from diverse fields, such as electrical engineering, chemical processes, biomedical systems or economic systems. In engineering a class of artificial neural networks (ANN) have been proven to be capable of representing a class of unknown nonlinear input–output mappings with arbitrary small approximation error capability [1,2]. The ANN are well structured for adaptive learning, have provable learning and convergence conditions (for at least linear-in-the-parameters networks), and have the capability of parallel processing and good generalisation properties for unseen data. An historically alternative approach is a fuzzy logic (FL) system [3,4] which has found significant practical applications in domestic products, FL controllers, pattern recognition and information processing. The advantage of the FL approach is its logicality and transparency, where it is easy to incorporate *a priori* knowledge about a system into an explicit fuzzy rule base. Additionally FL represents a form of imprecision and uncertainty common in linguistics through expressions of vagueness. A neuro-fuzzy theory [4–9] brings the ideas of ANN and FL together in a cohesive framework, such that the resultant model has a linear-in-the-parameters networks structure for learning properties, and is associated with a fuzzy rule base about the generated data knowledge.

Fuzzy systems [3,10–12] involve the procedures of fuzzification of a deterministic input signal with membership function, reasoning in a fuzzy rule set using a proper inference method and a defuzzification process to produce a deterministic output, each process while non-unique having an influence on the system's input–output mapping and overall performance. Usually in the fuzzy system both the antecedent and consequent part of the fuzzy rules are fuzzy subsets. Also particular choice of the membership sets, logical operators and defuzzification scheme lead to the fuzzy logic system having an equivalence to a class of analytical ANNs [5,9] (see also Section 7.2). However, in general FL systems' input–output relationships are highly nonlinear and there exist few mathematical methods to deal with them analytically. Takagi and Sugeno [13] have introduced a crisp type rule model in which the consequent parts of the fuzzy rules are crisp functional representations, or crisp real numbers in the simplified case, instead of fuzzy sets. The Takagi–Sugeno (T–S) fuzzy model has found widespread application in control and local modelling due to its nice analytical properties [14] for representing input–output processes.

The class of fuzzy logic systems which have an equivalence to certain ANNs (usually linear-in-the-parameters, single layer networks such as radial basis function (RBF), B-spline, ...) are called neuro-fuzzy systems. Most conventional neural networks lead only to 'black-box' model representation, yet the neuro-fuzzy network has an inherent model transparency that helps users to understand the system behaviours, oversee critical system operating regions, and/or extract physical laws or relationships that underpin the system. By using the B-spline functional based neuro-fuzzy approach [5,6], unknown dynamical systems are modelled in the form of a set of fuzzy rules in which numerical and linguistic data (usually *a priori* knowledge) are readily integrated. The basis functions are interpreted as a fuzzy membership function of individual rules, and the model output is decomposed into a convex combination (via the fuzzy membership function) of individual rules' outputs. This property is critically desirable for problems requiring insight into the underlying phenomenology, i.e. internal system behaviour interpretability and/or knowledge (rule) representation of the underlying process. Moreover, based on the fuzzy rules inference and model representation of Takagi and Sugeno [13], the neuro-fuzzy model can be functionally expressed as an operating point dependent fuzzy model with a local linear description that lends itself directly to conventional estimation and control synthesis [6,15,16]. In particular, the operating point dependent structure can be represented by a state–space model, enabling the ready applications of neuro-fuzzy networks for data fusion [17] or data based controller design via the well-known Kalman filtering state estimator [17,18].

The problem of the curse of dimensionality [19] has been a main obstacle in nonlinear modelling using associative memory networks or FL, whereby the network complexity increases exponentially as the number of dependent variables increases constraining practical applications of fuzzy logic to two-dimensional problems. Networks or knowledge representations that suffer from the curse of dimensionality include all lattice-based networks such as FL, RBF, Karneva distributed memory maps and all neuro-fuzzy networks (e.g. adaptive network-based fuzzy inference system (ANFIS) [7], Takagi and Sugeno model [13], etc.). This problem also militates against model transparency for high-dimensional systems since they generate massive rule sets, or require too many parameters, making it impossible for a human to comprehend the resultant rule set. Consequently the major purpose of neuro-fuzzy model construction algorithms is to select a parsimonious model structure that resolves the bias/variance dilemma (for finite training data), has a smooth prediction surface (e.g. parameter control via regularisation), produces good generalisation (for unseen data), and with an interpretable representation – often in the form of (fuzzy) rules. An obvious approach to dealing with a multivariate function is to decompose it into a sum of lower-dimensional functions. A model construction algorithm, called adaptive spline modelling of observational data (ASMOD) has been derived [20,21] based on the analysis of variance (ANOVA). For general linear in the parameter systems, an orthogonal least squares (OLS) algorithm based on Gram–Schmidt orthogonal decomposition can be used to determine the model's significant elements and associated parameter estimates and the overall model structure [22]. Regularisation techniques have been incorporated into the OLS algorithm to produce a regularised orthogonal least squares (ROLS) algorithm that reduces the variance of parameter estimates [23,24]. To produce a model with good generalisation capabilities, model selection criteria such as the Akaike information criterion (AIC) [25] are usually incorporated into the procedure to determine the model construction process. Yet the use of AIC or other information based criteria, if used in forward regression, only affects the stopping point of the model selection, but does not penalise regressors that might cause poor model performance, e.g. too large parameter variance or ill-posedness of the regression matrix, if this is selected. This is due to the fact that AIC or other information based criteria are usually simplified measures derived as an approximation formula that is particularly sensitive to model complexity.

In order to achieve a model structure with improved model generalisation, it is natural that a model generalisation capability cost function should be used in the overall model searching process, rather than only being applied as a measure of model complexity. In optimum experimental design [26], model adequacy is evaluated by design criteria that are statistical measures of goodness of experimental designs by virtue of design efficiency and experimental effort. Quantitatively, model adequacy is measured as function of the eigenvalues of the design matrix. In recent studies [27,28], composite cost functions have been introduced to optimise the model approximation ability using the forward OLS algorithm [22], and simultaneously determine model adequacy using an A-optimality design criterion (i.e. minimises the variance of the parameter estimates), or a D-optimality criterion (i.e. optimises the parameter efficiency and model robustness via the maximisation of the determinant of the design matrix). It was shown that the resultant models can be improved based on A- or D-optimality. Combining a locally regularised orthogonal least squares (LROLS) model selection [29] with D-optimality experimental design further enhances model robustness [30].

Due to the inherent transparency properties of a neuro-fuzzy network, a parsimonious model construction approach should lead also to a logical rule extraction process that increases model transparency, as simpler models inherently involve fewer rules which are in turn easier to interpret. One drawback of most current neuro-fuzzy learning algorithms is that learning is based upon a set of one-dimensional regressors, or basis functions (such as B-splines, Gaussians, etc.), but not upon a set of fuzzy rules (usually in the form of multidimensional input variables), resulting in opaque models during the learning process. Since modelling is inevitably iterative it can be greatly enhanced if the modeller can interpret or interrogate the derived rule base during learning itself, allowing him or her to terminate the process when his or her object-ives are achieved. There are valuable recent developments on rule based learning and model construction, including a linear approximation approach combined with uncer-tainty modelling [31], and various fuzzy similarity measures combined with genetic algorithms [32,33]. Recently we introduced a new neuro-fuzzy model construction and parameter estimation algorithm based on a T–S inference mechanism [34,35]. Following establishing a one to one mapping between a fuzzy rule base and a model matrix feature subspace, an extended Gram–Schmidt orthogonal decomposi-tion algorithm combined with D-optimality and parameter regularisation is introduced for the structure determination and parameter estimation of *a priori* unknown dynam-ical systems in the form of a set of fuzzy rules [35]. The proposed algorithm enhances the previous algorithm [35] via the combined LOLS and D-optimality for robust rule selection, and extends the combined LOLS and D-optimality algorithm [30] from conventional regressor regression to orthogonal subspace regression.

This chapter is organised as follows. Section 7.2 briefly introduces the concepts of neuro-fuzzy system and the important T–S local neuro-fuzzy model. Section 7.3 reviews some concepts on neuro-fuzzy network design, construction and estimation algorithms. Following the introduction of the ASMOD algorithm, further develop-ments on combining T–S local neuro-fuzzy model with ASMOD algorithms are briefly reviewed, and finally we introduce the recently developed algorithm, the extended Gram–Schmidt orthogonal decomposition algorithm combined with D-optimality and parameter regularisation [35]. Section 7.4 provides an illustrative example followed by conclusions in Section 7.5.

## 7.2    Neuro-fuzzy system concepts

Fuzzy logic and fuzzy systems [3,10–12] have received considerable attention both in scientific and popular media, yet the basic concept of vagueness goes back to at least the 1920s. Zadeh's seminal paper in 1965 on fuzzy logic introduced much of the terminology that is now used in fuzzy logic [36]. The considerable success of fuzzy logic products, as in automobiles, cameras, washing machines, rice cookers, etc. has done much to temper much of the scorn poured out by the academic commun-ity on the ideas first postulated by Zadeh [36]. The existing fuzzy logic literature, number of international conferences and academic journals, together with a rapidly increasing number and diversity of applications is a testament to the vitality and importance of this subject. Many engineering applications require a fuzzy system that simply operates as a functional mapping, mapping real-valued input vector $\mathbf{x}$ to real-valued output $y$. The task is to approximate a function $\hat{y} = f(\mathbf{x})$ on a bounded (*compact*) area of the input space by using real output $y$ as a target for the model output $\hat{y}$. In contrast to the data-driven methods used to train artificial neural networks, fuzzy systems are designed using human-centred engineering techniques where the system is used to encode the heuristic knowledge articulated by a domain-specific expert.

The main reason for the criticism of fuzzy logic is that the resultant models are mathematically opaque, or there is no formal mathematical representation of the system's behaviour, which prevents application of conventional empirical modelling techniques to fuzzy systems, making model validation and benchmarking difficult to perform. Equally carrying out stability analysis on fuzzy logic controllers (the main area of applications) had been until recently impossible [4]. Also fuzzy logic is claimed to represent a degree of uncertainty or imprecision associated with variables, yet there is no distribution theory associated with fuzzy logic that allows the propagation of uncertainty to be tracked through the various stages of information processing in a fuzzy reasoning system. While this is true (despite efforts to link fuzzy logic to possibility theory and probability theory) [37], fuzzy logic is primarily about linguistic vagueness through its ability to allow an element to be a partial member of a set, so that its membership value can lie between 0 and 1 and be interpreted as: the degree to which an event may be classified as something. Neuro-fuzzy systems have emerged in recent years as researchers [5,7,9] have tried to combine the natural linguistic/symbolic transparency of fuzzy systems with the provable learning and representation capability of linear in the weights ANNs. The combination of qualitative based reasoning via fuzzy logic and quantitative adaptive numeric/data processing via ANNs, is a potentially powerful concept, since it allows within a single framework intelligent qualitative and quantitative reasoning ($IQ^2$) to be achieved. Truly intelligent systems must make use of all available knowledge: numerical (e.g. sensory derived data), expert or heuristic rules (including safety jacket rules to constrain network learning or adaptability), and known functional relationships such as physical laws (so called mechanistic knowledge). Neuro-fuzzy systems allow all these knowledge sources to be incorporated in a single information framework. A neuro-fuzzy system initially contains all the components necessary to implement a fuzzy algorithm with basic elements:

- A knowledge base which contains definitions of fuzzy sets and the fuzzy operators;
- An inference engine which performs all the output calculations;
- A fuzzifier which represents the real-valued (crisp) inputs as fuzzy sets;
- A defuzzifier which transforms the fuzzy output set into a real-valued output (crisp).

The fundamental structure of a fuzzy system is shown in Figure 7.1, in which the inputs–outputs are usually crisp or deterministic variables. Central to fuzzy logic is the concept of fuzzy sets [36], since they give the designer a method for providing a precise representation of vague natural language terms such as hot, cold, warm, etc. Fuzzy logic generalises the concept of a characteristic function in Boolean logic, whereby a fuzzy membership function $\mu_A(x) \in [0, 1]$ is used to represent vague statements such as $x$ is $A$, in which $A$ is the fuzzy label describing the variable $x$. Fuzzy logic provides the necessary mechanisms for manipulating and inferring conclusions based upon this vague information.

A rule base in a fuzzy algorithm represents expert knowledge in the form of IF–THEN production rules to relate vague input statements to vague output decisions/actions. The input or rule antecedent (IF) defines imprecisely the system

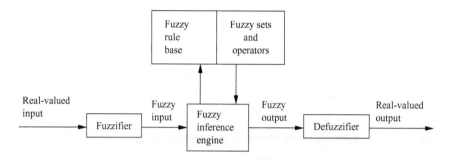

*Figure 7.1    Basic components of a fuzzy system*

states, and the consequent part (THEN) represents actions which could be taken to remedy the state condition. The knowledge base of a fuzzy system includes a rule base to describe a (usually) nonlinear system behaviour by using a set of linguistic rules, $r_{ij}$s, such as:

$$r_{ij} : \text{IF } (x_1 \text{ is } A_1^{i_1} \text{ AND } \cdots \text{ AND } x_n \text{ is } A_n^{i_n}) \text{ THEN } (\hat{y} \text{ is } B^j), \quad c_{ij}, \quad (7.1)$$

where $[x_1, \ldots, x_n]^T = \mathbf{x}$ is system input vector. Let $m_k$s be integers denoting the number of fuzzy sets for $x_k$s, and $i_k = 1, 2, \ldots, m_k$, $k = 1, \ldots, n$. An index corresponding to the sequence $\{i_1, \ldots, i_n\}$ is denoted by $i = 1, 2, \ldots, p$, with $p = \prod_{k=1}^n m_k$. Thus $A^i = [A_1^{i_1}, \ldots, A_n^{i_n}]^T$ denotes a fuzzy set in the $n$-dimensional input space. The output $B^j$, $j = 1, \ldots, q$, is a fuzzy set of the output with $q$ representing the number of fuzzy sets for both model output $\hat{y}$ and real output $y$. Each $c_{ij} \in [0, 1]$ denotes the degree of confidence on how the input fuzzy set $A^i$ is related to the output $B^j$.

A finite number of vague or fuzzy rules forms the basis for the fuzzy system's knowledge base and to generalise or interpolate between these rules, the inference engine weights each rule according to its firing strength, which in turn is determined by both the shape of the fuzzy membership functions $\mu_A(x)$ and the logical operators used by the inference engine. A standard fuzzy logic system utilises the logical functions of AND, OR, IF($\cdot$) and THEN($\cdot$) [5,7,9]. If the algebraic operators of product and sum, rather than truncation operators min and max are used as logical functions, then the resultant system (1) produces smoother interpolation, (2) provides an equivalence between ANNs and fuzzy logic if $\mu_A(x)$ are radial basis functions, or B-splines and (3) enables fuzzy system to be readily analysed [5,9]. In the following it is shown how to use a centre of gravity defuzzification algorithm in conjunction with algebraic operators, and a centre of gravity defuzzification algorithm to achieve an equivalence between a fuzzy system and a neural network [5].

With a centre of gravity defuzzification algorithm, the network's output is given by

$$\hat{y}(\mathbf{x}) = \frac{\int_Y \mu_B(y) y \, dy}{\int_Y \mu_B(y) \, dy}, \quad (7.2)$$

where $\mu_B(y) \in [0, 1]$, in which $B$ is the fuzzy label describing the variable $y$. When the $T$ norm and $S$-norm [5] operators are implemented using product and sum functions, respectively, then the centre of gravity defuzzification algorithm becomes

$$\hat{y}(\mathbf{x}) = \frac{\int_Y \int_X \mu_A(\mathbf{x}) \sum_{ij} \mu_{A^i}(\mathbf{x}) \mu_{B^j}(y) c_{ij} y \, d\mathbf{x} \, dy}{\int_Y \int_X \mu_A(\mathbf{x}) \sum_{ij} \mu_{A^i}(\mathbf{x}) \mu_{B^j}(y) c_{ij} \, d\mathbf{x} \, dy}. \tag{7.3}$$

But for bounded and symmetric fuzzy output sets (such as B-splines) the integrals $\int_Y \mu_{B^j}(y) dy$, for all $j$, are equal and so the following relationship holds:

$$\frac{\int_Y \mu_{B^j}(y) y \, dy}{\int_Y \mu_{B^j}(y) \, dy} = y_j^c,$$

where $y_j^c$ is the centre of the $j$th output set, and (7.3) therefore reduces to

$$\hat{y}(\mathbf{x}) = \frac{\int_X \mu_A(\mathbf{x}) \sum_i \mu_{A^i}(\mathbf{x}) \sum_j c_{ij} y_j^c \, d\mathbf{x}}{\int_X \mu_A(\mathbf{x}) \sum_i \mu_{A^i}(\mathbf{x}) \sum_j c_{ij} \, d\mathbf{x}}.$$

Suppose that the multivariate fuzzy input sets form a partition of unity [5], i.e. $\sum_i \mu_{A^i}(x) = 1$ and that the $i$th rule confidence vector $\mathbf{c}_i = [c_{i1}, \ldots, c_{iq}]^T$ is normalised, i.e. $\sum_j c_{ij} = 1$, then the defuzzified output becomes

$$\hat{y}(\mathbf{x}) = \frac{\int_X \mu_A(\mathbf{x}) \sum_i \mu_{A^i}(\mathbf{x}) w_i \, d\mathbf{x}}{\int_X \mu_A(\mathbf{x}) \, d\mathbf{x}}, \tag{7.4}$$

where $w_i = \sum_j c_{ij} y_j^c$ is the weight associated with the $i$th fuzzy membership function. The transformation from the weight $w_i$ to the vector of rule confidence $\mathbf{c}_i$ is a one-to-many mapping, although for fuzzy sets defined by symmetric B-splines of order $k \geq 2$, it can be inverted in the sense that for a given $w_i$ there exists a *unique* $\mathbf{c}_i$ that will generate the desired output (see Reference 5, for confidence learning rules). That is, $c_{ij} = \mu_{B^j}(w_i)$, or the rule confidence represents the different grade of membership of the weight to the various fuzzy output sets. Clearly we can alternate between the weight or rule confidence space while preserving the encoded information. This is the only method which is consistent within the fuzzy methodology for mapping between weights and rule confidences.

When the fuzzy input set $\mu_A(\mathbf{x})$ is a singleton, the numerator and denominator integrals in (7.4) cancel to give

$$y^s(\mathbf{x}) = \sum_i \mu_{A^i}(\mathbf{x}) w_i, \tag{7.5}$$

where $y^s(\mathbf{x})$ is called the fuzzy singleton output. This is an important observation since $y^s(\mathbf{x})$ is a *linear* combination of the fuzzy input sets and does not depend on the choice of fuzzy output sets. It also provides a useful link between fuzzy and neural networks and allows both approaches to be treated within a unified framework. The reduction in the computational cost of implementing a fuzzy system in this manner and the overall algorithmic simplification is illustrated in Figure 7.2.

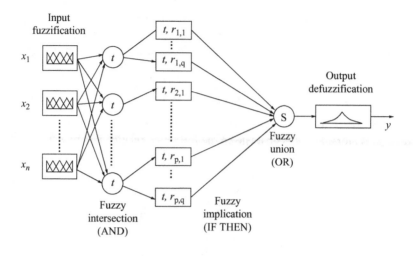

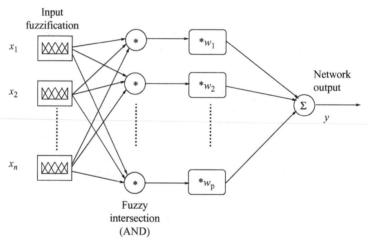

*Figure 7.2    An illustration of the information flow through a fuzzy system (top) and the resulting simplification (bottom) when algebraic operators are used in conjunction with a centre of gravity defuzzification algorithm, and the singleton input is represented by a crisp fuzzy set*

The analysis also illustrates how the centre of gravity defuzzification procedure implicitly imposes a partition of unity on the fuzzy input membership functions. Consider the above system when the fuzzy input sets do not sum to unity, which could be due to their univariate shape or the operator used to represent fuzzy intersection. The output is then given by

$$y^s(\mathbf{x}) = \frac{\sum_i \mu_{A^i}(\mathbf{x}) w_i}{\sum_j \mu_{A^j}(\mathbf{x})} = \sum_i \mu_{\hat{A}^i}(\mathbf{x}) w_i, \tag{7.6}$$

where the normalised fuzzy input membership functions $\mu_{\hat{A}^i}(\mathbf{x}) = \mu_{A^i}(\mathbf{x})/$ $\sum_j \mu_{A^j}(\mathbf{x})$ form a partition of unity. This normalisation step is very important because it determines the *actual* influence of the fuzzy set on the system's output and can make previously convex sets non-convex.

Substitute (7.6) into (7.4) giving

$$\hat{y}(\mathbf{x}) = \frac{\int_X \mu_A(\mathbf{x}) y^s(\mathbf{x}) \, d\mathbf{x}}{\int_X \mu_A(\mathbf{x}) \, d\mathbf{x}}. \tag{7.7}$$

Note that when the input to the fuzzy system is a fuzzy distribution rather than a singleton, the effect of (7.7) is to smooth or low pass filter the neuro-fuzzy system's output $\hat{y}$. This is illustrated in Figure 7.3. It can be seen that as the width of the fuzzy input set increases, the overall output of the system becomes less sensitive to the shape of either the input set or the sets used to represent the linguistic terms. However, this is not always desirable as the output also becomes less sensitive to individual rules and the input variable, and in the limit as the input set shape has an arbitrarily large width (representing complete uncertainty about the measurement) the system's output will be constant everywhere.

Using algebraic operation to calculate equation (7.7) leads to the following fundamental neuro-fuzzy modelling theorem that links fuzzy logic with neural network representations:

**Theorem 1**  [5] *When algebraic operators are used to implement the fuzzy logic function, crisp inputs are fuzzified using singleton fuzzification and centre of gravity defuzzification is employed, the input–output functional mapping can be represented by a linear combination of the input fuzzy membership functions:*

$$\hat{y}(\mathbf{x}) = \sum_{i=1}^{p} \left( \frac{\mu_{A^i}(\mathbf{x})}{\sum_j \mu_{A^j}(\mathbf{x})} \right) w_i \equiv \sum_{i=1}^{p} \psi_i(\mathbf{x}) w_i.$$

The defuzzified output is a weighted average of the fuzzy singleton outputs over the support of the fuzzy input set $\mu_A(\mathbf{x})$. Figure 7.4 illustrates the basic structure of a generalised neuro-fuzzy network. This *neuro-fuzzy network* is simply a weighted sum of nonlinear normalised basis functions $\psi_i(\mathbf{x})$, for which the weights can be trained by a linear optimisation training algorithm. This neuro-fuzzy model satisfies the Stone–Weirstrass theorem and as such is a universal approximator and can hence approximate any continuous nonlinear function $f(\mathbf{x})$ defined on a compact domain with arbitrary accuracy.

### 7.2.1  Takagi–Sugeno local neuro-fuzzy model

A very popular approach to neuro-fuzzy modelling and control is to use local fuzzy modelling approaches [13], in which fuzzy rules $r_i$s are in the form:

$$r_i: \text{IF } (x_1 \text{ is } A_1^{i_1} \text{ AND } \cdots \text{ AND } x_n \text{ is } A_n^{i_n}) \text{ THEN } (y \text{ is } \hat{y}_i(\mathbf{x})), \tag{7.8}$$

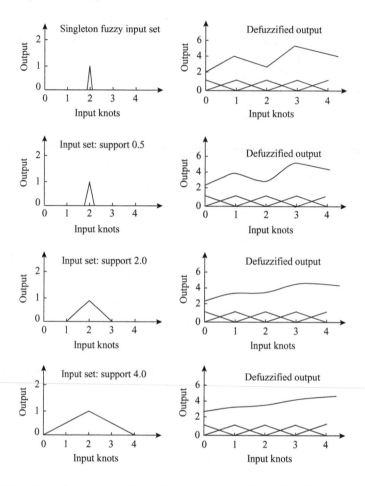

*Figure 7.3    Four fuzzy input sets and their corresponding defuzzified outputs, when the fuzzy rule base consists of triangular membership functions. The original triangular membership functions used to represent the linguistic terms are shown on the bottom of the graphs on the right and it can clearly be seen that as the width of the input set increases the system becomes less sensitive to the input variable and the set shapes*

where the output $\hat{y}_i(\mathbf{x})$ is a deterministic function. In practice $\hat{y}_i(\mathbf{x})$ is chosen as a linear combination of components of the input vector, i.e.

$$\hat{y}_i(\mathbf{x}) = w_{1i}x_1 + w_{2i}x_2 + \cdots + w_{ni}x_n, \tag{7.9}$$

allowing linear stability control theory to be used [6,15,16]. Using algebraic operators (product/sum), the truth value of the antecedent part of the fuzzy rule (7.8) is

$$\mu_{A^i}(\mathbf{x}) = \prod_{k=1}^{n} \mu_{A_k^{i_k}(x_k)}.$$

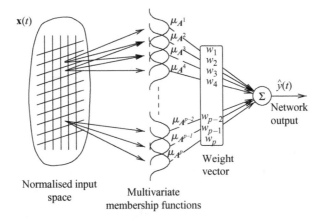

*Figure 7.4    The basic structure of a neuro-fuzzy network, resulting from Theorem 1*

Using centre of gravity defuzzification of the rule set (7.8) leads to the real output

$$\hat{y}(\mathbf{x}) = \frac{\sum_{i=1}^{p} \mu_{A^i}(\mathbf{x})\hat{y}_i(\mathbf{x})}{\sum_{i=1}^{p} \mu_{A^i}(\mathbf{x})} = \frac{\sum_{i=1}^{p} \prod_{k=1}^{n} \mu_{A_k^{i_k}}(x_k)\hat{y}_i(\mathbf{x})}{\sum_{i=1}^{p} \prod_{k=1}^{n} \mu_{A_k^{i_k}}(x_k)}. \tag{7.10}$$

Consider a neuro-fuzzy network using a B-spline function, which has been widely used in surface fitting applications, as membership functions [38–40]. The univariate B-spline basis functions are defined on a real-valued measurement, and parameterised by the *order* of a piecewise polynomial $l$ and a knot vector. Suppose there exist $m$ linguistic terms (and hence fuzzy membership functions) in the fuzzy variable $x$. The knot vector $\lambda = [\lambda_{-l+1}, \ldots, \lambda_j, \ldots, \lambda_m]^T$ breaks into a number of intervals on the universe of discourse of $x$ by using a set of values, $\lambda_i$s, which satisfy the following relationship:

$$x_{\min} < \lambda_1 \leq \lambda_2 \leq \cdots \leq \lambda_{m-l} < x_{\max}, \tag{7.11}$$

where $x_{\min}, x_{\max}$ are the minimum/maximum value of $x$. A set of the extra knot values which define the basis functions at each end must also be specified and these should satisfy:

$$\lambda_{-l+1} \leq \cdots \leq \lambda_0 = x_{\min}, \tag{7.12}$$

$$x_{\max} = \lambda_{m-l+1} \leq \cdots \leq \lambda_m, \tag{7.13}$$

and this is illustrated in Figure 7.5 for order 2 (triangular) fuzzy membership functions. Then the membership functions are defined on the interior space $m - k + 1$ intervals wide (defined by equation (7.11)). For simplicity, denote the knot vector for all $x_k$s as $\lambda$. The output of B-spline membership functions for $x_k$ can be calculated using

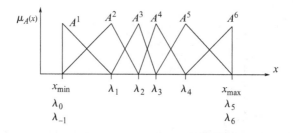

*Figure 7.5    Six B-spline fuzzy membership functions of order $k = 2$ where a non-uniform knot placement strategy is used*

the following simple and stable recursive relationship:

$$\mu_{A_{k,l}^j}(x_k) = \left( \frac{x_k - \lambda_{j-l}}{\lambda_{j-1} - \lambda_{j-l}} \right) \mu_{A_{k-1,l-1}^j}(x_k) + \left( \frac{\lambda_j - x}{\lambda_j - \lambda_{j-k+1}} \right) \mu_{A_{k,l-1}^j}(x_k),$$

$$\mu_{A_{k,1}^j}(x_k) = \begin{cases} 1, & \text{if } x_k \in [\lambda_{j-1}, \lambda_j), \\ 0, & \text{otherwise}, \end{cases} \tag{7.14}$$

where $\mu_{A_{k,l}^j}(x_k)$ is the $j$th membership function of order $l$, with $j = i_k$ and $m = m_k$ for notational simplicity.

Multidimensional B-spline basis functions are formed by a direct multiplication of univariate basis functions via

$$N_i(\mathbf{x}) = \prod_{k=1}^{n} \mu_{A_{k,l}^{i_k}}(x_k), \tag{7.15}$$

for $i = 1, \ldots, p$. Since B-splines have a partition of unity, i.e. $\sum_{i=1}^{p} \mu_{A^i}(\mathbf{x}) = 1$, (7.10) is simplified to

$$\hat{y}(\mathbf{x}) = \sum_{i=1}^{p} \prod_{k=1}^{n} \mu_{A_k^{i_k}(x_k)} \hat{y}_i(\mathbf{x}) = \sum_{i=1}^{p} N_i(\mathbf{x}) \hat{y}_i(\mathbf{x}). \tag{7.16}$$

Substituting (7.9) into (7.16) gives the following weighted linear model:

$$\hat{y}(\mathbf{x}) = \sum_{k=1}^{n} a_k(\mathbf{x}) x_k, \tag{7.17}$$

where the coefficients or weights $a_k(\mathbf{x}) = \sum_{i=1}^{p} N_i(\mathbf{x}) w_{ki}$ are effectively subneuro-fuzzy networks.

Takagi–Sugeno modelling has been proven popular due to the use of local linear models and the almost universal use of triangular (order 2) basis functions owing to their analytical tractability. While local linear modelling is justified for many dynamic processes, restricting basis functions to order 2 is not justified unless the fundamental relationship is piecewise linear. One of the fundamental problems of T–S models is how to decompose an *a priori* unknown global nonlinear system into a series of local models (7.17), which are minimum in number as well as with prescribed model approximation error. An obvious application of the local neuro-fuzzy model (7.17) is to time-series modelling where $\mathbf{x}(t)$ is the set of model regressors $[y(t-1), \ldots,$ $y(t-n_y), u(t-1), \ldots, u(t-n_u)]^\mathrm{T}$. In this case, (7.17) becomes an ARMA type model.

## 7.3   Neuro-fuzzy network design, construction and estimation

Neuro-fuzzy models are essentially adaptive fuzzy systems, where the adaptation is based on the fundamentals of feedback learning and weight adjustment found in conventional parametric optimisation so that the model produces a good approximation to the true system. Given a data set $D_N = \{\mathbf{x}(t), y(t)\}_{t=1}^N$, where $N$ is the number of data samples and $t$ is the label for data samples, consider an unknown process

$$y(\mathbf{x}(t)) = f(\mathbf{x}(t), \mathbf{w}) + e(t), \tag{7.18}$$

where $e(t)$ is the modelling error. One of the objectives is to find some model structure $f(\mathbf{x}(t), \mathbf{w})$ and parameter estimator of $\mathbf{w}$ to minimise modelling errors. Neuro-fuzzy models as described above are within the class of approximation models that are a particular form of generalised linear models that form an ideal framework for performing nonlinear system identification. In system identification there are a number of modelling attributes that need to be considered when evaluating a particular algorithm, such as

- *Model accuracy.* The model should accurately approximate the derived system across the training data.
- *Generalisation.* The model should be able to generalise, this is the model's ability to model the desired system accurately for unseen inputs.
- *Interpretability* or *transparency.* Often the prior understanding of the underlying system is limited and it is beneficial to the modeller if the model process provides knowledge about the underlying physical process, i.e. qualitative physics (or grey-box modelling as against black-box modelling); such knowledge is useful in validating the behaviour of the model (in say diagnostic problems, and in problems of knowledge discovering).
- *Ability to encode a priori knowledge.* *A priori* knowledge is often available describing certain aspects of the system's operation. Any such knowledge should be exploited in modelling via say rule/weight initialisation.
- *Efficient implementation.* The modelling technique must use computational resources (i.e. data, speed, memory, ...) efficiently as in many applications these are inherently limited.
- *Fast adaptation.* This property is critical when the model is employed on line, where process adaptation is carried out as new system knowledge is acquired.

A model construction algorithm ASMOD was initially derived by Kavli [20] to identify an ANOVA model decomposition from training data. This decomposition describes the additive decomposition in simpler subfunctions, which solves the curse of dimensionality in two ways: (1) for many functions certain interactions and inputs are redundant; (2) correlations between many inputs are within the error bars of model approximation. This iterative algorithm generates a globally partitioned B-spline model, which is directly applicable to neuro-fuzzy models. Complementary approaches [41] have led to the following improved form of ASMOD. For neuro-fuzzy modelling based on a lattice structure, interactions usually should be less than or equal to four input variables to avoid a rule explosion and retain network transparency. The ANOVA decomposition reduces a globally partitioned, lattice-based neuro-fuzzy system into a sum of smaller lattice-based neuro-fuzzy systems, retaining network transparency and linearity in the weights for parameter training.

Alternatively, due to the inherent transparency properties of a neuro-fuzzy network, a parsimonious model construction approach should lead also to a logical rule extraction process that increases model transparency, as simpler models inherently involve fewer rules, which are in turn easier to interpret. One drawback of most current neuro-fuzzy learning algorithms is that learning is based upon a set of one-dimensional regressors, or basis functions (such as B-splines, Gaussians, etc.), but not upon a set of fuzzy rules (usually in the form of multidimensional input variables), resulting in opaque models during the learning process. Since modelling is inevitably iterative it can be greatly enhanced if the modeller can interpret or interrogate the derived rule base during learning itself, allowing him or her to terminate the process when his or her objectives are achieved. Following the introduction of ASMOD in the next section, a robust extended Gram–Schmidt method is then presented [34,35]. Model rule bases are decomposed into orthogonal subspaces [34], so as to enhance model transparency with the capability of interpreting the derived rule base energy level. A locally regularised orthogonal least-squares algorithm, combined with a D-optimality used for subspace based rule selection, has been extended for fuzzy rule regularisation and subspace based information extraction [35].

### 7.3.1   The ASMOD construction modelling algorithm

As with fuzzy/expert/associative memory systems, neuro-fuzzy systems become computationally inefficient as the input space dimension increases due to the curse of dimensionality. The curse of dimensionality was a phrase introduced by Bellman referring to the exponential increase in resources (data, memory, ...) required by a system as the input dimension increases [19]. For example, for a complete base fuzzy system with $n$ inputs, with $m_k$ fuzzy sets on each input, the total number of rules or combinations of linguistic input terms is $p = \prod_{k=1}^{n} m_k$.

An obvious approach to dealing with a multivariate function $f(\mathbf{x})$ is to decompose it into a sum of lower-dimensional functions. One such decomposition is the ANOVA expansion given by

$$f(\mathbf{x}) = f_0 + \sum_{i=1}^{n} f_i(x_i) + \sum_{i=1}^{n-1} \sum_{j=i+1}^{n} f_{ij}(x_i, x_j) + \cdots + f_{1,2,\ldots,n}(x_1, \ldots, x_n).$$

$$(7.19)$$

The ASMOD algorithm is an offline iterative construction algorithm that uses *growing* and *pruning* in which a more complex model can always reproduce the current model exactly, a distinct advantage that B-splines have over alternatives such as RBFs. Within the improved ASMOD algorithm there are three refinement processes.

*Univariate addition/submodel deletion.* A prior model is given as the ability to additively include a new input variable, by the addition of a new univariate submodel. Univariate submodels are only incorporated into the existing model if the new input is not in the current model. These univariate submodels are drawn from an initial model set or external store, which may be the encoding of prior knowledge about the process being modelled, e.g. output smoothness requiring third-order B-splines. Note that the univariate models can be of differing order to reflect prior knowledge about a particular input variable. For $n_s$ submodels in the external store, with $n$ inputs, the number of possible univariate additions is $n_s(n - n_m)$, for $n_m$ the number of inputs of the current model.

The corresponding pruning step is submodel deletion, where a superfluous submodel is removed from the current model. This is achieved by deleting each of the submodels in the current model, producing a set of candidate evaluation models, each containing one less submodel than the current model. These are trained on the training data set to see if there is any submodel that can be deleted.

*Tensor multiplication/tensor submodel splitting.* Input variable interaction necessary to construct model submodels $f(x_1, \ldots, x_j)$ is incorporated by tensor multiplication (fuzzy AND) of an existing submodel with a univariate submodel from the external store. This refinement step is computationally expensive as for every iteration there are many different candidate models. If the submodels are tensor multiplied with the different fuzzy set distributions in the external store, the number of possible candidate refinements for replacing an existing submodel with $n_u$ inputs is $n_s(\sum_{i=1}^{n-n_u} C_{n-n_u}^i)$. The original ASMOD algorithm overcame this computational problem by allowing tensor multiplication of existing submodels. This approach generates submodels with large numbers of inputs in the early stage of construction as well as preventing the same input occurring in different submodels. This is avoided in the improved ASMOD algorithm by restricting the search so that a model dimension can only increase by one at each iteration, and that all input variables present in the current model, but not present in a given submodel, are considered for tensor multiplication (provided that the candidate model is not redundant). This gives the number of candidate models of the order of $n_s(n_m - n_u)$.

The corresponding pruning step is to replace a submodel with $n_u \geq 2$ inputs by submodels that depend on combinations of less than $n_u$ inputs. As with tensor multiplication, tensor splitting leads to a large number of possible splits; however, restricting the dimension reduction to one per iteration of any submodel significantly reduces the computational load. After splitting a submodel, any redundant submodels are removed.

*Knot insertion/knot removal.* Model flexibility of a submodel can be increased by the introduction of a new fuzzy set (or basis function) to one or more of the inputs. This is achieved by inserting a new knot into one of its univariate knot vectors, creating a new set of fuzzy membership functions for that input variable. The shape of the new fuzzy set is predetermined by the order of the existing fuzzy sets and their distribution on the domain of the input variable. All fuzzy sets which have non-zero grade of membership at the new knot location are automatically modified to ensure that a partition of unity is preserved. To reduce computation, the new knot is restricted to lie halfway between existing interior knots present in the submodel. The number of new rules produced by knot insertion is equal to the number of multivariate fuzzy sets produced from the tensor product of all the other fuzzy variables in the submodel, i.e. $\prod_{i=1,\,i\neq j}^{n_u} p_i$, where $p_i$ is the number of fuzzy sets defined on the $i$th input, and $j$ is the input index on which the new knot is inserted.

Knot deletion prunes out redundant submodel flexibility by removal of a knot from one of the submodel knot vectors (or equivalently removing a fuzzy membership function). Every interior knot in the entire ASMOD model is considered for knot dele-tion, resulting in several redundant multidimensional basis functions being removed.

Clearly even the improved ASMOD has potentially serious computational problems, which can be resolved by:

*Reducing basis function order.* (Sub)models can be simplified by reducing the order of the B-spline membership functions defined on an input. In the limit the order can be reduced to 1, allowing the introduction of crisp decision boundaries, which may be useful in classification/identification applications where a system changes abruptly between operating conditions (e.g. as in a direct fuel injected engine). If there are no interior points, then for order 1 basis functions the dependency of this submodel on this input can be removed.

*Stagewise construction.* Restricting the number of refinements and their ordering reduces the number of possible candidate models at each iteration. Also by assuming that the external stores are sufficiently representative knot insertions can be avoided until the overall model additive structure has been identified. Hence, we recommended the following stagewise construction strategy in implementing ASMOD:

1   Identify the additive structure by using the univariate addition, tensor product and tensor splitting refinements (i.e. as in the original ASMOD algorithm).
2   Delete any redundant submodels by performing submodel deletion refinement.
3   Optimise the fuzzy set distributions by insertion and deletion of knots.
4   Check for redundant membership functions by performing the reduced order refinement.

This iterative model construction algorithm allows the user to review models as the refinement proceeds, allowing premature termination if a more complex but acceptable model is generated.

ASMOD has been extensively used in practical data analysis problems, includ-ing autonomous underwater vehicle modelling [42], automobile MPG modelling [43], engine torque modelling [43], Al–Zn–Mg–Cu materials modelling [44], car brake sys-tems modelling [45], aircraft flight control [46], gas turbine engine diagnostics [47], breast cancer diagnostics [48], multisensor data fusion [49], driver modelling [50,51] and command and control [52].

There are a few further developments combining T–S local neuro-fuzzy model with ASMOD, including some structural learning algorithms for T–S local neuro-fuzzy models with applications in state estimation for data fusion and controller design [17,18,53–55]. A neuro-fuzzy local linearisation (NFLL) model has been decomposed into submodels in an ANOVA form as a modified ASMOD (MASMOD) algorithm for automatic fuzzy partitioning of the input space, in which B-splines, as membership functions, were automatically constructed, that is, the number of decomposed submodels, the number of B-splines required in each submodel and their positions and shapes were automatically determined by the MASMOD algorithm based on observational data and other *a priori* process knowledge. Most early work in local linearisation has been based on the first-order T–S fuzzy model [13]. Least squares (LS) and least mean squares (LMS) are widely used conventional methods for optimisation. Expectation-maximisation (EM) is a general technique for maximum likelihood or maximum *a posteriori* estimation and has become an alternative to LS and LMS techniques in solving many estimation problems [56–60]. In order to automatically partition the input space using fuzzy sets in an effective (parsimonious) manner, a hybrid learning scheme for NFLL model has been proposed, which combines the MASMOD and EM algorithms [53].

Effectively neuro-fuzzy local linearisation enables a system input–output model to be transformed into a state–space model, which is necessary for applications such as state estimation, state feedback controller or state vector data fusion. Gan and Harris [61] have developed an adaptively constructed recurrent neuro-fuzzy network and its learning scheme for feedback linearisation (NFFL for short), which enlarges the class of nonlinear systems that can be feedback linearised using neuro-fuzzy networks. The NFLL and NFFL models of a nonlinear system can be easily reformulated as a time-varying linear system model whose time-varying coefficients depend on both membership functions and the parameters in local linear models. If the underlying nonlinear system is locally or feedback linearisable, then the time-varying coefficients will change slowly with the system state vector, and linear techniques for state estimation and adaptive control, such as Kalman filter, can be directly applicable to the NFLL or NFFL model, resulting in neuro-fuzzy state estimators [54,55,62]. Alternatively the neuro-fuzzy local linearisation scheme also results in a Kalman filtering-based state estimation with coloured noise for a controller design applicable to some unknown dynamical process [18].

### 7.3.2 The extended Gram–Schmidt orthogonal decomposition algorithm with regularisation and D-optimality in orthogonal subspaces

The basic idea to enhance improved model sparsity, robustness and transparency is to use rule-based learning and model construction with some goodness measure embodying the modelling attributes listed above. Due to the inherent transparency properties of a neuro-fuzzy network, a one to one mapping between a fuzzy rule base and a model matrix feature subspace is readily established [35]. By incorporating the other modelling attributes of sparsity and model robustness, an extended Gram–Schmidt orthogonal decomposition algorithm combined with D-optimality and parameter regularisation is introduced for the structure determination and parameter estimation of *a priori* unknown dynamical systems in the form of a set of fuzzy rules [35]. The algorithm is presented below.

Consider a general representation combining local fuzzy modelling approaches [13] with ANOVA decomposition. Model (7.18) can be simplified by decomposing it into a set of $K$ local models $f_i(\mathbf{x}^{(i)}(t), \mathbf{w}_i)$, $i = 1, \ldots, K$, where $K$ is to be determined, each of which operates on a local region depending on the sub-measurement vector $\mathbf{x}^{(i)} \in \Re^{n_i}$, a subset of the input vector $\mathbf{x}$, i.e. $\mathbf{x}^{(i)} \in \mathcal{X}_i \in \Re^{n_i}$, $(n_i < n)$, $\mathcal{X}_1 \cup \cdots \cup \mathcal{X}_K = \mathcal{X}$. Each of the local models $f_i(\mathbf{x}^{(i)}(t), \mathbf{w}_i)$ can be represented by a set of linguistic rules

$$
\begin{aligned}
&\text{IF} \qquad \mathbf{x}^{(i)} \text{ is } A^{(i)} \\
&\text{THEN} \quad \hat{y}(t) = f_i(\mathbf{x}^{(i)}(t), \mathbf{w}_i),
\end{aligned}
\tag{7.20}
$$

where the fuzzy set $A^{(i)} = [A_1^{(i)}, \ldots, A_{n_i}^{(i)}]^T$ denotes a fuzzy set in the $n_i$-dimensional input space, $\Re^{n_i}$ and is given as an array of linguistic values, based on a predetermined input spaces partition into fuzzy sets via some prior system knowledge of the operating range of the data set. Usually if $\mathbf{x}^{(j)} = \mathbf{x}^{(k)}$, for $j \neq k$, then $A^{(j)} \cap A^{(k)} = \emptyset$, where $\emptyset$ denotes empty set. $\cup_{i=1}^{K} A^{(i)}$ defines a complete fuzzy partition of the input space $\mathcal{X}$. For an appropriate input space decomposition, the local models can have essentially local linear behaviour. In this case, using the well-known T–S fuzzy inference mechanism [13], the output of system (7.18) can be represented by

$$
f(\mathbf{x}(t), \mathbf{w}) = \sum_{i=1}^{K} N_i(\mathbf{x}^{(i)}(t)) f_i(\mathbf{x}^{(i)}(t), \mathbf{w}_i),
\tag{7.21}
$$

where $f_i(\mathbf{x}^{(i)}(t), \mathbf{w}_i)$ is a linear function of $\mathbf{x}^{(i)}$, given by

$$
f_i(\mathbf{x}^{(i)}(t), \mathbf{w}_i) = \mathbf{x}^{(i)}(t)^T \mathbf{w}_i,
\tag{7.22}
$$

and $\mathbf{w}_i \in \Re^{n_i}$ denotes parameter vector of the $i$th fuzzy rule or local model. $N_i(\mathbf{x}^{(i)})$ is a fuzzy membership function of the rule (7.20). These are computed from (7.14) and (7.15) by using sub-measurement vector $\mathbf{x}^{(i)}$ instead of $\mathbf{x}$.

Substitute (7.21) and (7.22) into (7.18) to yield

$$
y(t) = \sum_{i=1}^{K} \psi_i(\mathbf{x}^{(i)}(t))^T \mathbf{w}_i + e(t) = \psi(\mathbf{x}(t))^T \mathbf{w} + e(t),
\tag{7.23}
$$

where $\psi_i(\mathbf{x}^{(i)}(t)) = [\psi_{i1}(t), \ldots, \psi_{in_i}(t)]^T = N_i(\mathbf{x}^{(i)}(t))\mathbf{x}^{(i)} \in \Re^{n_i}$ · $\psi(\mathbf{x}(t)) = [\psi_1(\mathbf{x}^{(1)}(t))^T, \ldots, \psi_K(\mathbf{x}^{(K)}(t))^T]^T \in \Re^p$ · $\mathbf{w} = [\mathbf{w}_1^T, \ldots, \mathbf{w}_K^T]^T \in \Re^p$, where $p = \sum_{i=1}^{K} n_i$.

For the finite data set $D_N = \{\mathbf{x}(t), y(t)\}_{t=1}^N$, equation (7.23) can be written in a matrix form as

$$\mathbf{y} = \sum_{i=1}^K \Psi^{(i)} \mathbf{w}^{(i)} + \mathbf{e} = \Psi \mathbf{w} + \mathbf{e}, \tag{7.24}$$

where $\mathbf{y} = [y(1), y(2), \ldots, y(N)]^T \in \Re^N$ is the output vector, $\Psi^{(i)} = [\psi_i(\mathbf{x}(1)), \ldots, \psi_i(\mathbf{x}(N))]^T \in \Re^{N \times n_i}$ is the regression matrix associated with the $i$th fuzzy rule, $\mathbf{e} = [e(1), \ldots, e(N)]^T \in \Re^N$ is the model residual vector. $\Psi = [\Psi^{(1)}, \ldots, \Psi^{(K)}] \in \Re^{N \times p}$ is the full regression matrix.

An effective way of overcoming the curse of dimensionality is to start with a moderate sized rule base according to the actual data distribution. $K$ local models are selected as an initial model base based on model identifiability via an A-optimality design criterion [26] with the advantage of enhanced model transparency to quantify and interpret fuzzy rules and their identifiability.

*Rule-based learning and initial model base construction.* Rule-based knowledge, i.e. information associated with a fuzzy rule, is highly appropriate for users to understand a derived data-based model. Most current learning algorithms in neuro-fuzzy modelling are based on an ordinary $p$-dimensional linear in the parameter model. Model transparency during learning cannot be automatically achieved unless these regressors have a clear physical interpretation, or are directly associated with physical variables. Alternatively, a neuro-fuzzy network is inherently transparent for rule-based model construction. In (7.24), each of $\Psi^{(i)}$ is constructed based on a unique fuzzy membership function $N_i(\cdot)$, providing a link between a fuzzy rule base and a matrix feature subspace spanned by $\Psi^{(i)}$. Rule-based knowledge can be easily extracted by exploring this link.

**Definition 1**   *Basis of a subspace: If $n_i$ vectors $\psi_j^{(i)} \in \Re^N$, $j = 1, 2, \ldots, n_i$, satisfy the nonsingular condition that $\Psi^{(i)} = [\psi_1^{(i)}, \ldots, \psi_{n_i}^{(i)}] \in \Re^{N \times n_i}$ has a full rank of $n_i$, they span an $n_i$-dimensional subspace $S^{(i)}$, then $\Psi^{(i)}$ is the basis of the subspace $S^{(i)}$.*

**Definition 2**   *Fuzzy rule subspace: Suppose $\Psi^{(i)}$ is nonsingular, clearly $\Psi^{(i)}$ is the basis of an $n_i$-dimensional subspace $S^{(i)}$, which is a functional representation of the fuzzy rule (7.20) by using T–S fuzzy inference mechanism with a unique label $N_i(\cdot)$. $S^{(i)}$ is defined as a fuzzy rule subspace of the $i$th fuzzy rule.*

$\Psi^{(i)}$, the sub-matrix associated with the $i$th rule, can be expanded as

$$\Psi^{(i)} = \mathbf{N}^{(i)} X^{(i)}, \tag{7.25}$$

where   $\mathbf{N}^{(i)} = \text{diag}\{N_i(1), \ldots, N_i(N)\} \in \Re^{N \times N}$,   $X^{(i)} = [\mathbf{x}^{(i)}(1), \mathbf{x}^{(i)}(2), \ldots, \mathbf{x}^{(i)}(N)]^T \in \Re^{N \times n_i}$.

Equation (7.25) shows that each rule base is simply constructed by a weighting matrix multiplied to the regression matrix of original input variables. The weighting matrix $\mathbf{N}^{(i)}$ can be regarded as a data-based spatial prefiltering over the input region. Without loss of generality, it is assumed that $X^{(i)}$ is nonsingular, and $N > n_i$, as $\text{rank}(X^{(i)}) = n_i$. As

$$\text{rank}(\Psi^{(i)}) = \min[\text{rank}(\mathbf{N}^{(i)}), \text{rank}(X^{(i)})]. \tag{7.26}$$

For $\Psi^{(i)}$ to be nonsingular, then $\text{rank}(\mathbf{N}^{(i)}) > n_i$, this means that for the input region denoted by $N_i(\cdot)$, its basis function needs to be excited by at least $n_i$ data points.

The A-optimality design criteria for the weighting matrix $\mathbf{N}^{(i)}$ which is given by Reference 26

$$J_A(\mathbf{N}^{(i)}) = \frac{1}{N} \sum_{t=1}^{N} N_i(t), \tag{7.27}$$

provides an indication for each fuzzy rule on its identifiability and hence a metric for selecting appropriate model rules. The full model rule set can then be rearranged in descending order of identifiability, followed by utilising only the first $K$ rules with identifiability to construct a model rule base set.

*Orthogonal subspace decomposition and regularisation in orthogonal subspace.* For ease of exposition, we initially introduce some notations and definitions that are used in the development of the new extended Gram–Schmidt orthogonal decomposition algorithm.

**Definition 3**   *Orthogonal subspaces: For a p-dimensional matrix space $S \in \Re^{N \times p}$, two of its subspaces $Q^{(i)} \in \Re^{N \times n_i} \subset S$ and $Q^{(j)} \in \Re^{N \times n_j} \subset S$, $(n_i < p, n_j < p)$ are orthogonal if and only if any two vectors $\mathbf{q}^{(i)}$ and $\mathbf{q}^{(j)}$ that are located in the two subspaces, respectively, i.e. $\mathbf{q}^{(i)} \in Q^{(i)}$ and $\mathbf{q}^{(j)} \in Q^{(j)}$, are orthogonal, that is, $[\mathbf{q}^{(i)}]^T\mathbf{q}^{(j)} = 0$, for $i \neq j$.*

The $p$-dimensional space $S$ $(p = \sum_{i=1}^{K} n_i)$, can be decomposed by $K$ orthogonal subspaces $Q^{(i)}, i = 1, \ldots, K$, given by References 63 and 64

$$Q^{(1)} \oplus \cdots \oplus Q^{(K)} = S \in \Re^{p \times N}, \tag{7.28}$$

where $\oplus$ denotes sum of orthogonal sets. From Definition 1, if there are any linear uncorrelated $n_i$ vectors located in $Q^{(i)}$, denoted as $\mathbf{q}_j^{(i)} \subset Q^{(i)}, j = 1, \ldots, n_i$, then the matrix $\mathbf{Q}^{(i)} = [\mathbf{q}_1^{(i)}, \ldots, \mathbf{q}_{n_i}^{(i)}]$, forms a basis of $Q^{(i)}$. Note that these $n_i$ vectors need not be mutually orthogonal, i.e. $[\mathbf{Q}^{(i)}]^T\mathbf{Q}^{(i)} = \mathbf{D}^{(i)} \in \Re^{n_i \times n_i}$, where $\mathbf{D}^{(i)}$ is not required to be diagonal.

Clearly if two matrix subspaces $Q^{(i)}, Q^{(j)}$ have the basis of full rank matrices $\mathbf{Q}^{(i)} \in \Re^{N \times n_i}, \mathbf{Q}^{(j)} \in \Re^{N \times n_j}$, then they are orthogonal if and only if

$$[\mathbf{Q}^{(i)}]^T\mathbf{Q}^{(j)} = \mathbf{0}_{n_i \times n_j}, \tag{7.29}$$

where $\mathbf{0}_{n_i \times n_j} \in \Re^{n_i \times n_j}$ is a zero matrix.

**Definition 4** *Vector decomposition to subspace basis: If $K$ orthogonal subspaces $\mathcal{Q}^{(i)}$, $i = 1, \ldots, K$, are defined by a series of $K$ matrices $\mathbf{Q}^{(i)}$, $i = 1, \ldots, K$ as subspace basis based on Definition 3, then an arbitrary vector $\hat{\mathbf{y}} \in \Re^N \in S$ can be uniquely decomposed as*

$$\hat{\mathbf{y}} = \sum_{i=1}^{K} \sum_{j=1}^{n_i} \theta_{i,j} \mathbf{q}_j^{(i)} = \sum_{i=1}^{K} \mathbf{Q}^{(i)} \boldsymbol{\theta}_i, \tag{7.30}$$

*where $\theta_{i,j}$s are combination coefficients. $\boldsymbol{\theta}_i = [\theta_{i,1}, \ldots, \theta_{i,n_i}]^{\mathrm{T}} \in \Re^{n_i}$.*

As the result of the orthogonality of $[\mathbf{q}^{(i)}]^{\mathrm{T}} \mathbf{q}^{(j)} = 0$ (for $i \neq j$), from (7.30),

$$\hat{\mathbf{y}}^{\mathrm{T}} \hat{\mathbf{y}} = \sum_{i=1}^{K} \boldsymbol{\theta}_i^{\mathrm{T}} \mathbf{D}^{(i)} \boldsymbol{\theta}_i. \tag{7.31}$$

Clearly the variance of the vector $\hat{\mathbf{y}}$ projected into each subspace can be computed as $\boldsymbol{\theta}_i^{\mathrm{T}} \mathbf{D}^{(i)} \boldsymbol{\theta}_i$, for $i = 1, \ldots, K$.

Consider the nonlinear system (7.23) given as a vector form by (7.24). By introducing an orthogonal subspace decomposition $\boldsymbol{\Psi} = \mathbf{QR}$, (7.24) can be written as

$$\mathbf{y} = \mathbf{Q}\boldsymbol{\theta} + \mathbf{e} = \sum_{i=1}^{K} \mathbf{Q}^{(i)} \boldsymbol{\theta}_i + \mathbf{e}, \tag{7.32}$$

where $\mathbf{Q} = [\mathbf{Q}^{(1)}, \ldots, \mathbf{Q}^{(K)}]$ spans a $p$-dimensional space $S$ with $\mathbf{Q}^{(i)}$, $i = 1, \ldots, K$ spanning its subspaces $\mathcal{Q}^{(i)}$, as defined via Definition 3. The auxiliary parameter vector $\boldsymbol{\theta} = \mathbf{Rw} = [\boldsymbol{\theta}_1^{\mathrm{T}}, \ldots, \boldsymbol{\theta}_K^{\mathrm{T}}]^{\mathrm{T}} \in \Re^p$, where $\mathbf{R}$ is a block upper triangular matrix

$$\mathbf{R} = \begin{bmatrix} R_{1,1} & R_{1,2} & \cdots & R_{1,K} \\ 0 & R_{2,2} & \cdots & R_{2,K} \\ & \cdots & \cdots & \\ 0 & \cdots & R_{i,j} & \cdots \\ & \cdots & \cdots & \\ 0 & \cdots & \cdots & R_{K,K} \end{bmatrix} \in \Re^{p \times p}, \tag{7.33}$$

in which $R_{i,j} \in \Re^{n_i \times n_j} \cdot R_{i,i} = \mathbf{I}_{n_i \times n_i}$, a unit matrix $\in \Re^{n_i \times n_i}$.

**Definition 5** *The extended Gram–Schmidt orthogonal decomposition algorithm [34]: An orthogonal subspace decomposition for model (7.32) can be realised based on an extended Gram–Schmidt orthogonal decomposition algorithm as follows. Set $\mathbf{Q}^{(1)} = \boldsymbol{\Psi}^{(1)}$, $R_{1,1} = \mathbf{I}_{n_1 \times n_1}$ and, for $j = 2, \ldots, K$, set $R_{j,j} = \mathbf{I}_{n_j \times n_j}$,*

$$\mathbf{Q}^{(j)} = \boldsymbol{\Psi}^{(j)} - \sum_{i=1}^{j-1} \mathbf{Q}^{(i)} * R_{i,j}, \tag{7.34}$$

*where*

$$R_{i,j} = [\mathbf{D}^{(i)}]^{-1}[\mathbf{Q}^{(i)}]^T \Psi^{(j)} \in \Re^{n_i \times n_j}, \tag{7.35}$$

*for $i = 1, \ldots, j - 1$.*

From (7.30) and (7.32), if the system output vector $\mathbf{y}$ is decomposed as a term $\hat{\mathbf{y}}$ by projecting onto orthogonal subspaces $\mathbf{Q}^{(i)}$, $i = 1, \ldots, K$, and an uncorrelated term $e(t)$ that is unexplained by the model, such that the projection onto each subspace basis (or a percentage energy contribution of these subspaces towards the construction of $\mathbf{y}$) can be readily calculated via

$$[\text{err}]_i = \frac{\boldsymbol{\theta}_i^T \mathbf{D}^{(i)} \boldsymbol{\theta}_i}{\mathbf{y}^T \mathbf{y}} \tag{7.36}$$

the output variance projected onto each subspace can be interpreted as the contribution of each fuzzy rule in the fuzzy system, subject to the existence of previous fuzzy rules. To include the most significant subspace basis with the largest $[\text{err}]_i$ as a forward regression procedure is a direct extension of conventional forward OLS algorithm [22]. The output variance projected into each subspace can be interpreted as the output energy contribution explained by a new rule demonstrating the significance of the new rule towards the model. At each regression step, a new orthogonal subspace basis is formed by using a new fuzzy rule and the existing fuzzy rules in the model, with the rule basis with the largest $[\text{err}]_i$ to be included in the final model until

$$1 - \sum_{i=1}^{n_f} [\text{err}]_i < \rho \tag{7.37}$$

satisfies for an error tolerance $\rho$ to construct a model with $n_f < K$ rules. The parameter vectors $\mathbf{w}_i$, $i = 1, \ldots, n_f$ can be computed by the following back substitution procedure: Set $\mathbf{w}_{n_f} = \boldsymbol{\theta}_{n_f}$, and, for $i = n_f - 1, \ldots, 1$

$$\mathbf{w}_i = \boldsymbol{\theta}_i - \sum_{j=i+1}^{n_f} R_{i,j} * \mathbf{w}_j. \tag{7.38}$$

The concept of orthogonal subspace decomposition based on fuzzy rule bases is illustrated in Figure 7.6. This figure illustrates (7.34) that forms the orthogonal bases. Because of the one-to-one mapping of a fuzzy rule to a matrix subspace, a series of orthogonal subspace bases are formed by using fuzzy rule subspace basis $\Psi^{(i)}$ in a forward regression manner, such that $\{Q^{(1)} \oplus Q^{(2)} \oplus \cdots \oplus Q^{(i)}\} = \{S^{(1)} \cup S^{(2)} \cup \cdots \cup S^{(i)}\}$, $\forall i$, while maximising the output variance of the model at each regression step $i$. Note that the well-known orthogonal schemes such as the classical Gram–Schmidt method construct orthogonal vectors as bases based on regression vectors (one-dimensional), but the new algorithm extends the classical Gram–Schmidt orthogonal decomposition scheme to the orthogonalisation of subspace bases (multidimensional). The extended Gram–Schmidt orthogonal

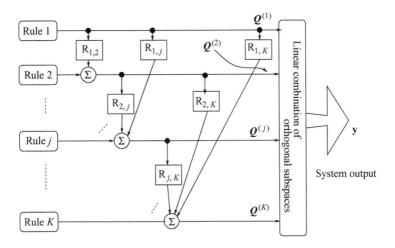

*Figure 7.6    Orthogonal subspace decomposition based on fuzzy rule bases*

decomposition algorithm is not only an extension from classical Gram–Schmidt orthogonal axis decomposition to orthogonal subspace decomposition, but also an extension from basis function regression to matrix subspace regression, introducing a significant advantage of model transparency to interpret fuzzy rule energy level.

Note that the use of [err]$_i$ aims to optimise the model in terms of approximation capability, but not in terms of model robustness. The extended Gram–Schmidt orthogonal subspace decomposition algorithm given below is based on two concepts: (1) parameter regularisation and (2) D-optimality for rule-based model construction.

Regularisation can be used as an effective mechanism to overcome overfitting to noise. The concept of parameter regularisation may be incorporated into a forward orthogonal least-squares algorithm as a locally regularised orthogonal least-square estimator for subspace selection by defining a regularised error reduction ratio due to the submatrix $\mathbf{Q}_i$ as follows.

**Definition 6** *Locally regularised least-squares cost function in orthogonal subspaces: The orthogonal subspace-based regularised least squares uses the following error criterion:*

$$J_R(\boldsymbol{\theta}, \boldsymbol{\tau}) = \mathbf{e}^{\mathrm{T}}\mathbf{e} + \boldsymbol{\theta}^{\mathrm{T}}\mathcal{T}\boldsymbol{\theta}, \tag{7.39}$$

*where $\boldsymbol{\tau} = [\tau_1, \ldots, \tau_K]^{\mathrm{T}}$, $\tau_k > 0$, $k = 1, 2, \ldots, K$ are regularisation parameters, and the diagonal matrix $\mathcal{T} = \mathrm{diag}\{\tau_1 \mathbf{I}_{n_1 \times n_1}, \tau_2 \mathbf{I}_{n_2 \times n_2}, \ldots, \tau_K \mathbf{I}_{n_K \times n_K}\}$, $\mathbf{I}$ is a unit matrix. The regularised least-squares estimate of $\boldsymbol{\theta}$, is given by [65]*

$$\boldsymbol{\theta} = (\mathbf{Q}^{\mathrm{T}}\mathbf{Q} + \mathcal{T})^{-1}\mathbf{Q}^{\mathrm{T}}\mathbf{y}. \tag{7.40}$$

After some simplification, it can be shown that the criterion (7.39) can be expressed as

$$\mathbf{e}^{\mathrm{T}}\mathbf{e} + \boldsymbol{\theta}^{\mathrm{T}}\mathcal{T}\boldsymbol{\theta} = \mathbf{y}^{\mathrm{T}}\mathbf{y} - \sum_{i=1}^{K} \boldsymbol{\theta}_i^{\mathrm{T}}(\mathbf{D}^{(i)} + \tau_i \mathbf{I})\boldsymbol{\theta}_i, \tag{7.41}$$

where $\mathbf{D}^{(i)} = [\mathbf{Q}^{(i)}]^{\mathrm{T}}\mathbf{Q}^{(i)}$. Normalising (7.41) by $\mathbf{y}^{\mathrm{T}}\mathbf{y}$ yields

$$\frac{\mathbf{e}^{\mathrm{T}}\mathbf{e} + \boldsymbol{\theta}^{\mathrm{T}}\mathcal{T}\boldsymbol{\theta}}{\mathbf{y}^{\mathrm{T}}\mathbf{y}} = 1 - \sum_{i=1}^{K} \frac{\boldsymbol{\theta}_i^{\mathrm{T}}(\mathbf{D}^{(i)} + \tau_i \mathbf{I})\boldsymbol{\theta}_i}{\mathbf{y}^{\mathrm{T}}\mathbf{y}}. \tag{7.42}$$

The regularised error reduction ratio $[\mathrm{rerr}]_i$ due to the sub-matrix $\mathbf{Q}^{(i)}$ is given by

$$[\mathrm{rerr}]_i = \frac{\boldsymbol{\theta}_i^{\mathrm{T}}(\mathbf{D}^{(i)} + \tau_i \mathbf{I})\boldsymbol{\theta}_i}{\mathbf{y}^{\mathrm{T}}\mathbf{y}}. \tag{7.43}$$

An appropriate choice of $\tau$ can smooth parameter estimates (noise rejection), and $\tau$ can be optimised by using a separate procedure, such as Bayesian hyper-parameter optimisation [30], or a genetic algorithm. For simplicity of illustration, it is assumed that an appropriate $\tau$ is predetermined to simplify the procedure. The regularised least-squares solution of (7.32) is given by

$$\boldsymbol{\theta}_i = [\mathbf{D}^{(i)} + \tau_i \mathbf{I}]^{-1}[\mathbf{Q}^{(i)}]^{\mathrm{T}}\mathbf{y}, \tag{7.44}$$

which follows from the fact that $\mathbf{Q}^{(i)}$, $i = 1, \ldots, K$ are mutually orthogonal subspace bases, and $\mathbf{D}^{(i)} = [\mathbf{Q}^{(i)}]^{\mathrm{T}}\mathbf{Q}^{(i)}$.

The introduction of D-optimality enhances model robustness and simplifies the model selection procedure [30]. In addition to parameter regularisation, it is natural to use a cost function to search rule base by including a composite cost function based on the D-optimality experimental design criterion [35].

**Definition 7** *D-optimality experimental design cost function in orthogonal subspaces: In experimental design, the data covariance matrix $(\boldsymbol{\Psi}^{\mathrm{T}}\boldsymbol{\Psi})$ is called the design matrix. The D-optimality design criterion maximises the determinant of the design matrix for the constructed model. Consider a model with orthogonal subspaces with design matrix as $(\mathbf{Q}^{\mathrm{T}}\mathbf{Q})$, and a subset of these subspaces is selected in order to construct an $n_{\mathrm{f}}$-subspace ($n_{\mathrm{f}} \ll K$) model that maximises the D-optimality $\det(\mathbf{Q}_{n_{\mathrm{f}}}^{\mathrm{T}}\mathbf{Q}_{n_{\mathrm{f}}})$, where $\mathbf{Q}_{n_{\mathrm{f}}}$ is a column subset of $\mathbf{Q}$ representing a constructed subset model with $n_{\mathrm{f}}$ sub-matrices selected from $\mathbf{Q}$ (consisting of $K$ sub-matrices). It is straightforward to verify that the maximisation of $\det(\mathbf{Q}_{n_{\mathrm{f}}}^{\mathrm{T}}\mathbf{Q}_{n_{\mathrm{f}}})$ is equivalent to the minimisation of $J_{\mathrm{D}} = -\log(\det(\mathbf{Q}_{n_{\mathrm{f}}}^{\mathrm{T}}\mathbf{Q}_{n_{\mathrm{f}}}))$ [35].*

Clearly

$$
\begin{aligned}
J_{\mathrm{D}} &= -\log(\det(\mathbf{Q}_{n_{\mathrm{f}}}^{\mathrm{T}}\mathbf{Q}_{n_{\mathrm{f}}})) \\
&= -\log\left[\prod_{i=1}^{n_{\mathrm{f}}}\det(\mathbf{D}^{(i)})\right] \\
&= -\sum_{i=1}^{n_{\mathrm{f}}}\log[\det(\mathbf{D}^{(i)})].
\end{aligned}
\tag{7.45}
$$

It can be easily verified that the maximisation of $\det(\mathbf{Q}_{n_{\mathrm{f}}}^{\mathrm{T}}\mathbf{Q}_{n_{\mathrm{f}}})$ is identical to the maximisation of $\det(\mathbf{\Psi}_{n_{\mathrm{f}}}^{\mathrm{T}}\mathbf{\Psi}_{n_{\mathrm{f}}})$, where $\mathbf{\Psi}_{n_{\mathrm{f}}}$ is a column subset of $\mathbf{\Psi}$, representing a constructed subset model with $n_{\mathrm{f}}$ sub-matrices selected from $\mathbf{\Psi}$ (consisting of $K$ sub-matrices) [34].

Finally aiming at improved model robustness and rule-based learning, the following algorithm combines the two separate previous works, the subspace based rule-based model construction [34] and the combined LOLS and D-optimality algorithm [30] for robust rule-based model construction. The composite cost function as defined in Definition 8 below is used in the identification algorithm.

**Definition 8** *Combined locally regularised cost function and D-optimality in orthogonal subspaces: the combined LROLS and D-optimality algorithm based on orthogonal subspace decomposition is based on the combined criterion*

$$
J_{\mathrm{c}}(\boldsymbol{\theta},\boldsymbol{\tau},\beta) = J_{\mathrm{R}}(\boldsymbol{\theta},\boldsymbol{\tau}) + \beta J_{\mathrm{D}},
\tag{7.46}
$$

*for model selection, where $\beta$ is a fixed small positive weighting for the D-optimality cost. Equivalently, a combined error reduction ratio defined as*

$$
[\mathrm{cerr}]_i = \frac{\boldsymbol{\theta}_i^{\mathrm{T}}(\mathbf{D}^{(i)} + \tau_i\mathbf{I})\boldsymbol{\theta}_i + \beta\log[\det(\mathbf{D}^{(i)})]}{\mathbf{y}^{\mathrm{T}}\mathbf{y}},
\tag{7.47}
$$

*is used for model selection, and the selection is terminated with an $n_{\mathrm{f}}$-subspace model when*

$$
[\mathrm{cerr}]_i \le 0 \quad \text{for } n_{\mathrm{f}} + 1 \le i \le K.
\tag{7.48}
$$

Given a proper $\boldsymbol{\tau}$ and a small $\beta$, the new extended Gram–Schmidt orthogonal subspace decomposition algorithm with regularisation and D-optimality for rule-based model construction is given below.

### 7.3.3   The identification algorithm

An extended classical Gram–Schmidt scheme combined with parameter regularisation and D-optimality selective criterion in orthogonal subspaces can be summarised as the following procedure:

1.  At the $j$th forward regression step, where $j \geq 1$, for $j \leq l \leq K$, compute

$$R_{i,j}^{(l)} = \begin{cases} [\mathbf{D}^{(i)}]^{-1}(\mathbf{Q}^{(i)})^T \mathbf{\Psi}^{(l)} & \text{for } i = 1, \ldots, j-1 \text{ and if } j \neq 1 \\ \mathbf{I}_{n_l \times n_l} & \text{if } i = j \end{cases}$$

$$\mathbf{Q}_{(l)}^{(j)} = \begin{cases} \mathbf{\Psi}^{(l)} & \text{if } j = 1 \\ \mathbf{\Psi}^{(l)} - \sum_{i=1}^{j-1} \mathbf{Q}^{(i)} R_{i,j}^{(l)} & \text{if } j > 1 \end{cases}$$

$$\mathbf{D}_{(l)}^{(j)} = [\mathbf{Q}_{(l)}^{(j)}]^T \mathbf{Q}_{(l)}^{(j)}, \tag{7.49}$$

$$\boldsymbol{\theta}_j^{(l)} = [\mathbf{D}_{(l)}^{(j)} + \tau_j \mathbf{I}]^{-1} [\mathbf{Q}_{(l)}^{(j)}]^T \mathbf{y} \tag{7.50}$$

$$[\text{cerr}]_j^{(l)} = \frac{[\boldsymbol{\theta}_j^{(l)}]^T (\mathbf{D}_{(l)}^{(j)} + \tau_j \mathbf{I}) \boldsymbol{\theta}_j^{(l)} + \beta \log\left[\det\left(\mathbf{D}_{(l)}^{(j)}\right)\right]}{\mathbf{y}^T \mathbf{y}}. \tag{7.51}$$

Find

$$[\text{cerr}]_j^{(l_j)} = \max\{[\text{cerr}]_j^{(l)}, \; j \leq l \leq K\}.$$

(NB: for rule base selection) \hfill (7.52)

and select

$$R_{i,j} = R_{i,j}^{(l_j)}, \quad \text{for } i = 1, \ldots, j$$

$$\mathbf{Q}^{(j)} = \mathbf{Q}_{(l_j)}^{(j)} = \mathbf{\Psi}^{(l_j)} - \sum_{i=1}^{j-1} \mathbf{Q}^{(i)} R_{i,j}$$

$$[\text{cerr}]_j = [\text{cerr}]_j^{(l_j)}$$

$$\boldsymbol{\theta}_j = \boldsymbol{\theta}_j^{(l_j)} \tag{7.53}$$

$$\mathbf{D}^{(j)} = [\mathbf{Q}^{(j)}]^T \mathbf{Q}^{(j)}$$

$$[\text{err}]_j = \frac{[\boldsymbol{\theta}_j]^T \mathbf{D}^{(j)} \boldsymbol{\theta}_j}{\mathbf{y}^T \mathbf{y}}.$$

(NB: for selected rule base energy level information extraction)

The selected sub-matrix $\mathbf{\Psi}^{(l_j)}$ exchanges columns with sub-matrix $\mathbf{\Psi}^{(j)}$. For notational convenience, all the sub-matrices will still be referred as $\mathbf{\Psi}^{(j)}$, $j = 1, \ldots, K$, according to the new column sub-matrix order $j$ in $\mathbf{\Psi}$, even if some of the column sub-matrices have been interchanged.

2.  The procedure is monitored and terminated at the derived $j = n_f$ step, when $[\text{cerr}]_{n_f} \leq 0$, for a predetermined $\beta > 0$. Otherwise, set $j = j + 1$, goto step 1.
3.  Calculate the original parameters according to (7.38).

## 7.4   Illustrative example

*Nonlinear two-dimensional surface modelling.* The MATLAB logo was generated by the first eigenfunction of the L-shaped membrane. A $51 \times 51$ meshed data set is generated by using MATLAB commands

$$x = \text{linspace}(0, 1, 51);$$

$$y = \text{linspace}(0, 1, 51);$$

$$[X, Y] = \text{meshgrid}(x, y); \qquad (7.54)$$

$$Z = \text{membrane}(1, 25);$$

such that output $Z$ is defined over a unit square input region $[0, 1]^2$. The data set $z(x, y)$, shown in Figure 7.8(a), is used to model the target function (the first eigenfunction of the L-shaped membrane function).

For both $x, y$, define a knot vector $[-0.4, -0.2, 0, 0.25, 0.5, 0.75, 1, 1.2, 1.4]$, and use a piecewise quadratic B-spline fuzzy membership function to build a one-dimensional model, resulting in $M = 6$ basis functions. These basis functions, as shown in Figure 7.7, correspond to 6 fuzzy rules: (1) IF ($x$ or $y$) is (very small) (VS);

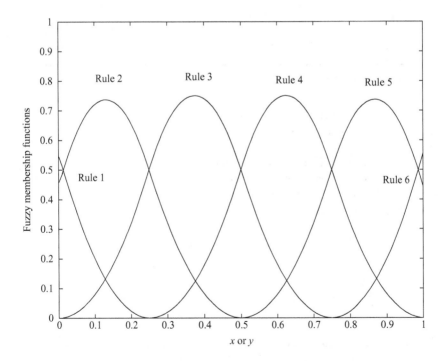

*Figure 7.7   The fuzzy membership functions for x or y. ©2004 IEEE*

*Table 7.1   Fuzzy rules identifiability in illustrative example: rules about
x and y*

| $\frac{1}{N}\sum_{t=1}^{N}N_j(t)$ | VS | S | MS | ML | L | VL |
|---|---|---|---|---|---|---|
| Rules ($x$) | 0.0510* | 0.2039* | 0.2451* | 0.2451 | 0.2039* | 0.0510 |
| Rules ($y$) | 0.0510* | 0.2039 | 0.2451 | 0.2451* | 0.2039* | 0.0510* |

The star '*' indicates rules included in the final model.

*Table 7.2   Fuzzy rules identifiability in illustrative example: rules about
x and y*

| $\frac{1}{N}\sum_{t=1}^{N}N_j(t)$ | Rules ($x$) | | | | | |
|---|---|---|---|---|---|---|
| | VS | S | MS | ML | L | VL |
| Rules ($y$) | | | | | | |
| VS | 0.0026* | 0.0104 | 0.0125 | 0.0125 | 0.0104 | 0.0026* |
| S | 0.0104 | 0.0416 | 0.0500* | 0.0500 | 0.0416 | 0.0104 |
| MS | 0.0125 | 0.0500* | 0.0601* | 0.0601* | 0.0500 | 0.0125* |
| ML | 0.0125 | 0.0500* | 0.0601* | 0.0601* | 0.0500* | 0.0125 |
| L | 0.0104 | 0.0416 | 0.0500 | 0.0500 | 0.0416 | 0.0104 |
| VL | 0.0026 | 0.0104* | 0.0125 | 0.0125 | 0.0104 | 0.0026 |

The star '*' indicates rules included in the final model.

(2) IF ($x$ or $y$) is (small) (S); (3) IF ($x$ or $y$) is (medium-small) (MS); (4) IF ($x$ or $y$) is (medium-large) (ML); (5) IF ($x$ or $y$) is (large) (L), and (6) IF ($x$ or $y$) is (very large) (VL).

The univariate and bivariate membership functions (interaction between univariate membership function $x$ and $y$ via tensor product) are used as model set and shown in Tables 7.1 and 7.2, in which, the identifiability of fuzzy rules are listed based on (7.27). From Tables 7.1 and 7.2, it is seen that all the rules have been uniformly excited. There are 48 rules.

By using the fuzzy model (7.21) for the modelling of $Z(x, y)$, the neuro-fuzzy model is simply given as

$$\hat{Z}(t) = \sum_{j=1}^{48} N_j(\mathbf{x}(t))\mathbf{x}^T\mathbf{w}_j, \tag{7.55}$$

where $t$ denotes the data label, and $\mathbf{x}(t) = [x, y]^T$ is given by the meshed values of $[x, y]$ in the input region $[0, 1]^2$. Hence each of the fuzzy rule $\Psi^{(j)} = N_j(\mathbf{x}(t))\mathbf{x}(t)$ spans a two-dimensional space, i.e. $n_j = 2$, $\forall j$. The proposed algorithm based on

*Table 7.3    System error reduction ratio by the selected rules*

| Selected rule(s) [err]$_j(t)$ | (x is ML) and (y is MS) 0.7044 | (x is MS) and (y is S) 0.1181 | (x is ML) and (y is ML) 0.0954 | (x is MS) 0.0292 | (y is ML) 0.0193 |
|---|---|---|---|---|---|
| Selected rule(s) [err]$_j(t)$ | (x is VS) 0.0086 | (x is VL) and (y is MS) 0.0056 | (x is MS) and (y is MS) 0.0040 | (x is S) and (y is MS) 0.0028 | (y is L) 0.0021 |
| Selected rule(s) [err]$_j(t)$ | (y is VL) 0.0037 | (x is S) 0.0016 | (x is VS) and (y is VS) 0.0011 | (x is MS) and (y is ML) 0.0004 | (x is S) and (y is ML) 0.0003 |
| Selected rule(s) [err]$_j(t)$ | (x is L) and (y is ML) 0.0002 | (x is VL) and (y is VS) 0.0005 | (x is L) 0.0000 | (y is VS) 0.0002 | (x is S) and (y is VL) 0.0001 |

the extended Gram–Schmidt orthogonal decomposition has been applied, in which each rule subspace being spanned by a two-dimensional rule basis is mapped into orthogonal matrix subspaces. The modelling results contain rule-based information of percentage energy increment (or the model error reduction ratio) by the selected rule to the model as shown in Table 7.3 for $\tau_j = 10^{-4}$, $\beta = 0.01$. The MSE of the resultant 20-rule model is $3.4527 \times 10^{-4}$. In Table 7.3, the selected rules are ordered in the sequence of being selected, and the model selection automatically terminates at a 20-rule model ($[cerr]_{21} < 0$). The model prediction of the 20-rule model is shown in Figure 7.8(b). For this example, the modelling results are insensitive to the value of $\tau_j$. It has been shown that by using a weighting for the D-optimality cost function, the entire model construction procedure becomes automatic. It can be seen that the model has some limitations over the modelling of corner and edge of the surface due to the data being only piecewise smooth and piecewise nonlinear. This factor may contribute to the fact that regularisation may not help in reducing misfit in some strong nonlinear behaviour region. Global nonlinear modelling using B-spline for strong nonlinear behaviour such as piecewise smooth and piecewise nonlinear data is under investigation.

## 7.5   Conclusions

Neuro-fuzzy model uses a set of fuzzy rules to model unknown dynamical systems with a B-spline functional [5,6]. T–S inference mechanism is useful to produce an operating point dependent structure facilitating the use of state–space models for data fusion or data-based controller design. This chapter aims to introduce some neuro-fuzzy model construction, design and estimation algorithms that are developed to deal with fundamental problems in data modelling, such as model sparsity, robustness, transparency and rule-based learning process. Some work

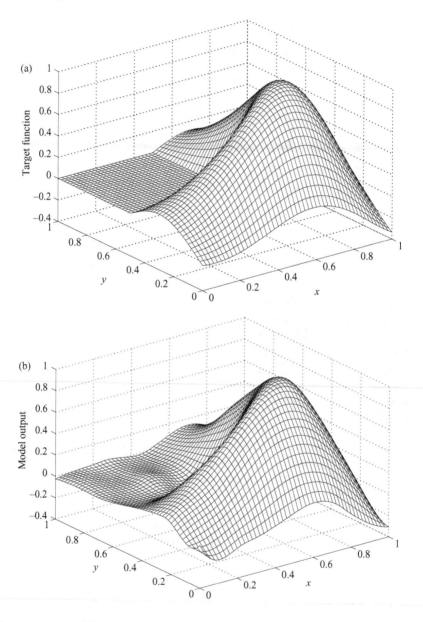

*Figure 7.8    Modelling results. ©2004 IEEE*

on ASMOD that has been derived based on ANOVA are reviewed initially and then a locally regularised orthogonal least-squares algorithm, based on T–S and ANOVA and combined with a D-optimality used for subspace-based rule selection, has been proposed for fuzzy rule regularisation and subspace-based information extraction.

# References

1 LEONTARITIS, I. J., and BILLINGS, S. A.: 'Input–output parametric models for nonlinear systems – part 1: deterministic nonlinear systems – part 2: stochastic nonlinear systems', *International Journal of Control*, 1985, **41** (1), pp. 303–44

2 WARWICK, K.: 'Dealing with complexity: a neural network approach' (Springer-Verlag, London, 1997)

3 HARRIS, C. J., MOORE, C. G., and BROWN, M.: 'Intelligent control: some aspects of fuzzy logic and neural networks' (World Scientific Press, Singapore, 1993)

4 WANG, L. X.: 'A course in fuzzy systems and control' (Prentice-Hall, New York, 1997)

5 BROWN, M., and HARRIS, C. J.: 'Neurofuzzy adaptive modelling and control' (Prentice-Hall, Hemel Hempstead, 1994)

6 HARRIS, C. J., HONG, X., and GAN, Q.: 'Adaptive modelling, estimation and fusion from data: a neurofuzzy approach' (Springer-Verlag, Heidelberg, 2002)

7 JANG, J. S. R., SUN, C. T., and MIZUTANI, E.: 'Neuro-fuzzy and soft computing: a computational approach to learning and machine intelligence' (Prentice-Hall, Upper Saddle River, NJ, 1997)

8 SINHA, N. K., and GUPTA, M. M. (Eds): 'Soft computing & intelligent systems: theory & applications' (Academic Press, New York, 2000)

9 WANG, L. X.: 'Adaptive fuzzy systems and control: design and analysis' (Prentice-Hall, New York, 1994)

10 COX, E.: 'The fuzzy systems handbook' (AP Professional, Boston, MA, 1994)

11 DRIANKOV, D., HELLENDOORN, H., and REINFRANK, M.: 'An introduction to fuzzy control' (Springer-Verlag, Berlin, 1993)

12 ZIMMERMANN, H. J.: 'Fuzzy set theory and its applications' (Kluwer Academic Publishers, Boston, MA, 1996)

13 TAKAGI, T., and SUGENO, M.: 'Fuzzy identification of systems and its applications to modelling and control', *IEEE Transactions on Systems, Man, and Cybernetics*, 1985, **15**, pp. 116–32

14 MURRAY-SMITH, R., and JOHANSEN, T. A.: 'Multiple model approaches to modelling and control' (Taylor & Francis, London, 1997)

15 FENG, M., and HARRIS, C. J.: 'Adaptive neurofuzzy control for a class of state-dependent nonlinear processes', *International Journal of Systems Science*, 1998, **29** (7), pp. 759–71

16 WANG, H., BROWN, M., and HARRIS, C. J.: 'Modelling and control of nonlinear, operating point dependent systems via associative memory networks', *Journal of Dynamics and Control*, 1996, **6**, pp. 199–218

17 GAN, Q., and HARRIS, C. J.: 'Comparison of two measurement fusion methods for Kalman-filter-based multisensor data fusion', IEEE Transactions. Aerospace and electronics systems. 2001, **37** (1), pp. 273–80

18 HARRIS, C. J., and HONG, X.: 'Neurofuzzy control using Kalman filtering state feedback with coloured noise for unknown nonlinear processes'. *Proceedings of*

the *IMechE Part I, Journal of Systems and Control Engineering*, 2001, **215** (5), pp. 423–35

19   BELLMAN, R.: 'Adaptive control processes' (Princeton University Press, Princeton, NJ, 1966)

20   KAVLI, T.: 'ASMOD – an algorithm for adaptive spline modelling of observation data', *International Journal of Control*, 1993, **58** (4), pp. 947–67

21   BOSSLEY, K. M.: 'Neurofuzzy modelling approaches in system identification'. Ph.D. thesis, Department of ECS, University of Southampton, 1997

22   CHEN, S., BILLINGS, S. A., and LUO, W.: 'Orthogonal least squares methods and their applications to non-linear system identification', *International Journal of Control*, 1989, **50**, pp. 1873–96

23   CHEN, S., WU, Y., and LUK, B. L.: 'Combined genetic algorithm optimization and regularized orthogonal least squares learning for radial basis function networks', *IEEE Transactions on Neural Networks*, 1999, **10**, pp. 1239–43

24   ORR, M. J. L.: 'Regularisation in the selection of radial basis function centers', *Neural Computation*, 1995, **7** (3), pp. 954–75

25   AKAIKE, H.: 'A new look at the statistical model identification', *IEEE Transactions on Automatic Control*, 1974, **AC-19**, pp. 716–23

26   ATKINSON, A. C., and DONEV, A. N.: 'Optimum experimental designs' (Clarendon Press, Oxford, 1992)

27   HONG, X., and HARRIS, C. J.: 'Nonlinear model structure detection using optimum experimental design and orthogonal least squares', *IEEE Transactions on Neural Networks*, 2001, **12** (2), pp. 435–9

28   HONG, X., and HARRIS, C. J.: 'Nonlinear model structure design and construction using orthogonal least squares and d-optimality design', *IEEE Transactions on Neural Networks*, 2001, **13** (5), pp. 1245–50

29   CHEN, S.: 'Locally regularised orthogonal least squares algorithm for the construction of sparse kernel regression models'. Proceedings of the sixth international conference on *Signal Processing*, Beijing, China, 26–30 August, 2002, pp. 1229–32

30   CHEN, S., HONG, X., and HARRIS, C. J.: 'Sparse kernel regression modelling using combined locally regularised orthogonal least squares and d-optimality experimental design', *IEEE Transactions on Automatic Control*, 2003, **48** (6), pp. 1029–36

31   TANIGUCHI, T., TANAKA, K., OHTAKE, H., and WANG, H. O.: 'Model construction, rule reduction and rubust compensation for generalised form of Takagi–Sugeno fuzzy systems', *IEEE Transactions on Fuzzy Systems*, 2001, **9** (4), pp. 525–38

32   JIN, Y.: 'Fuzzy modelling of high dimensional systems: complexity reduction and interpretability improvement', *IEEE Transactions on Fuzzy Systems*, 2000, **8** (2), pp. 212–21

33   ROUBOS, H., and SETNES, M.: 'Compact and transparent fuzzy models and classifiers through iterative complexity reduction', *IEEE Transactions on Fuzzy Systems*, 2001, **9** (4), pp. 516–24

34  HONG, X., and HARRIS, C. J.: 'A neurofuzzy network knowledge extraction and extended Gram–Schmidt algorithm for model subspace decomposition', *IEEE Transactions on Fuzzy Systems*, 2003, **11** (4), pp. 528–41

35  HONG, X., HARRIS, C. J., and CHEN, S.: 'Robust neurofuzzy rule base knowledge extraction and estimation using subspace decomposition combined with regularization and d-optimality', *IEEE Transactions on System, Man, and Cybernetics, Part B*, 2004, **34** (1), pp. 598–608

36  ZADEH, L. A.: 'Fuzzy sets', *Information and Control*, 1965, **8**, pp. 338–53

37  WOLKENHAUER, O.: 'Possibility theory with application to data analysis' (Research Studies Press Ltd., John Wiley and Sons, London, 1998)

38  DIERCKX, P.: 'Curve and surface fitting with splines' (Clarendon Press, Oxford, 1995)

39  HUNT, K. J., HAAS, R., and BROWN, M.: 'On the functional equivalence of fuzzy inference systems and spline-based networks', *International Journal of Neural Systems*, 1995, **6** (2), pp. 171–84

40  LANE, S. H.: 'Theory and development of higher order cmac neural networks', *IEEE Control Systems Magazine*, 1992, **12** (2), pp. 23–30

41  BOSSLEY, K. M., BROWN, M., and HARRIS, C. J.: 'Neurofuzzy model construction for the modelling of non-linear processes'. Proceedings of the third European control conference, vol. 3, Rome, Italy, 1995, pp. 2438–43

42  BOSSLEY, K. M., BROWN, M., and HARRIS, C. J.: 'Neurofuzzy identification of an autonomous underwater vehicle', *International Journal of Systems Science*, 1999, **30** (9), pp. 901–13

43  BROWN, M., GUNN, S. R., NG, C. N., and HARRIS, C. J.: 'Neurofuzzy modelling: a transparent approach', in WARWICK, K. (Ed.): 'Dealing with complexity: a neural network approach' (Springer-Verlag, Berlin, 1997) pp. 110–24

44  FEMMINELLA, O. P., STARINK, M. J., BROWN, M., SINCLAIR, I., HARRIS, C. J., and REED, P. A. S.: 'Data preprocessing/model initialisation in neurofuzzy modelling of structure property relationships in Al–Zn–Mg–Cu alloys', *ISIJ*, 1999, **39** (10), pp. 1027–37

45  FERADAY, S.: 'Intelligent approaches to modelling and interpreting disc brake squeal data'. Technical report, Ph.D. thesis, Department of ECS, University of Southampton, 2001

46  ALFORD, A., and HARRIS, C. J.: 'Using b-splines to represent optimal actuator demands for systems featuring position/rate limited actuators'. Proceedings of the fifth IFAC workshop, *AIARTC'98*, Mexico, 1998, pp. 271–6

47  BRIDGETT, N. A., and HARRIS, C. J.: 'Neurofuzzy identification and control of a gas turbine jet engine'. Proceedings of the international symposium on *Signal processing, robotics and neural networks*, Lille, France, 1994, pp. 462–7

48  BRIDGETT, N. A., BRANDT, J., and HARRIS, C. J.: 'A neural network for use in the diagnosis and treatment of breast cancer'. Proceedings of the fourth International Conference on *ANN*, Cambridge, UK, 1995, pp. 448–53

49  HARRIS, C. J., BAILEY, A., and DODD, T. J.: 'Multi-sensor data fusion in defence and aerospace', *The Aeronautical Journal*, 1998, **102** (1015), pp. 229–44

50  AN, P. E., and HARRIS, C. J.: 'An intelligent driver warning system for vehicle collision avoidance', *IEEE Transactions on Systems, Man, and Cybernetics*, 1995, **26** (2), pp. 254–61

51  HARRIS, C. J., ROBERTS, J. M., and AN, P. E.: 'An intelligent driver warning system', in JAIN, J. C., JOHNSON, R. P., TAKEFUA, Y., ZADEH, L.A. (Eds): 'Knowledge based intelligence techniques in industry' (CRC Press, Boca Raton, FL, 1998) pp. 1–52

52  HALL, M., and HARRIS, C. J.: 'Neurofuzzy systems for command and control'. Proceedings of the $C^2$ *research and technology symposium*, Montercy, Naval PGS, 1998, pp. 461–71

53  GAN, Q., and HARRIS, C. J.: 'A learning scheme combining EM and MASMOD algorithms for nonlinear system modelling', *IEEE Transactions on Neural Networks*, 2001, **12** (1), pp. 43–53

54  GAN, Q., and HARRIS, C. J.: 'Fuzzy local linearisation and local basis function expansion in nonlinear system modelling', *IEEE Transactions on Systems, Man, and Cybernetics, Part B*, 1999, **29** (4), pp. 559–65

55  GAN, Q., and HARRIS, C. J.: 'Neurofuzzy state estimators using a modified ASMOD and Kalman filter algorithm'. Proceedings of the international conference on *Computational intelligence for modelling control and automation*, vol. 1, Vienna, Austria, 1999, pp. 214–19

56  DEMPSTER, A. P., LAIRD, N. M., and RUBIN, D. B.: 'Maximum likelihood from incomplete data via the EM algorithm (with discussions)', *Journal of the Royal Statistical Society: B*, 1977, **39**, pp. 1–39

57  JORDAN, M., and JACOBS, R. A.: 'Hierarchical mixtures of experts and the EM algorithm', *Neural Computation*, 1994, **6**, pp. 181–214

58  JORDAN, M., and XU, L.: 'Convergence results for the EM approach to mixtures of experts architectures', *Neural Networks*, 1995, **8** (9), pp. 1409–31

59  XU, L., and JORDAN, M. I.: 'On convergence properties of the EM algorithm for Gaussian mixtures', *Neural Computation*, 1996, **8**, pp. 129–51

60  GERSHENFELD, N., SCHONER, B., and METOIS, E.: 'Cluster-weighted modeling for time-series analysis', *Nature*, 1999, **397**, pp. 329–32

61  GAN, Q., and HARRIS, C. J.: 'Linearisation and state estimation of unknown discrete-time nonlinear dynamic systems using recurrent neurofuzzy networks', *IEEE Transactions on Systems, Man, and Cybernetics, Part B*, 1999, **29** (6), pp. 802–17

62  WU, Z.-Q., and HARRIS, C. J.: 'A neurofuzzy network structure for modelling and state estimation of unknown nonlinear systems', *International Journal of Systems Science*, 1997, **28**, pp. 335–45

63  GRUENBERG, K. W., and WEIR, A. J.: 'Linear geometry' (D. Van Nostrand Company, Inc., New York, 1967)

64  SODERSTRÖM, T., and STOICA, P.: 'System identification' (Prentice-Hall, New York, 1989)

65  MARQUARDT, D. W.: 'Generalised inverse, ridge regression, biased linear estimation and nonlinear estimation', *Technometrics*, 1970, **12** (3), pp. 591–612

*Chapter 8*

# A neural network approach for nearly optimal control of constrained nonlinear systems

*Murad Abu-Khalaf and Frank L. Lewis*

## 8.1   Introduction

The design of control systems requires one to consider various types of constraints and performance measures. Constraints encountered in control system design are due to physical limitations imposed on the controller and the plant. This includes actuator saturation, constraints on the states and others. Performance measures on the other hand are related to optimality issues. This includes minimum fuel, minimum energy, minimum time and other optimal control problems. In this chapter, a unified framework for constructing nearly optimal closed-loop controllers for constrained control systems is presented.

The control of systems with saturating actuators has been the focus of many researchers for many years. Several methods for deriving control laws considering the saturation phenomena are found in Saberi *et al.* [1] and Sussmann *et al.* [2]. Other methods that deal with constraints on the states of the system as well as the control inputs are found in Bitsoris and Gravalou [3], Hu *et al.* [4], Henrion *et al.* [5] and Gilbert and Tan [6]. This line of research is based on mathematical programming and the set invariance theory to design controllers that satisfy the required constraints. The controllers developed are not necessarily in closed-loop form. Moreover, optimality issues are not the main concern in this theme of work. Most of these methods do not consider finding optimal control laws for general nonlinear systems.

The optimal control of constrained input systems is well established. The controller can be found by applying the Pontryagin's minimum principle. This usually requires solving a split boundary differential equation and the result is an open-loop optimal control [7].

There have been several studies to derive and solve for closed-loop optimal control laws for constrained input systems. Bernstein [8] studied the performance optimisation of saturated actuators control. Lyshevski and Meyer [9] and Lyshevski [10] presented a general framework for the design of optimal state feedback control laws. They propose the use of nonquadratic performance functionals that enables encoding various kinds of constraints on the control system. These performance functionals are used along with the famous Hamilton–Jacobi–Bellman (HJB) equation that appears in optimal control theory [7]. The controllers are closed loop and are in terms of the value function which satisfies the HJB equation. However, it remains unclear how to solve the HJB equations formulated using nonquadratic performance functionals.

In this chapter, we study the constrained optimal control problem through the framework of the HJB equation. We show how to systematically obtain approximate solutions of the HJB equation when nonquadratic performance functions are used. The solution of the HJB equation is a challenging problem due to its inherently nonlinear nature. For linear systems, this equation results in the well-known Riccati equation used to derive a linear state feedback control. But even when the system is linear, the saturated control requirement makes the required control nonlinear, and makes the solution of the HJB equation more challenging.

In the general nonlinear case, the HJB equation generally cannot be solved for explicitly. There has been a great deal of effort to confront this issue. Approximate HJB solutions have been found using many techniques such as those developed by Saridis and Lee [11], Beard *et al.* [12,13], Beard [14], Murray *et al.* [15], Lee *et al.* [16], Bertsekas and Tsitsiklis [17,18], Munos *et al.* [19], Kim *et al.* [20], Han and Balakrishnan [21], Liu and Balakrishnan [22,23], Lyshevski [10,24–26], Lyshevski and Meyer [9] and Huang and Lin [27].

In this presentation, we focus on solving the HJB solution using the so-called generalised HJB equation (GHJB) which we will refer to as Lyapunov equation (LE) [11,14]. Saridis and Lee [11] developed a successive approximation method that improves a given initial stabilising control. This method reduces to the well-known Kleinman iterative method for solving the Riccati equation for linear systems [28]. However, for nonlinear systems, it is unclear how to solve the LE. Therefore, successful application of the LE was limited until the novel work of Beard and co-workers [12–14]. He uses a Galerkin spectral approximation method to find approximate solutions to the LE at each iteration on a given compact set. The framework in which the algorithm is presented in Beard's work requires the computation of a large number of integrals and it is also not able to handle explicit constraints on the controls, which is what we are interested in.

In this chapter, we apply the successive approximation theory for performance functionals that are nonquadratic. We use neural networks to solve for the value function of the HJB equation, and to construct a nearly optimal constrained state feedback controller.

Neural networks have been used to control nonlinear systems. Werbos [29] first proposed using neural networks to find optimal control laws using the HJB equation. Parisini [30] used neural networks to derive optimal control laws for a discrete-time stochastic nonlinear system. Successful neural network controllers have been reported in References 31–36. It has been shown that neural networks can effectively extend adaptive control techniques to nonlinearly parameterised systems. The status of neural network control as of 2001 appears in Reference 37.

In summary, the objective of this chapter is to derive optimal control laws for systems with constraints, and construct those using neural networks that have been tuned offline. For constrained input systems, we present two optimal control problems. The first is a regular optimal saturated regulator, while the second is a minimum time optimal control problem. Therefore, in the next section of this chapter we introduce the HJB equation for constrained input systems using nonquadratic performance functions. We also show the LE that will be useful in implementing the successive approximation theory. In Section 8.3, we derive the successive solution theory for nonquadratic performance functions. It will be shown that instead of solving for the value function using the HJB directly, we can solve for a sequence of cost functions through the LE that converge uniformly to the value function that solves the HJB equation. In Section 8.4, we show how to construct nonquadratic performance functional to address minimum time and constrained state problems. In Section 8.5, we show how to use neural networks to solve for the cost function of the LE over a predefined region of the state space in a least squares sense. In Section 8.6 numerical examples are given to demonstrate the techniques presented in this chapter and that serve as a tutorial for other dynamical systems. The solution technique of this chapter combines the successive approximation with the method of weighted residuals to get a least-squares solution of the HJB that is formulated using a nonquadratic functional that encodes the constraints on the input.

## 8.2 Nearly optimal regulation of systems with actuator saturation

Consider an affine in the control nonlinear dynamical system of the form

$$\dot{x} = f(x) + g(x)u(x), \tag{8.1}$$

where $x \in \mathbb{R}^n$, $f(x) \in \mathbb{R}^n$, $g(x) \in \mathbb{R}^{n \times m}$. And the input $u \in U$, $U = \{u = (u_1, \ldots, u_m) \in \mathbb{R}^m : \alpha_i \leq u_i \leq \beta_i, \ i = 1, \ldots, m\}$, where $\alpha_i, \beta_i$ are constants. Assume that $f + gu$ is Lipschitz continuous on a set $\Omega \subseteq \mathbb{R}^n$ containing the origin, and that the system (8.1) is stabilisable in the sense that there exists a continuous control on $\Omega$ that asymptotically stabilises the system. It is desired to find $u$, which minimises a generalised nonquadratic functional

$$V(x_0) = \int_0^\infty [Q(x) + W(u)] \, dt, \tag{8.2}$$

where $Q(x)$ and $W(u)$ are positive definite functions on $\Omega$, $\forall x \neq 0$, $Q(x) > 0$ and $x = 0 \Rightarrow Q(x) = 0$. For unbounded control inputs, a common choice for $W(u)$ is $W(u) = u^T R u$, where $R \in \mathbb{R}^{m \times m}$. Note that the control $u$ must not only stabilise the system on $\Omega$, but also make the integral finite. Such controls are defined to be admissible [13].

**Definition 1**   *(Admissible controls) A control u is defined to be admissible with respect to equation (8.2) on $\Omega$, denoted by $u \in \Psi(\Omega)$, if u is continuous on $\Omega$; $u(0) = 0$; u stabilises (8.1) on $\Omega$; $\forall x_0 \in \Omega$, $V(x_0)$ is finite.*

Equation (8.2) can be expanded as follows:

$$V(x_0) = \int_0^T [Q(x) + W(u)]\, dt + \int_T^\infty [Q(x) + W(u)]\, dt$$

$$= \int_0^T [Q(x) + W(u)]\, dt + V(x(T)). \tag{8.3}$$

If the cost function $V$ is differentiable at $x_0$, then rewriting equation (8.3)

$$\lim_{T \to 0} \frac{V(x_0) - V(x(T))}{T} = \lim_{T \to 0} \frac{1}{T} \int_0^T [Q(x) + W(u)]\, dt,$$

$$\dot{V} = \frac{dV}{dx}^T (f + gu) = -Q(x) - W(u). \tag{8.4}$$

Equation (8.4) is the infinitesimal version of equation (8.2) and is known as a nonlinear Lyapunov equation,

$$LE(V, u) \triangleq \frac{dV}{dx}^T (f + gu) + Q + W(u) = 0, \quad V(0) = 0. \tag{8.5}$$

The LE becomes the well-known HJB equation [7], on substitution of the optimal control

$$u^*(x) = -\frac{1}{2} R^{-1} g^T \frac{dV^*}{dx}, \tag{8.6}$$

where $V^*(x)$ is the value function of the optimal control problem which solves the HJB equation

$$HJB(V^*) \triangleq \frac{dV^*}{dx}^T f + Q - \frac{1}{4} \frac{dV^*}{dx}^T g R^{-1} g^T \frac{dV^*}{dx} = 0, \quad V^*(0) = 0. \tag{8.7}$$

It is shown in Reference 9 that the value function obtained from (8.7) serves as a Lyapunov function on $\Omega$.

To confront bounded controls, Lyshevski and Meyer [9] and Lyshevski [10] introduced a generalised nonquadratic functional

$$W(u) = 2 \int_0^u (\phi^{-1}(v))^T R\, dv,$$

$$v \in \mathbb{R}^m, \quad \phi \in \mathbb{R}^m, \quad \phi(v) = \begin{bmatrix} \phi(v_1) \\ \vdots \\ \phi(v_m) \end{bmatrix}, \quad \phi^{-1}(u) = \begin{bmatrix} \phi^{-1}(u_1) \\ \vdots \\ \phi^{-1}(u_m) \end{bmatrix}, \tag{8.8}$$

where $\phi(\cdot)$ is a bounded one-to-one function that belongs to $C^p$ ($p \geq 1$) and $L_2(\Omega)$. Moreover, it is a monotonic odd function with its first derivative bounded by the constant $M$. An example of such a function is the hyperbolic tangent $\phi(\cdot) = \tanh(\cdot)$. $R$ is positive definite and assumed to be symmetric for simplicity of analysis. Note that $W(u)$ is positive definite because $\phi^{-1}(u)$ is monotonic odd and $R$ is positive definite.

The LE when (8.8) is used becomes

$$\frac{dV^T}{dx}(f + g \cdot u) + Q + 2\int_0^u \phi^{-T}(v)R\,dv = 0, \qquad V(0) = 0. \tag{8.9}$$

Note that the LE becomes the HJB equation upon substituting the constrained optimal feedback control

$$u^*(x) = -\phi\left(\frac{1}{2}R^{-1}g^T\frac{dV^*}{dx}\right), \tag{8.10}$$

where $V^*(x)$ solves the following HJB equation

$$\frac{dV^{*T}}{dx}\left(f - g\phi\left(\frac{1}{2}R^{-1}g^T\frac{dV^*}{dx}\right)\right) + Q$$
$$+ 2\int_0^{-\phi((1/2)R^{-1}g^T(dV^*/dx))} \phi^{-T}(v)R\,dv = 0, \qquad V^*(0) = 0. \tag{8.11}$$

This is a nonlinear differential equation for which there may be many solutions. Existence and uniqueness of the value function has been shown in Reference 26. This HJB equation cannot generally be solved. There is no current method for rigorously confronting this type of equation to find the value function for the system. Moreover, current solutions are not well defined over a specific region in the state space.

**Remark 1** *Optimal control problems do not necessarily have smooth or even continuous value functions [38,39]. In Reference 40, using the theory of viscosity solutions, it is shown that for infinite horizon optimal control problems with unbounded cost functionals and under certain continuity assumptions of the dynamics, the value function is continuous, $V^*(x) \in C(\Omega)$. Moreover, if the Hamiltonian is strictly convex and if the continuous viscosity is semiconcave, then $V^*(x) \in C^1(\Omega)$ [39], satisfying the HJB equation everywhere. Note that for affine in input systems, (8.1), the Hamiltonian is strictly convex if the system dynamics are not bilinear, and if the integrand of the performance functional (8.2) does not have cross terms of the states and the input. In this chapter, we perform all derivations under the assumption of smooth solutions to (8.9) and (8.11) with all that this requires of necessary conditions. See References 11 and 41 for similar framework of solutions. If this smoothness assumption is released, then one needs to use the theory of viscosity solutions [39], to show that the continuous cost solutions of (8.9) do converge to the continuous value function of (8.11).*

## 8.3    Successive approximation of HJB for constrained inputs systems

It is important to note that the LE is linear in the cost function derivative, while the HJB is nonlinear in the value function derivative. Solving the LE for the cost function requires solving a linear partial differential equation, while the HJB equation solution involves a nonlinear partial differential equation, which may be impossible to solve. This is the reason for introducing the successive approximation technique for the solution of the LE, which is based on a sound proof in Reference 11.

Successive approximation using the LE has not yet been rigorously applied for bounded controls. In this section, we will show that the successive approximation technique can be used for constrained controls when certain restrictions on the control input are met. Then, having the successive approximation theory well set, in the next section we will introduce a neural network approximation of the value function, and employ the successive solutions method in a least-squares sense over a mesh with certain size on $\Omega$. This is far simpler than the Galerkin approximation appearing in References 12 and 13.

The successive approximation technique is now applied to the new set of equations (8.9) and (8.10). The following lemma shows how equation (8.10) can be used to improve the control law. It will be required that the bounding function $\phi(\cdot)$ is non-decreasing.

**Lemma 1**    *If $u^{(i)} \in \Psi(\Omega)$, and $V^{(i)} \in C^1(\Omega)$ satisfies the equation $LE(V^{(i)}, u^{(i)}) = 0$ with the boundary condition $V^{(i)}(0) = 0$, then the new control derived as*

$$u^{(i+1)}(x) = -\phi\left(\frac{1}{2}R^{-1}g^T\frac{dV^{(i)}}{dx}\right), \tag{8.12}$$

*is an admissible control for the system on $\Omega$. Moreover, if the bounding function $\phi(\cdot)$ is a monotone odd function, and $V^{(i+1)}$ is the unique positive definite function satisfying equation $LE(V^{(i+1)}, u^{(i+1)}) = 0$, with the boundary condition $V^{(i+1)}(0) = 0$, then $V^*(x) \le V^{(i+1)}(x) \le V^{(i)}(x), \forall x \in \Omega$.*

**Proof**    To show the admissibility part, since $V^{(i)} \in C^1(\Omega)$, the continuity assumption on $g$ implies that $u^{(i+1)}$ is continuous. Since $V^{(i)}$ is positive definite it attains a minimum at the origin, and thus, $dV^{(i)}/dx$ must vanish there. This implies that $u^{(i+1)}(0) = 0$. Taking the derivative of $V^{(i)}$ along the system $f + gu^{(i+1)}$ trajectory we have,

$$\dot{V}^{(i)}(x, u^{(i+1)}) = \frac{dV^{(i)}}{dx}^T f + \frac{dV^{(i)}}{dx}^T gu^{(i+1)}, \tag{8.13}$$

$$\frac{dV^{(i)}}{dx}^T f = -\frac{dV^{(i)}}{dx}^T gu^{(i)} - Q - 2\int_0^{u^{(i)}} \phi^{-T}(v)R\,dv. \tag{8.14}$$

Therefore equation (8.13) becomes

$$\dot{V}^{(i)}(x, u^{(i+1)}) = -\frac{\mathrm{d}V^{(i)}}{\mathrm{d}x}^T g u^{(i)} + \frac{\mathrm{d}V^{(i)}}{\mathrm{d}x}^T g u^{(i+1)} - Q - 2\int_0^{u^{(i)}} \phi^{-T}(v) R \, \mathrm{d}v.$$

(8.15)

Since $\mathrm{d}V^{(i)}/\mathrm{d}x^T g(x) = -2\phi^{-T}(u^{(i+1)})R$, we get

$$\dot{V}^{(i)}(x, u^{(i+1)}) = -Q + 2\left[\phi^{-T}(u^{(i+1)})R(u^{(i)} - u^{(i+1)}) - \int_0^{u^{(i)}} \phi^{-T}(v) R \, \mathrm{d}v\right].$$

(8.16)

The second term in the previous equation is negative when $\phi^{-1}$, and hence $\phi$ is non-decreasing. To see this, note that the design matrix $R$ is symmetric positive definite, this means we can rewrite it as $R = \Lambda \Sigma \Lambda$, where $\Sigma$ is a triangular matrix with its values being the singular values of $R$ and $\Lambda$ is an orthogonal symmetric matrix. Substituting for $R$ in (8.16) we get,

$$\dot{V}^{(i)}(x, u^{(i+1)})$$
$$= -Q + 2\left[\phi^{-T}(u^{(i+1)})\Lambda \Sigma \Lambda(u^{(i)} - u^{(i+1)}) - \int_0^{u^{(i)}} \phi^{-T}(v)\Lambda \Sigma \Lambda \, \mathrm{d}v\right].$$

(8.17)

Applying the coordinate change $u = \Lambda^{-1}z$ to (8.17)

$$\dot{V}^{(i)}(x, u^{(i+1)})$$
$$= -Q + 2\phi^{-T}(\Lambda^{-1}z^{(i+1)})\Lambda \Sigma \Lambda(\Lambda^{-1}z^{(i)} - \Lambda^{-1}z^{(i+1)})$$
$$\quad - 2\int_0^{z^{(i)}} \phi^{-T}(\Lambda^{-1}\zeta)\Lambda \Sigma \Lambda \Lambda^{-1} \, \mathrm{d}\zeta$$
$$= -Q + 2\phi^{-T}(\Lambda^{-1}z^{(i+1)})\Lambda \Sigma(z^{(i)} - z^{(i+1)}) - 2\int_0^{z^{(i)}} \phi^{-T}(\Lambda^{-1}\zeta)\Lambda \Sigma \, \mathrm{d}\zeta$$
$$= -Q + 2\pi^T(z^{(i+1)})\Sigma(z^{(i)} - z^{(i+1)}) - 2\int_0^{z^{(i)}} \pi^T(\zeta)\Sigma \, \mathrm{d}\zeta,$$

(8.18)

where $\pi^T(z^{(i)}) = \phi^{-1}(\Lambda^{-1}z^{(i)})^T \Lambda$.

Since $\Sigma$ is a triangular matrix, we can now decouple the transformed input vector such that

$$\dot{V}^{(i)}(x, u^{(i+1)})$$

$$= -Q + 2\pi^T(z^{(i+1)})\Sigma(z^{(i)} - z^{(i+1)}) - 2\int_0^{z_k^{(i)}} \pi^T(\zeta)\Sigma\,d\zeta$$

$$= -Q + 2\sum_{k=1}^{m} \Sigma_{kk} \left[ \pi(z_k^{(i+1)})(z_k^{(i)} - z_k^{(i+1)}) - \int_0^{z_k^{(i)}} \pi(\zeta_k)\,d\zeta_k \right]. \qquad (8.19)$$

Since the matrix $R$ is positive definite, then we have the singular values $\Sigma_{kk}$ being all positive. Also, from the geometrical meaning of

$$\pi\left(z_k^{(i+1)}\right)\left(z_k^{(i)} - z_k^{(i+1)}\right) - \int_0^{z_k^{(i)}} \pi(\zeta_k)\,d\zeta_k,$$

this term is always negative if $\pi(z_k)$ is monotone and odd. Because $\phi(\cdot)$ is monotone and odd, and because it is a one-to-one function, it follows that $\phi^{-1}(\cdot)$ is odd and monotone. Hence, since $\pi^T(z^{(i)}) = \phi^{-1}(\Lambda^{-1}z^{(i)})^T\Lambda$, it follows that $\pi(z_k)$ is monotone and odd. This implies that $V^{(i)}(x, u^{(i+1)}) \leq 0$ and that $V^{(i)}(x)$ is a Lyapunov function for $u^{(i+1)}$ on $\Omega$. Following Definition 1, $u^{(i+1)}$ is admissible on $\Omega$.

For the second part of the lemma, along the trajectories of $f + gu^{(i+1)}$, and $\forall x_0$ we have

$$V^{(i+1)} - V^{(i)}$$

$$= \int_0^\infty \left\{ Q(x(\tau, x_0, u^{(i+1)})) + 2\int_0^{u^{(i+1)}(x(\tau, x_0, u^{(i+1)}))} \phi^{-T}(v)R\,dv \right\} d\tau$$

$$- \int_0^\infty \left\{ Q(x(\tau, x_0, u^{(i+1)})) + 2\int_0^{u^{(i)}(x(\tau, x_0, u^{(i+1)}))} \phi^{-T}(v)R\,dv \right\} d\tau$$

$$= -\int_0^\infty \frac{d(V^{(i+1)} - V^{(i)})^T}{dx}[f + g\,u^{(i+1)}]\,d\tau. \qquad (8.20)$$

Because $\text{LE}(V^{(i+1)}, u^{(i+1)}) = 0$, $\text{LE}(V^{(i)}, u^{(i)}) = 0$

$$\frac{dV^{(i)}}{dx}^T f = -\frac{dV^{(i)}}{dx}^T gu^{(i)} - Q - 2\int_0^{u^{(i)}} \phi^{-T}(v)R\,dv, \qquad (8.21)$$

$$\frac{dV^{(i+1)}}{dx}^T f = -\frac{dV^{(i+1)}}{dx}^T gu^{(i+1)} - Q - 2\int_0^{u^{(i+1)}} \phi^{-T}(v)R\,dv. \qquad (8.22)$$

Substituting (8.21) and (8.22) in (8.20) we get

$$V^{(i+1)}(x_0) - V^{(i)}(x_0)$$

$$= -2 \int_0^\infty \left\{ \phi^{-T}(u^{(i+1)}) R(u^{(i+1)} - u^{(i)}) - \int_{u^{(i)}}^{u^{(i+1)}} \phi^{-T}(v) R \, dv \right\} d\tau.$$

$$(8.23)$$

By decoupling equation (8.23) using $R = \Lambda \Sigma \Lambda$, it can be shown that $V^{(i+1)}(x_0) - V^{(i)}(x_0) \leq 0$ when $\phi(\cdot)$ is non-decreasing. Moreover, it can be shown by contradiction that $V^*(x_0) \leq V^{(i+1)}(x_0)$. ∎

The next theorem is a key result on which the rest of the chapter is justified. It shows that successive improvement of the saturated control law converges to the optimal saturated control law for the given actuator saturation model $\phi(\cdot)$. But first we need the following definition.

**Definition 2**  *Uniform convergence: a sequence of functions $\{f_n\}$ converges uniformly to $f$ on a set $\Omega$ if $\forall \varepsilon > 0$, $\exists N(\varepsilon): n > N \Rightarrow |f_n(x) - f(x)| < \varepsilon \ \forall x \in \Omega$, or equivalently $\sup|_{x \in \Omega} f_n(x) - f(x)| < \varepsilon$, where $||$ is the absolute value.*

**Theorem 1**  *If $u^{(0)} \in \Psi(\Omega)$, then $u^{(i)} \in \Psi(\Omega)$, $\forall i \geq 0$. Moreover, $V^{(i)} \to V^*$, $u^{(i)} \to u^*$ uniformly on $\Omega$.*

**Proof**  From Lemma 1, it can be shown by induction that $u^{(i)} \in \Psi(\Omega)$, $\forall i \geq 0$. Furthermore, Lemma 1 shows that $V^{(i)}$ is a monotonically decreasing sequence and bounded below by $V^*(x)$. Hence $V^{(i)}$ converges pointwise to $V^{(\infty)}$. Because $\Omega$ is compact, then uniform convergence follows immediately from Dini's theorem [42]. Due to the uniqueness of the value function [7,26], it follows that $V^{(\infty)} = V^*$. Controllers $u^{(i)}$ are admissible, therefore they are continuous having unique trajectories due to the locally Lipschitz continuity assumptions on the dynamics. Since (8.2) converges uniformly to $V^*$, this implies that system's trajectories converges to $\forall x_0 \in \Omega$. Therefore $u^{(i)} \to u^{(\infty)}$ uniformly on $\Omega$. If $dV^{(i)}/dx$ converges uniformly to $dV^*/dx$, we conclude that $u^{(\infty)} = u^*$. To prove that $dV^{(i)}/dx \to dV^*/dx$ uniformly on $\Omega$, note that $dV^{(i)}/dx$ converges uniformly to some continuous function $J$. Since $V^{(i)} \to V^*$ uniformly and $dV^{(i)}/dx$ exists $\forall i$, hence it follows that the sequence $dV^{(i)}/dx$ is term-by-term differentiable [42], and $J = dV^*/dx$. ∎

The following is a result from Reference 14 which we tailor here to the case of saturated control inputs. It basically guarantees that improving the control law does not reduce the region of asymptotic stability of the initial saturated control law.

**Corollary 1**  *If $\Omega^*$ denotes the region of asymptotic stability of the constrained optimal control $u^*$, then $\Omega^*$ is the largest region of asymptotic stability of any other admissible control law.*

**Proof** The proof is by contradiction. Lemma 1 showed that the saturated control $u^*$ is asymptotically stable on $\Omega^{(0)}$, where $\Omega^{(0)}$ is the stability region of the saturated control $u^{(0)}$. Assume that $u_{\text{Largest}}$ is an admissible controller with the largest region of asymptotic stability $\Omega_{\text{Largest}}$. Then, there is $x_0 \in \Omega_{\text{Largest}}$, $x_0 \notin \Omega^*$. From Theorem 1, $x_0 \in \Omega^*$ which completes the proof.    ∎

Note that there may be stabilising saturated controls that have larger stability regions than $u^*$, but are not admissible with respect to $Q(x)$ and the system $(f, g)$.

## 8.4  Nonquadratic performance functionals for minimum time and constrained states control

### 8.4.1  Minimum time problems

For a system with saturated actuators, we want to find the control signal required to drive the system to the origin in minimum time. This is done through the following performance functional

$$V = \int_0^\infty \left[ \tanh(x^T Q x) + 2 \int_0^u (\phi^{-1}(v))^T R \, dv \right] dt. \tag{8.24}$$

By choosing the coefficients of the weighting matrix $R$ very small, and for $x^T Q x \gg 0$, the performance functional becomes,

$$V = \int_0^{t_s} 1 \, dt, \tag{8.25}$$

and for $x^T Q x \approx 0$, the performance functional becomes,

$$V = \int_{t_s}^\infty \left[ x^T Q x + 2 \int_0^u (\phi^{-1}(v))^T R \, dv \right] dt. \tag{8.26}$$

Equation (8.13) represents usually performance functionals used in minimum-time optimisation because the only way to minimise (8.13) is by minimising $t_s$.

Around the time $t_s$, we have the performance functional slowly switching to a nonquadratic regulator that takes into account the actuator saturation. Note that this method allows an easy formulation of a minimum-time problem, and that the solution will follow using the successive approximation technique. The solution is a nearly minimum-time controller that is easier to find compared with techniques aimed at finding the exact minimum-time controller. Finding an exact minimum-time controller requires finding a bang–bang controller based on a switching surface that is hard to determine [7,43].

### 8.4.2  Constrained states

In literature, there exist several techniques that find a domain of initial states such that starting within this domain guarantees that a specific control policy will not violate the constraints [6]. However, we are interested in improving the given control laws so that they do not violate specific state–space constraints. For this we choose the following nonquadratic performance functional,

$$Q(x,k) = x^T Q x + \sum_{l=1}^{n_c} \left( \frac{x_l}{B_l - \alpha_l} \right)^{2k}, \tag{8.27}$$

where $n_c$, $B_l$, are the number of constrained states, the upper bound on $x_l$, respectively. The integer $k$ is positive, and $\alpha_l$ is a small positive number. As $k$ increases, and $\alpha_l \to 0$, the nonquadratic term will dominate the quadratic term when the state–space constraints are violated. However, the nonquadratic term will be dominated by the quadratic term when the state–space constraints are not violated. Note that in this approach, the constraints are considered soft constraints that can be hardened by using higher values for $k$ and smaller values for $\alpha_l$.

Having the successive approximation theory well set for nonquadratic functionals, in the next section we will introduce a neural network approximation of the value function, and employ the successive solutions method in a least-squares sense over a compact set of the stability region, $\Omega$. This is far simpler than the Galerkin approximation appearing in Reference 14.

## 8.5   Neural network least-squares approximate HJB solution

Although equation (8.9) is a linear differential equation, when substituting (8.10) into (8.9), it is still difficult to solve for the cost function $V^{(i)}(x)$. Therefore, Neural Nets are now used to approximate the solution for the cost function $V^{(i)}(x)$ at each successive iteration $i$. Moreover, for the approximate integration, a mesh is introduced in $\Re^n$. This yields an efficient, practical and computationally tractable solution algorithm for general nonlinear systems with saturated controls. We provide here a theoretically rigorous justification of this algorithm.

### 8.5.1   Neural network approximation of the cost function $V(x)$

It is well known that neural networks can be used to approximate smooth functions on prescribed compact sets [32]. Since our analysis is restricted to a set within the stability region, neural networks are natural for our application. Therefore, to successively solve (8.9) and (8.10) for bounded controls, we approximate $V^{(i)}$ with

$$V_L^{(i)}(x) = \sum_{j=1}^{L} w_j^{(i)} \sigma_j(x) = \mathbf{w}_L^{T^{(i)}} \boldsymbol{\sigma}_L(x), \tag{8.28}$$

which is a neural network with the activation functions $\sigma_j(x) \in C^1(\Omega)$, $\sigma_j(0) = 0$. The neural network weights are $w_j^{(i)}$ and $L$ is the number of hidden-layer neurons. Vectors

$$\boldsymbol{\sigma}_L(x) \equiv [\sigma_1(x)\sigma_2(x) \cdots \sigma_L(x)]^T, \qquad \mathbf{w}_L^{(i)} \equiv \left[ w_1^{(i)} w_2^{(i)} \cdots w_L^{(i)} \right]^T$$

are the vector activation function and the vector weight, respectively. The neural network weights will be tuned to minimise the residual error in a least-squares sense over a set of points within the stability region $\Omega$ of the initial stabilising control. The least-squares solution attains the lowest possible residual error with respect to the Neural Network weights.

For the $LE(V, u) = 0$, the solution $V$ is replaced with $V_L$ having a residual error

$$LE\left(V_L(x) = \sum_{j=1}^{L} w_j \sigma_j(x), u\right) = e_L(x). \qquad (8.29)$$

To find the least-squares solution, the method of weighted residuals is used [44]. The weights $\mathbf{w}_L$ are determined by projecting the residual error onto $de_L(x)/d\mathbf{w}_L$ and setting the result to zero $\forall x \in \Omega$ using the inner product, i.e.

$$\left\langle \frac{de_L(x)}{d\mathbf{w}_L}, e_L(x) \right\rangle = 0, \qquad (8.30)$$

where $\langle f, g \rangle = \int_\Omega fg \, dx$ is a Lebesgue integral [45]. Equation (8.30) becomes,

$$\langle \nabla \sigma_L(f + gu), \nabla \sigma_L(f + gu) \rangle \mathbf{w}_L$$

$$+ \left\langle Q + 2 \int_0^u (\phi^{-1}(v))^T R \, dv, \nabla \sigma_L(f + gu) \right\rangle = 0. \qquad (8.31)$$

The following technical results are needed.

**Lemma 2**   *If the set $\{\sigma_j\}_1^L$ is linearly independent and $u \in \Psi(\Omega)$, then the set*

$$\{\nabla \sigma_j^T(f + gu)\}_1^L \qquad (8.32)$$

*is also linearly independent.*

**Proof**   See Reference 13.   ∎

Because of Lemma 2, $\langle \nabla \sigma_L(f + gu), \nabla \sigma_L(f + gu) \rangle$ is of full rank, and thus is invertible. Therefore a unique solution for $\mathbf{w}_L$ exists and computed as

$$\mathbf{w}_L = - \langle \nabla \sigma_L(f + gu), \nabla \sigma_L(f + gu) \rangle^{-1}$$

$$\times \left\langle Q + 2 \int_0^u (\phi^{-1}(v))^T R \, dv, \nabla \sigma_L(f + gu) \right\rangle. \qquad (8.33)$$

Having solved for the neural net weights, the improved control is given by

$$u = -\phi\left(\tfrac{1}{2} R^{-1} g^T(x) \nabla \sigma_L^T \mathbf{w}_L\right). \qquad (8.34)$$

Equations (8.33) and (8.34) are successively solved at each iteration ($i$) until convergence.

### 8.5.2 Convergence of the method of least squares

In what follows, we show convergence results associated with the method of least-squares approach to solve for the cost function the LE using the Fourier series expansion (8.28). But before this, we want to consider the notation and definitions associated with convergence issues.

**Definition 3** *Convergence in the mean: a sequence of functions $\{f_n\}$ that is Lebesgue-integrable on a set $\Omega$, $L_2(\Omega)$, is said to converge in the mean to $f$ on $\Omega$ if $\forall \varepsilon > 0, \exists N(\varepsilon) : n > N \Rightarrow \|f_n(x) - f(x)\|_{L_2(\Omega)} < \varepsilon$, where $\|f\|^2_{L_2(\Omega)} = \langle f, f \rangle$.*

The convergence proof for the least-squares method is done in the Sobolev function space setting. This space allows defining functions that are $L_2(\Omega)$ with their partial derivatives.

**Definition 4** *Sobolev Space $H^{m,p}(\Omega)$: Let $\Omega$ be an open set in $\mathbb{R}^n$ and let $u \in C^m(\Omega)$. Define a norm on $u$ by*

$$\|u\|_{m,p} = \sum_{0 \le |\alpha| \le m} \left( \int_\Omega |D^\alpha u(x)|^p \, dx \right)^{1/p}, \quad 1 \le p < \infty.$$

*This is the Sobolev norm in which the integration is the Lebesgue integration. The completion of $\{u \in C^m(\Omega) : \|u\|_{m,p} < \infty\}$ with respect to $\|\,\|_{m,p}$ is the Sobolev space $H^{m,p}(\Omega)$. For $p = 2$, the Sobolev space is a Hilbert space [46].*

The LE can be written using the linear operator $A$ defined on the Hilbert space $H^{1,2}(\Omega)$

$$\overbrace{\frac{dV}{dx}^T (f + gu)}^{AV} = \overbrace{-Q - W(u)}^{P}.$$

In Reference 47, it is shown that if the set $\{\sigma_j\}_1^L$ is complete, and the operator $A$ and its inverse are bounded, then $\|AV_L - AV\|_{L_2(\Omega)} \to 0$ and $\|V_L - V\|_{L_2(\Omega)} \to 0$. However, for the LE, it can be shown that these sufficiency conditions are violated.

Neural networks based on power series have an important property that they are differentiable. This means that they can approximate uniformly a continuous function with all its partial derivatives of order $m$ using the same polynomial, by differentiating the series termwise. This type of series is $m$-uniformly dense. This is known as the High Order Weierstrass Approximation theorem. Other types of neural networks not necessarily based on power series that are $m$-uniformly dense are studied in Reference 48.

**Lemma 3** *High Order Weierstrass Approximation Theorem: Let $f(x) \in C^m(\Omega)$ in the compact set $\Omega$, then there exists a polynomial, $f_N(x)$, such that it converges uniformly to $f(x) \in C^m(\Omega)$, and such that all its partial derivatives up to order $m$ converge uniformly [44,48].*

**Lemma 4**   *Given N linearly independent set of functions $\{f_n\}$. Then $\|\alpha_N f_N\|^2_{L_2(\Omega)} \to 0 \Leftrightarrow \|\alpha_N\|^2_{l_2} \to 0$.*

**Proof**   To show the sufficiency part, note that the Gram matrix, $G = \langle f_N, f_N \rangle$, is positive definite. Therefore, $\alpha_N^T G_N \alpha_N \geq \underline{\lambda}(G_N) \|\alpha_N\|^2_{l_2}$, $\underline{\lambda}(G_N) > 0 \, \forall N$. If $\alpha_N^T G_N \alpha_N \to 0$, then $\|\alpha_N\|^2_{l_2} = \alpha_N^T G_N \alpha_N / \underline{\lambda}(G_N) \to 0$ because $\underline{\lambda}(G_N) > 0 \, \forall N$.

To show the necessity part, note that

$$\|\alpha_N\|^2_{L_2(\Omega)} - 2\|\alpha_N f_N\|^2_{L_2(\Omega)} + \|f_N\|^2_{L_2(\Omega)} = \|\alpha_N - f_N\|^2_{L_2(\Omega)},$$

$$2\|\alpha_N f_N\|^2_{L_2(\Omega)} = \|\alpha_N\|^2_{L_2(\Omega)} + \|f_N\|^2_{L_2(\Omega)} - \|\alpha_N - f_N\|^2_{L_2(\Omega)}.$$

Using the Parallelogram law

$$\|\alpha_N - f_N\|^2_{L_2(\Omega)} + \|\alpha_N + f_N\|^2_{L_2(\Omega)} = 2\|\alpha_N\|^2_{L_2(\Omega)} + 2\|f_N\|^2_{L_2(\Omega)}.$$

As $N \to \infty$

$$\|\alpha_N - f_N\|^2_{L_2(\Omega)} + \|\alpha_N + f_N\|^2_{L_2(\Omega)} = 2\overbrace{\|\alpha_N\|^2_{L_2(\Omega)}}^{\to 0} + 2\|f_N\|^2_{L_2(\Omega)},$$

$$\Rightarrow \|\alpha_N - f_N\|^2_{L_2(\Omega)} \to \|f_N\|^2_{L_2(\Omega)},$$

$$\Rightarrow \|\alpha_N + f_N\|^2_{L_2(\Omega)} \to \|f_N\|^2_{L_2(\Omega)}.$$

As $N \to \infty$

$$2\|\alpha_N f_N\|^2_{L_2(\Omega)} = \overbrace{\|\alpha_N\|^2_{L_2(\Omega)}}^{\to 0} + \|f_N\|^2_{L_2(\Omega)} - \overbrace{\|\alpha_N - f_N\|^2_{L_2(\Omega)}}^{\to \|f_N\|^2_{L_2(\Omega)}} \to 0.$$

Therefore, $\|\alpha_N\|^2_{l_2} \to 0 \Rightarrow \|\alpha_N f_N\|^2_{L_2(\Omega)} \to 0$.   ∎

Before discussing the convergence results for the method of least squares, the following four assumptions are needed.

**Assumption 1**   The LE solution is positive definite. This is guaranteed for stabilisable dynamics and when the performance functional satisfies zero-state observability.

**Assumption 2**   The system's dynamics and the performance integrands $Q(x) + W(u(x))$ are such that the solution of the LE is continuous and differentiable, therefore, belonging to the Sobolev space $V \in H^{1,2}(\Omega)$.

**Assumption 3**   We can choose a complete coordinate elements $\{\sigma_j\}_1^\infty \in H^{1,2}(\Omega)$ such that the solution $V \in H^{1,2}(\Omega)$ and its partial derivatives $\{\partial V / \partial x_1, \ldots, \partial V / \partial x_n\}$ can be approximated uniformly by the infinite series built from $\{\sigma_j\}_1^\infty$.

**Assumption 4**   The sequence $\{\psi_j = A\sigma_j\}$ is linearly independent and complete.

In general the infinite series, constructed from the complete coordinate elements $\{\sigma_j\}_1^\infty$, need not be differentiable. However, from Lemma 3 and Reference 48 we know that several types of neural networks can approximate a function and all its partial derivatives uniformly.

Linear independence of $\{\psi_j\}$ follows from Lemma 2, while completeness follows from Lemma 3 and Reference 48:

$$\forall V, \ \varepsilon \ \exists L, \mathbf{w}_L \ \therefore |V_L - V| < \varepsilon, \ \forall i \ |dV_L/dx_i - dV/dx_i| < \varepsilon.$$

This implies that as $L \to \infty$

$$\sup_{x \in \Omega} |AV_L - AV| \to 0 \Rightarrow \|AV_L - AV\|_{L_2(\Omega)} \to 0,$$

and therefore completeness of the set $\{\psi_j\}$ is established.

The next theorem uses these assumptions to conclude convergence results of the least-squares method which is placed in the Sobolev space $H^{1,2}(\Omega)$.

**Theorem 2**  *If assumptions 1–4 hold, then approximate solutions exist for the LE equation using the method of least squares and are unique for each L. In addition, the following results are achieved:*

(R1)   $\|LE(V_L(x)) - LE(V(x))\|_{L_2(\Omega)} \to 0,$
(R2)   $\|dV_L/dx - dV/dx\|_{L_2(\Omega)} \to 0,$
(R3)   $\|u_L(x) - u(x)\|_{L_2(\Omega)} \to 0.$

**Proof**   Existence of a least-squares solution for the LE can be easily shown. The least-squares solution $V_L$ is nothing but the solution of the minimisation problem

$$\|AV_L - P\|^2 = \min_{\tilde{V} \in S_L} \|A\tilde{V} - P\|^2 = \min_{\mathbf{w}} \|\mathbf{w}_L^T \mathbf{\psi}_L - P\|^2,$$

where $S_L$ is the span of $\{\varphi_1, \ldots, \varphi_L\}$.

Uniqueness follows from the linear independence of $\{\psi_1, \ldots, \psi_L\}$.

The first result R1, follows from the completeness of $\{\psi_j\}$.

To show the second result, R2, write the LE in terms of its series expansion on $\Omega$ with coefficients $c_j$

$$\mathrm{LE}\left(V_L = \sum_{i=1}^{N} w_i \sigma_i\right) - \mathrm{LE}\overbrace{\left(V = \sum_{i=1}^{\infty} c_i \sigma_i\right)}^{=0} = \varepsilon_L(x),$$

$$(\mathbf{w}_L - \mathbf{c}_L)^T \nabla \mathbf{\sigma}_L(f + gu) = \varepsilon_L(x) + \overbrace{\sum_{i=N+1}^{\infty} c_i \frac{d\varphi_i}{dx}(f + gu)}^{e_L(x)}.$$

Note that $e_L(x)$ converges uniformly to zero due to Lemma 3, and hence converges in the mean. On the other hand $\varepsilon_L(x)$ is shown to converge in the mean to zero using the least-squares method as seen in R1. Therefore,

$$\|(\mathbf{w}_L - \mathbf{c}_L)^T \nabla \sigma_L (f + gu)\|^2_{L_2(\Omega)} = \|\varepsilon_L(x) + e_L(x)\|^2_{L_2(\Omega)}$$
$$\leq 2\|\varepsilon_L(x)\|^2_{L_2(\Omega)} + 2\|e_L(x)\|^2_{L_2(\Omega)} \to 0.$$

Because $\nabla \sigma_L (f + gu)$ is linearly independent, using Lemma 4, we conclude that $\|\mathbf{w}_L - \mathbf{c}_L\|^2_{l_2} \to 0$. Therefore, because the set $\{d\sigma_i/dx\}$ is linearly independent, we conclude from Lemma 4 that $\|(\mathbf{w}_L - \mathbf{c}_L)^T \nabla \sigma_L\|^2_{L_2(\Omega)} \to 0$. Because the infinite series with $c_j$ converges uniformly it follows that $\|dV_L/dx - dV/dx\|_{L_2(\Omega)} \to 0$.

Finally, the third result, R3, follows by noting that $g(x)$ is continuous and therefore bounded on $\Omega$, this implies using R2 that

$$\left\| -\frac{1}{2} R^{-1} g^T \left( \frac{dV_L}{dx} - \frac{dV}{dx} \right) \right\|^2_{L_2(\Omega)}$$
$$\leq \left\| -\frac{1}{2} R^{-1} g^T \right\|^2_{L_2(\Omega)} \left\| \left( \frac{dV_L}{dx} - \frac{dV}{dx} \right) \right\|^2_{L_2(\Omega)} \to 0.$$

Denote

$$\alpha_{j,L}(x) = -\frac{1}{2} g_j^T \frac{dV_L}{dx}, \qquad \alpha_j(x) = -\frac{1}{2} g_j^T \frac{dV}{dx},$$

$$u_L - u = -\phi \left( \frac{1}{2} g^T \frac{dV_L}{dx} \right) + \phi \left( \frac{1}{2} g^T \frac{dV}{dx} \right)$$
$$= \begin{bmatrix} \phi(\alpha_{1,L}(x)) - \phi(\alpha_1(x)) \\ \vdots \\ \phi(\alpha_{m,L}(x)) - \phi(\alpha_m(x)) \end{bmatrix}.$$

Because $\phi(\cdot)$ is smooth, and under the assumption that its first derivative is bounded by a constant $M$, then we have $\phi(\alpha_{j,L}) - \phi(\alpha_j) \leq M(\alpha_{j,L}(x) - \alpha_j(x))$, therefore

$$\|\alpha_{j,L}(x) - \alpha_j(x)\|_{L_2(\Omega)} \to 0 \Rightarrow \|\phi(\alpha_{j,L}) - \phi(\alpha_j)\|_{L_2(\Omega)} \to 0,$$

hence R3 follows. ∎

**Corollary 2**   *If the results of Theorem 2 hold, then*

$$\sup_{x \in \Omega} |dV_L/dx - dV/dx| \to 0, \quad \sup_{x \in \Omega} |V_L - V| \to 0, \quad \sup_{x \in \Omega} |u_L - u| \to 0.$$

**Proof**   As the coefficients of the neural network, $w_j$, series converge to the coefficients of the uniformly convergent series, $c_j$, i.e. $\|\mathbf{w}_L - \mathbf{c}_L\|^2_{l_2} \to 0$, and since the mean error goes to zero in R2 and R3, hence uniform convergence follows. ∎

The next theorem is required to show the admissibility of the controller derived using the technique presented in this chapter.

**Corollary 3** *Admissibility of* $u_L(x)$:

$$\exists L_0 : L \geq L_0, \quad u_L \in \Psi(\Omega).$$

**Proof** Consider the following LE equation

$$\dot{V}^{(i)}(x, u_L^{(i+1)}) = \overbrace{-Q - 2 \int_{u^{(i+1)}}^{u^{(i)}} \phi^{-T}(v) R \, dv + 2\phi^{-T}(u^{(i+1)}) R(u^{(i)} - u^{(i+1)})}^{\leq 0}$$

$$- 2\phi^{-T}(u^{(i+1)}) R(u_L^{(i+1)} - u^{(i+1)}) - 2 \int_0^{u^{(i+1)}} \phi^{-T}(v) R \, dv.$$

Since $u_L^{(i+1)}$ is guaranteed to be within a tube around $u^{(i+1)}$ because $u_L^{(i+1)} \to u^{(i+1)}$ uniformly, therefore one can easily see that

$$\phi^{-T}(u^{(i+1)}) R u_L^{(i+1)} \geq \frac{1}{2} \cdot \phi^{-T}(u^{(i+1)}) R u^{(i+1)} + \alpha \int_0^{u^{(i+1)}} \phi^{-T} R \, dv$$

with $\alpha > 0$ is satisfied $\forall x \in \Omega \backslash \Omega_1(\varepsilon(L))$ where $\Omega_1(\varepsilon(L)) \subseteq \Omega$ containing the origin. Hence $\dot{V}^{(i)}(x, u_L^{(i+1)}) < 0$, $\forall x \in \Omega \backslash \Omega_1(\varepsilon(L))$. Given that $u_L^{(i+1)}(0) = 0$, and from the continuity of $u_L^{(i+1)}$, there exists $\Omega_2(\varepsilon(L)) \subseteq \Omega_1(\varepsilon(L))$ containing the origin for which $\dot{V}^{(i)}(x, u_L^{(i+1)}) < 0$. As $L$ increases, $\Omega_1(\varepsilon(L))$ gets smaller while $\Omega_2(\varepsilon(L))$ gets larger and the inequality is satisfied $\forall x \in \Omega$. Therefore, $\exists L_0 : L \geq L_0$, $\dot{V}^{(i)}(x, u_L^{(i+1)}) < 0$, $\forall x \in \Omega$ and hence $u_L \in \Psi(\Omega)$. ∎

**Corollary 4** *Positive definiteness of* $V_L^{(i)}(x)$

$$V_L^{(i)}(x) = 0 \Leftrightarrow x = 0, \quad \text{elsewhere } V_L^{(i)}(x) > 0.$$

**Proof** The proof is going to be by contradiction. Assuming that $u \in \Psi(\Omega)$, then Lemma 2 is satisfied. Therefore

$$\mathbf{w}_L = -\langle \nabla \sigma_L(f + gu), \nabla \sigma_L(f + gu) \rangle^{-1}$$

$$\cdot \left\langle Q + 2 \int (\phi^{-1}(u))^T R \, du, \nabla \sigma_L(f + gu) \right\rangle.$$

Assume also that

$$\exists x_a \neq 0, \quad \text{s.t.} \sum_{j=1}^{L} w_j \sigma_j(x_a) = \mathbf{w}_L^T \sigma_L(x_a) = 0.$$

Then,

$$-\left\langle Q + 2\int (\phi^{-1}(u))^T R\, du, \nabla_{\sigma L}(f + gu)\right\rangle^T$$

$$\cdot \left\langle \nabla_{\sigma L}(f + gu), \nabla_{\sigma L}(f + gu)\right\rangle^{-T} \sigma_L(x_a) = 0.$$

Note that because Lemma 2 is satisfied then $\langle \nabla_{\sigma L}(f + gu), \nabla_{\sigma L}(f + gu)\rangle^{-1}$ is a positive definite constant matrix. This implies that

$$\left\langle Q + 2\int (\phi^{-1}(u))^T R\, du, \nabla_{\sigma L}(f + gu)\right\rangle^T \sigma_L(x_a) = 0.$$

We can expand this matrix representation into a series form,

$$\left\langle Q + 2\int (\phi^{-1}(u))^T R\, du, \nabla_{\sigma L}(f + gu)\right\rangle^T \sigma_L(x_a)$$

$$= \sum_{j=1}^{L} \left\langle Q + 2\int (\phi^{-1}(u))^T R\, du, \frac{d\sigma_j}{dx}(f + gu)\right\rangle \sigma_j(x_a) = 0.$$

Note that,

$$\left\langle Q + 2\int (\phi^{-1}(u))^T R\, du, \frac{d\sigma_j}{dx}(f + gu)\right\rangle$$

$$= \int_{\Omega} \left\{ \left(Q + 2\int (\phi^{-1}(u))^T R\, du\right)\left(\frac{d\sigma_j}{dx}(f + gu)\right)\right\} dx.$$

Thus,

$$\sum_{j=1}^{L} \int_{\Omega} \left\{ \left(Q + 2\int \phi^{-T} R\, du\right)\left(\frac{d\sigma_j}{dx}(f + gu)\right)\right\} dx \cdot \sigma_j(x_a) = 0.$$

Using the mean value theorem, $\exists \xi \in \Omega$ such that,

$$\int_{\Omega} \left\{ \left[Q + 2\int \phi^{-T} R\, du\right] \times [\sigma_L^T(x_a)\nabla_{\sigma L}(f + gu)]\right\} dx$$

$$= \mu(\Omega)\left\{ \left[Q + 2\int \phi^{-T} R\, du\right] \times [\sigma_L^T(x_a)\nabla_{\sigma L}(f + gu)]\right\}(\xi),$$

where $\mu(\Omega)$ is the Lebesgue measure of $\Omega$.

This implies that,

$$
0 = \sum_{j=1}^{L} \mu(\Omega) \left[ \left( Q + 2 \int \phi^{-T} R \, du \right) \cdot \frac{d\sigma_j}{dx} (f + gu) \right] (\xi) \times \sigma_j(x_a)
$$

$$
= \mu(\Omega) \left[ Q + 2 \int \phi^{-T} R \, du \right] (\xi) \cdot \sum_{j=1}^{L} \left[ \frac{d\sigma_j}{dx} (f + gu) \right] (\xi) \times \sigma_j(x_a)
$$

$$
\Rightarrow \sum_{j=1}^{L} \left[ \frac{d\sigma_j}{dx} (f + gu) \right] (\xi) \times \sigma_j(x_a) = 0.
$$

Now, we can select a constant $\sigma_j(x_a)$ to be equal to a constant $c_j$. Thus we can rewrite the above formula as follows:

$$
\sum_{j=1}^{L} c_j \left[ \frac{d\sigma_j}{dx} (f + gu) \right] (\xi) = 0.
$$

Since $\xi$ depends on $\Omega$, which is arbitrary, this means that, $\nabla \sigma_j (f + gu)$ is not linearly independent, which contradicts our assumption. ∎

**Corollary 5** *It can be shown that* $\sup_{x \in \Omega} |u_L(x) - u(x)| \to 0$ *implies that* $|\sup_{x \in \Omega} J(x) - V(x)| \to 0$, *where* $\mathrm{LE}(J, u_L) = 0$, $\mathrm{LE}(V, u) = 0$.

### 8.5.3 Convergence of the method of least squares to the solution of the HJB equation

In this section, we would like to have a theorem analogous to Theorem 1 which guarantees that the successive least-squares solutions converge to the value function of the HJB equation (8.11).

**Theorem 3** *Under the assumptions of Theorem 2, the following is satisfied* $\forall i \geq 0$:

(i) $\sup_{x \in \Omega} \left| V_L^{(i)} - V^{(i)} \right| \to 0$,

(ii) $\sup_{x \in \Omega} \left| u_L^{(i+1)} - u^{(i+1)} \right| \to 0$,

(iii) $\exists L_0 : L \geq L_0, u_L^{(i+1)} \in \Psi(\Omega)$.

**Proof** The proof is by induction.

*Basis step.* Using Corollaries 2 and 3, it follows that for any $u^{(0)} \in \Psi(\Omega)$, one has

(I) $\sup_{x \in \Omega} \left| V_L^{(0)} - V^{(0)} \right| \to 0$,

(II) $\sup_{x \in \Omega} \left| u_L^{(1)} - u^{(1)} \right| \to 0$,

(III) $\exists L_0 : L \geq L_0, u_L^{(1)} \in \Psi(\Omega)$.

*Inductive step.* Assume that

(a)  $\sup_{x \in \Omega} \left| V_L^{(i-1)} - V^{(i-1)} \right| \to 0,$

(b)  $\sup_{x \in \Omega} \left| u_L^{(i)} - u^{(i)} \right| \to 0,$

(c)  $\exists L_0 : L \geq L_0, u_L^{(i)} \in \Psi(\Omega).$

If $J^{(i)}$ is such that $LE(J^{(i)}, u_L^{(i)}) = 0$, then from Corollary 2, $J^{(i)}$ can be uniformly approximated by $V_L^{(i)}$. Moreover, from assumption (b) and Corollary 5, it follows that as $u_L^{(i)} \to u^{(i)}$ uniformly then $J^{(i)} \to V^{(i)}$ uniformly. Therefore, $V_L^{(i)} \to V^{(i)}$ uniformly.

Because $V_L^{(i)} \to V^{(i)}$ uniformly, then $u_L^{(i+1)} \to u^{(i+1)}$ uniformly by Corollary 2. From Corollary 3, $\exists L_0 : L \geq L_0 \Rightarrow u_L^{(i+1)} \in \Psi(\Omega)$.

Hence the proof by induction is complete.   ∎

The next theorem is an important result upon which the algorithm proposed in Section 8.5.4 is justified.

**Theorem 4**   $\forall \varepsilon > 0, \exists i_0, L_0 : i \geq i_0, L \geq L_0$ *the following is satisfied:*

(A)  $\sup_{x \in \Omega} \left| V_L^{(i)} - V^* \right| < \varepsilon,$

(B)  $\sup_{x \in \Omega} \left| u_L^{(i)} - u^* \right| < \varepsilon,$

(C)  $u_L^{(i)} \in \Psi(\Omega).$

**Proof**   The proof follows directly from Theorems 1 and 3.   ∎

### 8.5.4   *Algorithm for nearly optimal neurocontrol design with saturated controls: introducing a mesh in $\mathbb{R}^n$*

Solving the integration in (8.33) is expensive computationally. However, an integral can be fairly approximated by replacing the integral with a summation series over a mesh of points on the integration region. This results in a nearly optimal, computationally tractable solution procedure.

By introducing a mesh on $\Omega$, with mesh size equal to $\Delta x$, we can rewrite some terms of (8.33) as follows:

$$X = \lfloor \nabla \sigma_L (f + gu)|_{x_1} \cdots \nabla \sigma_L (f + gu)|_{x_p} \rfloor^T, \tag{8.35}$$

$$Y = \left\lfloor Q + 2 \int_0^u \phi^{-T}(v) R \, dv \bigg|_{x_1} \cdots Q + 2 \int_0^u \phi^{-T}(v) R \, dv \bigg|_{x_p} \right\rfloor^T, \tag{8.36}$$

where $p$ in $x_p$ represents the number of points of the mesh. This number increases as the mesh size is reduced.

Note that

$$\langle \nabla \sigma_L(f + gu), \nabla \sigma_L(f + gu) \rangle = \lim_{\|\Delta x\| \to 0} (X^T X) \cdot \Delta x,$$

$$\left\langle Q + 2 \int_0^u \phi^{-T}(v) R \, dv, \nabla \sigma_L(f + gu) \right\rangle = \lim_{\|\Delta x\| \to 0} (X^T Y) \cdot \Delta x. \tag{8.37}$$

This implies that we can calculate $\mathbf{w}_L$ as

$$\mathbf{w}_{L,p} = -(X^T X)^{-1}(X^T Y). \tag{8.38}$$

We can also use Monte Carlo integration techniques in which the mesh points are sampled stochastically instead of being selected in a deterministic fashion [49]. This allows a more efficient numerical integration technique. In any case, however, the numerical algorithm at the end requires solving (8.38) which is a least-squares computation of the neural network weights.

Numerically stable routines that compute equations like (8.38) do exist in several software packages, such as MATLAB, which we use to perform the simulations in this chapter.

A flowchart of the computational algorithm presented in this chapter is shown in Figure 8.1. This is an offline algorithm run *a priori* to obtain a neural network feedback controller that is a nearly optimal solution to the HJB equation for the constrained control input case.

The neurocontrol law structure is shown in Figure 8.2. It is a neural network with activation functions given by $\sigma$, multiplied by a function of the system's state variables.

## 8.6 Numerical examples

We now show the power of our neural network control technique of finding nearly optimal nonlinear saturated controls for general systems. Four examples are presented.

### 8.6.1 Multi-input canonical form linear system with constrained inputs

We start by applying the algorithm obtained above for the linear system

$$\dot{x}_1 = 2x_1 + x_2 + x_3,$$

$$\dot{x}_2 = x_1 - x_2 + u_2,$$

$$\dot{x}_3 = x_3 + u_1.$$

It is desired to control the system with input constraints $|u_1| \le 3, |u_2| \le 20$. This system when uncontrolled has eigenvalues with positive real parts. This system is not asymptotically null controllable, therefore global asymptotic stabilisation cannot be achieved [2].

The algorithm developed in this chapter is used to derive a nearly optimal neurocontrol law for a specified region of stability around the origin. The following smooth

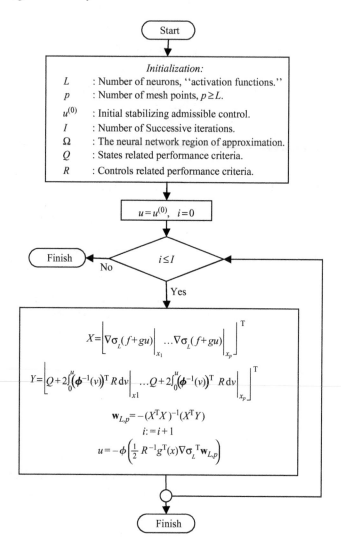

*Figure 8.1   Successive approximation algorithm for nearly optimal saturated neurocontrol*

function is used to approximate the value function of the system,

$$
\begin{aligned}
V_{21}(x_1, x_2, x_3) = {} & w_1 x_1^2 + w_2 x_2^2 + w_3 x_3^2 + w_4 x_1 x_2 + w_5 x_1 x_3 + w_6 x_2 x_3 + w_7 x_1^4 \\
& + w_8 x_2^4 + w_9 x_3^4 + w_{10} x_1^2 x_2^2 + w_{11} x_1^2 x_3^2 + w_{12} x_2^2 x_3^2 \\
& + w_{13} x_1^2 x_2 x_3 + w_{14} x_1 x_2^2 x_3 + w_{15} x_1 x_2 x_3^2 + w_{16} x_1^3 x_2 \\
& + w_{17} x_1^3 x_3 + w_{18} x_1 x_2^3 + w_{19} x_1 x_3^3 + w_{20} x_2 x_3^3 + w_{21} x_2^3 x_3.
\end{aligned}
$$

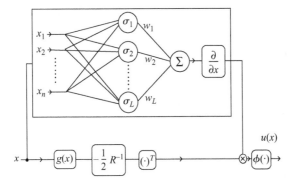

*Figure 8.2    Neural-network-based nearly optimal saturated control law*

Selecting the approximation for $V(x)$ is usually a natural choice guided by engineering experience and intuition. With this selection, we guarantee that $V(0) = 0$. This is a neural net with polynomial activation functions. It has 21 activation functions containing powers of the state variable of the system up to the fourth power. We selected neurons with fourth-order power of the states variables because for neurons with second-order power of the states, the algorithm did not converge. Moreover, we found that sixth power polynomials did not improve the performance over fourth power ones. The number of neurons required is chosen to guarantee the uniform convergence of the algorithm. If fewer neurons are used, then the algorithm might not properly approximate the cost function associated with the initial stabilising control, and thus the improved control using this approximated cost might not be admissible. The activation functions for the neural network neurons selected in this example satisfy the properties of activation functions discussed in Section 8.4.1 and Reference 32.

To initialise the algorithm, a stabilising control is needed. It is very easy to find this using linear quadratic regulator (LQR) for unconstrained controls. In this case, the performance functional is

$$\int_0^\infty (x_1^2 + x_2^2 + x_3^2 + u_1^2 + u_2^2) \, dt.$$

Solving the corresponding Riccati equation, the following stabilising unconstrained state feedback control is obtained:

$$u_1 = -8.31x_1 - 2.28x_2 - 4.66x_3,$$
$$u_2 = -8.57x_1 - 2.27x_2 - 2.28x_3.$$

However, when the LQR controller works through saturated actuators, the stability region shrinks. Further, this optimal control law derived for the linear case will not be optimal anymore working under saturated actuators. Figure 8.3 shows the

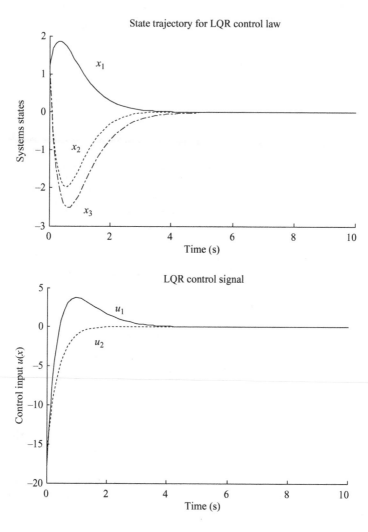

*Figure 8.3    LQR optimal unconstrained control*

performance of this controller assuming working with unsaturated actuators for the initial conditions $x_i(0) = 1.2$, $i = 1, 2, 3$. Figure 8.4 shows the performance when this control signal is bounded by $|u_1| \leq 3, |u_2| \leq 20$. Note how the bounds destroy the performance.

In order to model the saturation of the actuators, a nonquadratic cost performance term (8.8) is used as explained before. To show how to do this for the general case of $|u| \leq A$, we assume that the function $\phi(\text{cs})$ is given as $A \times \tanh(1/A \cdot \text{cs})$, where cs is assumed to be the command signal to the actuator. Figure 8.5 shows this for the case $|u| \leq 3$.

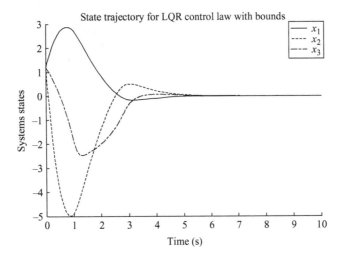

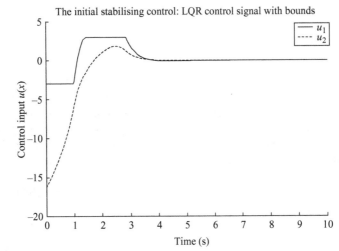

*Figure 8.4    LQR control with actuator saturation*

Following that, the nonquadratic cost performance is calculated to be

$$W(u) = 2 \int_0^u (\phi^{-1}(v))^T R \, dv$$

$$= 2 \int_0^u \left( A \tanh^{-1} \left( \frac{v}{A} \right) \right)^T R \, dv$$

$$= 2 \cdot A \cdot R \cdot u \cdot \tanh^{-1} \left( \frac{u}{A} \right) + A^2 \cdot R \cdot \ln \left( 1 - \frac{u^2}{A^2} \right).$$

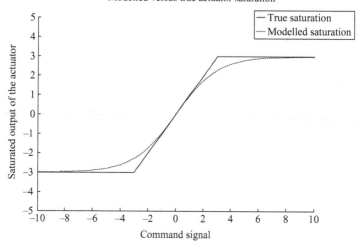

Figure 8.5    *Model of saturation*

This nonquadratic cost performance is then used in the algorithm to calculate the optimal bounded control. The improved bounded control law is found using the technique presented in the previous section. The algorithm is run over the region $-1.2 \leq x_1 \leq 1.2, -1.2 \leq x_2 \leq 1.2, -1.2 \leq x_3 \leq 1.2$ with the design parameters $R = I_{2x2}, Q = I_{3x3}$. This region falls within the region of asymptotic stability of the initial stabilising control. Methods to estimate the region of asymptotic stability are discussed in Reference 50.

After 20 successive iterations, the algorithm converges to the following nearly optimal saturated control,

$$
u_1 = -3 \tanh \left( \frac{1}{3} \left\{ \begin{array}{l} 7.7x_1 + 2.44x_2 + 4.8x_3 + 2.45x_1^3 + 2.27x_1^2x_2 \\ +3.7x_1x_2x_3 + 0.71x_1x_2^2 + 5.8x_1^2x_3 + 4.8x_1x_3^2 \\ +0.08x_2^3 + 0.6x_2^2x_3 + 1.6x_2x_3^2 + 1.4x_3^3 \end{array} \right\} \right),
$$

$$
u_2 = -20 \tanh \left( \frac{1}{20} \left\{ \begin{array}{l} 9.8x_1 + 2.94x_2 + 2.44x_3 - 0.2x_1^3 - 0.02x_1^2x_2 \\ +1.42x_1x_2x_3 + 0.12x_1x_2^2 + 2.3x_1^2x_3 + 1.9x_1x_3^2 \\ +0.02x_2^3 + 0.23x_2^2x_3 + 0.57x_2x_3^2 + 0.52x_3^3 \end{array} \right\} \right).
$$

This is the control law in terms of the state variables and a neural net following the structure shown in Figure 8.2. The suitable performance of this saturated control law is revealed in Figure 8.6.

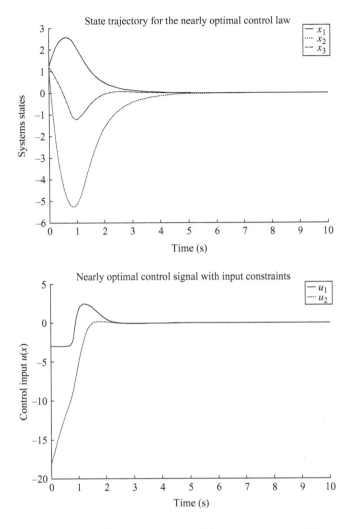

*Figure 8.6    Nearly optimal nonlinear neural control law considering actuator saturation*

### 8.6.2    Nonlinear oscillator with constrained input

We consider next a nonlinear oscillator having the dynamics

$$\dot{x}_1 = x_1 + x_2 - x_1(x_1^2 + x_2^2),$$
$$\dot{x}_2 = -x_1 + x_2 - x_2(x_1^2 + x_2^2) + u.$$

It is desired to control the system with control limits of $|u| \leq 1$. The following smooth function is used to approximate the value function of the system,

$$V_{24}(x_1, x_2) = w_1 x_1^2 + w_2 x_2^2 + w_3 x_1 x_2 + w_4 x_1^4 + w_5 x_2^4 + w_6 x_1^3 x_2 + w_7 x_1^2 x_2^2$$
$$+ w_8 x_1 x_2^3 + w_9 x_1^6 + w_{10} x_2^6 + w_{11} x_1^5 x_2 + w_{12} x_1^4 x_2^2 + w_{13} x_1^3 x_2^3$$
$$+ w_{14} x_1^2 x_2^4 + w_{15} x_1 x_2^5 + w_{16} x_1^8 + w_{17} x_2^8 + w_{18} x_1^7 x_2 + w_{19} x_1^6 x_2^2$$
$$+ w_{20} x_1^5 x_2^3 + w_{21} x_1^4 x_2^4 + w_{22} x_1^3 x_2^5 + w_{23} x_1^2 x_2^6 + w_{24} x_1 x_2^7.$$

This neural net has 24 activation functions containing powers of the state variable of the system up to the eighth power. In this example, the order of the neurons is higher than in the previous example, Section 8.6.1, to guarantee uniform convergence. The complexity of the neural network is selected to guarantee convergence of the algorithm to an admissible control law. When only up to the sixth-order powers are used, convergence of the iteration to admissible controls was not observed.

The unconstrained state feedback control $u = -5x_1 - 3x_2$, is used as an initial stabilising control for the iteration. This is found after linearising the nonlinear system around the origin, and building an unconstrained state feedback control which makes the eigenvalues of the linear system all negative. Figure 8.7 shows the performance of the bounded controller $u = \text{sat}_{-1}^{+1}(-5x_1 - 3x_2)$, when running it through a saturated actuator for $x_1(0) = 0$, $x_2(0) = 1$. Note that it is not good.

The nearly optimal saturated control law is now found through the technique presented in Section 8.3. The algorithm is run over the region $-1 \le x_1 \le 1$, $-1 \le x_2 \le 1$, $R = 1$, $Q = I_{2x2}$. After 20 successive iterations, the nearly optimal saturated control law is found to be,

$$u = -\tanh \begin{pmatrix} 2.6x_1 + 4.2x_2 + 0.4x_2^3 - 4.0x_1^3 - 8.7x_1^2 x_2 - 8.9x_1 x_2^2 \\ -5.5x_2^5 + 2.26x_1^5 + 5.8x_1^4 x_2 + 11x_1^3 x_2^2 + 2.6x_1^2 x_2^3 \\ +2.00x_1 x_2^4 + 2.1x_2^7 - 0.5x_1^7 - 1.7x_1^6 x_2 - 2.71x_1^5 x_2^2 \\ -2.19x_1^4 x_2^3 - 0.8x_1^3 x_2^4 + 1.8x_1^2 x_2^5 + 0.9x_1 x_2^6 \end{pmatrix}.$$

This is the control law in terms of a neural network following the structure shown in Figure 8.2. The suitable performance of this saturated control law is revealed in Figure 8.8. Note that the states and the saturated input in Figure 8.8 have less oscillations when compared to those of Figure 8.7.

### 8.6.3    Constrained state linear system

Consider the following system:

$$\dot{x}_1 = x_2,$$
$$\dot{x}_2 = x_1 + x_2 + u,$$
$$|x_1| \le 3.$$

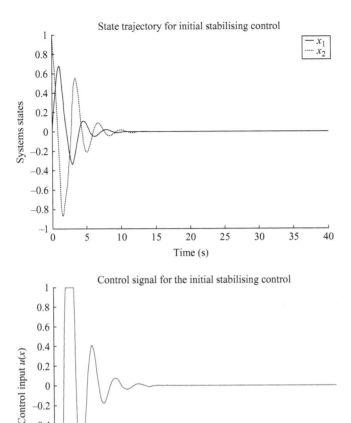

*Figure 8.7 Performance of the initial stabilising control when saturated*

For this we select the following performance functional:

$$Q(x, 14) = x_1^2 + x_2^2 + \left(\frac{x_1}{3-1}\right)^{10},$$

$$W(u) = u^2.$$

Note that, we have chosen the coefficient $k$ to be 10, $B_1 = 3$ and $\alpha_1 = 1$. A reason why we have selected $k$ to be 10 is that a larger value for $k$ requires using many activation functions in which a large number of them will have to have powers higher than the value $k$. However, since this simulation was carried on a double

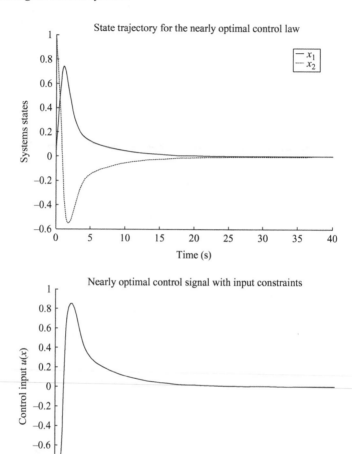

*Figure 8.8    Nearly optimal nonlinear control law considering actuator saturation*

precision computer, then power terms higher than 14 do not add up nicely and round-off errors seriously affect determining the weights of the neural network by causing a rank deficiency.

An initial stabilising controller, the LQR $-2.4x_1 - 3.6x_2$, that violates the state constraints is shown in Figure 8.9. The performance of this controller is improved by stochastically sampling from the region $-3.5 \leq x_1 \leq 3.5$, $-5 \leq x_2 \leq 5$, where $p = 3000$, and running the successive approximation algorithm for 20 times.

It can be seen that the nearly optimal control law that considers the state constraint tends not to violate the state constraint as the LQR controller does. It is important

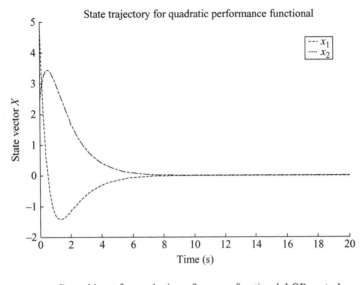

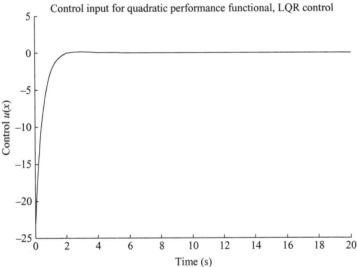

*Figure 8.9    LQR control without considering the state constraint*

to realise, that as we increase the order $k$ in the performance functional, then we get larger and larger control signals at the starting time of the control process to avoid violating the state constraints (Figure 8.10).

A smooth function of the order 45 that resembles the one used for the nonlinear oscillator in the previous example is used to approximate the value function of the system. The weights $W_o$ are found by successive approximation. Since $R = 1$, the final

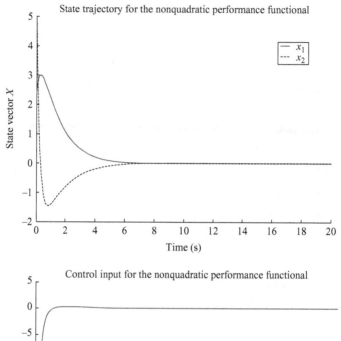

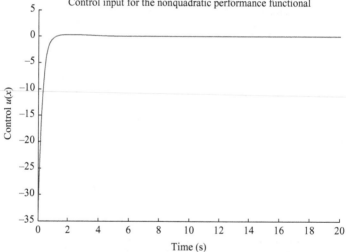

*Figure 8.10*   *Nearly optimal nonlinear control law considering the state constraint*

control law becomes,

$$u(x) = -\frac{1}{2} W_o^T \frac{\partial V}{\partial x_2}.$$

It was noted that the nonquadratic performance functional returns an overall cost of 212.33 when the initial conditions are $x_1 = 2.4$, $x_2 = 5.0$ for the optimal controller, while this cost increases to 316.07 when the linear controller is used. It is this increase in cost detected by the nonquadratic performance functional that causes the system to avoid violating the state constraints. If this difference in costs is made bigger, then we actually increase the set of initial conditions that do not violate the constraint. This however, requires a larger neural network, and high precision computing machines.

### 8.6.4   Minimum time control

Consider the following system:

$$\dot{x}_1 = x_2,$$
$$\dot{x}_2 = -x_2 + u.$$

It is desired to control the system with control limits of $|u| \leq 1$ to drive it to origin in minimum time. Typically, from classical optimal control theory [43], we find out that the control law required is a bang–bang controller that switches back and forth based on a switching surface that is calculated using Pontryagin's minimum principle. It follows that the minimum time control law for this system is given by

$$s(x) = x_1 - \frac{x_2}{|x_2|} \ln(|x_2| + 1) + x_2,$$

$$u^*(x) = \begin{cases} -1, & \text{for } x \text{ such that } s(x) > 0, \\ +1, & \text{for } x \text{ such that } s(x) < 0, \\ -1, & \text{for } x \text{ such that } s(x) = 0 \text{ and } x_2 < 0, \\ 0, & \text{for } x = 0. \end{cases}$$

The response to this controller is shown in Figure 8.11. It can be seen that this is a highly nonlinear control law, which requires the calculation of a switching surface. This is, however, a formidable task even for linear systems with state dimension larger than 3. However, when using the method presented in this chapter, finding a nearly minimum-time controller becomes a less complicated matter.

We use the following nonquadratic performance functional:

$$Q(x) = \tanh \left( \left( \frac{x_1}{0.1} \right)^2 + \left( \frac{x_2}{0.1} \right)^2 \right),$$

$$W(u) = 0.001 \times 2 \int_0^u \tanh^{-1}(\mu) \, d\mu.$$

A smooth function of the order 35 is used to approximate the value function of the system. We solve for this network by stochastic sampling. Let $p = 5000$ for $-0.5 \leq x_1 \leq 0.5$, $-0.5 \leq x_2 \leq 0.5$. The weights $W_o$ are found after iterating for 20 times. Since $R = 1$, the final control law becomes,

$$u(x) = -\tanh \left( \frac{1}{2} W_o^T \frac{\partial V}{\partial x_2} \right).$$

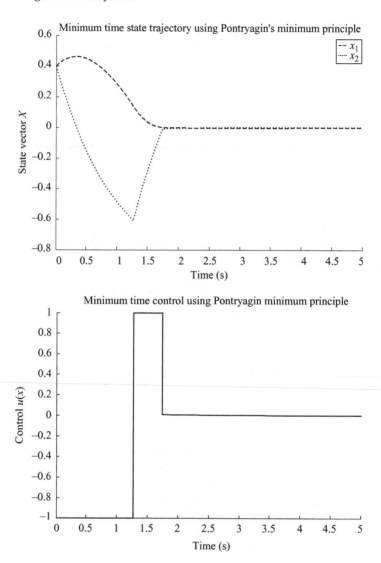

*Figure 8.11    Performance of the exact minimum-time controller*

Figure 8.12 shows the performance of the controller obtained using the algorithm presented in this chapter and compares it with that of the exact minimum-time controller. Figure 8.13 plots the state trajectory of both controllers. Note that the nearly minimum-time controller behaves as a bang–bang controller until the states come close to the origin when it starts behaving as a regulator.

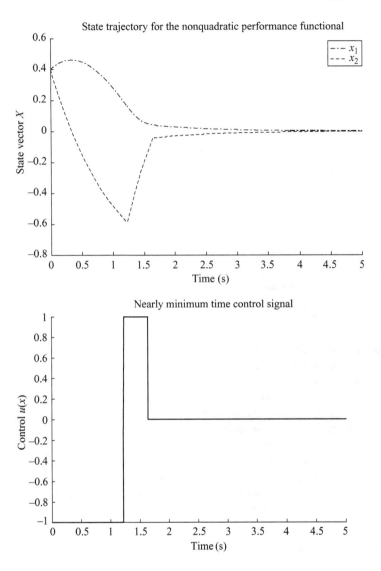

*Figure 8.12 Performance of the nearly minimum-time controller*

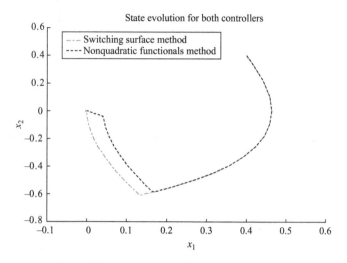

*Figure 8.13    State evolution for both minimum-time controllers*

## 8.7   Conclusion

A rigorous computationally effective algorithm to find nearly optimal controllers in state feedback form for general nonlinear systems with constraints is presented that approaches the problem of constrained optimisation from a practical engineering tractable point. The control is given as the output of a neural network. This is an extension of the novel work in References 14 and 24. Conditions under which the theory of successive approximation [28] applies were shown. Several numerical examples were discussed and simulated.

This algorithm requires further study into the problem of increasing the region of asymptotic stability. Moreover, adaptive control techniques can be blended to formulate adaptive optimal controllers for general nonlinear systems with constraints and unknown system dynamics $f, g$.

### Acknowledgements

This work has been supported by the National Science Foundation ECS-0140490 grant, and by the Army Research Office DAAD 19-02-1-0366 grant.

### References

1  SABERI, A., LIN, Z., and TEEL, A.: 'Control of linear systems with saturating actuators', *IEEE Transactions on Automatic Control*, 1996, **41** (3), pp. 368–78
2  SUSSMANN, H., SONTAG, E. D., and YANG, Y.: 'A general result on the stabilization of linear systems using bounded controls', *IEEE Transactions on Automatic Control*, 1994, **39** (12), pp. 2411–25

3 BITSORIS, G., and GRAVALOU, E.: 'Design techniques for the control of discrete-time systems subject to state and control constraints', *IEEE Transactions on Automatic Control*, 1999, **44** (5), pp. 1057–61

4 HU, T., LIN, Z., and CHEN, B. M.: 'An analysis and design method for linear systems subject to actuator saturation and disturbance', *Automatica*, 2002, **38** (2), pp. 351–9

5 HENRION, D., TARBOURIECH, S., and KUCERA, V.: 'Control of linear systems subject to input constraints: a polynomial approach', *Automatica*, 2001, **37** (4), pp. 597–604

6 GILBERT, E., and TAN, K. T.: 'Linear systems with state and control constraints: the theory and application of maximal output admissible sets', *IEEE Transactions on Automatic Control*, 1991, **36** (9), pp. 1008–20

7 LEWIS, F. L., and SYRMOS, V. L.: 'Optimal control' (John Wiley & Sons, Inc., New York, 1995)

8 BERNSTEIN, D. S.: 'Optimal nonlinear, but continuous, feedback control of systems with saturating actuators', *International Journal of Control*, 1995, **62** (5), pp. 1209–16

9 LYSHEVSKI, S. E., and MEYER, A. U.: 'Control system analysis and design upon the Lyapunov method'. Proceedings of the *American Control Conference*, June 1995, pp. 3219–23

10 LYSHEVSKI, S. E.: 'Optimal control of nonlinear continuous-time systems: design of bounded controllers via generalized nonquadratic functionals'. Proceedings of the *American Control Conference*, June 1998, pp. 205–9

11 SARIDIS, G., and LEE, C. S.: 'An approximation theory of optimal control for trainable manipulators', *IEEE Transactions on Systems, Man, and Cybernetics*, 1979, **9** (3), pp. 152–9

12 BEARD, R., SARIDIS, G., and WEN, J.: 'Approximate solutions to the time-invariant Hamilton–Jacobi–Bellman equation', *Journal of Optimization Theory and Application*, 1998, **96** (3), pp. 589–626

13 BEARD, R., SARIDIS, G., and WEN, J.: 'Galerkin approximations of the generalized Hamilton–Jacobi–Bellman equation', *Automatica*, 1997, **33** (12), pp. 2159–77

14 BEARD, R.: 'Improving the closed-loop performance of nonlinear systems'. Ph.D. thesis, Rensselaer Polytechnic Institute, Troy, NY 12180, 1995

15 MURRAY, J., COX, C., SAEKS, R., and LENDARIS, G.: 'Globally convergent approximate dynamic programming applied to an autolander'. Proceedings of the *American Control Conference*, June 2001, pp. 2901–6

16 LEE, H. W. J., TEO, K. L., LEE, W. R., and WANG, S.: 'Construction of sub-optimal feedback control for chaotic systems using B-splines with optimally chosen knot points', *International Journal of Bifurcation and Chaos*, 2001, **11** (9), pp. 2375–87

17 BERTSEKAS, D. P., and TSITSIKLIS, J. N.: 'Neuro-dynamic programming: an overview'. Proceedings of the 34th IEEE conference on *Decision and Control*, New Orleans, LA, December 1995, pp. 560–4

18  BERTSEKAS, D. P., and TSITSIKLIS, J. N.: 'Neuro-dynamic programming' (Athena Scientific, Belmont, MA, 1996)

19  MUNOS, R., BAIRD, L. C., and MOORE, A.: 'Gradient descent approaches to neural-net-based solutions of the Hamilton–Jacobi–Bellman equation', *International joint conference on Neural Networks IJCNN*, 1999, vol. 3, pp. 2152–7

20  KIM, Y. H., LEWIS, F. L., and DAWSON, D.: 'Intelligent optimal control of robotic manipulators using neural networks', *Automatica*, 2000, **36**, pp. 1355–64

21  HAN, D., and BALAKRISHNAN, S. N.: 'State-constrained agile missile control with adaptive-critic based neural networks'. Proceedings of the *American Control Conference*, June 2000, pp. 1929–33

22  LIU, X., and BALAKRISHNAN, S. N.: 'Convergence analysis of adaptive critic based optimal control'. Proceedings of the *American Control Conference*, June 2000, pp. 1929–33

23  LIU, X., and BALAKRISHNAN, S. N.: 'Adaptive critic based neuro-observer'. Proceedings of the *American Control Conference*, June 2001, pp. 1616–21

24  LYSHEVSKI, S. E.: 'Control systems theory with engineering applications' (Birkhauser, Boston, MA, 2001)

25  LYSHEVSKI, S. E.: 'Role of performance functionals in control laws design'. Proceedings of the *American Control Conference*, June 2001, pp. 2400–5

26  LYSHEVSKI, S. E.: 'Constrained optimization and control of nonlinear systems: new results in optimal control'. Proceedings of IEEE conference on *Decision and control*, December 1996, pp. 541–6

27  HUANG, J., and LIN, C. F.: 'Numerical approach to computing nonlinear $H_\infty$ control laws', *Journal of Guidance, Control, and Dynamics*, 1995, **18** (5), pp. 989–94

28  KLEINMAN, D.: 'On an iterative technique for Riccati equation computations', *IEEE Transactions on Automatic Control*, 1968, **13** (1), pp. 114–15

29  MILLER, W. T., SUTTON, R., and WERBOS, P.: 'Neural networks for control' (The MIT Press, Cambridge, MA, 1990)

30  PARISINI, T., and ZOPPOLI, R.: 'Neural approximations for infinite-horizon optimal control of nonlinear stochastic systems', *IEEE Transactions on Neural Networks*, 1998, **9** (6), pp. 1388–408

31  CHEN, F.-C., and LIU, C.-C.: 'Adaptively controlling nonlinear continuous-time systems using multilayer neural networks', *IEEE Transactions on Automatic Control*, 1994, **39** (6), pp. 1306–10

32  LEWIS, F. L., JAGANNATHAN, S., and YESILDIREK, A.: 'Neural network control of robot manipulators and nonlinear systems' (Taylor & Francis, London, UK, 1999)

33  POLYCARPOU, M. M.: 'Stable adaptive neural control scheme for nonlinear systems', *IEEE Transactions on Automatic Control*, 1996, **41** (3), pp. 447–51

34  ROVITHAKIS, G. A., and CHRISTODOULOU, M. A.: 'Adaptive control of unknown plants using dynamical neural networks', *IEEE Transactions on Systems, Man, and Cybernetics*, 1994, **24** (3), pp. 400–12

35 SADEGH, N: 'A perceptron network for functional identification and control of nonlinear systems', *IEEE Transactions on Neural Networks*, 1993, **4** (6), pp. 982–8

36 SANNER, R. M., and SLOTINE, J.-J. E.: 'Stable adaptive control and recursive identification using radial Gaussian networks'. Proceedings of IEEE conference on *Decision and control*, Brighton, 1991, pp. 2116–23

37 NARENDRA, K. S., and LEWIS, F. L. (Eds): 'Special issue on neural network feedback control', *Automatica*, 2001, **37** (8), pp. 1147–8

38 HUANG, C.-S., WANG, S., and TEO, K. L.: 'Solving Hamilton–Jacobi–Bellman equations by a modified method of characteristics', *Nonlinear Analysis*, 2000, **40**, pp. 279–93

39 BARDI, M., and CAPUZZO-DOLCETTA, I.: 'Optimal control and viscosity solutions of Hamilton–Jacobi–Bellman equations' (Birkhauser, Boston, MA, 1997)

40 LIO, F. D.: 'On the Bellman equation for infinite horizon problems with unbounded cost functional', *Applied Mathematics and Optimization*, 2000, **41**, pp. 171–97

41 VAN DER SCHAFT, A. J.: '$L_2$-gain analysis of nonlinear systems and nonlinear state feedback $H_\infty$ control', *IEEE Transactions on Automatic Control*, 1992, **37** (6), pp. 770–84

42 APOSTOL, T.: 'Mathematical analysis' (Addison-Wesley, Reading, MA, 1974)

43 KIRK, D.: 'Optimal control theory: an introduction' (Prentice-Hall, NJ, 1970)

44 FINLAYSON, B. A.: 'The method of weighted residuals and variational principles' (Academic Press, New York, 1972)

45 BURK, F.: 'Lebesgue measure and integration' (John Wiley & Sons, New York, 1998)

46 ADAMS, R., and FOURNIER, J.: 'Sobolev spaces' (Academic Press, Oxford, 2003, 2nd edn.)

47 MIKHLIN, S. G.: 'Variational methods in mathematical physics' (Pergamon, Oxford, 1964)

48 HORNIK, K., STINCHCOMBE, M., and WHITE, H.: 'Universal approximation of an unknown mapping and its derivatives using multilayer feedforward networks', *Neural Networks*, 1990, **3**, pp. 551–60

49 EVANS, M., and SWARTZ, T.: 'Approximating integrals via Monte Carlo and deterministic methods' (Oxford University Press, Oxford, 2000)

50 KHALIL, H.: 'Nonlinear systems' (Prentice-Hall, Upper Saddle River, NJ, 2003, 3rd edn.)

*Chapter 9*

# Reinforcement learning for online control and optimisation

*James J. Govindhasamy, Seán F. McLoone, George W. Irwin, John J. French and Richard P. Doyle*

## 9.1 Introduction

Current control methodologies can generally be divided into model based and model free. The first contains conventional controllers, the second so-called intelligent controllers [1–3]. Conventional control designs involve constructing dynamic models of the target system and the use of mathematical techniques to derive the required control law. Therefore, when a mathematical model is difficult to obtain, either due to complexity or the numerous uncertainties inherent in the system, conventional techniques are less useful. Intelligent control may offer a useful alternative in this situation.

An intelligent system learns from experience. As such, intelligent systems are adaptive. However, adaptive systems are not necessarily intelligent. Key features of adaptive control systems are that they

- continuously identify the current state of the system
- compare the current performance to the desired one and decide whether a change is necessary to achieve the desired performance
- modify or update the system parameters to drive the control system to an optimum performance.

These three principles, identification, decision and modification, are inherent in any adaptive system. A learning system on the other hand is said to be intelligent because it improves its control strategy based on past experience or performance. In other words, an adaptive system regards the current state as novel (i.e. localisation), whereas a learning system correlates experience gained at previous plant operating regions with the current state and modifies its behaviour accordingly [4] for a more long-term effect.

There are many examples in industrial control where conventional automatic control systems (e.g. self-tuning controllers) are not yet sufficiently advanced to cater for nonlinear dynamics across different operating regions or to predict the effects of current controller changes in the longer term. This is certainly true of very large, highly interconnected and complex systems. In these situations an intelligent approach for evaluating possible control alternatives, can be of value. One such framework, called the adaptive critic design (ACD), was proposed by Werbos [5,6]. The design of nonlinear optimal neurocontrollers using this ACD paradigm, is currently attracting much renewed interest in the academic community.

However, closer examination of the current literature suggests that a number of restrictive assumptions have had to be introduced which run counter to the original ACD concept. Wu and Pugh [7], Wu [8], Chan *et al.* [9], Zeng *et al.* [10] and Riedmiller [11] all assumed *a priori* knowledge of the plant in selecting the control action. Ernst *et al.* [12], Park *et al.* [13,14], Venayagamoorthy *et al.* [15–17], Iyer and Wunsch [18], Radhakant and Balakrishnan [19] and Sofge and White [20,21] trained their neurocontrollers offline using a plant model. While Hoskins and Himmelblau [22] successfully implemented an online model-free adaptive heuristic critic (AHC) [23,24], the control employed was constrained to be of bang–bang form.

Adaptive heuristic critic and Q-learning, variants of the ACD, will be discussed in Chapter 10 in the context of multi-agent control within an internet environment. In this chapter, the model-free, action-dependent adaptive critic (ADAC) design of Si and Wang [25] is extended to produce a fully online neurocontroller without the necessity to store plant data during a successful run. In particular, the potential limitations of Si and Wang's stochastic back-propagation training approach, in terms of poor convergence and parameter shadowing [26] are avoided. This is done by introducing a second-order training algorithm, based on the recursive Levenberg–Marquardt (RLM) [27,28] approach. This training algorithm incorporates the temporal difference (TD) strategy [29,30], and is hence designated TD-RLM.

The performance of our new ADAC scheme for reinforcement learning is validated using simulation results from an 'inverted pendulum' (pole-balancing) control task. This is often used as an example of the inherently unstable, multiple-output, dynamic systems present in many balancing situations, like a two-legged walking robot or the aiming of a rocket thruster. It has also been widely used to demonstrate many modern and classical control engineering techniques.

Finally, results from a collaboration with Seagate Technology Ltd on data from an actual industrial grinding process used in the manufacture of disk-drive platters suggest that the ADAC can achieve a 33 per cent reduction in rejects compared to a proprietary controller, one of the first reported industrial applications of this emerging technology.

The chapter is organised as follows. Section 9.2 gives a detailed description of the ADAC framework and its neural network implementation. The training algorithm for the neural implementation is then compared in simulation on the control of an inverted pendulum in Section 9.3. Section 9.4 introduces the industrial grinding process case study and presents results on the application of ADACs to identification and control of the process. Conclusions appear in Section 9.5.

## 9.2 Action-dependent adaptive critics

The algorithm used to illustrate the features of the ADAC shown in Figure 9.1, is based on Si and Wang [25] and belongs to the approximate dynamic programming family. Such methods were first introduced and formalised by Werbos [5,6]. It is useful to summarise how this method came to be used.

The fundamental solution to sequential optimisation problems relies on Bellman's principle of optimality [31]: 'an optimal trajectory has the property that no matter how the intermediate point is reached, the rest of the trajectory must coincide with an optimal trajectory as calculated with the intermediate point as the starting point'. This principle is applied in reinforcement learning by devising a 'primary' reinforcement function or reward, $r(k)$, that incorporates a control objective for a particular scenario in one or more measurable variables. A secondary utility is then formed, which incorporates the desired control objective through time. This is called the Bellman equation and is expressed as

$$J(k) = \sum_{i=0}^{\infty} \gamma^i r(k+i), \tag{9.1}$$

where $\gamma$ is a discount factor ($0 < \gamma < 1$), which determines the importance of the present reward as opposed to future ones. The reinforcement, $r(k)$, is binary with $r(k) = 0$ when the event is successful (the objective is met) and $r(k) = -1$ when failure occurs (when the objective is not met). The purpose of dynamic programming is then to choose a sequence of control actions to maximise $J(k)$, which is also called the cost-to-go. Unfortunately, such an optimisation is not feasible computationally due to the backward numerical solution process needed which requires future information

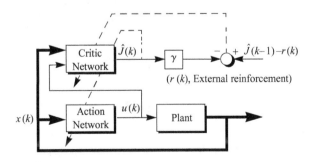

*Figure 9.1    Schematic of the ADAC scheme*

for real problems. There is thus a need for a more tractable approximation method which uses the Bellman recursion equation for the cost-to-go,

$$J(k) = r(k) + \gamma J(k+1). \tag{9.2}$$

Werbos proposed a variety of methods for estimating $J(k)$ using artificial neural networks (ANN) as function approximators. These methods are called ACDs. This term is generally applied to any module that provides learning reinforcement to a second, Action Network module (i.e. a controller) [5].

The standard classification of these ACDs is based on the critic's inputs and outputs. In heuristic dynamic programming (HDP), for example, the critic's output is an estimate of the value of $J(k)$ while in dual heuristic programming (DHP) the critic's output is an estimate of the derivative of $J(k)$ with respect to the states. Globalised dual heuristic programming (GDHP) approximates both $J(k)$ and its derivatives by adaptation of the critic network. In the action dependent versions of HDP and DHP, the critic's inputs are augmented with the controller's output (action), hence the names ADHDP, ADDHP and ADGDHP. The reader is referred to References 5 and 6 for more details of the characteristics and uses of these different methods. Only the ADHDP is discussed here as Si and Wang's [25] approach is closely related to this method.

For illustration, suppose a discrete nonlinear, time-varying system is defined as,

$$x(k+1) = f[x(k), u(k), k], \tag{9.3}$$

where $x \in \Re^n$ represents the state vector and $u \in \Re^m$ denotes the control action. The cost function is represented by

$$J[x(k)] = \sum_{i=k}^{\infty} \gamma^{i-k} r[x(i), u(i)], \tag{9.4}$$

where $r$ is the reinforcement signal and $\gamma$ is a discount factor $(0 < \gamma < 1)$. The objective is to choose the control action, $u(i)$, $i = k, k+1, \ldots$, so that the cost $J$ defined in equation (9.4) is minimised. However, in equation (9.2), the future value of the cost $J(k+1)$ is required which is not known *a priori*. In the ACDs, an adaptive critic network is used to produce the required estimate of $J(k+1)$.

The adaptive critic network is trained to minimise the following error measured over time:

$$E_c(k) = \sum_k [\hat{J}(k) - r(k) - \gamma \hat{J}(k+1)]^2, \tag{9.5}$$

$\hat{J}(k)$, the output of the critic network, is given by

$$\hat{J}(k) = \hat{J}[x(k), u(k), W_c], \tag{9.6}$$

where $W_c$ are the parameters of the critic network. Here the reinforcement signal indicates the performance of the overall system, i.e. failure $= -1$ or success $= 0$.

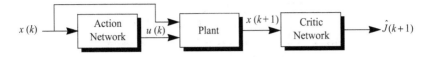

*Figure 9.2  Conventional ADHDP [5,6]*

When $E_c(k) = 0$ for all $k$, equation (9.5) can be simplified to

$$\hat{J}(k) = r(k) + \gamma \hat{J}(k+1)$$
$$= r(k) + \gamma[r(k+1) + \gamma \hat{J}(k+2)]$$
$$= \cdots$$
$$= \sum_{i=k}^{\infty} \gamma^{i-k} r(i), \qquad (9.7)$$

which is the same as equation (9.4). However, this approach to training would require a plant model to predict $x(k+1)$ and consequently the cost-to-go, $\hat{J}(k+1)$, as shown in Figure 9.2.

At this point it is worth mentioning that, in ACDs, there are two partial derivative terms $\partial \hat{J}(k)/\partial W_c(k)$ and $\partial \hat{J}(k+1)/\partial W_c(k)$ in the back-propagation path from the Bellman equation. When ACDs were implemented without a model network, the second partial derivative term was simply ignored. This can be detrimental as demonstrated in References 32–34. In previous implementations such as DHP and GDHP, a model network was employed to take into account the term $\partial \hat{J}(k+1)/\partial W_c(k)$.

Si and Wang [25] proposed a method which resolved the dilemma of either ignoring the $\partial \hat{J}(k+1)/\partial W_c(k)$ term, leading to poor learning accuracy, or including an additional system model network and hence more computation. These authors modified the Bellman recursion in equation (9.2) so that, instead of approximating $J(k)$, the Critic Network would approximate $J(k+1)$. This is done by defining the future accumulated reward-to-go at time $t$, as

$$Q(k) = r(k+1) + \gamma r(k+2) + \cdots, \qquad (9.8)$$

and using the Critic Network to provide $Q(k)$ as an estimate of $J(k+1)$, i.e. $Q(k) = J(k+1)$, as shown in Figure 9.3.

In their method, the Critic Network is trained by storing the estimated cost at $k-1$, $Q(k-1)$. The current estimated cost $Q(k)$, and the current reward, $r(k)$, are then used to determine the TD error (i.e. the error between two successive estimates of $Q$) as given by

$$e_c(t) = Q(k-1) - r(k) - \gamma Q(k), \qquad (9.9)$$

where

$$Q(k-1) = r(k) + \gamma Q(k) \qquad (9.10)$$

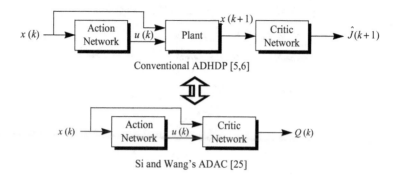

*Figure 9.3    Comparison of the conventional ADHDP and Si and Wang's ADAC*

is the recursive form of equation (9.8). Thus the cost for the Critic Network to be minimised during training is defined as

$$E_c(k) = \sum_t [Q(k-1) - r(k) - \gamma Q(k)]^2. \tag{9.11}$$

When $E_c(k) = 0$ for all $k$, equation (9.11) simplifies to

$$Q(k-1) = r(k) + \gamma Q(k)$$
$$= r(k) + \gamma[r(k+1) + \gamma Q(k+1)]$$
$$= \cdots$$
$$= \sum_{i=k+1}^{\infty} \gamma^{i-k-1} r(i), \tag{9.12}$$

where $Q(\infty) = 0$. Comparing equations (9.7) and (9.12), it can be seen that by minimising $E_c(k)$, the Critic Network output then provides an estimate of $J(k+1)$ in equation (9.4), i.e. the value of the cost function in the immediate future.

Training can either be performed 'backward-in-time' or 'forward-in-time' [35]. In the case of backward-in-time training, the target output of the Critic Network at time $t$, is computed from the previous network output, $Q(k-1)$ using equation (9.10), i.e.

$$Q(k) = \frac{1}{\gamma}[Q(k-1) - r(k)]. \tag{9.13}$$

Thus the network is trained to realise the mapping

$$\text{Critic: } \{x(k), u(k)\} \rightarrow \frac{1}{\gamma}[Q(k-1) - r(k)]. \tag{9.14}$$

The arrangement is as shown in Figure 9.4.

In the alternative forward-in-time approach at $k-1$, the Critic Network is trained to produce the output $Q(k-1)$ and the output at time $t$ is used to generate the required

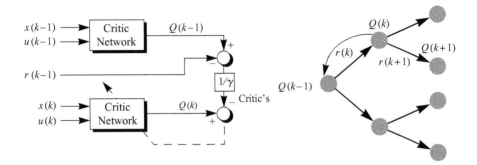

*Figure 9.4    Illustration of the 'backward-in-time' mapping*

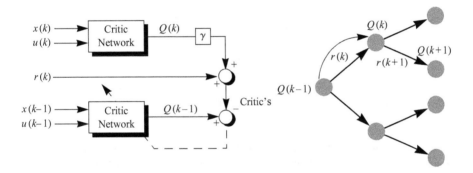

*Figure 9.5    Illustration of the 'forward-in-time' mapping*

target output, i.e. $r(k) + \gamma Q(k)$. Here the Critic Network realises the mapping

$$\text{Critic: } \{x(k-1), u(k-1)\} \rightarrow r(k) + \gamma Q(k) \tag{9.15}$$

and, $Q(k-1)$ is the network output. The forward-in-time arrangement is depicted in Figure 9.5.

The Action Network is trained after the Critic Network, with the objective of maximising the critic output, $Q(k)$. This strategy indirectly enables the Action Network to produce favourable control actions and zero reward, i.e. $r(k) = 0$. Thus, the target output from the Action Network for training purposes can be equated to zero, so that the Critic Network's output is as close to zero as possible. The mapping desired from the Action Network is given by

$$\text{Action: } \{x(k)\} \rightarrow \{Q(k) = 0\}. \tag{9.16}$$

The training of the Action Network requires that it be connected to the Critic Network so that the target mapping in equation (9.16), then refers to the output of the whole network as shown in Figure 9.6.

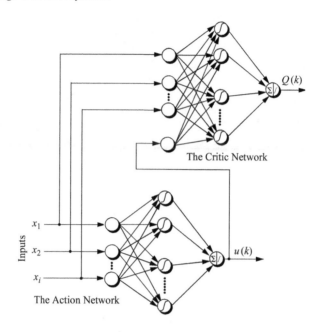

*Figure 9.6    The Action and Critic Networks in an ADAC implementation*

Having defined the network mappings and cost functions, stochastic gradient-based training algorithms can be used to train the networks. In Si and Wang's original ADAC design [25] networks were trained using a temporal difference error-based stochastic back-propagation algorithm (TD-SBP). This is a first-order gradient-descent method and therefore subject to poor convergence and parameter shadowing problems [26]. Here second-order training, in the form of a recursive implementation of the Levenberg–Marquardt [27,28] algorithm (TD-RLM), is introduced to address these deficiencies and compared with Si and Wang's original TD-SBP approach. In each case the Critic and Action Networks are implemented as multilayer perceptron neural networks.

### 9.2.1    First-order training algorithm: temporal difference stochastic back-propagation (TD-SBP)

The output of the Critic Network is $Q(k)$ and its prediction error is given by

$$e_{\mathrm{c}}(k) = \gamma Q(k) - [Q(k-1) - r(k)]. \tag{9.17}$$

Thus, the objective function to be minimised is the instantaneous estimate of the mean-squared prediction error

$$E_{\mathrm{c}}(k) = \tfrac{1}{2}e_{\mathrm{c}}^2(k). \tag{9.18}$$

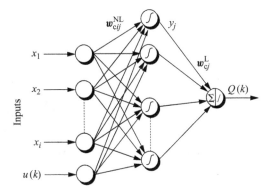

*Figure 9.7  Critic Network with the process states and control action, u(k) from the Action Network as inputs*

For a multilayer perceptron Critic Network (Figure 9.7), the output is calculated in a feedforward manner as follows:

$$g_j(k) = \sum_{i=1}^{N_i+1} w_{cij}^{NL}(k)x_i(k), \tag{9.19}$$

$$y_j(k) = \frac{1 - e^{-g_j(k)}}{1 + e^{-g_j(k)}}, \tag{9.20}$$

where $x$ is the input vector, $w_c^{NL}$ is the input to the hidden layer (or nonlinear) weights, $g$ is the input to the hidden layer nodes and $y$ is the output of the hidden layer nodes. Note that the index $N_i + 1$ is to include $u(k)$ (i.e. $x_{N_i+1} = u(k)$), the output of the Action Network, as shown in Figure 9.7. Finally, the output, $Q(k)$, is calculated as

$$Q(k) = \sum_{j=1}^{N_h} w_{cj}^{L}(k)y_j(k), \tag{9.21}$$

where $w_c^L$ is the linear (or hidden to output layer) weights vector.

The weights-update rule for the Critic Network is based on a combination of gradient-descent weight adaptation and temporal-difference (TD) learning [29,30]. The linear weights, $w_c^L$, are updated as:

$$w_{cj}^{L}(k+1) = w_{cj}^{L}(k) + \Delta w_{cj}^{L}(k), \tag{9.22}$$

where

$$\Delta w_{cj}^{L}(k) = -\beta_c(k)\left[\frac{\partial E_c(k)}{\partial w_{cj}^{L}(k)}\right] = -\beta_c(k)\left[\frac{\partial E_c(k)}{\partial Q(k)} \cdot \frac{\partial Q(k)}{\partial w_{cj}^{L}(k)}\right]. \tag{9.23}$$

The nonlinear weights, $w_c^{NL}$, are correspondingly updated as:

$$w_{cij}^{NL}(k+1) = w_{cij}^{NL}(k) + \Delta w_{cij}^{NL}(k) \tag{9.24}$$

and

$$\Delta w_{cij}^{NL}(k) = -\beta_c(k) \left[ \frac{\partial E_c(k)}{\partial w_{cij}^{NL}(k)} \right]$$

$$= -\beta_c(k) \left[ \frac{\partial E_c(k)}{\partial Q(k)} \cdot \frac{\partial Q(k)}{\partial y_j(k)} \cdot \frac{\partial y_j(k)}{\partial g_j(k)} \cdot \frac{\partial g_j(k)}{\partial w_{cij}^{NL}(k)} \right]. \tag{9.25}$$

In both cases, $\beta_c > 0$ is the learning rate.

The prediction error for the Action Network update is

$$e_a(k) = Q(k), \tag{9.26}$$

where the instantaneous estimate of the objective function to be minimised is given by

$$E_a(k) = \tfrac{1}{2} e_a^2(k). \tag{9.27}$$

The output of the Action Network shown in Figure 9.8 is calculated in a feedforward manner and is expressed as

$$f_j(k) = \sum_{i=1}^{N_i} w_{aij}^{NL}(k) x_i(k), \tag{9.28}$$

where $f$ is the input to the hidden layer nodes, $w_a^{NL}$ is the input to the hidden layer weights and $x$ is the input vector. For the MLP architecture, the output is given by

$$u(k) = \sum_{j=1}^{N_h} w_{aj}^{L}(k) z_j(k), \tag{9.29}$$

where $w_a^{L}$ is the linear weights vector and $z$ is the output layer input function given as

$$z_j(k) = \frac{1 - e^{-f_j(k)}}{1 + e^{-f_j(k)}}. \tag{9.30}$$

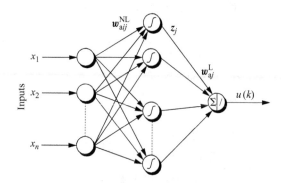

Figure 9.8   *Action Network topology receiving the process states*

The weight-update rule for the Action Network is again based on gradient-descent adaptation. The linear weights, $w_a^L$, adaptation rule is

$$w_{aj}^L(k+1) = w_{aj}^L(k) + \Delta w_{aj}^L(k), \tag{9.31}$$

where

$$\Delta w_{aj}^L(k) = -\beta_a(k)\left[\frac{\partial E_a(k)}{\partial w_{aj}^L(k)}\right] = -\beta_a(k)\left[\frac{\partial E_a(k)}{\partial Q(k)} \cdot \frac{\partial Q(k)}{\partial u(k)} \cdot \frac{\partial u(k)}{\partial w_{aj}^L(k)}\right], \tag{9.32}$$

while the nonlinear weights, $w_a^{NL}$ are adapted according to:

$$w_{aij}^{NL}(k+1) = w_{aij}^{NL}(k) + \Delta w_{aij}^{NL}(k) \tag{9.33}$$

and

$$\Delta w_{aij}^{NL}(k) = -\beta_a(k)\left[\frac{\partial E_c(k)}{\partial w_{aij}^{NL}(k)}\right]$$
$$= -\beta_a(k)\left[\frac{\partial E_c(k)}{\partial Q(k)} \cdot \frac{\partial Q(k)}{\partial u(k)} \cdot \frac{\partial u(k)}{\partial z_j(k)} \cdot \frac{\partial z_j(k)}{\partial f_j(k)} \cdot \frac{\partial f_j(k)}{\partial w_{aij}^{NL}(k)}\right]. \tag{9.34}$$

Again $\beta_a > 0$ is the learning rate.

### 9.2.2 Second-order training algorithms

The reason for considering second-order training methods is that the original ADAC algorithm by Si and Wang [25] used TD-SBP, which is known to have poor convergence properties [36]. In addition, it is susceptible to parameter shadowing when used online. This occurs when the network parameters continually adapt so that the network output tracks the desired output instead of converging to the desired model weights, $w^*$ [26]. It is therefore logical to develop a more robust and stable training algorithm based on existing second-order recursive schemes.

As with SBP in the previous section, the second-order recursive training schemes investigated here also seek to minimise a squared-error cost function, $E(k)$. Using the single data point available at the $k$th sample instant, an instantaneous estimate of $E(k)$ is derived as

$$E(k) = \frac{1}{2}e^2(k), \tag{9.35}$$

where $e(k)$ is defined as

$$e_c(k) = \gamma Q(k) - [Q(k-1) - r(k)] \quad \text{and} \quad e_a(k) = Q(k), \tag{9.36}$$

for the Critic and Action Networks, respectively. The use of the Levenberg–Marquardt algorithm here is motivated by the fact that it is often the best algorithm for offline training of neural networks. Hence the recursive version of this algorithm might be expected to yield similar performance benefits when training ADACs. There have been numerous studies demonstrating the superiority of second-order recursive estimation algorithms, such as recursive least squares (RLS), over first-order algorithms, such as least mean squares (LMS), for linear-in-the-parameter models [37–40]. These algorithms can also be applied for the identification of nonlinear-in-the-parameter models as shown in References 41–43. This is illustrated as follows. Assume that the predicted output at time $k$, for either of the Critic Network or Action Network, is represented by

$$\hat{y}(k) = a[w(k), x(k)], \tag{9.37}$$

where $a(\cdot)$ is the nonlinear mapping function produced by the MLP, while $w$ and $x$ are the parameters (i.e. weights) and state (i.e. input) vector, respectively. As in the linear-in-parameter technique, the nonlinear estimation requires the gradient vector $\nabla \psi[w(k)]$ to be derived, i.e.

$$\nabla \psi[w(k)] = \frac{\partial}{\partial w} a[w(k), x(k)], \tag{9.38}$$

where

$$
\begin{aligned}
\nabla \psi[w_{cj}^{L}(k)] &= \frac{\partial E_c(k)}{\partial Q(k)} \cdot \frac{\partial Q(k)}{\partial w_{cj}^{L}(k)}, \\
\nabla \psi[w_{cij}^{NL}(k)] &= \frac{\partial E_c(k)}{\partial Q(k)} \cdot \frac{\partial Q(k)}{\partial y_j(k)} \cdot \frac{\partial y_j(k)}{\partial g_j(k)} \cdot \frac{\partial g_j(k)}{\partial w_{cij}^{NL}(k)}, \\
\nabla \psi[w_{aj}^{L}(k)] &= \frac{\partial E_a(k)}{\partial Q(k)} \cdot \frac{\partial Q(k)}{\partial u(k)} \cdot \frac{\partial u(k)}{\partial w_{aj}^{L}(k)}, \\
\nabla \psi[w_{aij}^{NL}(k)] &= \frac{\partial E_c(k)}{\partial Q(k)} \cdot \frac{\partial Q(k)}{\partial u(k)} \cdot \frac{\partial u(k)}{\partial z_j(k)} \cdot \frac{\partial z_j(k)}{\partial f_j(k)} \cdot \frac{\partial f_j(k)}{\partial w_{aij}^{NL}(k)}.
\end{aligned}
\tag{9.39}
$$

The Gauss–Newton Hessian, $R(k)$, matrix approximation can be estimated recursively at each iteration by

$$R(k) = \alpha(k)R(k-1) + [1 - \alpha(k)](\nabla \psi[w(k)]\nabla \psi^{T}[w(k)]), \tag{9.40}$$

where $\alpha$ is called the forgetting factor and controls the rate at which $R(k)$ adapts to new inputs. It is usually set at $0.975 < \alpha < 1.0$. The weight update is then given by

$$w(k+1) = w(k) + R(k)^{-1}(\nabla \psi[w(k)])e(k), \tag{9.41}$$

where $e(k)$ is the temporal difference error, either $e_c(k)$ for the Critic Network or $e_a(k)$ for the Action Network (equation (9.36)). This recursive formulation is rarely used directly due to the $O(N_w^3)$ computational complexity of the inverse calculation which has to be performed at each iteration [27]. Instead the matrix inversion lemma,

$$(A + BC)^{-1} = A^{-1} - A^{-1}B(1 + CA^{-1}B)^{-1}CA^{-1} \tag{9.42}$$

is used to compute the inverse of $R(k)$. This leads to the following weight-update procedure, referred to as the recursive prediction error (RPE) algorithm [41,42]:

$$P(k) = \frac{1}{\alpha}[P(k-1) - P(k-1)(\nabla\psi[\mathbf{w}(k)])S^{-1}(k)(\nabla\psi^T[\mathbf{w}(k)])P(k-1)],$$
(9.43)

$$S(k) = \alpha(k) + (\nabla\psi^T[\mathbf{w}(k)])P(k-1)(\nabla\psi[\mathbf{w}(k)]),$$
(9.44)

$$\mathbf{w}(k+1) = \mathbf{w}(k) + P(k)(\nabla\psi[\mathbf{w}(k)])e(k).$$
(9.45)

The matrix $P(k)$ $(= R(k)^{-1})$ here can be interpreted as the covariance matrix of the weight estimate $\mathbf{w}(k)$. If the input signals to the plant are not well excited, the covariance matrix $P(k)$ tends to become large, a phenomenon known as covariance wind-up [41,42]. A constant trace adjustment is usually applied to the covariance matrix, as suggested by Salgado *et al.* [44], to overcome this problem. Thus

$$P(k) = \frac{\tau}{\text{trace}[P(k)]}P(k), \quad \tau > 0$$
(9.46)

and $\tau$ is a positive scalar normally set to value of one [23].

Ljung and Söderström [45], suggested that by using a time-varying forgetting factor, $\alpha(k)$, rather than a fixed value with $\alpha(k) < 1$ at the beginning, $P(k)$ converges faster from its initial value and $\alpha(k) \to 1$ as $k \to \infty$ will provide stability. This is achieved by using the following update rule for $\alpha(t)$:

$$\alpha(k) = \bar{\alpha}\alpha(k-1) + (1-\bar{\alpha}),$$
(9.47)

where $\bar{\alpha}$ is a scalar $(<1)$ which determines the rate of convergence of $\alpha(k)$ to 1. According to Gunnarsson [46,47], the RLM algorithm is simply the regularised form of the RPE method. Thus, RLM is obtained by incorporating a regularisation term in the Hessian update expressed in equation (9.40), giving

$$R(k) = \alpha(k)R(k-1) + [1-\alpha(k)]\{(\nabla\psi[\mathbf{w}(k)]\nabla\psi^T[\mathbf{w}(k)]) + \rho I_{N_w}\}.$$
(9.48)

Unfortunately, it is now impractical to directly apply the matrix inversion lemma. Ngia *et al.* [27] and Ngia and Sjöberg [28] proposed adding the regularisation term $\rho$ to one diagonal element of $R(k)$ at a time as a solution to this problem. This corresponds to rewriting equation (9.48) as

$$R(k) = \alpha(k)R(k-1) + [1-\alpha(k)]\{(\nabla\psi[\mathbf{w}(k)]\nabla\psi^T[\mathbf{w}(k)]) + \rho Z_{N_w}\},$$
(9.49)

where $Z_{N_w}$ is an $N_w \times N_w$ zero matrix, except with one of its diagonal elements set to one. The diagonal element $z_{ii}$ set equal to 1 changes from iteration to iteration as determined by the following expressions,

$$z_{ii} = 1 \quad \text{when } i = k \bmod(N_w) + 1 \text{ and } k > N_w,$$
(9.50)

$$z_{ii} = 0 \quad \text{otherwise.}$$
(9.51)

With this modification, equation (9.49) can be rewritten as

$$R(k) = \alpha(k)R(k-1) + [1 - \alpha(k)][\Omega(k)\Lambda(k)^{-1}\Omega^T(k)], \tag{9.52}$$

where $\Omega(k)$ is a $N_w \times 2$ matrix with the first column containing $\nabla\psi[w(k)]$ and the second column consisting of a $N_w \times 1$ zero vector with one element set to 1 in accordance with (9.50) and (9.51) above, i.e.

$$\Omega^T(k) = \begin{bmatrix} 0\ldots & \begin{matrix} \nabla\psi[w(k)] \\ 1 \\ \Uparrow \end{matrix} & \ldots 0 \end{bmatrix} \quad \text{and} \quad \Lambda(k)^{-1} = \begin{bmatrix} 1 & 0 \\ 0 & \rho \end{bmatrix}. \tag{9.53}$$

$$\text{position} = k \bmod (N_w) + 1$$

The matrix inversion lemma can now be applied to equation (9.52) which leads to the RLM formulation [27]

$$S(k) = \alpha(k)\Lambda(k) + \Omega^T(k)P(k-1)\Omega(k), \tag{9.54}$$

$$P(k) = \frac{1}{\alpha(k)}[P(k-1) - P(k-1)\Omega(k)S^{-1}(k)\Omega^T(k)P(k-1)]$$

$$\text{with } P(k) = \frac{1}{\text{trace}[P(k)]}P(k), \tag{9.55}$$

$$w(k+1) = w(k) + P(k)(\nabla\psi[w(k)])e(k). \tag{9.56}$$

$S(k)$ is now a $2 \times 2$ matrix, which is much more cost effective to invert than the $N_w \times N_w$ matrix that arises when the matrix inversion lemma is applied to equation (9.48).

The overall training procedure for the ADAC is summarised in Figure 9.9. Training begins with the weights being initialised randomly before each run. At time $k$, the Action Network and Critic Network both receive the input state vector, $x(k)$. The Action Network outputs a control action, $u(k)$, to the plant. At the same time the Critic Network outputs $Q(k)$ and it is stored. At time $k + 1$, both the Action and Critic Networks receive the next state vector, $x(k+1)$ and produce their corresponding outputs, $u(k + 1)$ and $Q(k + 1)$, respectively. The reward $r(k + 1)$ is obtained based on the outcome of the control action, i.e. $r = 0$ when it is a success or $r = -1$ when it is a failure. The Critic Network weights are updated once this reward value is obtained. Then the Action Network weights are updated. Once the updates are done, the cycle is repeated until the stopping criteria have been met.

In the next section the proposed TD-RLM algorithm will be evaluated in simulation on the well-known inverted pendulum case study and its performance compared to the TD-SBP algorithm used by Si and Wang [25].

## 9.3   Inverted pendulum

The inverted pendulum or pole-balancing task is representative of the inherently unstable, multiple-output, dynamic systems present in many balancing situations,

---

ADAC training summary

1. At the start of every run initialise all network weights to random values in the range [−1.0,1.0].

2. At time *t*, retrieve the state vector, $x(k)$, from the process or plant being controlled.

3. Generate a control action, $u(k)$ and apply it to the plant.

4. Generate and store the Critic Network output $Q(k)$ (to be used as '$Q(k-1)$' to calculate the cost function).

5. Retrieve the new state vector, $x(k+1)$, from plant and generate the next control action, $u(k+1)$.

6. Generate the next Critic Network's output $Q(k+1)$.

7. Calculate the external reinforcement, $r(k+1)$ resulting from the control action.

8. Adjust the Critic Network and the Action Network weights using equations (9.22)–(9.34) for TD-SBP, or using equations (9.54)–(9.56) for TD-RLM.

9. Repeat from Step 3 until the stopping criteria have been met.

---

*Figure 9.9    Summary of ADAC algorithm*

such as two-legged walking robots and the aiming of a rocket thruster, and is frequently used as a benchmark for modern and classical control-engineering techniques. The inverted pendulum problem, depicted in Figure 9.10, has historical importance for reinforcement learning as it was one of the first successful applications of an algorithm based on model-free action policy estimation, as described in 1983 by Barto *et al.* [48] in their pioneering paper. Systems that learn to control inverted pendulums were first developed over 30 years ago by Michie and Chambers [49] and have been the subject of much research since then. See e.g. References 23–25, 43, 50–57.

Control-engineering techniques involve detailed analyses of the system to be controlled. When the lack of knowledge about a task precludes such analyses, a control system must adapt as information is gained through experience with the task. To investigate the use of learning methods for such cases, it is assumed that very little is known about the inverted pendulum system, including its dynamics. The system is viewed as a black-box generating as output the system's state and accepts as input a control action.

The inverted pendulum task considered here involves balancing a pole hinged to the top of a wheeled cart that travels along a track, as shown in Figure 9.10. The control objective is to apply a sequence of right and left forces of fixed magnitude to the cart so that the pole balances and the cart does not hit the end of the track (for this simulation a −10 N and 10 N correspond to the left and right forces,

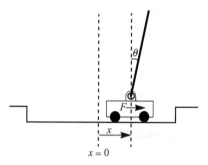

*Figure 9.10   The inverted pendulum system*

respectively). A zero magnitude force is not permitted. The state of the cart–pole system has to remain outside certain regions of the state–space to be considered successfully controlled. There is no unique solution and any state–space trajectory that does not pass through these regions is considered acceptable. The only information provided regarding the control goal is the reinforcement signal, $r(k)$, which signals a failure when either the pole falls outside $\pm 12°$ or the cart hits the bounds of the track at $\pm 2.4$ m.

The cart–pole simulation model used in this study is defined by the equations

$$\ddot{\theta}(t) = \left\{ g \sin\theta(t) - \cos\theta(t) \left[ \frac{-F(t) + ml\dot{\theta}^2(t)\sin\theta(t) + \mu_c \operatorname{sgn}[\dot{x}(t)]}{m_c + m} \right] - \frac{\mu_p \dot{\theta}(t)}{ml} \right\}$$

$$/ \left\{ l \left[ \frac{4}{3} - \frac{m\cos^2\theta(t)}{m_c + m} \right] \right\}, \tag{9.57}$$

$$\ddot{x}(t) = \frac{F(t) + ml[\dot{\theta}^2(t)\sin\theta(t) - \ddot{\theta}(t)\cos\theta(t)] - \mu_c \operatorname{sgn}[\dot{x}(t)]}{m_c + m}, \tag{9.58}$$

$$\operatorname{sgn}(x) = \begin{cases} 1, & \text{if } x > 0, \\ 0, & \text{if } x = 0, \\ -1, & \text{if } x < 0. \end{cases} \tag{9.59}$$

This includes all the nonlinearities and reactive forces of the physical system such as friction. The system is constrained to move within the vertical plane. Here $x(t)$ and $\theta(t)$ are the horizontal position of the cart and the angle of the pole, respectively and

$l$, the length of the pole $= 0.5$ m
$m$, the mass of the pole $= 0.1$ kg
$m_c$, the mass of the pole and cart $= 1.1$ kg
$F$, the magnitude of the force $= \pm 10$ N
$g$, the acceleration due to gravity $= 9.8\,\text{ms}^{-2}$
$\mu_c$, friction of cart on track coefficient $= 5 \times 10^{-4}$
$\mu_p$, friction of pole on cart coefficient $= 2 \times 10^{-6}$.

This set of nonlinear differential equations was solved numerically using fourth-order Runge–Kutta in the simulation and sampled every 0.02 s to produce four discrete state variables defined as follows:

$x(k)$ = the horizontal position of the cart, relative to the track, in metres,
$\dot{x}(k)$ = the horizontal velocity of the cart, in metres/second,
$\theta(k)$ = the angle between the pole and vertical, in degrees, clockwise being
    positive,
$\dot{\theta}(k)$ = the angular velocity of the pole, in degrees/second.

### 9.3.1   Inverted pendulum control results

The ADAC used to implement the inverted pendulum controller is illustrated in Figure 9.11. Both the Action and Critic Networks were implemented as single hidden layer MLP neural networks. Each network had six neurons in its hidden layer. Thus the Action Network was a 4-6-1 MLP architecture and the Critic Network was a 5-6-1 architecture.

The simulation studies were based on 100 runs, each consisting of 10 trials, during which the Action Network had to control the inverted pendulum within set boundaries. Here the external reinforcement signal was defined as

$$r(k) = \begin{cases} -1, & \text{if } |\theta| > 12° \text{ or } |x| > 2.4\,\text{m,} \\ 0, & \text{otherwise.} \end{cases} \tag{9.60}$$

The controller was considered to be successful if it managed to balance the inverted pendulum for $6 \times 10^5$ time steps of 0.02 s each (i.e. for 3 h 20 min). If after ten trials

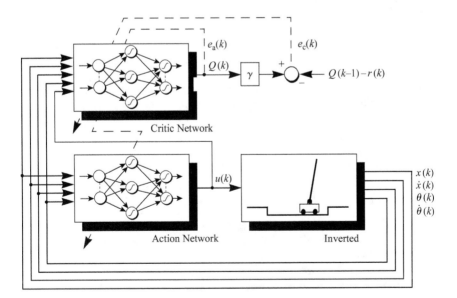

*Figure 9.11   The ADAC control strategy for the inverted pendulum system*

the controller still failed to control the pendulum, that run was considered a failure and a new run was initiated with the pendulum states set to zero and all the network weights initialised randomly. Figure 9.12 depicts the result of a typical successful run of the ADAC controller.

Figure 9.12 clearly shows that the TD-RLM algorithm converges almost immediately compared to the ADAC which was trained using TD-SBP. Also note that the ADAC controller balances the pendulum, using a force, of a fixed magnitude and alternating sign applied to the cart (see Figure 9.13). This bang–bang control leads to the zig-zag oscillation observed in the graphs.

The performance of both training algorithms was measured in terms of the frequency of successful runs and the average computational time for a successful run. The actual computational cost was also considered as a function of Action and Critic Network sizes, to get a better measure of overall performance, in terms of the training efficiency.

Figure 9.14 shows that the TD-RLM training algorithm gave the best overall performance in terms of the frequency of successful runs. It can also be seen that

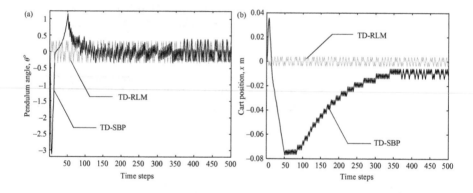

*Figure 9.12    An example of a typical pendulum angle and cart position trajectory for a successful run using TD-SBP and TD-RLM*

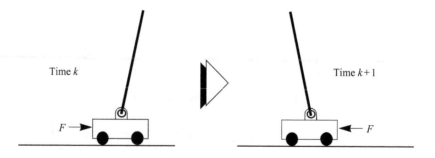

*Figure 9.13    Forces applied to the cart to balance the inverted pendulum between a sample interval*

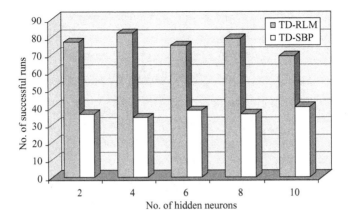

*Figure 9.14   Variation in number of successful runs with the size of the ADAC networks*

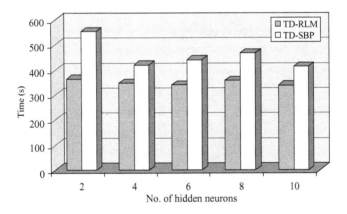

*Figure 9.15   Variation in average computation time per successful run with the size of the ADAC networks*

the original TD-SBP algorithm by Si and Wang [25] was quite poor in terms of the number of successful runs, being at times 50 per cent lower than the alternative second-order method.

Furthermore, as shown in Figure 9.15, the TD-SBP implementation also proved less efficient compared to TD-RLM, in terms of average computation time per successful run, with the second-order method being the most consistent for different network sizes. The overall average computation time per successful run as the ADAC network sizes varied from 2 to 10 hidden neurons was 459.75 s for TD-SBP and 350.41 s for TD-RLM.

Figure 9.16 plots the time variations in the squared training error for the Critic Network obtained with each algorithm. These illustrate the speed of convergence

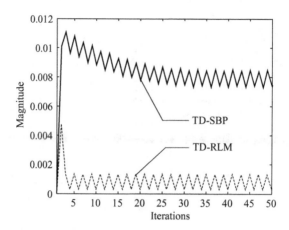

*Figure 9.16    Comparison of the training cost $E_c$ of the Critic Network obtained with TD-RLM and TD-SBP*

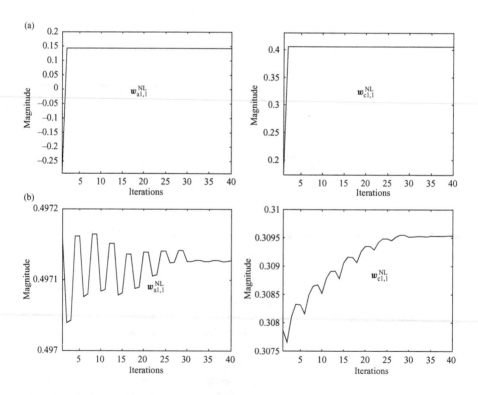

*Figure 9.17    The trajectory of the first components of the weight vector of the Action Network, $w_{a1,1}^{NL}$ (left) and of the Critic Network, $w_{c1,1}^{NL}$ (right) for (a) TD-RLM and (b) TD-SBP*

of the neurocontroller weights. Again, the zig-zag oscillation is due to the nature of the bang-bang control strategy employed. Further evidence of the superior convergence speed produced by second-order training is provided in Figure 9.17. This shows the variation in the first component of the weight vector for the Action Network and the Critic Network, $w^{NL}_{a1,1}$ and $w^{NL}_{c1,1}$, respectively, with each algorithm.

## 9.4 Ring grinding process

The second application is the ADAC modelling and control of an industrial ring grinding process used in the manufacture of disk-drive platters. Here, aluminium subtrate disks are ground in batches of 12 between two grindstones, as shown in Figures 9.18 and 9.19. The stones can be moved apart to allow loading and unloading of the disks using a pick-and-place unit. During operation the grindstones are rotated in opposite directions with pressure applied to the upper one. This causes the subtrate disks between them to rotate, thereby ensuring uniform grinding of their surfaces. The rate at which the disks are ground, called the removal rate, is the critical variable. It varies depending on a number of parameters including stone wear, exerted pressure, lubricant viscosity and coolant flow rate. The initial thickness of the disks also varies, although the disks in any one batch are sorted to be approximately the same thickness.

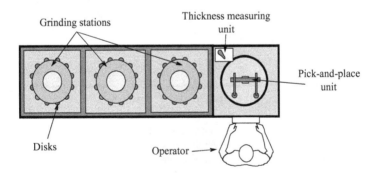

*Figure 9.18    Layout of the ring grinding process*

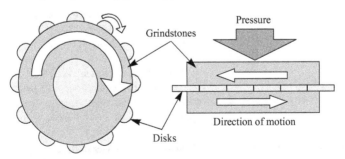

*Figure 9.19    The ring grinding process*

The thickness of one disk from each batch is measured before the batch is ground. The system controller determines the actual removal rate from the previous batch and estimates the current value of removal rate using a proprietary control law. It predicts how much material has to be removed by subtracting the target thickness from the input thickness and then calculates the necessary grinding duration for the current batch.

When the grinding is completed, the selected disk is measured again. If it is within specification, then the whole batch is passed. If the disk is too thick (above the upper specification limit), the disks are ground again (i.e. reworked) but if the disk is too thin (below the lower specification limit), the batch is rejected. When a grindstone is newly installed (i.e. replaced due to wear), the pressure is initially set to a low value and then gradually increased to an upper limit to counteract the stone deterioration, which in turn increases the removal rate. The removal rate subsequently decreases until a stage is reached where it is so low that the grindstone has to be resurfaced which is done by slicing off the worn part of the grindstone. Once re-installed the whole process is repeated.

Various process variables are logged for each grind cycle as part of the company's own process performance monitoring procedure. These include the current removal rate, the pressure between the grindstones and the cumulative cycle time. Cumulative cycle time is logged as it is an indication of wear and ageing of the grindstones, which in turn impacts on the removal rate. A summary of these variables and the identifiers used for them in this chapter is provided in Table 9.1.

*Table 9.1   Grinding process variables*

| Variables | Definition |
|---|---|
| Removal rate, $rr(k)$ | Rate of material removal from a disk during the grind cycle. Units are in microinch per min |
| Previous removal rate, $rr(k-1)$ | Removal rate from the previous grind cycle |
| Cycle time, $c(k)$ | Grind cycle time. Units are in seconds |
| Cumulative cycle time, $cct(k)$ | Sum of all previous cycle times since the grindstone was last resurfaced |
| Pressure, $p(k)$ | Pressure between the grindstones. Units are in p.s.i. |
| Loading thickness, $T_L(k)$ | Thickness of the disk before the grinding process begins. Units are in mil(s) |
| Unloading thickness, $T(k)$ | Thickness of the disk after the completion of the grinding process. Units are in mil(s) |
| Target thickness, $T_{SP}(k)$ | Desired thickness required for each grind cycle. Units are in mil(s) |
| Upper control limit, $UCL(k)$ | Upper control thickness limit specification. Units are in mil(s) |
| Lower control limit, $LCL(k)$ | Lower control thickness limit specification. Units are in mil(s) |

The grindstone data used in this investigation was detrended and normalised to lie within the interval $[-1, 1]$. A sample of the data showing the typical variations found in all the variables is plotted in Figure 9.20.

The main aim here is to achieve accurate thickness control in order to minimise the number of out-of-specification disks produced by the grinding process. This process optimisation can be achieved through manipulation of the grind cycle time as illustrated in Figure 9.21. Neural network-based direct inverse control has been shown to provide an effective solution to this problem [58] and was therefore chosen as the basis for the ADAC investigation. The ADAC framework was considered for two elements of the controller design, namely developing a process model and fine-tuning the final controller. A process model is needed as this forms the basis for the direct inverse controller implementation. The recommended model is one which predicts the removal rate, for each grind cycle, on the basis of the current state of the process [58].

The existing proprietary controller was used as a reference for evaluating the performance of the model and subsequent controller design. Since the proprietary controller was a model-free implementation, it did not generate an explicit removal rate prediction, $\hat{rr}(k)$. Rather, this was inferred from the generated cycle time,

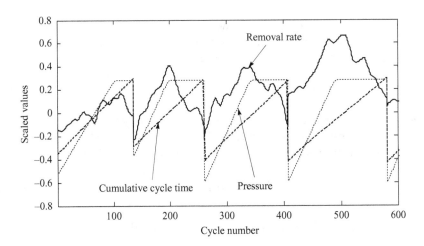

*Figure 9.20    Variables used in modelling the ring grinding process*

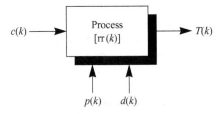

*Figure 9.21    The ring grinding process block diagram*

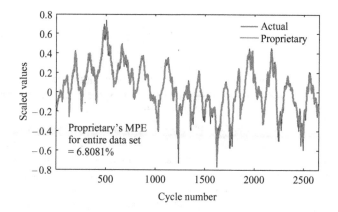

*Figure 9.22    The removal rate prediction from the proprietary scheme*

$c(k)$, as

$$\hat{rr}(k) = \frac{T_L(k) - T_{SP}(k)}{c(k)}, \tag{9.61}$$

where $T_L(k)$ is the loading thickness and $T_{SP}(k)$ is the setpoint or target thickness. Note that the actual removal rate, $rr(k)$, was obtained by replacing $T_{SP}(k)$ by the measured unloading thickness, $T(k)$, in equation (9.61), i.e.

$$rr(k) = \frac{T_L(k) - T(k)}{c(k)}. \tag{9.62}$$

Figure 9.22 compares the predicted removal rate, $\hat{rr}(k)$, with the actual removal rate, $rr(k)$, over the life of a typical grindstone. The accuracy of the prediction is measured in terms of the percentage normalised mean prediction error (MPE), defined as

$$MPE = \frac{1}{n} \sum_{j=1}^{n} \frac{|rr(k) - \hat{rr}(k)|}{\sigma(k)} \times 100\%, \tag{9.63}$$

where $\sigma(k)$ is the standard deviation of $rr(k)$. In this case the MPE for the grindstone was 6.8 per cent.

### 9.4.1    Model development

The process model to be developed here predicts the grindstone removal rate and this is used to calculate the cycle time for the grinding machine. Accurate prediction of removal rate will therefore lead to an improved cycle time estimate for the grind process. In the ADAC framework, the Action Network is trained to form the process model.

Based on previous experience [58], it was decided to incorporate error feedback to compensate for the low-frequency offsets in the Action Network prediction, in order to further enhance the accuracy of the process model. This 'predict–correct' technique uses past plant outputs and the corresponding model predictions to generate a correction to the current estimate $\hat{rr}^*(k)$ and successful applications have been reported in Rovnak and Corlis [59], Willis *et al.* [60], Lightbody [61] and Irwin *et al.* [62]. The predict–correct scheme is implemented as follows:

$$\hat{rr}^*(k) = \hat{rr}(k) + \frac{1}{N} \sum_{j=1}^{N} [rr(k-j) - \hat{rr}(k-j)]. \tag{9.64}$$

A first-order, predict–correct term was incorporated into the Action Network predictor, as shown in Figure 9.23. The complete ADAC identification strategy is then as shown in Figure 9.24.

For this study a nonlinear ARX modelling strategy was employed where the removal rate was estimated as a function of previous removal rates, $rr(k-1)$, $rr(k-3)$ and $rr(k-5)$, current pressure, $p(k)$, and the current and past cumulative cycle times, $cct(k)$ and $cct(k-1)$. Thus, the Action Network was trained to learn the unknown mapping

$$\hat{rr}(k) = f[rr(k-1), rr(k-3), rr(k-5), p(k), cct(k), cct(k-1)]. \tag{9.65}$$

The goal was to minimise the absolute tracking error between the desired removal rate, $rr(k)$, and the predicted, $rr^*(k)$. The external reinforcement signal, $r(k)$, for the ADAC model was chosen as

$$r(k) = \begin{cases} -1, & |\text{Prediction error}| \geq 5 \times 10^{-4}, \\ 0, & \text{otherwise.} \end{cases} \tag{9.66}$$

After experimenting with different architectures, the ADAC networks that were found to produce the minimum MPE error used five hidden neurons (i.e. a 6-5-1 MLP for the Action Network and a 7-5-1 MLP for the Critic Network). Figure 9.25 compares the ADAC model obtained for the grindstone with the corresponding proprietary model and clearly shows the benefit of nonlinear modelling using the ADAC framework as the MPE has been reduced by 42 per cent compared to the proprietary model.

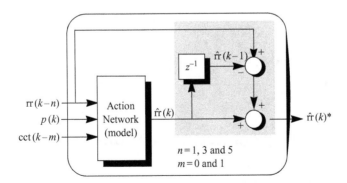

*Figure 9.23*   *The augmented Action Network prediction model for the grinding process (model + PC)*

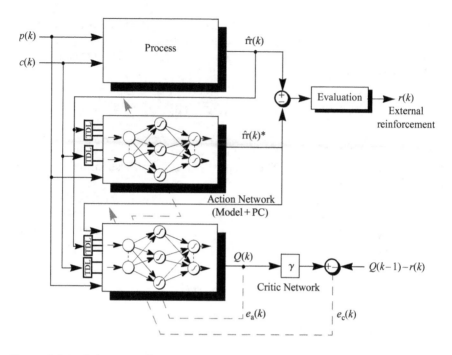

*Figure 9.24    Schematic of modelling strategy using the ADAC*

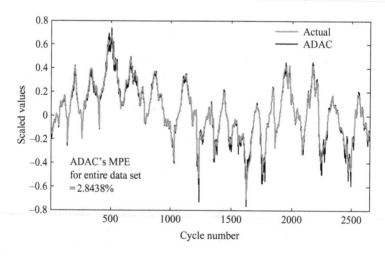

*Figure 9.25    The removal rate prediction from the ADAC scheme*

## 9.4.2    Disk thickness control results

The model identified previously was used to provide an accurate estimate of $rr(k)$ at each iteration to produce the open-loop thickness controller depicted in Figure 9.26.

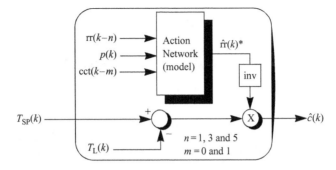

*Figure 9.26    Open-loop control using the Action Network removal rate predictor*

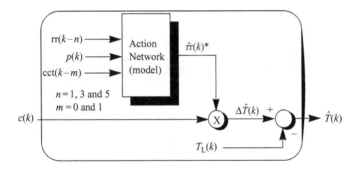

*Figure 9.27    Forward process ADAC model*

In fact this is a direct inverse control implementation, which can be seen as follows. First, note that the removal rate model can also be used to generate a $c(k)$-to-$T(k)$ forward process model, as shown in Figure 9.27. Lee and Shin [63] pointed out that, this particular formulation allows the inverse plant model to be obtained without having to invert an artificial neural network (ANN) model, as is usually the case with a neural-control scheme [64]. Thus, Figure 9.26 represents an exact inverse of the forward process model and therefore a direct inverse controller. The final complete direct inverse model control scheme is shown in Figure 9.28.

The ADAC-model direct inverse controller can be applied to the process as a fixed parameter controller. Alternatively, it can be fine tuned online within an ADAC control framework, with the reinforcement signal now defined in terms of control specifications, namely the upper and lower control limits (UCLs and LCLs) for the unloading thicknesses of the disks. This gives

$$r(k) = \begin{cases} -1, & \text{UCL} < \text{ULT} < \text{LCL}, \\ 0, & \text{otherwise.} \end{cases} \tag{9.67}$$

Figure 9.29 shows the unloading thickness prediction obtained with the resulting ADAC control strategy while Table 9.2 provides a comparison with the fixed parameter ADAC model-based controller and the proprietary controller.

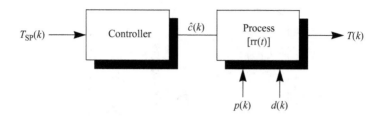

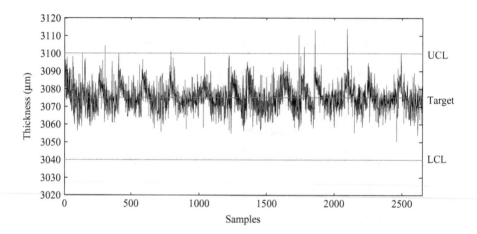

*Figure 9.28    Open-loop inverse model control using the ADAC modelling method of the grinding process*

*Figure 9.29    Unloading thickness prediction of the ADAC controller*

*Table 9.2    Performance comparison between the actual and the ADAC schemes*

|  | Target thickness | Unloading thickness | | Number of rejects |
|---|---|---|---|---|
|  |  | Mean | Variance |  |
| Proprietary |  | 3075.85 | 7.095 | 12 |
| ADAC-mod. | 3075 | 3076.12 | 6.71 | 10 |
| ADAC-cont. |  | 3076.28 | 6.407 | 8 |

It can be seen that the ADAC model direct inverse controller (ADAC-mod.) and the online tuned ADAC controller (ADAC-cont.) both outperform the proprietary scheme with the tuned ADAC controller yielding a 33.33 per cent reduction in the number of rejects. Figure 9.30 compares the unloading thickness distribution

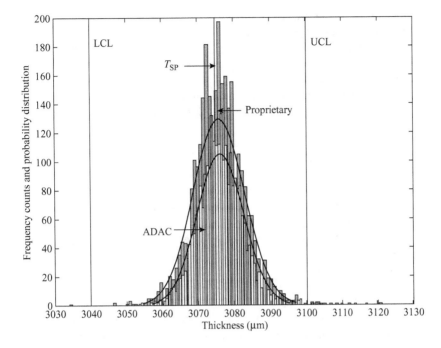

*Figure 9.30   Unloading thickness distribution plot of the ADAC method compared to the proprietary scheme*

of the proprietary control scheme with that obtained from the ADAC controller and clearly shows than tighter thickness control was achieved with the ADAC control scheme.

## 9.5   Conclusions

This chapter extends the model-free action dependent adaptive critic (ADAC) of Si and Wang by presenting a fully online, intelligent neurocontroller which avoids the necessity to store plant data during a successful run. In particular, the potential limitations of their stochastic back-propagation training, in terms of poor convergence and parameter shadowing, are avoided by introducing a modified version of recursive Levenberg–Marquardt (RLM) called the temporal difference RLM (TD-RLM) algorithm. This was demonstrated in an initial simulation study on inverted pendulum control.

The performance of the new ADAC scheme, for both identification and control, has been validated using data from an actual industrial grinding process used in the manufacture of aluminium substrates for disk drives. The results suggest that the ADAC can achieve a 33 per cent reduction in rejects compared to a proprietary controller. The authors believe that this is one of the first reported industrial applications of this emerging technology.

For identification the ADAC still requires the incorporation of the predict correct strategy to achieve convergence and there are many parameters, such as the learning rates, which need to be tuned by trial-and-error. These are both areas for further work. However, the ability of the ADAC to converge satisfactorily from scratch far outweighs these limitations.

The control performance of the ADAC is most encouraging and clearly demonstrates the feasibility of successfully using reinforcement learning for process control applications.

# References

1  PASSINO, K.: 'Bridging the gap between conventional and intelligent control', *IEEE CSM*, June 1993, pp. 12–18
2  ANTSAKLIS, P.: 'Final report of task force in intelligent control'. Technical committee on *Intelligent control*, IEEE Control Systems Society, 1993
3  ANTSAKLIS, P.: 'Intelligent control', in WEBSTER, J.G., (Ed.): 'Encyclopedia of electrical and electronics engineering' (John Wiley & Sons, Inc., New York, 1999), vol. 10, pp. 493–503.
4  GUPTA, M. M., and RAO, D. H. (Eds): 'Neuro-control systems: a tutorial', in 'Neuro-control systems: theory and application' (IEEE Press, Piscataway, NJ, 1994) pp. 1–43
5  WERBOS, P. J.: 'A menu of designs for reinforcement learning over time', in MILLER, W. T., SUTTON, R. S., and WERBOS, P. J. (Eds): 'Neural networks for control' (MIT Press, Cambridge, MA, 1990) pp. 67–95
6  WERBOS, P. J.: 'Approximate dynamic programming for real-time control and neural modelling', in WHITE, D. A., and SOFGE, D. A. (Eds): 'Handbook of intelligent control: neural, fuzzy and adaptive approaches' (Van Nostrand Reinhold, New York, 1992) pp. 493–525
7  WU, Q. H., and PUGH, A. C.: 'Reinforcement learning control of unknown dynamic systems', *Proceedings of the IEE – Part D: Control Theory and Applications*, 1993, **140**, pp. 313–22
8  WU, Q. H.: 'Reinforcement learning control using interconnected learning automata', *International Journal of Control*, 1995, **62** (1), pp. 1–16
9  CHAN, K. H., JIANG, L., TILOSTON, P., and WU, Q. H.: 'Reinforcement learning for the control of large-scale power systems'. Proceedings of second international symposium on *Engineering of intelligent systems* (EIS'2000), Paisley, UK, 2000
10  ZENG, X., ZHOU, J., and VASSEUR, C.: 'A strategy for controlling nonlinear systems using a learning automation', *Automatica*, 2000, **36**, pp. 1517–24
11  RIEDMILLER, M.: 'Concepts and facilities of a neural reinforcement learning control architecture for technical process control', *Neural Computation and Application Journal*, 1999, **8**, pp. 323–38
12  ERNST, D., GLAVIC, M., and WEHENKEL, L.: 'Power system stability control: reinforcement learning framework', accepted for publication in *IEEE Transaction on Power Systems*, vol. 19, no. 1, pp. 427–436, Feb 2004

13 PARK, J. W., HARLEY, R. G., and VENAYAGAMOORTHY, G. K.: 'Adaptive critic designs and their implementations on different neural network architectures'. Proceedings of the international joint conference on *Neural networks*, 2003, vol. 3, pp. 1879–84

14 PARK, J. W., HARLEY, R. G., and VENAYAGAMOORTHY, G. K.: 'Adaptive critic based optimal neurocontrol for synchronous generator in power system using MLP/RBF neural networks'. *Conference Record – IAS Annual Meeting (IEEE Industry Applications Society)*, 2002, vol. 2, pp. 1447–54

15 VENAYAGAMOORTHY, G. K., HARLEY, R. G., and WUNSCH, D. C.: 'A nonlinear voltage controller with derivative adaptive critics for multimachine power systems'. *IEEE Power Industry Computer Applications Conference*, 2001, pp. 324–9

16 VENAYAGAMOORTHY, G. K., HARLEY, R. G., and WUNSCH, D. C.: 'Excitation and turbine neurocontrol with derivative adaptive critics of multiple generators on the power grid'. Proceedings of the international joint conference on *Neural networks*, 2001, vol. 2, pp. 984–9

17 VENAYAGAMOORTHY, G. K., HARLEY, R. G., and WUNSCH, D. C.: 'Comparison of heuristic dynamic programming and dual heuristic programming adaptive critics for neurocontrol of a turbogenerator', *IEEE Transactions on Neural Networks*, 2002, **13** (3), pp. 764–73

18 IYER, M. S., and WUNSCH, D. C. II: 'Dynamic re-optimization of a fed-batch fermentor using adaptive critic designs', *IEEE Transactions on Neural Networks*, 2001, **12** (6), pp. 1433–44

19 RADHAKANT, P., and BALAKRISHNAN, S. N.: 'Proper orthogonal decomposition based optimal neurocontrol synthesis of a chemical reactor process using approximate dynamic programming', *Neural Networks*, 2003, **16**, pp. 719–28

20 SOFGE, D. A., and WHITE, D. A.: 'Neural network based process optimization and control'. Proceedings of the IEEE conference on *Decision and Control*, 1990, vol. 6, pp. 3270–6

21 WHITE, D. A., and SOFGE, D. A.: 'Neural network based control for composite manufacturing', American Society of Mechanical Engineers, Materials Division (Publication) MD, *Intelligent Processing of Materials*, 1990, **21**, pp. 89–97

22 HOSKINS, J. C., and HIMMELBLAU, D. M.: 'Process control via artificial neural networks and reinforcement learning', *Computers and Chemical Engineering*, 1992, **16** (4), pp. 241–51

23 ANDERSON, C. W.: 'Learning and problem solving with multilayer connectionist systems'. Doctoral dissertation, Department of Computer and Information Science, University of Massachusetts, Amherst, MA, 1986

24 ANDERSON, C. W.: 'Strategy learning with multilayer connectionist representations'. Technical report TR87-507.3, GTE Laboratories, Waltham, MA, 1987. Revision of the fourth international workshop on *Machine learning*, June 1987, pp. 103–14

25 SI, J., and WANG, Y. T.: 'Online learning control by association and reinforcement', *IEEE Transactions on Neural Networks*, 2001, **12** (2), pp. 264–76

26   McLOONE, S. F.: 'Neural network identification: a survey of gradient based methods'. Digest 98/521, IEE Colloquium Optimization in Control: Methods and Applications, London, November 1998

27   NGIA, L. S. H., SJÖBERG, J., and VIBERG, M.: 'Adaptive neural networks filter using a recursive Levenberg–Marquardt search direction'. Conference record of the 32nd Asilomar conference on *Signals, system and computers*, Pacific Grove, CA, November 1998, pp. 697–701

28   NGIA, L. S. H., and SJÖBERG, J.: 'Efficient training of neural nets for non-linear adaptive filtering using a recursive Levenberg–Marquardt algorithm', *IEEE Transactions on Signal Processing*, 2000, **48** (7), pp. 1915–26

29   SUTTON, R. S.: 'Learning to predict by the method of temporal differences', *Machine Learning*, 1988, **3**, pp. 9–44

30   SUTTON, R. S.: 'Implementation details of the TD($\lambda$) procedure for the case of vector predictions and backpropagation'. GTE Laboratories Technical Note TN87-509.1, August 1989

31   BELLMAN, R. E.: 'Dynamic programming' (Princeton University Press, Princeton, NJ, 1957)

32   PROKHOROV, D., SANTIAGO, R., and WUNSCH, D.: 'Adaptive critic designs: a case study for neurocontrol', *Neural Networks*, 1995, **8** (9), pp. 1367–72

33   PROKHOROV, D., and WUNSCH, D.: 'Adaptive critic designs', *IEEE Transactions on Neural Networks*, 1997, **8**, pp. 997–1007

34   WERBOS, P. J.: 'Consistency of HDP applied to a simple reinforcement learning problem', *Neural Networks*, 1990, **3** (2), pp. 179–89

35   LIU, D., XIONG, X., and ZHANG, Y.: 'Action-dependant adaptive critic designs'. Proceedings of the INNS–IEEE international joint conference on *Neural networks*, July 2001, pp. 990–5

36   SUTTON, R. S., and WHITEHEAD, S. D.: 'Online learning with random representations'. Proceedings of the tenth international conference on *Machine learning*, University of Massachusetts, Amherst, MA, 27–29 June 1993, Morgan Kaufmann Publishers, pp. 314–21

37   GOODWIN, G. C., and PAYWE, R. L.: 'Dynamic system identification: experiment design and data analysis' (Academic Press, New York, 1977)

38   JOHANSSON, R.: 'System modelling identification' (Prentice-Hall Information and System Science Series, NJ, 1993)

39   LJUNG, L.: 'System identification: theory for the user' (Prentice-Hall, Englewood Cliffs, NJ, 1987)

40   YOUNG, P. C.: 'Recursive estimation and time series analysis' (Springer-Verlag, Berlin, 1984)

41   CHEN, S., BILLINGS, S. A., and GRANT, P. M.: 'Non-linear system identification using neural networks', *International Journal of Control*, 1990, **51** (6), pp. 1191–214

42   CHEN, S., COWAN, C. F. N., BILLINGS, S. A., and GRANT, P. M.: 'Parallel recursive prediction error algorithm for training layered networks', *International Journal of Control*, 1990, **51** (6), pp. 1215–28

43 CONNELL, M., and UTGOFF, P.: 'Learning to control a dynamic physical system'. Proceedings *AAAI-87*, American Association for Artificial Intelligence, Seattle, WA, 1987, vol. 2, pp. 456–60

44 SALGADO, M. E., GOODWIN, G. C., and MIDDLETON, R. H.: 'Modified least squares algorithm incorporating exponential resetting and forgetting', *International Journal of Control*, 1988, **47** (2), pp. 477–91

45 LJUNG, L., and SÖDERSTRÖM, T.: 'Theory and practice of recursive identification' (MIT, Cambridge, MA, 1983)

46 GUNNARSSON, S.: 'On covariance modification and regularization in recursive least square identification'. 10th IFAC symposium on *System Identification*, SYSID 94, 1994, pp. 661–6

47 GUNNARSSON, S.: 'Combining tracking and regularization in recursive least square identification'. Proceedings of the 35th IEEE conference on *Decision and control*, IEEE Press, Piscataway, NJ, 1996, vol. 3, pp. 2551–2

48 BARTO, A. G., SUTTON, R. S., and ANDERSON, C. W.: 'Neuronlike elements that can solve difficult learning control problems', *IEEE Transactions on Systems, Man and Cybernetics*, 1983, **13**, pp. 835–46

49 MICHIE, D., and CHAMBERS, R.: 'Boxes: an experiment in adaptive control', in DALE, E., and MICHIE, D. (Eds): 'Machine intelligence' (Oliver and Boyd, Edinburgh, 1968)

50 ANDERSON, C. W.: 'Learning to control an inverted pendulum using neural networks', *IEEE Control Systems Magazine*, 1989, **9**, pp. 31–7

51 GUEZ, A., and SELINSKY, J.: 'A trainable neuromorphic controller', *Journal of Robotic System*, 1988, **5** (4), pp. 363–88

52 GUEZ, A., and SELINSKY, J.: 'A neuromorphic controller with a human teacher'. *IEEE International Conference on Neural networks*, 1988, **2**, pp. 595–602

53 HANDELMAN, D., and LANE, S.: 'Fast sensorimotor skill acquisition based on rule-based training of neural networks', in BEKEY, G., and GOLDBERG, K. (Eds): 'Neural networks in robotics' (Kluwer Academic Publishers, Boston, MA, 1991)

54 HOUGEN, D.: 'Use of an eligibility trace to self-organize output', in RUCK, D. (Ed.): 'Science of artificial neural networks II' (SPIE 1966, 1993) pp. 436–47

55 SAMMUT, C.: 'Experimental results from an evaluation of algorithms that learn to control dynamic systems'. Proceedings of the fifth international conference on *Machine learning*, Morgan Kaufman, San Mateo, CA, 1988, pp. 437–43

56 WIDROW, B.: 'The original adaptive neural net broom-balancer'. International symposium on *Circuits and Systems*, 1987, pp. 351–7

57 WIDROW, B., GUPTA, N., and MAITRA, S.: 'Punish/reward: learning with a critic in adaptive threshold systems', *IEEE Transactions on Systems, Man and Cybernetics*, 1973, **3** (5), pp. 445–65

58 GOVINDHASAMY, J. J., McLOONE, S. F., IRWIN, G. W., DOYLE, R. P., and FRENCH, J. J.: 'Neural modelling, control and optimisation of an industrial grinding process', accepted for publication in *Control Engineering Practice*,

2003. Article in press is available online at the Control Engineering Practice Journal Website

59  ROVNAK, J. A., and CORLIS, R.: 'Dynamic matrix based control of fossil power plants', *IEEE Transactions on Energy Conversion*, 1991, **6** (2), pp. 320–6

60  WILLIS, M. J., DI MASSIMO, C. D., MONTAGUE, G. A., THAM, M. T., and MORRIS, A. J.: 'Artificial neural networks in process engineering', *IEE Proceedings – Part D: Control Theory and Applications*, 1991, **138** (3), pp. 256–66

61  LIGHTBODY, G.: 'Identification and control using neural networks', Ph.D. dissertation, The Intelligent Systems and Control Group, The Queen's University of Belfast, Northern Ireland, UK, 1993

62  IRWIN, G. W., O'REILLY, P., LIGHTBODY, G., BROWN, M., and SWIDENBANK, E.: 'Electrical power and chemical process applications', in IRWIN, G. W., WARWICK, K., and HUNT, K. J. (Eds): 'Neural network applications in control' (IEE Control Engineering Series 53, The Institution of Electrical Engineers, London, UK, 1995)

63  LEE, C. W., and SHIN, Y. C.: 'Intelligent modelling and control of computer hard disk grinding processes', in MEECH, J. A., VEIGA, S. M., VEIGA, M. V., LECLAIR, S. R., and MAGUIRE, J. F. (Eds): Proceedings of the third international conference on *Intelligent Processing and Manufacturing of Materials*, CD-ROM, Vancouver, Canada, 2001, pp. 829–38

64  HUNT, K. J., SBARBARO-HOFER, D., ZBIKOWSKI, R., and GAWTHROP, P. J.: 'Neural networks for control systems – a survey', *Automatica*, 1992, **28**, pp. 1083–112

*Chapter 10*

# Reinforcement learning and multi-agent control within an internet environment

*P. R. J. Tillotson, Q. H. Wu and P. M. Hughes*

## 10.1 Introduction

A multi-agent system consists of many individual computational agents, distributed throughout an environment, capable of learning environmental management strategies, environmental interaction and inter-agent communication. Multi-agent controllers offer attractive features for the optimisation of many real world systems. One such feature is that agents can operate in isolation, or in parallel with other agents and traditional control systems. This allows multi-agent controllers to be implemented incrementally as funds allow. The distrusted nature of multi-agent control allows local control to be integrated into a global system. Each of the individual agent controllers can be comparatively simple and optimised for local problems. Through agent communication, the agents can take into consideration the global goals of a system when they are developing their local behaviour strategies. Multi-agent controllers have adaptive behaviour, developed by the artificial intelligence community, to learn and optimise local behaviour and at the same time contribute to the global system performance.

Many real world problems are both non-deterministic and non-stationary in nature. In complex environments, there may be many variables that need to be considered, making classical analytic methods difficult to implement. Multi-agent techniques allow complex problems to be deconstructed into simpler problems, and harness expert knowledge about subproblems to develop local solutions.

There are many different flavours of intelligent behaviour developed by the artificial intelligence community. Expert systems were among the earliest solutions. These use the knowledge of domain experts to construct deterministic behaviours based on environmental observation and existing knowledge. Later the behaviour selection mechanisms were made adaptive. Genetic algorithms take an evolutionary approach to system optimisation. The behaviour of a system is encoded in a genetic model which is optimised through interactions with the environment. In this chapter, we consider the reinforcement learning approach for intelligent behaviour.

In reinforcement learning the environment and possible actions are quantised into sets known as the state and action space, respectively. Reinforcement learning's task is to develop mappings between observed states and actions that optimise agents' behaviour. An agent observes a state, selects and implements an action and receives an associated reward. The reward or reinforcement signal, as it is more formally known, is a feedback signal from the environment which the agent uses to determine the value of selecting particular actions given the observed environmental state.

Reinforcement learning techniques are based on the Markov decision process (MDP). This chapter introduces the MDP and presents two commonly used reinforcement learning techniques. These are Watkin's $Q(\lambda)$ and Sutton's generalised temporal difference algorithm Sarsa($\lambda$). Both of these algorithms are then used, based on the multi-agent system with agents distributed throughout a network, in the context of a TCP/IP forwarding engine to control routing within an internet. The results of simulation studies are presented to provide an example usage of reinforcement learning in a multi-agent system to route Internet packets from source nodes to destinations in an optimal manner.

## 10.2   Markov decision processes

Markov decision processes originated in the study of stochastic optimal control [1]. In the 1980s and 1990s, incompletely known MDPs were recognised as a natural problem formulation for reinforcement learning [2]. MDPs have become widely studied within artificial intelligence and are particularly suitable in planning problems, e.g. as in decision-theoretic planning [3], and in conjunction with structured Bayes nets [4]. In robotics, artificial life and evolutionary methods, it is less common to use the language and mathematics of MDPs, but again these problems are well expressed in terms of MDPs. MDPs provide a simple, precise, general and relatively neutral way of talking about a learning or planning agent, interacting with its environment to achieve a goal. As such, MDPs are starting to provide a bridge to biological efforts to understand the mind, with analysis in MDP-like terms being found in both neuroscience [5] and psychology [6]. MDPs are stochastic processes that describe the evolution of dynamically controlled systems by sequences of decisions or actions. In essence, MDPs present a model in which an agent interacts synchronously with a world [7]. In Figure 10.1 the agent observes the state of the environment, from which it determines actions. The state is a representation of all the environmental properties relevant to an agent's decision-making process. As such there is never an uncertainty about an agent's current state.

Under MDPs a discrete-time stochastic dynamical system consists of a state–space, an action set and a probability distribution governing the next state of the system given the past state of the system and the actions performed. In the

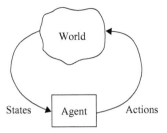

*Figure 10.1   An MDP models the synchronous interactions between agent and world*

following, it is assumed that the state and action sets are finite. The state–space can be represented by the set $S = s_1, s_2, \ldots, s_n$ of possible system states with an action set being represented by $A = a_1, a_2, \ldots, a_m$, the set of possible actions. At any particular time a system will be in some state $s \in S$. An MDP's job is to select a suitable action $a \in A$ according to its probability distribution and policy. Assuming an agent is performing actions according to an optimal policy, then for any given state the agent's course of action is fixed.

A general discrete-time stochastic and dynamic system, with Markovian properties and $T$ time stages, can be represented using $T$ transition matrices of size $N \times N$ [8]. Each matrix at instant $t$ can be represented as:

$$
\begin{bmatrix} p_{11}^1 & p_{12}^1 & p_{13}^1 \\ p_{21}^1 & p_{22}^1 & p_{23}^1 \\ p_{31}^1 & p_{32}^1 & p_{33}^1 \end{bmatrix}, \begin{bmatrix} p_{11}^2 & p_{12}^2 & p_{13}^2 \\ p_{21}^2 & p_{22}^2 & p_{23}^2 \\ p_{31}^2 & p_{32}^2 & p_{33}^2 \end{bmatrix}, \ldots, \begin{bmatrix} p_{11}^T & p_{12}^T & p_{13}^T \\ p_{21}^T & p_{22}^T & p_{23}^T \\ p_{31}^T & p_{32}^T & p_{33}^T \end{bmatrix}, \tag{10.1}
$$

and their components are constructed from probabilistic elements $p_{ij}$, which describe how likely the system will occupy state $s_j$ at time $t + 1$, given that at time $t$ the system occupies state $s_i$, or more formally:

$$
p_{ij} = \text{Prob}(S^{t+1} = s_j | S^t = s_i). \tag{10.2}
$$

Formally, an action $a_k$ maps each state $s_i$ into a distribution over $S$ that characterises the possible outcomes of that action. Therefore actions can be represented by a set of $T$ transition matrices with elements $p_{ij}^{kt}$. This is the probability of entering state $s_j$ at time $t + 1$ given that at time $t$ the system was in state $s_i$ and action $a_k$ was executed, or mathematically:

$$
p_{ij}^{kt} = \text{Prob}(S^{t+1} = s_j | S^t = s_i, A^t = a_k). \tag{10.3}
$$

This describes a probability distribution for the future state of the environment over all combinations of current state and actions selected. The effects of an action may depend on any aspect of the current state. When applying MDPs to real situations, it is important to determine the key characteristics of an environment. These need to be encoded in the model developed to represent the state–space. Often agents use sensors to monitor their environment. These sensors provide a quantised observation of the environmental state on which the agent bases its decision-making process. It is important for designers to understand the complexity of their problem environments and the imperfect view of them provided by the available sensing capability.

### 10.2.1   Value functions

The MDPs use value functions to estimate the benefit of being in a particular state, or even the benefit of performing a particular action in a given state. How beneficial it is to an agent is defined in terms of the expected future rewards. However, the rewards an agent can expect to receive in the future are dependent on what actions it will take. Accordingly, the value functions are defined with respect to particular policies.

A policy, $\pi$, is a mapping from each state, $s \in S$, and action, $a \in A$, to the probability $\pi(s, a)$ of taking action $a$ when in state $s$. The value of a state $s$ under a policy $\pi$, denoted by $V^\pi(s)$, is the expected return when starting in state $s$ and following $\pi$ thereafter. For MDPs $V^\pi(s)$ is formally defined as follows:

$$V^\pi(s) = E_\pi\{R_t | s_t = s\} = E_\pi\left\{\sum_{k=0}^{\infty} \gamma^k r_{t+k+1} \bigg| s_t = s\right\},\qquad(10.4)$$

where $E_\pi\{\}$ denotes the expected value given that the agent follows policy $\pi$. The function $V^\pi$ is called the state-value function for policy $\pi$. Similarly an action-value function for policy $\pi$ exists. This function is denoted as $Q^\pi(s, a)$, and is defined as the value of taking action $a$ in state $s$ under a policy $\pi$. This quantity is the expected return having started in state $s$, taking action $a$, and thereafter following policy $\pi$. Or more formally:

$$Q^\pi(s, a) = E_\pi\{R_t | s_t = s, a_t = a\} = E_\pi\left\{\sum_{k=0}^{\infty} \gamma t_{t+k+1} \bigg| s_t = s, a_t = a\right\}.$$

$$(10.5)$$

Value functions are estimated from experience. If an agent follows policy $\pi$ and maintains an average, for each state encountered, of the actual returns following a specific state, then the average will converge to the states value, $V^\pi(s)$, as the number of times that state is visited. Separate averages are maintained for each action taken in a specific state. In a similar way these averages converge to the action values, $Q^\pi(s, a)$.

A fundamental property of value functions used throughout reinforcement learning and dynamic programming is that they satisfy particular recursive relationships. For any policy $\pi$ and any state $s$, the following condition holds between the values of $s$ and the values of its possible successor states:

$$V^{\pi}(s) = E_{\pi}\{R_t | s_t = s\}$$

$$= E_{\pi}\left\{\sum_{k=0}^{\infty}\gamma^k r_{t+k+1} \,\middle|\, s_t = s\right\}$$

$$= E_{\pi}\left\{r_{t+1} + \gamma \sum_{k=0}^{\infty}\gamma^k r_{t+k+2} \,\middle|\, s_t = s\right\}$$

$$= \sum_a \pi(s,a) \sum_{s'} \mathcal{P}^a_{ss'}\left[\mathcal{R}^a_{ss'} + \gamma E_{\pi}\left\{\sum_{k=0}^{\infty}\gamma^k r_{t+k+2} \,\middle|\, s_{t+1} = s'\right\}\right]$$

$$= \sum_a \pi(s,a) \sum_{s'} \mathcal{P}^a_{ss'}[\mathcal{R}^a_{ss'} + \gamma V^{\pi}(s')], \tag{10.6}$$

where

$$\mathcal{P}^a_{ss'} = \Pr\{s_{t+1} = s | s_t = s, a_t = a\},$$
$$\mathcal{R}^a_{ss'} = E\{r_{t+1} | s_t = s, a_t = a, s_{t+1} = s'\}. \tag{10.7}$$

Equation (10.6) is the Bellman equation for $V^{\pi}$. It expresses a relationship between the value of a state and the values of its successor states, as seen in Figure 10.2(a).

Each open circle represents a state and each solid circle represents a state–action pair. Starting from $s$, the root node at the top, the agent could take any set of actions; three are shown in Figure 10.2(a). From each of these, the environment could respond with one of several next states, $s'$, along with a reward $r$. The Bellman equation (10.6) averages over all possibilities, weighting each by its probability of occurring. It states that the value of the starting state must equal the discounted value of the expected next states, plus the reward expected along the way. The value function $V^{\pi}$ is the unique solution to its Bellman equation. Figure 10.2(b) depicts the similar case for $Q^{\pi}$.

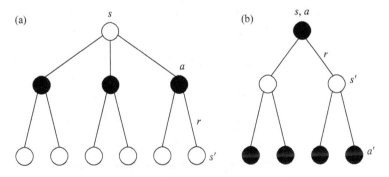

*Figure 10.2    Backup diagrams for (a) $V^{\pi}$ and (b) $Q^{\pi}$*

## 10.2.2   *Acting optimally*

Ideally agents based on MDPs should act in such a way, so as to maximise some measure of the long-term reward received. One such strategy is called finite horizon optimality, in which the agent attempts to maximise the expected sum of rewards that it gets in the next $n$ steps. Hence it should maximise the following:

$$E\left[\sum_{t=0}^{n-1} r_t\right]. \tag{10.8}$$

As mentioned previously, $r_t$ is the reward received on time step $t$. However, in most real world situations, the appropriate horizon length $n$ will not be known. The horizon length can be considered to be the lifetime of the agent, and in most cases this is assumed to be infinite. The most straightforward strategy, developed for undefined agent lifetime, is the infinite horizon discounted model, in which the rewards are summed over the infinite lifetime, and then geometrically discounted using the discount factor $0 < \gamma < 1$. Hence the agent should implement a policy which optimises the following [7]:

$$E\left[\sum_{t=0}^{\infty} \gamma r_t\right]. \tag{10.9}$$

Using this model, rewards received earlier in an agent's lifetime have more value than those received later. The sum is ensured to be finite because of the discount factor. This sum is also the expected amount of reward received if a decision to terminate the run is made on each step with probability $1 - \gamma$. Hence, the larger the discount factor, the more effect future rewards have on current decision-making [9].

A policy is a description of an agent's behaviour. Two types of policy can be considered, the stationary and non-stationary cases. For the stationary case the choice of action depends only on the state and is independent of the time step [10]. A non-stationary policy is a sequence of situation-action mappings, indexed by time. The policy $\pi_t$ is used to choose the actions on the $t$th time step as a function of the current state, $i_t$. In the finite horizon model, the optimal policy is not typically stationary. However, in the infinite horizon discounted model the agent always has a constant expected amount of time remaining, so there is no reason to change the action strategy. Hence, there exists a stationary optimal policy.

The agent's objective is to learn an optimal Markov policy, a mapping from states to probabilities of taking each available primitive action, $\pi : S \times A \mapsto [0, 1]$, which maximises the expected discounted future reward from each state $s$:

$$\begin{aligned}
V^{\pi}(s) &= E\{r_{t+1} + \gamma r_{t+1} + \gamma^2 r_{t+1} + \cdots | s_t = s, \pi\} \\
&= E\{r_{t+1} + \gamma V^{\pi}(s_{t+1}) | s_t = s, \pi\} \\
&= \sum_{a \in A_s} \pi(s, a)\left[r_s^a + \gamma \sum_{s'} p_{ss'} V^{\pi}(s')\right],
\end{aligned} \tag{10.10}$$

where $\pi(s,a)$ is the probability with which the policy $\pi$ chooses action $a \in A_s$ in state $s$, and $\gamma \in [0,1]$ is the discount rate. The quantity, $V^\pi(s)$, is called the value of state $s$ under policy $\pi$, and $V^\pi$ is called the state-value function for $\pi$. The optimal state-value function gives the value of a state under an optimal policy:

$$V^*(s) = \max_\pi V^\pi(s)$$

$$= \max_{a \in A_s} E\{r_{t+1} + \gamma V^*(s_{t+1}) | s_t = s, a_t = a\}$$

$$= \max_{a \in A_s} \left[ r_s^a + \gamma \sum_{s'} p_{ss'}^a V^*(s) \right]. \tag{10.11}$$

Any policy that achieves the maximum in (10.11) is by definition an optimal policy. Thus, given $V^*$, an optimal policy is easily formed by choosing in each state $s$ any action that achieves the maximum in (10.11). Planning in reinforcement learning refers to the use of models of the environment to compute value functions and thereby to optimise or improve policies. Particularly useful in this regard are Bellman equations, such as (10.10) and (10.11), which recursively relate the value functions to themselves. If we treat the values, $V^\pi(s)$ or $V^*(s)$, as unknowns, then a set of Bellman equations, for all $s \in S$, forms a system of equations whose unique solution is in fact $V^\pi$ or $V^*$ as given by (10.10) or (10.11). This fact is key to the way in which all temporal difference and dynamic programming methods estimate value functions.

Particularly important for learning methods is a parallel set of value functions and Bellman equations for state–action pairs rather than for states. The value of taking action $a$ in state $s$ under policy $\pi$, denoted as $Q^\pi(s,a)$, is the expected discounted future reward starting in $s$, taking $a$, and henceforth following $\pi$:

$$Q^\pi(s,a) = E\{r_{t+1} + \gamma r_{t+1} + \gamma^2 r_{t+1} + \cdots | s_t = s, a_t = a, \pi\}$$

$$= r_s^a + \gamma \sum_{s'} p_{ss'}^a V^\pi(s')$$

$$= r_s^a + \gamma \sum_{s'} p_{ss'}^a \sum_{a'} \pi(s,a') Q^\pi(s',a'). \tag{10.12}$$

This is known as the action-value function for policy $\pi$. The optimal action-value function is:

$$Q^*(s,a) = \max_\pi Q^\pi(s,a)$$

$$= r_s^a + \gamma \sum_{s'} p_{ss'}^a \max_a Q^*(s',a'). \tag{10.13}$$

Finally, many tasks are episodic in nature, involving repeated trials, or episodes, each ending with a reset to a standard state or state distribution. In these episodic tasks, we include a single special terminal state, arrival in which terminates the current episode. The set of regular states plus the terminal state (if there is one) is denoted as $S^+$. Thus, the $s'$ in $p_{ss'}^a$ in general ranges over the set $S^+$ rather than just $S$. In an episodic task, values are defined by the expected cumulative reward up until termination rather than over the infinite future (or, equivalently, we can consider the terminal state to transition to itself forever with a reward of zero).

## 10.3   Partially observable Markov decision processes

A partially observable Markov decision process (POMDP) is an MDP that works in environments in which the observations alone do not differentiate the current state of the system [11]. Hence the MDP is only capable of partially mapping the observation it takes into actual system states [7]. The POMDP is a development of the general MDP framework presented previously, with the following additional assumptions:

1.  At any decision epoch the state $i \in I$ may not be known.
2.  If the system is in state $i \in I$ at any decision epoch, if action $a \in A$ is taken and the resultant state is $j \in J$, we also receive some observation from a finite set $D$ whose realisation will be represented by $d$.
3.  There is a probability $q_{jd}^a$ if action $a \in A$ has been taken at the current decision epoch and the state at the next epoch is $j \in J$, that the realised observations $d \in D$ will be obtained.
4.  There is a reward $r_{ijd}^a$ given $(i, j, a, d)$, assumed to be received at the beginning of the time unit. We let:

$$r_i^a = \sum_{j \in J, d \in D} p_{ij}^a q_{jd}^a r_{ijd}^a. \tag{10.14}$$

5.  In order to model this problem, our new state variable will be:

$$\mu = (\mu_1, \mu_2, \ldots, \mu_i, \ldots, \mu_m) \in R_+^m \tag{10.15}$$

with

$$\sum_{i \in I} \mu_i = 1, \tag{10.16}$$

where $\mu_i$ is the current probability that the system is in state $i \in I$. The new state–space is $M$.

Initially we will have, for $t = 1$

$$\mu = \mu^1. \tag{10.17}$$

6.  Because $i \in I$ is not generally known, it is assumed that action selected in state $i$, $A(i)$ is independent of $i \in I$, i.e.

$$A(i) = a, \quad \forall i \in I. \tag{10.18}$$

7.  A general policy set $\Pi$ can be defined analogously to that of the MDP case, where the history is made up of the initial $\mu = \mu^1$ plus all subsequent observations and actions.

A POMDP is really just an MDP; we have a set of states, a set of actions, transitions and immediate rewards. An action's effect on the state in a POMDP is exactly the same as in an MDP. The only difference is in whether or not we can observe the current state of the process. In a POMDP, a set of observations is added to the model. Therefore instead of directly observing the current state, the state gives us an observation which provides a hint about what state it is in [12]. The observations can be probabilistic; so we need to also specify the observation model. This observation model simply tells us the probability of each observation for each state in the model.

Although the underlying dynamics of the POMDP are still Markovian, since there is no direct access to the current state, decisions require access to the entire history of the process, making this a non-Markovian process [13]. The history at a given point in time comprises our knowledge about our starting situation, all actions performed and all observations seen.

Maintaining a probability distribution over all of the states provides the same information as maintaining a complete history. When an action is performed and an observation taken, the probability distributions need to be updated. A POMDP is a sequential decision model for agents which act in stochastic environments with only partial knowledge about the state of the environment [14].

## 10.4 Q-learning

Q-learning is an on-policy form of model-free reinforcement learning. It provides agents with the capability of acting optimally in Markovian domains by experiencing the consequences of actions without requiring the agents to build explicit maps of their environment [15]. The Q-learning algorithm works by estimating the values of state–action pairs. The task facing the agent is that of determining an optimal policy, $\pi^*$; one that selects actions that maximise the long-term measure of reinforcement, given the current state [16]. Normally the measure used is the total discounted expected reward. By discounted reward, it is meant that future rewards are worth less than rewards received now, by a factor of $\gamma^s (0 < \gamma < 1)$. Under policy $\pi$, the value of state $s$ is:

$$V^\pi \equiv r_a + \gamma \sum_{s_{t+1}} P_{s_t s_{t+1}} [\pi(s_t)] V^\pi (s_{t+1}) \tag{10.19}$$

because the agent expects to receive reward $r$ immediately for performing the action $\pi$ recommends, and then moves to a state that is 'worth' $V^\pi (s_{t+1})$ to it, with probability $P_{s_t s_{t+1}} [\pi(s_{t+1})]$. The theory assures us that there is at least one stationary policy $\pi^*$ such that:

$$V^*(s_t) \equiv V^{\pi^*}(s_t) = \max_a \left\{ r_a + \gamma \sum_{s_{t+1}} P_{s_t s_{t+1}} [a] V^{\pi^*}(s_{t+1}) \right\}. \tag{10.20}$$

This is the best an agent can do from state $s$. Assuming that $r_a$ and $P_{s_t s_{t+1}} [a]$ are known, dynamic programming techniques provide a number of ways to calculate $V^*$ and $\pi^*$. The task faced by Q-learning is to determine $\pi^*$ without initially knowing these values. As such, Q-learning is a form of incremental dynamic programming, because of its stepwise method of determining the optimal policy.

For a policy $\pi$, $Q$-values are defined as follows:

$$Q^\pi (s_t, a) = r_a + \gamma \sum_{s_{t+1}} P_{s_1 s_{t+1}} [\pi (s_t)] V^\pi (s_{t+1}). \tag{10.21}$$

Hence the $Q$-value is the discounted reward for executing action $a$ at state $s_t$ and following policy $\pi$ thereafter. Therefore, the aim of Q-learning is to estimate the $Q$-values for an optimal policy. If these are defined as $Q^*(s_t, a) \equiv Q^{\pi^*}(s_t, a), \forall s, a$ it can be shown that $V^*(s_t) = \max_a Q^*(s_t, a)$ and if $a^*$ is an action at which the maximum is attained, then the optimal policy can be formulated as $\pi^*(s_t) \equiv a^*$. Herein lies the utility of Q-learning; if an agent can learn them, it can easily decide its best action.

Q-learning is applied to problems as an incremental algorithm, see Algorithm 10.1 in which $Q$-values are updated using the previous values. This process is performed using the error function, $\delta$, which calculates the sum of the immediate reward and the discounted value of the expected optimal state transition while negating the $Q$-value attributable to the current state–action pair. The error function is then used to update

---

**Algorithm 10.1** Tabular version of Watkin's $Q(\lambda)$
    Initialise $Q(\lambda)$ arbitrarily and $e(s, a) = 0 \; \forall s, a$
    Initialise $s, a$
    Repeat (for each step of episode):
        Take action $a$, observe $r, s'$
        Choose $a'$ from $s'$ using policy
            derived from $Q$ (e.g. $\varepsilon$-greedy)

    $a^* \leftarrow \arg\max_b Q(s', b)$

        (if $a'$ ties for the max, then $a^* \leftarrow a'$)

    $\delta \leftarrow r + \gamma Q(s', a^*) - Q(s, a)$
    $e(s, a) \leftarrow e(s, a) + 1$

    For all $s, a$

        $Q(s, a) \leftarrow Q(s, a) + \alpha \delta e(s, a)$
        If $a' = a^*$, then $e(s, a) \leftarrow \gamma \lambda e(s, a)$

            else $e(s, a) \leftarrow 0$
    $s \leftarrow s'; a \leftarrow a'$
    until $s$ is terminal

$a \to$ action $a' \to$ new action $a^* \to$ action with the maximal expected reward given the state $s \to$ state at time $n$ $s' \to$ state at time $n + 1$ $Q(s, a) \to$ value function $\alpha \to$ learning factor $\delta \to$ error function $e(s, a) \to$ eligibility trace function $\lambda \to$ trace decay parameter $\gamma \to$ discount rate.

---

the $Q$-value for the current state–action pair. Hence the error function incrementally constructs $Q$-values, which are the discounted reward for executing action $a$ at state $s_t$ and following the estimated optimal policy $\pi$ thereafter. Over time the estimated optimal policy, $\pi$, converges to the actual optimal policy, $\pi^*$.

## 10.5 Adaptive heuristic critic and temporal difference

The adaptive heuristic critic algorithm is an adaptive version of policy iteration, in which the value function computation is no longer implemented by solving a set of linear equations, but is instead computed by an algorithm called TD(0). A block diagram for this approach is given in Figure 10.3. It consists of two components: a critic (labelled AHC), and a reinforcement learning component (labelled RL). The reinforcement learning component can be modified to deal with multiple states and non-stationary rewards. But instead of acting to maximise instantaneous reward, it will be acting to maximise the heuristic value, $v$, that is computed by the critic. The critic uses the external reinforcement signal to learn to map states to their expected discounted values given that the policy being executed is the one currently instantiated in the RL component. As such the general TD($\lambda$) algorithm builds a model of its environment.

The policy $\pi$ implemented by RL is fixed and the critic learns the value function $V_\pi$ for that policy. Then the critic is fixed allowing the RL component to learn a new policy $\pi'$ that maximises the new value function, and so on.

It remains to be explained how the critic can learn the value of a policy. We define $\{s, a, r, s'\}$ to be an experience tuple summarising a single transition in the environment. Here $s$ is the agent's state before the transition, $a$ is its choice of action, $r$ the instantaneous reward it receives and $s'$ its resulting state. The value of a policy is learned using Sutton's TD(0) algorithm [17] which uses the following update rule.

$$V(s) = V(s) + \alpha(r + \gamma V(s') - V(s)). \qquad (10.22)$$

Whenever a state, $s$, is visited, its estimated value is updated to be closer to $r + \gamma V(s')$ since $r$ is the instantaneous reward received and $V(s')$ is the estimated value of the actually occurring next state. The key idea is that $r + \gamma V(s')$ is a sample of the value of $V(s)$, and it is more likely to be correct because it incorporates the real $r$. If the learning rate $\alpha$ is adjusted properly (it must be slowly decreased) and the policy holds fixed, TD(0) is guaranteed to converge to the optimal value function.

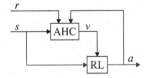

*Figure 10.3    Architecture for the adaptive heuristic critic*

The TD(0) rule as presented above is really an instance of a more general class of algorithms called TD($\lambda$), with $\lambda = 0$. TD(0) looks only one step ahead when adjusting value estimates; although it will eventually arrive at the correct answer, it can take quite a while to do so. The general TD($\lambda$) rule is similar to the TD(0) rule given earlier,

$$V(u) = V(u) + \alpha(r + \gamma V(s' - V(s)))e(u), \tag{10.23}$$

but it is applied to every state according to its eligibility $e(u)$, rather than just to the immediately previous state, $s$. One version of the eligibility trace is defined to be:

$$e(s) = \sum_{k=1}^{t} (\lambda\gamma)^{t-k} \delta_{s,s_k}, \qquad \text{where } \delta_{s,s_k} = \begin{cases} 1, & \text{if } s = s_k, \\ 0, & \text{otherwise.} \end{cases} \tag{10.24}$$

The eligibility of a state $s$ is the degree to which it has been visited in the recent past; when a reinforcement signal is received, it is used to update all the states that have been recently visited, according to their eligibility. When $\lambda = 0$ this is equivalent to TD(0). When $\lambda = 1$, it is roughly equivalent to updating all the states according to the number of times they were visited by the end of a run. Note that we can update the eligibility online as follows:

$$e(s) = \begin{cases} \gamma\lambda e(s) + 1, & \text{if } s = \text{current state}, \\ \gamma\lambda e(s), & \text{otherwise.} \end{cases} \tag{10.25}$$

It is computationally more expensive to execute the general TD($\lambda$), though it often converges considerably faster for large $\lambda$.

The general TD($\lambda$) algorithm is an off-policy method. It cannot update the policy it is currently using. Instead the critic and RL component alternate developing their respective models and policies offline. The Sarsa($\lambda$) algorithm, see Algorithm 10.2, provides an online implementation of the TD control method. Sarsa($\lambda$) applies the TD($\lambda$) prediction method to state–action pairs rather than to states, substituting state–action variables for state variable. Hence $V(s)$ becomes $Q(s, a)$ and $e(s)$ becomes $e(s, a)$. The Sarsa($\lambda$) approximates $Q^\pi(s, a)$, the action values for the current policy, $\pi$, then it improves the policy gradually based on the approximate value for the current policy.

## 10.6   Internet environment

The Internet can be considered to be an unreliable, best effort, connectionless packet delivery system. It is unreliable because delivery is not guaranteed. Packets may be lost, duplicated, delayed or delivered out of order. Unreliability also arises when resources fail or underlying networks fail. The service is called connectionless because each packet is routed independently. A sequence of packets sent from one machine to another may travel over different paths, some may be lost while others are delivered. Finally, the service is said to be best effort delivery because the Internet software makes an earnest effort to deliver packets.

**Algorithm 10.2** Tabular version of Sarsa($\lambda$)
  Initialise $Q(\lambda)$ arbitrarily and $e(s,a) = 0 \; \forall s, a$
  Initialise $s, a$
  Repeat (for each step of episode):
    Take action $a$, observe $r$, $s'$
    Choose $a'$ from $s'$ using policy
      derived from $Q$ (e.g. $\varepsilon$-greedy)

$$\delta \leftarrow r + \gamma Q(s',a') - Q(s,a)$$
$$e(s,a) \leftarrow e(s,a) + 1$$

    For all $s$, $a$

$$Q(s,a) \leftarrow Q(s,a) + \alpha \delta e(s,a)$$
$$e(s,a) \leftarrow \gamma \lambda e(s,a)$$

$$s \leftarrow s'; a \leftarrow a'$$
    until $s$ is terminal

$a \rightarrow$ action $a' \rightarrow$ new action $a^* \rightarrow$ action with the maximal expected reward given the state $s \rightarrow$ state at time $n$ $s' \rightarrow$ state at time $n + 1$ $Q(s, a) \rightarrow$ value function $\alpha \rightarrow$ learning factor $\delta \rightarrow$ error function $e(s,a) \rightarrow$ eligibility trace function $\lambda \rightarrow$ trace decay parameter $\gamma \rightarrow$ discount rate.

The Internet Protocol (IP) defines the unreliable, connectionless delivery mechanism. IP provides three basic definitions. First, the IP defines the basic unit of data transfer used throughout the Internet. These packets are called 'datagrams'. Second, IP software performs the routing function, choosing the paths over which data will be sent. Third, IP includes a set of rules that embody the idea of unreliable packet delivery [18]. These rules characterise how hosts and routers should process packets, how and when an error message should be generated and the conditions under which packets can be discarded.

The Internet interconnects multiple physical networks using routers. Each router has a direct connection to two or more physical networks. Routing is performed at a software level and any computer with multiple network connections can act as a router. Hence multi-homed hosts and dedicated routers can both route datagrams from one network to another. However, the TCP/IP standards define strict distinctions between the function of hosts and routers [19].

Routing is the method by which the host or router decides where to send the datagram. It may be able to send the datagram directly to the destination, if that destination is on one of the directly connected networks. However, the interesting case is when the destination is not directly reachable. In this case, the host or router attempts to send the datagram to a router that is nearer to the destination. A routing protocol's purpose is very simple: it is to supply the information that is needed to efficiently route datagrams [20].

The routing of IP datagrams is a table driven process, shown in Figure 10.4. As such the information held within routing tables must be kept up to date and valid. Indeed it is through this process that routes from source to destination are maintained. A class of algorithms known as 'Distance vector algorithms' perform exactly this purpose, see Algorithm 10.3. These simply use the distance to remote destinations as a measure of quality for routes. However, they are primarily intended for use in reasonably homogeneous networks of moderate size.

Routing in Internet environments is based on the information stored in routing tables at each router. The distance vector algorithm maintains the routes from source to destination. When a datagram arrives at a router, the router is responsible for selecting the next hop router for that datagram. One of the most commonly used protocols used in TCP/IP systems for propagating routing information is Routing Information Protocol (RIP). This protocol is intended for use in small- to medium-sized autonomous systems. As the networks grow in size, the amount of time for the

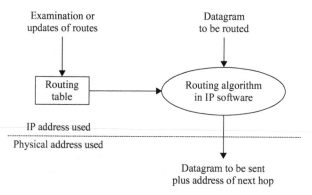

*Figure 10.4    Overview of the table driven routing process*

---

**Algorithm 10.3** The basic routing information algorithm

- Keep a table with an entry for every possible destination in the system. The entry contains the distance $D$ to the destination, and the first router $G$ on the route to that network.
- Periodically, send a routing update to every neighbour. The update is a set of messages that contain all of the information from the routing table. It contains an entry for each destination, with the distance to that destination.
- When a routing update arrives from neighbours $G'$, add the cost associated with the network that is shared with $G'$. Call the resulting distance $D'$. Compare the resulting distance with the current routing table entries. If the new distance $D'$ is smaller than the existing value $D$, adopt the new route. That is, change the table entry to have metric $D'$ and router $G'$. If $G'$ is the router from which the existing route came, then use the new metric; even if it is larger than the old one.

---

routing information to converge on the system becomes prohibitive. RIP has been found to be efficient on networks with no more than about 40 routers. RIP maintains a routing table at each router, with an entry for every possible destination within the autonomous system. The entry consists of both primary and secondary routes, which contain information relating to the destination address, the next router address en route and a metric giving the expected distance to the destination address. In addition to this, RIP maintains the address of a neighbouring default router.

## 10.7   Agent learning for Internet control

This following example compares the uses of two reinforcement learning techniques as the basis for multi-agent controllers that manage the forwarding engine at each node of a TCP/IP system. A traditional control technique is also implemented.

The forwarding engine is responsible for how each packet is treated. The goal is to ensure that as many packets as possible are routed from their source to destination nodes. For this tutorial a simple TCP/IP simulation was constructed that used RIP to maintain routes between nodes. RIP implements the distances vector algorithm presented above, and in this simulation both primary and secondary routes were maintained for all source–destination pairs.

The traditional forwarding engine uses Algorithm 10.4 to determine how each packet was to be treated. In this algorithm, the environments state is considered

---

**Algorithm 10.4** A basic routing algorithm
RouteDatagram (Datagram, RoutingTable)

    Extract destination address, D, from the datagram;
    else if the table contains routes for D
        else if first choice route has spare capacity
            send datagram via first choice route
        else if second choice route has spare capacity
            send datagram via second choice route
        else if the default route has spare capacity
            send datagram via default route
        else if internal store has spare capacity
            store datagram and send later
        else if any network has spare capacity
            send via that network
        else drop datagram
    else if the table contains a default route
        send datagram send via default route
    else drop datagram.

---

in terms of the available capacities on connected links and internal storage queues. This state maps deterministically to particular actions. Three routes are identified in the algorithm; however, these may in fact map to a single link. Each node has six available actions, but in essence it can either forward the packet via a known route, store the packet on the internal queue, send via any link or drop the packet.

Reinforcement learning techniques optimise their behaviour by learning their own internal state–action mappings. In this study, the first choice route, second choice route, default route, internal memory and other connections were tested for spare capacity, each returning a Boolean (TRUE/FALSE) response, with miscellaneous spare capacity on connections not being recognised as a specific route but returning a single Boolean response for any spare capacity. Hence at any time the agent receives five Boolean variables identifying whether there is spare capacity on its connection and internal storage. Considering these five variables to be a binary number, 32 states have been identified. The task of a reinforcement learning algorithm is to develop optimal mappings of these 32 states into the 6 identified actions.

Reinforcement learning agents were implemented at every node. The goal of these agents was the same as for the traditional control. This introduces a bit of a problem from the perspective of this multi-agent controller. One of the characteristics of multi-agent control is that agents can communicate to manage the global goals of the system. In this system, the goal was to maximise the network throughput. Implementing an explicit mechanism for agents' communication would increase the traffic conditions in the network and may compromise their ability to achieve their goal. Instead, these agents' communicate indirectly through environmental properties. The most obvious properties they monitor are the available capacities. In addition to this the hop count of each packet is used to form a composite reinforcement signal. Hence, packets that are routed most efficiently result in the highest reward. A measure of reward is also associated with each successfully routed packet.

## 10.8 Results and discussion

Multi-agent learning and traditional control techniques were both examined using a simulation of the Internet environment. For the purpose of this work a 20-node network was constructed, as shown in Figure 10.5. The topology of this network was chosen first to be asymmetrical and second to include a number of traffic bottlenecks. Hence, it is ensured that each agent is responsible for optimising datagram routing in different control environments.

The size of the network is by no means comparable to the Internet as a whole. This size was chosen because the distance vector algorithm was designed to operate in small-scale autonomous systems, usually with a maximum size of 40 nodes. However, the speed of convergence to optimal routes using the distance vector algorithm is dependent on the size of the network. Therefore the 20-node network was chosen to be a compromise between the rapid convergence of routing information and network complexity.

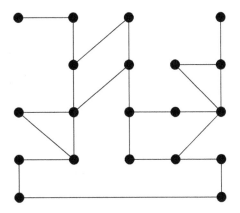

*Figure 10.5    Topology of a 20-node network*

In these simulations, heavy traffic conditions were selected. The TCP/IP forwarding mechanism is already optimised for lower traffic conditions. It is only under heavy traffic that intelligent routing could provide an improvement over the traditional controller.

Agent learning was implemented at every node, with Sarsa($\lambda$), $Q(\lambda)$ and traditional control systems being tested against two performance measures. The first was the percentage of datagrams dropped per time cycle. This can be considered to be a measure of the reliability of the network. The lower the percentage for datagrams dropped, the greater the percentage of datagrams successfully reaching their destinations. The second measure was the percentage of datagrams routed per time cycle. This can be considered to be a measure of network efficiency. The higher the percentage of datagrams routed the higher the network throughput. Hence, the network is making a more efficient use of its resources.

The results for the percentage of calls dropped per time cycle are shown in Figure 10.6. For the traditional controller, the percentage of calls dropped starts off at zero, rising rapidly, as network traffic builds up. However, this increase stabilises after about 20 cycles at around 13.5 per cent.

Whereas for $Q(\lambda)$ the percentage of calls dropped starts off at around 12 per cent, and oscillates between 8 and 14 per cent until the 50th time cycle, when it starts to fall; levelling off at around 8 per cent by the 100th time cycle, where on balance it remains. However, after the 250th time cycle the results for $Q(\lambda)$ oscillate wildly; peaking at 12.7 per cent and dropping to a minimum of 4.5 per cent.

While for Sarsa($\lambda$) the percentage of calls dropped starts off at 19 per cent this drops below the stabilised value for the traditional control after 20 time cycles and continues to drop to a value of around 5.5 per cent after the first 125 time cycles, after which the performance deteriorates mildly as the results pass through a local maximum of 7 per cent. However, by the 200th time cycle, the percentage of calls dropped has again begun to drop; and by the 250th time cycle has stabilised at 4.5 per cent.

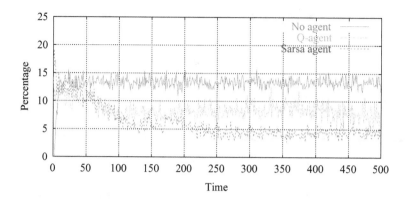

*Figure 10.6   Percentage of datagrams dropped per unit time for the traditional and agent controllers*

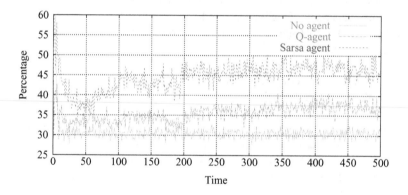

*Figure 10.7   Percentage of datagrams routed per unit time for the traditional and agent controllers*

As far as dropping datagram packets is concerned, the results presented in Figure 10.6 show intelligent multi-agent control outperforming the traditional control. The agent based Sarsa($\lambda$) algorithm achieved the best performance. While the $Q(\lambda)$ agents show a considerable improvement over the traditional control, they do not achieve the performance of Sarsa($\lambda$), whose early convergence characteristics are very similar to $Q(\lambda)$, but continue to improve beyond the point at which $Q(\lambda)$ stops.

With respect to the routing of datagram packets, the results presented in Figure 10.7 show that multi-agent controllers again outperform the traditional method, with the Sarsa($\lambda$) agents displaying the best performance overall. Indeed the Sarsa($\lambda$) agent control scheme levels out more rapidly, and then displays the fastest adaptation to its environment. While the $Q(\lambda)$ control scheme displays similar adaptation characteristics, they are in no way as pronounced. The $Q(\lambda)$ agents provide an improvement over the traditional control.

The improved performance of the Sarsa($\lambda$) agents over the $Q(\lambda)$ agents in these simulations is attributed to the use of more historical information in Sarsa($\lambda$). The eligibility mechanism in $Q(\lambda)$ only maintains traces while the agent is following its perceived optimal policy, whereas the Sarsa($\lambda$) mechanism maintains traces for all state–action combination. Assuming a relatively slow moving quasi stationary environment, the performance of these two algorithms would be expected to converge.

## 10.9 Conclusions

Reinforcement learning is a technique for adaptive behaviour that is based on the Markov decision process (MDP). It provides an intuitive model on which intelligent control can be developed. One of the main features of this and other artificial techniques is that the problems, for which they are most attractive, are characterised by complex multivariate dynamic environments. In implementing reinforcement learning, these complex problems are mapped into the comparatively limited variable space of reinforcement learning. There is no reason to believe that the resulting solution is stable. The success of applying reinforcement techniques is highly dependent on the developer's understanding of the underlying problem.

Reinforcement learning provides an attractive technique for developing intelligent solutions because although the resulting solutions are difficult to analyse qualitatively, the state and action space models are intuitive. Effectively problem domain information maintained is readily available through the state and action space models, which makes this a good technique for iterative development and gaining understanding of a problem space. Through an improved understanding of the problem, it may be possible to develop a fully optimised solution.

In some scenarios an agent's ability to be further developed may be more important than optimal behaviour. This is generally the case in non-stationary environments. Here the adaptability of learning techniques allows the agents to outperform traditional control paradigms. The paradox between acting suboptimally to determine optimal control strategies and optimal behaviour becomes less important where suboptimal action is an improvement over traditional control.

As an agent is able to perform locally without requiring the information of other agents directly, it can be seen that the multi-agent controllers have a great potential for the Internet, which has a complex and distributed control environment. In this chapter, reinforcement learning has been implemented for routing control in such an environment. In heavy traffic conditions, the TCP/IP forwarding engine for a network becomes a highly non-stationary environment. Both reinforcement learning techniques discussed in this chapter have improved characteristics over traditional control. However, the technique that used historical information more effectively was shown to provide the best solution. This may be expected where the underlying fundamentals of the problem are quasi stationary. In problems where the historical information is fundamentally outdated, and the new optimal solution has changed, maintaining eligibility for optimal policy alone may prove beneficial.

# References

1  BELLMAN, R. E.: 'A Markov decision process', *Journal of Mathematics and Mechanics*, 1957, **6**, pp. 679–93
2  WITTEN, I. H.: 'Exploring, modelling and controlling discrete sequential environments', *International Journal of Man–Machine Studies*, 1977, **9** (6), pp. 715–35
3  DEAN, T. L., KAELBLING, L. P., KIRMAN, J., and NOCHOLSON, A.: 'Planning under time constraints in stochastic domains', *Artificial Intelligence*, 1995, **76** (1–2), pp. 35–74
4  BOUTILIER, C., DEARDEN, R., and GOLDSZMIDT, M.: 'Exploiting structure in policy construction'. Proceedings of the fourth international joint conference on *Artificial intelligence*, 1995
5  SCHULTZ, W., DAYAN, P., and MONTAGUE, P. R.: 'A neural substrate of prediction and reward', *Science*, 1997, **275**, pp. 1593–8
6  BARTO, A. G., SUTTON, R. S., and WATKINS, C. J. C. H.: 'Learning and sequential decision making', in GABRIEL, M., and MOORE, J. (Eds): 'Learning and computational neuroscience: foundations of adaptive networks' (MIT Press, Cambridge, MA, 1990) pp. 539–602
7  KAELBLING, L. P., LITTMAN, M. L., and CASSANDRA, A. R.: 'Planning and acting in partially observable stochastic domains'. Technical report, Department of Computer Science, Brown University, 13, 1997
8  WHITE, D. J.: 'Markov decision processes' (John Wiley and Sons, New York, 1993)
9  CASSANDRA, A. R., KAELBLING, L. P., and LITTMAN, M. L.: 'Acting optimally in partially observable stochastic domains'. Technical report, Department of Computer Science, Brown University, 1997
10  HOWARD, A.: 'Dynamic programming and Markov decision processes' (John Wiley and Sons, New York, 1960)
11  LITTMAN, M. L., CASSANDRA, A. R., and KAELBLING, L. P.: 'Learning policies for partially observable environments: scaling up'. Technical report, Department of Computer Science, Brown University, 1997
12  CASSANDRA, A. R.: 'Exact and approximate algorithms for partially observable Markov decision processes'. Ph.D. thesis, Department of Computer Science, Brown University, 1998
13  LIU, W., and ZHANG, N. L.: 'A model approximation scheme for planning in partially observable stochastic domains', *Journal of Artificial Intelligence Research*, 1997, **7**, pp. 199–230
14  ZHANG, W., and ZHANG, N. L.: 'Speeding up the convergence of value iteration in partially observable Markov decision processes', *Journal of Artificial Intelligence Research*, 2001, **14**, pp. 29–51

15 WATKINS, C., and DAYAN, P.: 'Technical note: Q-learning', *Machine Learning*, 1992, **8**, pp. 279–92

16 WU, Q. H.: 'Reinforcement learning control using interconnected learning automata', *International Journal of Control*, 1995, **62** (1), pp. 1–16

17 SUTTON, R. S.: 'Learning to predict by the methods of temporal differences', *Machine Learning*, 1988, **3**, pp. 9–44

18 COMER, D. E.: 'Internetworking with TCP/IP, vol. 1' (Prentice-Hall, New York, 1995)

19 WRIGHT, G. R., and STEVENS, W. R.: 'TCP/IP illustrated, vol. 2' (Addison-Wesley, Reading, MA, 1995)

20 HEDRICK, C.: 'Routing information protocol', 1998, http://www.ietf.org/rfc/rfc1058.txt

*Chapter 11*

# Combined computational intelligence and analytical methods in fault diagnosis

*Ronald J. Patton, Józef Korbicz, Martin Witczak and Faisel Uppal*

## 11.1  Introduction

There is an increasing demand for manmade dynamical systems to become safer, more reliable, more cost-effective and less polluting to the environment. These requirements extend beyond normally accepted safety-critical systems of nuclear reactors, chemical plants or aircraft, to new systems such as autonomous vehicles or fast rail systems. An early detection of faults can help avoid the system shutdown, breakdown and even catastrophes involving human fatalities and material damage. A system that includes the capacity of detecting, isolating, identifying or classifying faults is called a fault diagnosis system and the procedure of fault detection and isolation is referred to as an FDI (fault detection and isolation) system. During the 1980s and 1990s many investigations have been made using analytical approaches, based on quantitative models, especially in the control engineering literature [1–3]. The idea is to generate signals that reflect inconsistencies between the nominal and faulty system operation. Such signals, termed residuals, are usually generated using analytical modelling methods, providing analytical (or functional) redundancy.

Considerable attention has been given to both research and application studies of real processes using analytical redundancy as this is a powerful alternative to the use of repeated hardware (hardware or software redundancy). During the last 20 years, researchers have directed their attention to the use of data-driven FDI techniques and a variety of new improved designs have emerged. The most common approach for model construction for FDI is to use the well-known tools for linear systems [4], based on robustness analysis. In spite of the simplicity of linear models, they introduce a degree of approximation that may not be fully acceptable in many cases. This is the main motivation for the development of the robust technique for FDI as well as nonlinear system identification methods. A few decades ago, nonlinear system identification consisted of methods that were applicable to very restricted classes of systems. The most popular classical nonlinear identification employs various kinds of polynomials [5]. However, there are many applications where these models do not give satisfactory results. To overcome these problems, the so-called computational intelligence methods (neural networks, neuro-fuzzy models, etc.) are being investigated, which can model a much wider class of nonlinear systems. Mathematical models used in the traditional FDI methods are potentially sensitive to modelling errors, parameter variation, noise and disturbance [4,6]. Process modelling has limitations, especially when the system is uncertain and the data are ambiguous.

Computational intelligence (CI) methods (e.g. neural networks, fuzzy logic and evolutionary algorithms) are known to overcome these problems to some extent [7,8]. Neural networks are known for their approximation, generalisation and adaptive capabilities and they can be very useful when analytical models are not available. However, the neural network operates as a black-box with no qualitative information available. Fuzzy logic systems, on the other hand, have the ability to model a nonlinear system and to express it in the form of linguistic rules making it more transparent (i.e. easier to interpret). They also have the inherent abilities to deal with imprecise or noisy data.

A neuro-fuzzy (NF) model [9,10] can be used as a powerful combination of neural networks and fuzzy logic techniques, incorporating the robust decoupling potential of the unknown input observer (UIO) (as a robust tool for FDI design) with the identification and modelling capabilities of a neural network. Many NF models have been successfully applied to a wide range of applications. However, none of these approaches have made use of this unique combination of CI with robust observer design.

This chapter provides a background and framework for futher studies in this important and rapidly developing field of research – combined analytical and CI methods in fault diagnosis as an important approach for enhancing the robustness and reliability of FDI methods for real engineering applications. Starting with brief consideration of the principles of quantitative model-based fault diagnosis, via the unknown input observer, the ideas behind the alternative use of artificial neural networks, fuzzy logic, NF structures and evolutionary algorithms are summarised to give the reader a feeling of their potential advantages and disadvantages in the FDI role. To illustrate the effectiveness of the extended unknown input observer (EUIO) approach and GMDH (group method of data handling) neural networks, a comprehensive simulation study regarding the nonlinear model of an induction motor and the DAMADICS benchmark problem [11] is performed.

### 11.1.1 Notation

| | |
|---|---|
| $t$ | time |
| $k$ | discrete time |
| $x_k, \hat{x}_k \in \mathbb{R}^n$ | state vector and its estimate |
| $y_k, \hat{y}_k \in \mathbb{R}^m$ | output vector and its estimate |
| $e_k \in \mathbb{R}^n$ | state estimation error |
| $u_k \in \mathbb{R}^r$ | input vector |
| $d_k \in \mathbb{R}^q$ | unknown input vector, $q \leq m$ |
| $w_k, v_k$ | process and measurement noise |
| $Q_k, \mathbb{R}k$ | covariance matrices of $w_k$ and $v_k$ |
| $p$ | parameter vector |
| $f_k \in \mathbb{R}^s$ | fault vector |
| $g(\cdot), h(\cdot)$ | nonlinear functions |
| $E_k \in \mathbb{R}^{n \times q}$ | unknown input distribution matrix |
| $L_{1,k}, L_{2,k}$ | fault distribution matrices |

## 11.2 Classical FDI methods based on multi-objective design

Model-based FDI techniques have been demonstrated to be capable of detecting and isolating the so-called abrupt faults very quickly and reliably. On the other hand, the detection of the so-called incipient faults, which are small and slowly developing, e.g. a sedimentation of a valve, constitutes a serious challenge to model-based FDI methods due to the inseparable effects of faults, model uncertainty, disturbances as well as that of noise. This means that the detection of incipient faults is not very straightforward without a suitable analysis and design concerning the FDI scheme.

One of the most popular approaches that can be applied for the purpose of robust FDI is to employ the so-called unknown input decoupling techniques [2,12]. Undoubtedly, the most common one is to use robust observers, such as the UIO [2,3,12] or a special modification of the Kalman filter [13], which can tolerate a degree of model uncertainty and hence increase the reliability of fault diagnosis. In these approaches, the model–reality mismatch is represented by the so-called unknown input and hence the state estimate and, consequently, the output estimates, are obtained taking into account model uncertainty. The key challenge is to minimise the effect of the uncertain input while maximising the effect of any fault signal. When the distributions of these two types of signals into the dynamical system are different, then decoupling methods can be used to achieve robustness in FDI. When the uncertain effects (the unknown inputs) act in the same way as the fault(s), this discrimination becomes much more difficult and can only be achieved to any degree using frequency domain filtering methods.

Unfortunately, much of the work in this subject is based on the use of linear systems principles. This is mainly because of the fact that the theory of observers (or filters in the stochastic case) is especially well developed for linear systems. The main disadvantage of UIO is that the distribution of an unknown input is required to facilitate the design procedure. Thus, the aim of the UIO approach to FDI is to completely eliminate the effect of an unknown input from the signal used for FDI (the residual); the approach may be limited to system operation considered sufficiently close enough to a stable point of system operation for linearity to be considered acceptable.

A direct consequence of the above discussion is that the design criteria of robust FDI techniques should take into account the effect of an unknown input as well as that of the faults. Indeed, each fault effect needs to be taken into account as well as each unknown input. Thus, there is a trade-off between fault sensitivity and robustness to model uncertainty. This leads directly to multi-objective design of fault detection systems, i.e. the minimisation of the effect of an unknown input (and noise in the stochastic case) and the maximisation of the fault sensitivity. While model-based FDI methods using these robustness ideas are elegant, they are very dependent on the ability to model the system and to identify or model the unknown inputs and fault input distributions. One direction may be to use nonlinear system methods (in an attempt to provide more global modelling of the system). An alternative approach is to use various combinations of quantitative model-based methods with soft-computing techniques (evolutionary-computing, genetic algorithms, neural networks, NF modelling, etc.). The ideas of maximising the joint advantages of model-based and soft-computing methods for FDI form the main motivation for this chapter.

This section presents an optimal residual generator design [14,15] that is based on the combination of multi-objective optimisation and the evolutionary algorithms [16], mainly genetic algorithms [17]. In this approach, observer-based residual generation is employed. In order to make the residual insensitive to model uncertainty and sensitive to faults, a number of performance indices are defined in the frequency domain to account for the fact that model uncertainty effects and faults occupy different frequency bands. The remaining part of this section presents an outline of the approach introduced by Chen *et al.* [14] and its modification that was proposed by Kowalczuk and Bialaszewski [15].

## 11.2.1    Designing the robust residual generator using multi-objective optimisation and evolutionary algorithms

Let us consider the system described by the following equations:

$$\dot{x}(t) = Ax(t) + Bu(t) + d(t) + L_1 f(t) + w(t), \tag{11.1}$$

$$y(t) = Cx(t) + Du(t) + L_2 f(t) + v(t). \tag{11.2}$$

Depending upon the type of fault that has occurred, the matrices $L_1$ and $L_2$ have the following forms:

$$L_1 = \begin{cases} 0, & \text{sensor faults,} \\ B, & \text{actuator faults,} \end{cases} \qquad L_2 = \begin{cases} I_m, & \text{sensor faults,} \\ D, & \text{actuator faults.} \end{cases}$$

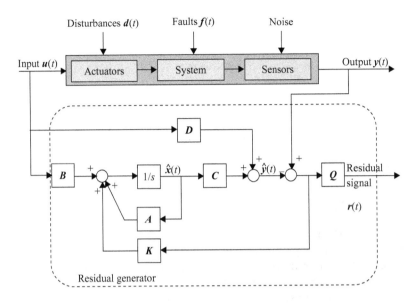

*Figure 11.1   Robust residual generator*

The observer for the system (11.1)–(11.2) can be described as follows (Figure 11.1):

$$\dot{\hat{x}}(t) = (A - KC)\hat{x}(t) + (B - KD)u(t) + Ky(t), \qquad (11.3)$$

$$\hat{y}(t) = C\hat{x}(t) + Du(t), \qquad (11.4)$$

$$r(t) = Q[y(t) - \hat{y}(t)]. \qquad (11.5)$$

The matrix $Q \in \mathbb{R}^{p \times m}$ is the weight factor of the residual signals, which is constant in most cases, but, generally, it can be varying.

Using the model (11.1)–(11.2) and the observer descriptions (11.3)–(11.5), it can be shown that the residual vector (neglecting noise effects) has the form

$$r(s) = Q\{L_2 + C(sI - A + KC)^{-1}(L_1 - KL_2)\}$$
$$\times f(s) + QC(sI - A + KC)^{-1}[d(s) + e(0)], \qquad (11.6)$$

where $e(0)$ is the initial state estimation error. The residual signal is also affected by measurement and process noise signals, too. In (11.1) and (11.2), the influence of the measurement noise $v(t)$ on the residual vector has the same character as the factor $L_2 f(t)$. The analogous conclusion can be drawn for the process noise $w(t)$ and the factor $L_1 f(t)$. These factors can be distinguished only in the frequency domain. For an incipient fault signal, the fault information is contained within a low frequency band as the fault development is slow. However, the noise signals comprise mainly high frequency signals. Both noise and fault effects can be separated by using different frequency-dependent weighting penalties. Finally, four performance indices can be defined [12]:

$$J_1(W_1, K, Q) = \sup_{\omega \in [\omega_1, \omega_2]} \bar{\sigma}\{W_1(j\omega)[QL_2 + QC(j\omega I - A$$

$$+ KC)^{-1}(L_1 - KL_2)]^{-1}\}, \tag{11.7}$$

$$J_2(W_2, K, Q) = \sup_{\omega \in [\omega_1, \omega_2]} \bar{\sigma}\{W_2(j\omega)QC(j\omega I - A + KC)^{-1}\}, \tag{11.8}$$

$$J_3(W_3, K, Q) = \sup_{\omega \in [\omega_1, \omega_2]} \bar{\sigma}\{W_3(j\omega)Q[I - C(j\omega I - A + KC)^{-1}K]\}, \tag{11.9}$$

$$J_4(K) = \|(A - KC)^{-1}\|_\infty, \tag{11.10}$$

where $\bar{\sigma}\{\cdot\}$ is the maximal singular value, and $(W_i(j\omega) \mid i = 1, 2, 3)$ are the weighting penalties, which separate the effects of noise and faults in the frequency domain. They represent the prior knowledge about the spectral properties of the process and introduce additional degrees of freedom of the detector design procedure.

The task is to maximise (11.7)–(11.10) simultaneously. Unfortunately, conventional optimisation methods used for solving the above-formulated multi-objective problem (11.7)–(11.10) are ineffective. An alternative solution is genetic algorithm implementation [12]. In this case, a string of real values, which represent the elements of matrices $(W_i \mid i = 1, 2, 3)$, $K$ and $Q$, is chosen as an individual that represents the potential solution to the optimisation problem being considered.

The approach developed by Chen *et al.* [14] has been further investigated by Kowalczuk and Bialaszewski [15]. They proposed to use a modified description of the system that is given as follows:

$$\dot{x}(t) = Ax(t) + Bu(t) + Ed(t) + L_1 f(t) + w(t), \tag{11.11}$$

$$y(t) = Cx(t) + Du(t) + L_2 f(t) + v(t). \tag{11.12}$$

A set of optimisation indices for (11.11) and (11.12) is given as follows:

$$J_1(K, Q) = \sup_{\omega} \bar{\sigma}\{\|W_1(j\omega)G_{rf}(j\omega)\|\}^{-1}, \tag{11.13}$$

$$J_2(K, Q) = \sup_{\omega} \bar{\sigma}\{\|W_2(j\omega)G_{rd}(j\omega)\|\}, \tag{11.14}$$

$$J_3(K, Q) = \sup_{\omega} \bar{\sigma}\{\|W_3(j\omega)G_{rw}(j\omega)\|\}, \tag{11.15}$$

$$J_4(K, Q) = \sup_{\omega} \bar{\sigma}\{\|W_4(j\omega)G_{rv}(j\omega)\|\}, \tag{11.16}$$

$$J_5(K, Q) = \bar{\sigma}\{(A - KC)^{-1}\}, \tag{11.17}$$

$$J_6(K, Q) = \bar{\sigma}\{(A - KC)^{-1}K\}, \tag{11.18}$$

where

$$G_{rf}(j\omega) = Q\{C(Ij\omega - (A - KC))^{-1}(L_1 - KL_2) + L_2\}, \tag{11.19}$$

$$G_{rd}(j\omega) = QC(Ij\omega - (A - KC))^{-1}E, \tag{11.20}$$

$$G_{rw}(j\omega) = QC(Ij\omega - (A - KC))^{-1}, \tag{11.21}$$

$$G_{rv}(j\omega) = Q\{I - C(Ij\omega - (A - KC))^{-1}K\}. \tag{11.22}$$

Contrary to Reference 14, it is proposed to minimise (11.13)–(11.16) in the whole frequency domain. Moreover, the authors suggest to fix the weighting matrices $(W_i(j\omega)|i = 1, 2, 3, 4)$ before the optimisation process is started. In order to maximise the influence of faults at low frequencies and minimise the noise effect at high frequencies, the matrix $W_1$ should have a low-pass property. The weighting function $W_2$ should have the same properties, while the spectral effect of $W_3$ and $W_4$ should be opposite to that of $W_1$. Since $(W_i(j\omega)|i = 1, 2, 3, 4)$ are fixed, the synthesis of the detection filter reduces to multi-objective optimisation with respect to $K$ and $Q$. Another contribution is that the spectral synthesis of the matrix $(A - KC)$ incorporates the additional task of robust stabilisation of the observer that is realised with the genetic algorithm. As has already been mentioned, Chen and Patton [12] utilised the method of sequential inequalities and the genetic algorithm for solving the design problem of a fault detection observer. Here the multi-objective optimisation problem is solved with a method that incorporates the Pareto-optimality and genetic algorithm-based optimisation in the whole frequency domain.

The above-described approach clearly shows that the fusion of classical and soft-computing techniques, i.e. the genetic algorithm here results in a new reliable fault detection technique. The objective of the subsequent part of this section is to outline the possible approaches of evolutionary algorithms to designing fault diagnosis systems.

## 11.2.2 Evolutionary algorithms in FDI design

There are two steps of signal processing in model-based FDI systems: symptom extraction (residual generation) and, based on these residuals, decision-making about the appearance of faults, their localisation and range (residual evaluation) (Figure 11.2).

There are relatively scarce publications on applications of evolutionary algorithms to the design of FDI systems. The proposed solutions [12,13,15,16,18] (see also the references therein) allow one to obtain a high efficiency of diagnostic systems. Apart from the application of a relatively simple genetic algorithm for designing fault detection observers, special attention should be paid to the latest development regarding genetic programming [19] based approaches to the modelling and FDI concerning dynamic nonlinear systems by designing nonlinear state–space models for diagnostic observers [18,20] or by designing the EUIO [18,21].

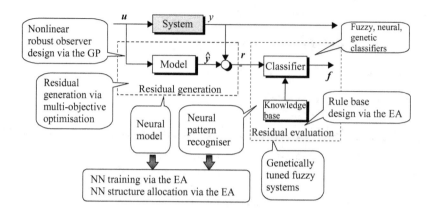

*Figure 11.2*   *Evolutionary algorithms in the process of designing an FDI system*
           *(EA – evolutionary algorithm; GP – genetic programming; NN – neural*
           *network)*

Among artificial intelligence methods applied to fault diagnosis systems, artificial neural networks are very popular, and they are used to construct neural models as well as neural classifiers [2,8,22]. Neural network approaches for FDI constitute the subject of a subsequent point in this chapter. But the construction of the neural model corresponds to two basic optimisation problems: the optimisation of a neural network architecture and its training process, i.e. searching for the optimal set of network free parameters. Evolutionary algorithms are very useful tools that are used to solve both problems, especially in the case of dynamic neural networks [8,16].

One can pin hopes on applications of evolutionary algorithms to the design of the residual evaluation module. The most interesting applications of genetic algorithms can be found in initial signal processing [23], in fuzzy systems tuning [24,25] and in rule base construction of an expert system [19,22].

All the above-mentioned possibilities of fusing the evolutionary algorithms with other classical and soft-computing techniques clearly show that such hybrid solutions may lead to very attractive developments in both the theory and practice of FDI. It is worth noticing that the process of designing fault diagnosis systems, can be reduced to a set of complex, global and usually multi-objective optimisation problems. This means that the conventional local optimisation routines cannot efficiently and reliably be used to tackle such a challenging task.

## 11.3   Nonlinear techniques in FDI

The main objective of this section is to provide a condensed overview of the classical nonlinear methods that can be applied for the purpose of fault diagnosis. The attention is restricted to the most popular approaches that can be divided into three groups, i.e. parameter estimation, parity relation and observers. For a description and references of other techniques the reader is referred to the following excellent books [2,3,12].

### 11.3.1 Parameter estimation

The task consists of detecting faults in a system by measuring the input $u_k$ and the output $y_k$, and then estimating the parameters of the model of the system [26]. In discrete time, the nonlinear input–output dynamics can be generally described by

$$y_k = g(\phi_k, p_k), \tag{11.23}$$

where $\phi_k$ may contain the previous or current system input $u_k$, the previous system or model output ($y$ or $\hat{y}$), and the previous prediction error. The model (11.23) can also be expressed in the state–space form; however, this does not change the general framework. If a fault now occurs in the system, this causes a change $\Delta p_k$ (residual) in the parameter vector $p_k$. Such a residual can then be used to detect and isolate the faults.

The model parameters should have physical meaning, i.e. they should correspond to the parameters of the system and if this is not the case the approach is severely limited. When the model parameters replicate those of their physical counterparts, the detection and isolation of faults is very straightforward. In a practical situation it can be difficult to distinguish a fault from a change in the parameter vector $p_k$ resulting from time-varying properties of the system. Moreover, the process of fault isolation may become extremely difficult because model parameters do not uniquely correspond to those of the system. Apart from the above-mentioned difficulties there are many classes of systems for which it is possible to derive models whose parameters have physical meaning. Distributed parameter systems [27] constitute such an important class. In order to increase the accuracy of parameter estimation and, consequently, the reliability of fault diagnosis, Uciński [27] proposed and developed various procedures that can be utilised for the development of an experimental design that facilitates high accuracy parameter estimation.

It should also be pointed out that the detection of faults in sensors and actuators is possible, but rather complicated [3] with the parameter estimation approach. Robustness with respect to model uncertainty can be tackled relatively easily (especially for linear systems) by employing robust parameter estimation techniques, e.g. the bounded-error approach [28].

### 11.3.2 Parity relation approach for nonlinear systems

An extension of the parity relation concept [1,2,12] to nonlinear polynomial dynamic systems was proposed by Guernez et al. [29]. In order to describe this approach, let us consider a system modelled by the state–space equations

$$x_{k+1} = g(x_k, u_k, f_k), \tag{11.24}$$

$$y_k = h(x_k, u_k, f_k), \tag{11.25}$$

where $g(\cdot)$ and $h(\cdot)$ are assumed to be polynomials. Equations (11.24)–(11.25) can always be expressed on a time window $[k - s, k]$. As a result, the following structure can be obtained:

$$y_{k-s,k} = H(x_{k-s}, u_{k-s,k}, f_{k-s,k}), \tag{11.26}$$

where $\boldsymbol{u}_{k-s,k} = \boldsymbol{u}_{k-s}, \ldots, \boldsymbol{u}_k$ and $\boldsymbol{f}_{k-s,k} = \boldsymbol{f}_{k-s}, \ldots, \boldsymbol{f}_k$. In order to check the consistency of the model equations, the state variables have to be eliminated. This results in the following equation:

$$\Phi(\boldsymbol{y}_{k-s,k}, \boldsymbol{u}_{k-s,k}, \boldsymbol{f}_{k-s,k}) = 0. \tag{11.27}$$

Since $\boldsymbol{g}(\cdot)$ and $\boldsymbol{h}(\cdot)$ are assumed to be polynomials, elimination theory can be applied to transform (11.26) into (11.27). Knowing that the $\Phi_i(\cdot)$ are polynomials and therefore they are expressed as sums of monomials, it seems natural to split the expression (11.27) into two parts, i.e.

$$\boldsymbol{r}_k = \Phi_1(\boldsymbol{y}_{k-s,k}, \boldsymbol{u}_{k-s,k}), \tag{11.28}$$

$$\boldsymbol{r}_k = \Phi_2(\boldsymbol{y}_{k-s,k}, \boldsymbol{u}_{k-s,k}, \boldsymbol{f}_{k-s,k}). \tag{11.29}$$

The right-hand side of (11.28) contains all the monomials in $\boldsymbol{y}_{k-s,k}$ and $\boldsymbol{u}_{k-s,k}$ only, while (11.29) contains all the monomials involving at least one of the components of $\boldsymbol{f}_{k-s,k}$. The above condition ensures that $\boldsymbol{r}_k = 0$ in the fault-free case. Since the fault signal $\boldsymbol{f}_{k-s,k}$ is not measurable, only equation (11.28) can be applied to generate the residual signal $\boldsymbol{r}_k$ and, consequently, to detect faults.

One drawback to this approach is that it is limited to polynomial models or, more precisely, to models for which the state vector $\boldsymbol{x}_k$ can be eliminated. Another drawback is that it is assumed that a perfect model is available, i.e. there is no model uncertainty. This may cause serious problems while applying the approach to real systems.

Another interesting approach was developed for bilinear systems [30]. By including bilinear terms into the system matrix, a linear time-varying model with a known time-varying feature was obtained and then a classical parity relation approach technique was utilised.

Parity relations for a more general class of nonlinear systems was proposed by Krishnaswami and Rizzoni [31]. The FDI scheme considered is shown in Figure 11.3.

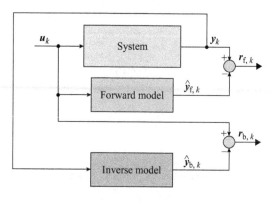

*Figure 11.3   Nonlinear parity relation-based FDI*

*Table 11.1    Principle of fault isolation with nonlinear parity relation*

| Fault location | Non-zero element of $r_{f,k}$ | Non-zero element of $r_{b,k}$ |
|---|---|---|
| $i$th sensor | $r_f^i$ | All elements dependent on $y_i$ |
| $i$th actuator | All elements dependent on $u_i$ | $r_b^i$ |

There are two residual vectors, namely, the forward $r_{f,k}$ residual vector and the backward $r_{b,k}$ residual vector. These residuals are generated using the forward and inverse (backward) models, respectively. Based on these residual vectors, fault detection can (theoretically) be easily performed while fault isolation should be realised according to Table 11.1. The authors suggest an extension of the proposed approach to cases where model uncertainty is considered. Undoubtedly, strict existence conditions for an inverted model as well as possible difficulties with the application of the known identification techniques make the usefulness of this approach, for a wide class of nonlinear systems, questionable.

Another parity relation approach for nonlinear systems was proposed by Shumsky [32]. The concepts of parity relation and parameter estimation fault detection techniques are combined. In particular, the parity relation is used to detect offsets in the model parameters. The necessary condition is that there exists a transformation $x_k = \xi(u_k, \ldots, u_{k+s}, y_k, \ldots, y_{k+s})$, which may cause serious problems in many practical applications. Another inconvenience is that the approach inherits most drawbacks concerning parameter estimation-based fault detection techniques.

### 11.3.3    Use of observers for FDI

Model linearisation is a straightforward way of extending the applicability of linear techniques to nonlinear systems. On the other hand, it is well known that such approaches work well when there is no large mismatch between the linearised model and the nonlinear system. Two types of linearisation can be distinguished, i.e. linearisation around the constant state and linearisation around the current state estimate. It is obvious that the second type of linearisation usually yields better results. Unfortunately, during such linearisation the influence of terms higher than linear is usually neglected (as in the case of the extended Luenberger observer and the extended Kalman filter (EKF)). This disqualifies such approaches for most practical applications. Such conditions have led to the development of linearisation-free observers for nonlinear systems.

This section briefly reviews the most popular observer-based residual generation techniques for nonlinear systems [33,34]. Their advantages, drawbacks as well as robustness to model uncertainty are discussed. It should be pointed out that the observers to be described are presented either in a discrete- or continuous-time form depending on their original form in the literature.

### 11.3.3.1   Extended Luenberger observers and Kalman filters

Let us consider a nonlinear discrete-time system modelled by the following state–space equations:

$$x_{k+1} = g(x_k, u_k) + L_{1,k} f_k, \tag{11.30}$$

$$y_{k+1} = h(x_{k+1}) + L_{2,k+1} f_{k+1}. \tag{11.31}$$

In order to apply the Luenberger observer [2] it is necessary to linearise equations (11.30) and (11.31) around either a constant value (e.g. $x = 0$) or the current state estimate $\hat{x}_k$. The latter approach seems to be more appropriate as it improves its approximation accuracy as $\hat{x}_k$ tends to $x_k$. In this case approximation can be realised as follows:

$$A_k = \left. \frac{\partial g(x_k, u_k)}{\partial x_k} \right|_{x_k = \hat{x}_k}, \qquad C_k = \left. \frac{\partial h(x_k)}{\partial x_k} \right|_{x_k = \hat{x}_k}. \tag{11.32}$$

As a result of using the Luenberger observer, the state estimation error takes the form

$$e_{k+1} = [A_{k+1} - K_{k+1} C_k] e_k + L_{1,k} f_k - K_{k+1} L_{2,k} f_k + o(x_k, \hat{x}_k), \tag{11.33}$$

where $o(x_k, \hat{x}_k)$ stands for the linearisation error caused by the approximation (11.32).

Because of a highly time-varying nature of $A_{k+1}$ and $C_k$ as well as the linearisation error $o(x_k, \hat{x}_k)$, it is usually very difficult to obtain an appropriate form of the gain matrix $K_{k+1}$. This is the main reason why this approach is rarely used in practice. As the Kalman filter constitutes a stochastic counterpart of the Luenberger observer, the EKF can also be designed for the following class of nonlinear systems:

$$x_{k+1} = g(x_k, u_k) + L_{1,k} f_k + w_k, \tag{11.34}$$

$$y_{k+1} = h(x_{k+1}) + L_{2,k+1} f_{k+1} + v_{k+1}, \tag{11.35}$$

where, similar to the linear case, $w_k$ and $v_k$ are zero-mean white noise sequences. Using the linearisation (11.32) and neglecting the influence of the linearisation error, it is straightforward to use the Kalman filter algorithm [2,3,35]. The main drawback to such an approach is that it works well only when there is no large mismatch between the model linearised around the current state estimate and the nonlinear behaviour of the system.

The EKF can also be used for deterministic systems, i.e. as an observer for the system (11.30)–(11.31) (see Reference 36 and the references therein). In this case, the noise covariance matrices can be set almost arbitrarily. As was proposed in Reference 36, this possibility can be used to increase the convergence of an observer.

Apart from the difficulties regarding linearisation errors, similar to the case of linear systems, the presented approaches do not take model uncertainty into account. This drawback disqualifies those techniques for most practical applications, although, there are applications for which such techniques work with acceptable efficiency e.g. Reference 2.

### 11.3.3.2 Thau observer

The observer proposed by Thau [37] can be applied to a special class of nonlinear systems which can be modelled by the following state–space equations:

$$\dot{x}(t) = Ax(t) + Bu(t) + L_1 f(t) + g(x(t), u(t)), \tag{11.36}$$

$$y(t) = Cx(t) + L_2 f(t). \tag{11.37}$$

This special model class can represent systems with both linear and nonlinear parts. The nonlinear part is continuously differentiable and locally Lipschitz, i.e.

$$\|g(x(t), u(t)) - g(\hat{x}(t), u(t))\| \le \gamma \|x(t) - \hat{x}(t)\|. \tag{11.38}$$

The structure of the Thau observer can be given as

$$\dot{x}(t) = Ax(t) + Bu(t) + g(\hat{x}(t), u(t)) + K(y(t) - \hat{y}(t)), \tag{11.39}$$

$$\hat{y}(t) = C\hat{x}(t), \tag{11.40}$$

where $K = P_\theta^{-1} C^T$, and $P_\theta$ is the solution to the Lyapunov equation:

$$A^T P_\theta + P_\theta A - C^T C + \theta P_\theta = 0, \tag{11.41}$$

where $\theta$ is a positive parameter, chosen in such a way so as to ensure a positive definite solution of (11.41). In order to satisfy the above condition, the Lipschitz constant $\gamma$ should satisfy the following condition:

$$\gamma < \frac{1}{2} \frac{\underline{\sigma}(C^T C + \theta P_\theta)}{\bar{\sigma}(P_\theta)}, \tag{11.42}$$

where $\bar{\sigma}(\cdot)$ and $\underline{\sigma}(\cdot)$ stand for the maximum and minimum singular values, respectively.

In spite of the fact that the design procedure does not require any linearisation, the conditions regarding the Lipschitz constant $\gamma$ are rather restrictive. This may limit any practical application of such an approach. Another difficulty arises from the lack of robustness to model uncertainty.

### 11.3.3.3 Observers for bilinear and polynomial systems

A polynomial (and, as a special case, bilinear) system description is a natural extension of linear models. Designs of observers for bilinear and (up to degree three) systems [38–41] involve only solutions of nonlinear algebraic or Ricatti equations. This allows online residual generation.

Let us consider a bilinear continuous-time system modelled by the following state–space equations:

$$\dot{x}(t) = Ax(t) + \sum_{i=1}^{r} B_i u_i(t) + E_1 d(t), \tag{11.43}$$

$$y(t) = Cx(t) + E_1 d(t). \tag{11.44}$$

With a slight abuse of notation, the influence of faults is neglected. However, faults can be very easily introduced without changing the design procedure.

An observer for the system (11.43)–(11.44) can be given as

$$\dot{\zeta}(t) = F\zeta(t) + Gy(t) + \sum_{i=1}^{r} L_i u_i(t)y(t), \tag{11.45}$$

$$\hat{x}(t) = H\zeta(t) + Ny(t). \tag{11.46}$$

Hou and Pugh [42] established the necessary conditions for the existence of the observer (11.45)–(11.46). Moreover, they proposed a design procedure involving a transformation of the original system (11.43)–(11.44) into an equivalent, quasi-linear one.

An observer for systems which can be described by state–space equations consisting of both linear and polynomial terms was proposed by Shields and Yu [40]. Similar to the case of the observer (11.43)–(11.44), here robustness to model uncertainty is tackled by means of an unknown input $d(t)$.

### 11.3.3.4   Nonlinear UIOs

This section presents an extension of the UIO for linear systems [12,13]. Such an extension can be applied to systems which can be modelled by the following state–space equations:

$$\dot{x}(t) = a(x(t)) + B(x(t))u(t) + E_1(x(t), u(t))d(t) + K_1(x(t), u(t))f(t), \tag{11.47}$$

$$y(t) = c(x(t)) + E_2(u(t))d(t) + K_2(x(t))f(t). \tag{11.48}$$

For notational convenience, the dependence of time $t$ is neglected (e.g. $u = u(t)$).

The underlying idea is to design an unknown input observer for the system (11.47)–(11.48) without model linearisation. For that purpose, the following observer structure is proposed by Alcorta and Frank [43] and Seliger and Frank [44]:

$$\dot{\hat{z}} = l(\hat{z}, y, u, \dot{u}), \tag{11.49}$$

$$r = m(\hat{z}, y, u), \tag{11.50}$$

where

$$z = T(x, u). \tag{11.51}$$

The unknown input decoupling condition can be stated as

$$\forall x, u \quad \frac{\partial T(x, u)}{\partial x} E_1(x, u) = 0. \tag{11.52}$$

The unknown input decoupling problem can now be realised by analytically solving a set of linear first-order partial differential equations (11.52). Moreover, if any fault

$f$ is to be reflected by the transformed model, it must be required that

$$\forall x, u \quad \mathrm{rank}\left(\frac{\partial T(x, u)}{\partial x}K_1(x, u)\right) = \mathrm{rank}(K_1(x, u)). \tag{11.53}$$

The effect of an unknown input can be decoupled from the output signal (11.48) in a similar way [44].

The main drawback to the proposed approach is that it requires a relatively complex design procedure, even for simple laboratory systems [45]. This may limit most practical applications of nonlinear input observers. Other problems may arise from the application of the presented observer to nonlinear discrete-time systems.

### 11.3.3.5 Extended unknown input observer

As has already been mentioned, the complex design procedure of nonlinear UIOs does not encourage engineers to apply them in practice. Bearing in mind such difficulties, the so-called EUIO was proposed [18,20]. The approach can be employed for the following class of nonlinear systems:

$$x_{k+1} = g(x_k) + h(u_k + L_{1,k}f_k) + E_k d_k, \tag{11.54}$$

$$y_{k+1} = C_{k+1}x_{k+1} + L_{2,k+1}f_{k+1}. \tag{11.55}$$

It should also be pointed out that the necessary condition for the existence of a solution to the problem of an unknown input decoupling is $\mathrm{rank}(C_{k+1}E_k) = \mathrm{rank}(E_k)$ ([12], p. 72, Lemma 3.1). The EUIO extends the approach developed by Chen *et al.* [13] for linear systems to nonlinear systems. It is proven [18,20,21] that the observer is convergent under some, not restrictive, conditions.

### 11.3.3.6 An illustrative example

The purpose of this section is to show the reliability and effectiveness of the observer-based fault detection scheme presented in Section 11.3.3.5. The numerical example considered here is a fifth-order two-phase nonlinear model of an induction motor, which has already been the subject of a large number of various control design applications (see Reference 36 and the references therein).

The complete discrete-time model in a stator-fixed $(a, b)$ reference frame is

$$x_{1,k+1} = x_{1,k} + h\left(-\gamma x_{1k} + \frac{K}{T_\mathrm{r}}x_{3k} + Kp x_{5k}x_{4k} + \frac{1}{\sigma L_\mathrm{s}}u_{1k}\right), \tag{11.56}$$

$$x_{2,k+1} = x_{2,k} + h\left(-\gamma x_{2k} - Kp x_{5k}x_{3k} + \frac{K}{T_\mathrm{r}}x_{4k} + \frac{1}{\sigma L_\mathrm{s}}u_{2k}\right), \tag{11.57}$$

$$x_{3,k+1} = x_{3,k} + h\left(\frac{M}{T_r}x_{1k} - \frac{1}{T_r}x_{3k} - px_{5k}x_{4k}\right), \tag{11.58}$$

$$x_{4,k+1} = x_{4,k} + h\left(\frac{M}{T_r}x_{2k} + px_{5k}x_{3k} - \frac{1}{T_r}x_{4k}\right), \tag{11.59}$$

$$x_{5,k+1} = x_{5,k} + h\left(\frac{pM}{JL_r}(x_{3k}x_{2k} - x_{4k}x_{1k}) - \frac{T_L}{J}\right), \tag{11.60}$$

$$y_{1,k+1} = x_{1,k+1}, \qquad y_{2,k+1} = x_{2,k+1}, \tag{11.61}$$

where $x_k = (x_{1,k}, \ldots, x_{n,k}) = (i_{sak}, i_{sbk}, \psi_{rak}, \psi_{rbk}, \omega_k)$ represents the currents, the rotor fluxes, and the angular speed, respectively, while $u_k = (u_{sak}, u_{sbk})$ is the stator voltage control vector, $p$ is the number of the pairs of poles and $T_L$ is the load torque. The rotor time constant $T_r$ and the remaining parameters are defined as

$$T_r = \frac{L_r}{R_r}, \qquad \sigma = 1 - \frac{M^2}{L_sL_r}, \qquad K = \frac{M}{\sigma L_sL_r^2}, \qquad \gamma = \frac{R_s}{\sigma L_s} + \frac{R_rM^2}{\sigma L_sL_r^2}, \tag{11.62}$$

where $R_s$, $R_r$ and $L_s$, $L_r$ are stator and rotor per-phase resistances and inductances, respectively, and $J$ is the rotor moment inertia.

The numerical values of the above parameters are as follows: $R_s = 0.18\,\Omega$, $R_r = 0.15\,\Omega$, $M = 0.068\,\text{H}$, $L_s = 0.0699\,\text{H}$, $L_r = 0.0699\,\text{H}$, $J = 0.0586\,\text{kg}\,\text{m}^2$, $T_L = 10\,\text{Nm}$, $p = 1$ and $h = 0.1\,\text{ms}$. The initial conditions for the observer and the system are $\hat{x}_k = (200, 200, 50, 50, 300)$ and $x_k = 0$. The unknown input distribution matrix is

$$E_k = \begin{bmatrix} 0.1 & 0 & 1 & 0 & 0 \\ 0 & 0.1 & 0 & 1 & 0 \end{bmatrix}^T. \tag{11.63}$$

The input signals are

$$u_{1,k} = 300\cos(0.03k), \qquad u_{2,k} = 300\sin(0.03k). \tag{11.64}$$

The unknown input is defined as

$$d_{1,k} = 0.09\sin(0.5\pi k)\cos(0.3\pi k), \qquad d_{2,k} = 0.09\sin(0.01k) \tag{11.65}$$

and $P_0 = 10^3 I$. The objective of presenting the next example is to show the effectiveness of the proposed observer as a residual generator in the presence of an unknown input. For that purpose, the following fault scenarios were considered:

*Case 1.*   An abrupt fault of $y_{1,k}$ sensor:

$$f_{s,k} = \begin{cases} 0, & k < 140, \\ -0.1y_{1,k}, & \text{otherwise.} \end{cases} \tag{11.66}$$

*Case 2.*   An abrupt fault of $u_{1,k}$ actuator:

$$f_{a,k} = \begin{cases} 0, & k < 140, \\ -0.2u_{1,k}, & \text{otherwise.} \end{cases} \tag{11.67}$$

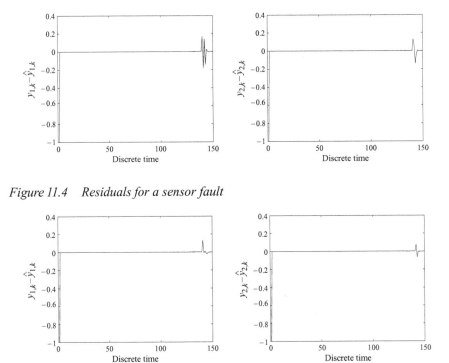

*Figure 11.4    Residuals for a sensor fault*

*Figure 11.5    Residuals for an actuator fault*

From Figures 11.4 and 11.5 it can be observed that the residual signal is sensitive to the faults under consideration. This, together with unknown input decoupling, implies that the process of fault detection becomes a relatively easy task.

## 11.4    Neural network approaches to FDI

A common disadvantage of analytical approaches to the FDI system is the fact that a precise mathematical model of the diagnosed plant is required. As no system can be modelled precisely, analytical approaches to FDI are associated with the robustness challenge defined in Section 12.1. An alternative solution can be obtained using soft-computing techniques [8,46,47], i.e. artificial neural networks [48], fuzzy logic, expert systems and evolutionary algorithms [49] or their combination as NF networks [50]. To apply soft-computing modelling [51], empirical data, principles and rules which describe the diagnosed process and other accessible qualitative and quantitative knowledge are required.

One of the most important classes of FDI methods, especially dedicated to non-linear processes, is the use of artificial neural networks [7,52–55]. There are many neural structures that can be effectively used for residual generation as well as residual evaluation. For residual generation, the neural network replaces the analytical

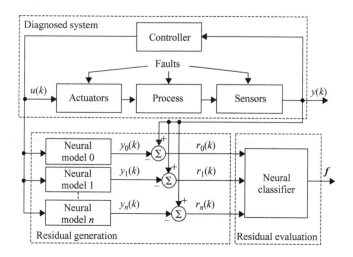

*Figure 11.6   Neural fault diagnosis scheme based on a bank of process models*

model [2,3], which describes the diagnosed process. To solve these neural modelling problems, dynamic neural networks such as a multilayer perceptron with tapped delay lines, with the dynamic model of neurons or recurrent and GMDH (group method of data handling) networks can be applied. Then the residuals generated by a bank of neural models (Figure 11.6) are evaluated by means of pattern classification [51]. In Figure 11.6 each neural model represents one class of system behaviour. One model represents the system under its normal conditions and each successive one – in a given faulty situation, representatively. To carry out the classification task several neural structures can be used, including static multilayer perceptron, Kohonen's self-organising map, radial basis and GMDH networks, as well as the multiple network structure. Similarly, as many other data-based techniques, the effectiveness of neural networks in FDI systems strongly depends on the quality of training data [48,56].

Learning data can be collected directly from the process if possible or from a simulation model that is as realistic as possible. The latter possibility is of special interest for data acquisition in different faulty situations, in order to test the residual generator, as those data are not generally available in the real process.

## 11.4.1   Neural structures for modelling

To date, many neural structures with dynamic characteristic have been developed [48,54,57], which are effective in modelling nonlinear processes. In general, all known structures can be divided into two groups: neural networks with external dynamics and with internal dynamics.

### 11.4.1.1   Neural networks with external dynamics

Neural networks with external dynamics can be designed by introducing explicit dynamics into the stand static network, i.e. the multilayer perceptron [48,56]

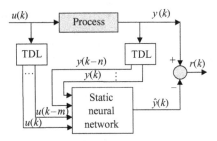

*Figure 11.7    Neural residual generator with external tapped delay lines (TDL)*

(Figure 11.7). In this case, the delay lines for the input $u(k)$ and the output $y(t)$ are linear filters. In general, the structure of such dynamic networks depends upon the format of the mathematical representation of the diagnosed process, i.e. state–space or input–output representation. It has been pointed out that the most suitable structure of the network, in order to model a nonlinear process, is the input–output format:

$$y(k) = \hat{f}[u(k), u(k-1), \ldots, u(k-m), y(k), y(k-1), \ldots, y(k-n)],$$

$$(11.68)$$

where $n$ and $m$ are the input and output signal delays, respectively, and $\hat{f}(\cdot)$ denotes the nonlinear function of the network which approximates the nonlinear function of the diagnosed process $f(\cdot)$. Moreover, $u(k)$ and $y(k)$ are the input and output signals, respectively. By comparing the process output $y(k)$ with the network output $\hat{y}(k)$, the residual $r(k) = y(k) - \hat{y}(k)$ can be obtained.

Such an input–output structure possesses some advantages over the state–space format, i.e. only one network is required in order to approximate the nonlinear function $f(\cdot)$. However, such an approach based on the static network (Figure 11.7) leads to quasi-dynamic models and the network used remains a static approximator. Moreover, the dimension of the input space $\{u(k), \ldots, u(k-m), y(k), \ldots, y(k-n)\}$ of the network increases, depending on the number of the available process data and the number of the past values used. The maximum delays $n$ and $m$ describe the dynamic orders of the diagnosed process.

### 11.4.1.2   Neural networks with internal dynamics

The main characteristic of the dynamic neural network is the fact that it possesses memory by introducing global or local recurrence. Introducing the recurrence into the network architecture, it is possible to memorise the information and use it later. Generally, globally recurrent neural networks in spite of their usefulness in control theory have some disadvantages. These architectures suffer from a lack of stability; for a given set of initial values the activations of the linear output neurons may grow unlimited. An alternative solution, which provides the dynamic behaviour of the neural model, is a network designed using dynamic neuron models. Such networks have an architecture that is somewhere in-between a feed-forward and a globally recurrent architecture. Based on the well-known classical static neuron model, different dynamic neuron models can be designed by using the various locations of the internal feedback. The best-known dynamic models are [48]: those with local activation feedback, local synapse feedback, local output feedback and the IIR filter [54,57,58].

In models with the IIR filter, dynamics are introduced into the neuron in such a way that neuron activation depends on its internal states. It can be done by introducing the IIR filter into the neuron structure [54,58]. The behaviour of the model considered can be described by the following set of equations:

$$\varphi(k) = \sum_{i=1}^{m} w_i u_i(k), \tag{11.69}$$

$$x(k) = -\sum_{j=1}^{n} a_j x(k-j) + \sum_{j=0}^{n} b_j \varphi(k-j), \tag{11.70}$$

$$y(k) = F(x(k) + c), \tag{11.71}$$

where $w_i$, $i = 1, \ldots, m$, are the input weights, $u_i(k)$ are the neuron inputs, $m$ is the number of inputs, $x(k)$ is the filter output, $a_j$ and $b_j$, $j = 1, \ldots, n$, are the feedback and feed-forward filter parameters, respectively, $F(\cdot)$ is a nonlinear activation function that produces the neuron output $y(k)$ and $c$ is the bias of the activation function.

Due to the dynamic characteristics of neurons, a neural network of the feed-forward structure can be designed. Taking into account the fact that this network has no recurrent link between the neurons, to adapt the network parameters, a training algorithm based on the back-propagation idea can be elaborated. The calculated output is propagated back to the inputs through hidden layers containing dynamic filters. As a result, extended dynamic back-propagation (EDBP) is defined [54]. This algorithm can have both online and offline forms, and therefore it can be widely used in FDI and control theory. The choice of a proper mode depends on problem specification.

### 11.4.1.3 Dynamic neural networks of the GMDH type

A disadvantage of most known neural networks is the fact that their architecture is arbitrarily defined [48,56]. An alternative approach is based on the integration of process training with the choice of the network optimal architecture. Such a designing procedure can be obtained by the GMDH approach [59–61].

The idea of GMDH is based on replacing the complex model of the process with partial models (neurons) by using the rules of variable selection. As usual, partial models have a small number of inputs $u_i(k)$, $i = 1, 2, \ldots, m$, and are implemented by GMDH neurons. The synthesis process of the GMDH network [61] is based on the iterative processing of a sequence of operations. This process leads to the evolution of the resulting model structure in such a way so as to obtain an approximation of the optimal degree of model complexity. The quality of the model can be measured with the application of various performance indexes [62]. The principle of GMDH

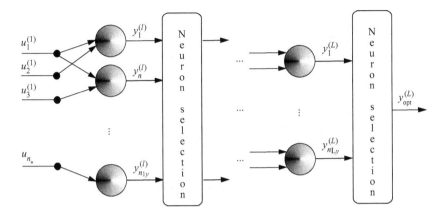

*Figure 11.8    Principle of the GMDH algorithm*

algorithm design is shown in Figure 11.8, where the multilayer GMDH network is constructed through the connection of a given number of neurons.

For such a network, the dynamics problem has been solved by introducing the dynamic model of the GMDH neuron, which can be given by Witczak *et al.* [63]:

$$x(k) = -a_1 x(k-1) - \cdots - a_{n_a} x(k - n_a) + \boldsymbol{b}_0^{\mathrm{T}} \boldsymbol{u}(k)$$
$$+ \boldsymbol{b}_1^{\mathrm{T}} \boldsymbol{u}(k-1) + \cdots + \boldsymbol{b}_{n_b}^{\mathrm{T}} \boldsymbol{u}(k - n_b), \tag{11.72}$$

$$y(k) = \xi(x(k)), \tag{11.73}$$

where $a_a, \ldots, a_n$ denote the feedback parameters of the introducing filter; $\boldsymbol{b}_0^{\mathrm{T}}, \boldsymbol{b}_1^{\mathrm{T}}, \ldots, \boldsymbol{b}_{n_b}^{\mathrm{T}}$ are the parameter vectors of the neuron; $n_a$ and $n_b$ denote delays of the filter outputs $x$ and the filter inputs $\boldsymbol{u}$, respectively. Moreover, $y(k)$ is the neuron output and $\xi$ denotes the nonlinear activation function.

To estimate the parameters of such a dynamic neuron model (11.72)–(11.73) with a nonlinear activation function, linear parameter estimation methods can be applied [64]. In these methods it is assumed that the discrete-time output from the identification system is represented as

$$y(k) = \boldsymbol{q}(k)^{\mathrm{T}} \boldsymbol{\theta} + \varepsilon(k), \tag{11.74}$$

where $\boldsymbol{q}(k) = [x(k-1), \ldots, u_{nu}(k - n_b)]$ is the regression vector, $\varepsilon(k)$ denotes disturbances and $\boldsymbol{\theta} \in \mathbb{R}^{n_\theta}$ is the parameter vector.

An outline of the GMDH algorithm can be described by a three-step procedure [61]:

1.  Determine all neurons (estimate their parameter vectors $\boldsymbol{\theta}_n^{(l)}$ with the training data set $\mathcal{T}$) whose inputs consist of all possible couples of input variables, i.e. $(n_a - 1)n_a/2$ couples (neurons).
2.  Use a validation data set $\mathcal{V}$, not employed during the parameter estimation phase, select several neurons which are best-fitted in terms of the chosen criterion.

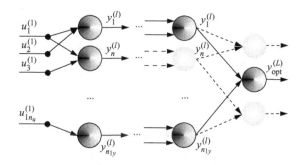

*Figure 11.9   Final structure of the GMDH neural network*

3. If the termination condition is fulfilled (the network fits the data with desired accuracy or the introduction of new neurons did not induce a significant increase in the approximation abilities of the neural network), then STOP, otherwise use the outputs of the best-fitted neurons (selected in Step 2) to form the input vector for the next layer, and then go to Step 1.

To obtain the final structure of the network (Figure 11.9), all unnecessary neurons are removed, leaving only those which are relevant to the computation of the model output. The procedure of removing unnecessary neurons is the last stage of the synthesis of the GMDH neural network. The feature of the above-mentioned algorithm is that the techniques for the parameter estimation of linear-in-parameter models can be used. Indeed, since $\xi(\cdot)$ is invertible, the neuron (11.72)–(11.73) can relatively easily be transformed into a linear-in-parameter one.

The important problem of GMDH network synthesis is a proper choice of the parameter estimation of the single neuron (11.72)–(11.73). Unfortunately, the least-squares method gives biased parameter estimates for the GMDH model. In order to solve this problem [63], the bounded-error approach as well as the outer bounding ellipsoid algorithm were applied.

## 11.4.2   Neural structures for classification

The residual evaluation is a logical decision-making process that transforms quantitative knowledge into qualitative ('Yes/No') statements. It can also be seen as a classification problem. The task is to match each pattern of the symptom vector with one of the pre-assigned classes of faults and the fault-free case. A variety of well-established approaches and techniques (thresholds, adaptive thresholds, statistical and classification methods) can be used for residual evaluation [46,47]. Among these approaches, fuzzy and neural classification methods are very attractive and more and more frequently used in FDI systems [8,47]. Different structures of static neural networks can be applied, including the multilayer perceptron, Kohonen networks, RBF networks and multiple network structures. In all cases, neural classifiers are fed with residuals which can be generated by analytical or soft-computing methods [2,3].

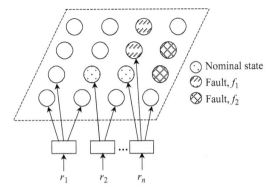

*Figure 11.10    Kohonen two-dimensional classifier*

### 11.4.2.1    Kohonen classifier

The Kohonen network is a self-organising map in which parameters are adapted by an unsupervised learning procedure based only on input patterns. Contrary to the standard supervised training methods, the unsupervised ones use input signals to extract knowledge from data. Inputs (residuals) and neurons in the competitive layer are connected entirely [56,65] (see the two-dimensional map in Figure 11.10). During the training process the weight parameters are adapted using the winner-takes-all rule. However, instead of adapting the winning neuron only, all neurons within a certain neighbourhood of the winner are adapted as well. The concept of the neighbourhood is extremely important during network processing. A typical neighbourhood function is of the form of the Mexican hat. After designing the network, a very important problem is how to assign the clustering results generated by the network to the desired results of a given problem. It is necessary to determine which regions of the feature map will be active during the occurrence of a given fault. A good result of the training process is shown in Figure 11.10, where three states are considered, and each of them (two faults $f_1$ and $f_2$, and a normal state) is separated. If after the training process there are overlapping regions, then the size of the map (the number of neurons) should be increased and the process training repeated again from the beginning.

### 11.4.2.2    Multiple network structure

In many cases, a single neural network of a finite size, i.e. a multilayer perceptron or radial basic network, does not assure the required mapping or its generalisation ability is not sufficient. To improve the quality of neural classifiers, multiple network schemes are proposed [66]. The idea of the multiple network scheme is to combine $n$ independently trained networks for $n$ working points and to classify a given input pattern by a decision module. The decomposition of a complex classification problem can be performed using independently trained neural classifiers (Figure 11.11) designed in such a way that each of them is able to recognise only few classes. The decision module in connection with the gate decide which neural classifier should classify a given pattern. This task can be carried out using a suitable rule and fuzzy

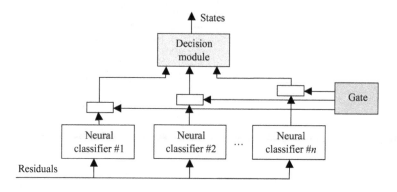

*Figure 11.11   Multiple neural classifier*

logic approach. Moreover, each neural classifier may have an optional structure and should be trained with a convenient algorithm based on different training sets. The only condition is that each neural classifier should have the same number of outputs.

### 11.4.3   Robust FDI with GMDH models

Based on the actuator benchmark definition [11], two structural models can be defined [67]:

$$F = f_F(X, P_1, P_2, T_1),\tag{11.75}$$

$$X = f_X(C_V, P_1, P_2, T_1),\tag{11.76}$$

where $f_F(\cdot)$ and $f_X(\cdot)$ denote unknown nonlinear functions of the flow rate and displacement, respectively. Using these functional relations, GMDH neural dynamic models have been developed. For research purposes, 19 faults ($f_1, f_2, \ldots, f_{19}$) have been selected and grouped into four sets: the faults of the control value, the pneumatic actuator, the positioner and the general/external faults. For the purpose of fault detection, two GMDH neural models corresponding to the relations (11.75) and (11.76) have been built. During the synthesis process of these networks, the so-called selection method was employed [63], and the final structures are shown in Figure 11.12.

Figure 11.13(a) and (b) present the modelling abilities of the obtained models $F = f_F(\cdot)$ and $X = f_X(\cdot)$ as well as the corresponding system output uncertainty. The thick solid line represents the real system output, the thin solid lines correspond to the system output uncertainty and the dashed line denotes the model response. From Figure 11.13, it is clear that the system response is contained within the system's output bounds, which have been designed with the estimated output error bounds [63].

The main objective of this application study was to develop a fault detection scheme for the valve actuator. On employing the models $F = f_F(\cdot)$ and $X = f_X(\cdot)$ for robust fault detection with the approach proposed in Reference 63, the selected results are shown in Figures 11.14 and 11.15. These figures present the residuals (the solid lines) and their bounds given by the adaptive thresholds (the dashed lines). It is clear that both faults $f_{18}$ and $f_{19}$ are detected.

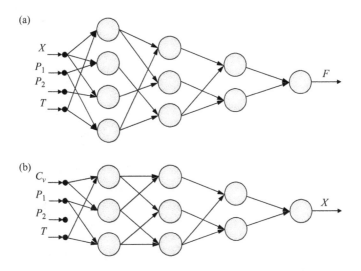

*Figure 11.12   Final structure of GMDH models: (a) for $F = f_F(\cdot)$ and (b) for $X = f_X(\cdot)$*

## 11.5   FDI using NF identification and design

### 11.5.1   NF modelling for FDI

In the last few years an increasing number of authors ([9,68–72] and the references therein) have been using integrated neural–fuzzy models in order to benefit from the advantages of both. NF techniques are now being investigated as powerful modelling and decision-making tools for FDI, along with the more traditional use of nonlinear and robust observers, parity space methods and hypothesis-testing theory.

The NF model combines, in a single framework, both numerical and symbolic knowledge. Automatic linguistic rule extraction is a useful aspect of NF, especially when little or no prior knowledge about the process is available [9,73]. For example, an NF model of a nonlinear dynamical system can be identified from the empirical data. This model may give us some insight about the nonlinearity and dynamical properties of the system [74]. Many researchers claim that NF-based FDI is advantageous when:

- enough system information is not available
- physical or semi-physical models are difficult to obtain
- sensor measurements are incomplete, assumed or missing
- training data are difficult to obtain because of a lack of control of some input variable
- the system exhibits strong nonlinear static and dynamic behaviour
- during the normal operation the frequency components of the input signal are not complete enough to build a model that can be reliable for all possible frequencies.

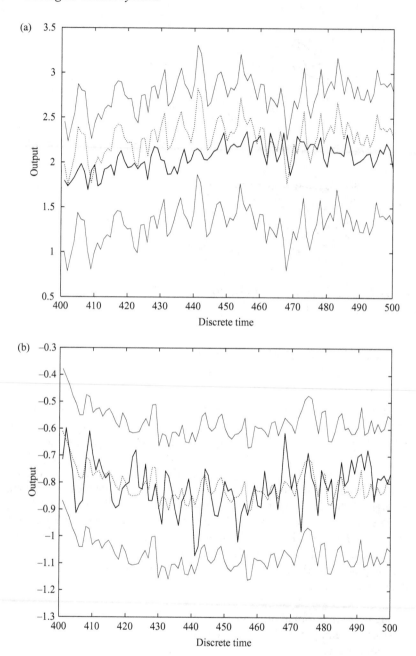

*Figure 11.13    Model and system outputs with the corresponding uncertainty range*

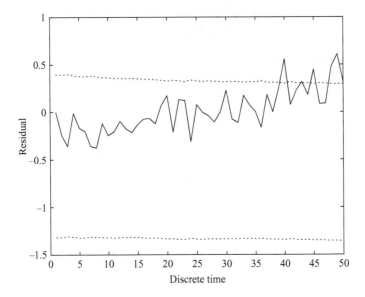

*Figure 11.14    Residual for fault $f_{18}$*

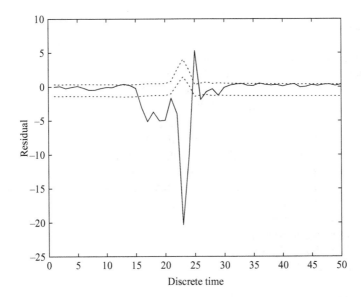

*Figure 11.15    Residual for fault $f_{19}$*

The two most important characteristics of NF models are generalising and reasoning capabilities. Depending on the application requirement, usually a compromise is made between the two. Transparency in NF modelling accounts for a more meaningful description of the process, i.e. fewer rules with appropriate membership functions. In the adaptive NF inference system known as the ANFIS [70,71], usually a fixed structure with grid partition is used. Antecedent and consequent parameters are identified by a combination of the least squares estimate and gradient-based method, called the hybrid-learning rule. This method is fast and easy to implement for low dimension input spaces. However, it is more prone to lose the transparency and the local model accuracy because of the use of error back-propagation, which is generally a global nonlinear optimisation procedure. A way to overcome this is to find the antecedents and rules separately, e.g. cluster and constrain the antecedents, and then find the consequents.

Neural networks and fuzzy logic systems can be combined in a number of different ways. However, there are two main frameworks: (1) NF based on neural network and (2) NF based on fuzzy logic. The simplest NF networks of this type (combination hybrid systems) use a 'fuzzifier' to combine fuzzy logic and the neural network. Simpson [75,76] proposed a structure for combining neural networks with fuzzy logic, called fuzzy min–max neural networks. Many other different structures have been established, integrating neural networks with the fuzzy logic with different training algorithms, e.g. References 71 and 77–79 proposed a five-layer NF network in which each layer represents a specific fuzzy logic functionality.

Many different NF models have been developed based on different architectures and training schemes, e.g. additive and multiplicative NF modelling [80], NF controller based on generic fuzzy-perceptron to learn fuzzy sets and rules by a reinforcement learning algorithm (NEuro-Fuzzy CONtrol: NEFCON) (variants: NEFCLASS: NEuro-Fuzzy CLASSification, NEFPROX: NEuro-Fuzzy apPROXimation) [81], fuzzy adaptive learning control network (FALCON) (and its variants) [77], fuzzy basis function network (FBFN) with orthogonal LSE learning [78], adaptive neuro-fuzzy inference systems (ANFIS) [71], rule extraction from local cluster neural nets (RULEX) [82], generalised approximate reasoning based intelligent controller (GARIC) [83], neural network driven fuzzy reasoning (NNFDR) [84], fuzzy-neural methods for complex data analysis problems (FuNe) [85], feedforward network model with a supervised learning algorithm and a dynamic architecture (Fuzzy RuleNet) [86], neural network based fuzzy logic design (NeuFuz) [87], fuzzy neurons [88], fuzzy ART models [89], fuzzy CMAC [90] and fuzzy restricted coulomb energy network (FRCE) [91].

*B-spline NF network.* The study of the B-spline NF model is inspired by the discovery of a connection between the radial basis neural network (similar in structure to B-spline neural network) and the fuzzy logic system [9]. The large number of rules needed, fixed grid partition and constant consequents generally cause the low interpretability of these networks [9,92]. This approach was applied to FDI in a sugar plant [72,93,94].

*Hierarchical NF networks.* These can be used to overcome the dimensionality problem by decomposing the system into a series of MISO and/or SISO systems called 'hierarchical systems' [95]. Ozyurt and Kandel [96] used a hybrid diagnostic methodology for fault diagnosis based on a hierarchical multilayer perceptron–elliptical neural network structure and a fuzzy expert system.

*NF models based on two major fuzzy systems.* Two major classes of knowledge representation in fuzzy modelling are TSK (Takagi, Sugeno and Kang) and Mamdani, proposed in References 97–99. In the linguistic or the Mamdani fuzzy model both antecedents and consequents are linguistic fuzzy sets. This model is mainly used to give a more linguistic description of the process as compared to the TSK fuzzy model. Linguistic models are interpreted by using a natural language and qualitative terms. This kind of system representation might be less suitable for control design and mathematical analysis but it is more compatible with the natural language. Based on the two classes described above, there are two principal types of NF networks preferred by most authors in the field of NF integration. In this network, the number of trainable parameters is large, defuzzification is more complicated and is more computationally involved. It is more transparent (more transparent than the TSK model in the sense of linguistic consequent MFs) and close to human reasoning, but the complexity is high and the training is difficult.

In TSK fuzzy models the antecedents are similar to the Mamdani fuzzy system. The consequents can be any function describing the response of the model within the fuzzy region. The local models can be exploited for control design and fault diagnosis. TSK models with linear consequents can be interpreted in terms of local linear models of the system and in terms of changes of the model parameters with respect to the antecedent variables [100]. These relationships can give valuable insights into the nonlinearities, and can be exploited in control design. TSK fuzzy models are suitable for the accurate modelling and identification.

An FDI framework can be used based on the TSK model for residual generation and the Mamdani model for the classification. Application studies carried out using this framework can be found in References 74 and 101–103.

### 11.5.2 Structure identification of NF models for FDI

The term structure identification (SI) introduced by Sugeno and Yasukawa [104] is widely used in the fuzzy literature and is considered as the task of finding the optimal number of input membership functions, number of rules, number of output membership functions and initial membership function parameters. One idea to use this is to identify a general structure of the NF network and then, based on this structure, to identify a series of NF models, which correspond to different points in the NF spectrum.

Various techniques can be applied to the process data to perform SI, including classification and regression trees (CART) [105], data clustering, evolutionary algorithms (Reference 106 and the references therein) and unsupervised learning (e.g. Kohonen features maps [65]). Using structure identification techniques, an initial TSK or Mamdani NF model can be constructed. The initial TSK model can be converted to other NF fuzzy models, which are at different points of the NF spectrum (i.e. different levels of transparency and accuracy). The product space clustering approach [107] can be used for structure identification of TSK fuzzy models as it facilitates the representation of a nonlinear system by several local linear dynamic models. Various algorithms have been developed to perform this, e.g. the local linear model tree algorithm (LOLIMOT) [108]. However, in structure identification, a major issue is sensitivity to uneven distribution of data. Another issue is the transparency of models. Tuning the rules and membership functions can enhance the transparency of NF models. These are often referred to as structure simplification techniques [107]. Another issue in NF structure identification is to find the optimal number of rules, for which a number of cluster validity measures [109,110] and complexity reduction methods, such as compatible cluster merging (CCM) [111], have been suggested.

### 11.5.3   NF-based FDI methods

Many NF structures have been successfully applied to a wide range of FDI applications. A general conceptual diagram for a fault diagnosis system is depicted in Figure 11.16. The diagnosis consists of two sequential steps: residual generation and residual evaluation. Many neural network and NF-based FDI schemes use simple networks in these two stages. In the first step $m$ number of residual signals are generated from $p$ measurements in order to determine the state of the process. The objective of fault isolation is to determine if a fault has occurred ($n$ is the total number of faults) and to find the location of the fault, by analysing the residual vector.

The four main NF-based FDI schemes are listed below:

*Residual generation and fault isolation.* The scheme presented in Figure 11.17 is one of the most common schemes used for FDI, in which NF networks with high approximation and generalisation capabilities are used to construct faulty and fault-free models of the system in order to generate residuals ($r$). If the system is large, it may be divided into subsystems and identified by a number of NF networks. Each residual is ideally sensitive to one particular fault in the system. In practice, however, because of noise and disturbances, residuals are sensitive to more than one fault. To take into account the sensitivity of residuals to various faults and noise, an NF classifier is used. A linguistic style (Mamdani) NF network is suitable to process the residuals

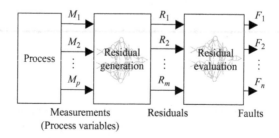

*Figure 11.16   General structure of NF diagnosis system*

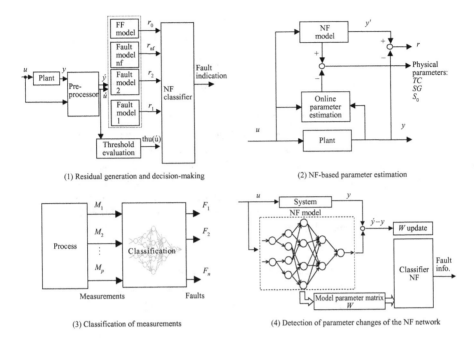

*Figure 11.17   NF-based fault detection and isolation schemes*

to indicate the fault as it is more transparent. Fuzzy threshold evaluation can be employed to take into account the imprecision of the residual generator in different regions in the input space. This approach depends heavily on the availability of the fault and fault-free data and it is more difficult to isolate faults that appear mainly in system dynamics.

*Parameter identification.* In this scheme, NF networks are employed to predict some system parameters. When faults occur, these parameters change and from the difference in the predicted and actual parameter values, fault isolation is carried out. One possibility to generate residuals is to use an NF model which is a nonlinear dynamic model of the system and which approximates it by local linear models. Such a model can be obtained by product space clustering [107] or tree-like algorithms (CART [105], LOLIMOT [108]). Each local model is a linear approximation of the process in an input–output subspace and fuzzy logic is employed for the selection of local models [108]. From the parameters of the local models, physical parameters, such as time constants, static gains, offsets, etc. [100] can be extracted for each operating point and compared with the parameters estimated online. This scheme is suitable especially where some prior knowledge is available about the parameters and the physical model of the system, and it heavily depends on the accuracy of the nonlinear dynamic model and requires sufficient excitation at each operating point for online parameter estimation.

*Classification of measurements.* Figure 11.17 describes an FDI scheme based on classification of measurements. In this method, measurements from the process are fed directly to NF classifiers and these are trained using the data to detect and isolate faults. This method appears to be relatively more convenient for static systems. For dynamic systems, multiple FD systems can be built for a number of operating conditions of interest.

*Detection of parameter changes of the NF model.* Another fault isolation scheme based on the detection of parameter changes of the NF model is shown in Figure 11.17. This scheme is based on an adaptive NF network such that its internal parameters are updated continuously. When faults occur in the system, it can be expected that the internal parameters of the NF system will change, and from these changes the faults can be predicted. Another classifier NF network is used to isolate the faults. This scheme also depends heavily on the availability of the fault and fault-free data but can be faster and comparatively more automatic (to acquire the fault information and use it for online isolation). However, it is difficult to use prior knowledge, if the data are missing in some input regions.

### 11.5.3.1   NF-based decoupling fault diagnosis scheme

For a system described by equations (11.1) and (11.2), the robust observer approach such as the UIO is undoubtedly the most commonly used model-based FDI approach. As has been mentioned above, much work in this field is directed towards linear systems. The existing linear extensions of this approach to stochastic systems [12] require a relatively complex design procedure and are restricted to a limited class of systems. The extension of this approach for stochastic systems (similar to that of the classic Kalman filter) is also imperfect in the sense of restrictive assumptions

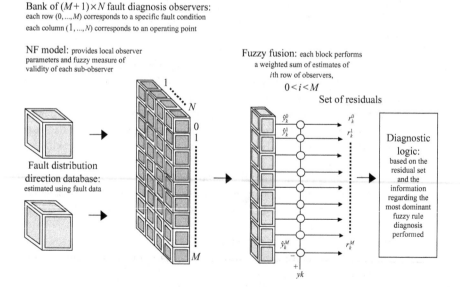

Bank of $(M+1) \times N$ fault diagnosis observers:
each row $(0, ..., M)$ corresponds to a specific fault condition
each column $(1, ..., N)$ corresponds to an operating point

NF model: provides local observer parameters and fuzzy measure of validity of each sub-observer

Fuzzy fusion: each block performs a weighted sum of estimates of $i$th row of observers,
$0 < i < M$

Set of residuals

Fault distribution direction database: estimated using fault data

Diagnostic logic: based on the residual set and the information regarding the most dominant fuzzy rule diagnosis performed

*Figure 11.18    NFDFDS overview*

concerning noise distribution (the noise is assumed to be zero-mean white noise sequences) [35].

Nonlinear extensions of the UIO can only be applied to nonlinear deterministic systems. For nonlinear stochastic systems, the only model-based FDI approach is based on the EKF, which is rather complex. The application of such robust model-based FDI techniques is usually limited due to the requirement of a nonlinear state–space model of a system.

Computational intelligence approaches, particularly NF approaches, are known to overcome some of the above-mentioned problems. However, due to the lack of current theory on NF-based FDI, and in particular a lack of effective online fault diagnosis methods, NF-based FDI had not attracted serious research and development attention until recently.

Bearing in mind the above combination of the model-based and NF approaches seems attractive especially for nonlinear systems. This scheme presented by Uppal and Patton [112] (Figure 11.18) comprises a bank of $(M + 1) \times N$ NF-based decoupling observers, where $M$ is the number of fault scenarios considered and $N$ is the number of operation points. It generates a set of residuals $(r_k^0 \cdots r_k^M)$ at the $k$th sampling instant, in the form of a structured residual set [12], while the diagnostic logic unit performs an analysis of the residuals to determine the nature and location of the faults. In this scheme, each residual is designed to be sensitive to a subset of faults (or the residuals are designed to have different responses to different faults). Ideally, each residual is sensitive to all but one fault, called a generalised residual set (based on the generalised observer scheme of Frank [33]). Each of the $M + 1$ fault diagnosis observers is a nonlinear system comprising $N$ linear (integrated by fuzzy fusion) sub-observers, each one corresponding to a different operating point of the process. Computing the difference between the actual and estimated outputs generates a set of residuals.

The NF-based decoupling fault diagnosis scheme (NFDFDS) is based on the identification of a particular type of the NF model (a simpler form of the TS (Takagi–Sugeno) NF model [98]). A method to generate locally optimal decoupled observers automatically from the NF model is proposed. In this proposed scheme, the computational limitations and stability constraints associated with the conventional fuzzy observer [113,114] are alleviated. The removal of these stability constraints simplifies and speeds up the joint identification and FDI problem. Moreover, the reasoning and approximation capabilities of NF networks can be exploited. The results obtained show that this scheme performs better than the conventional neural network- and NF-based FDI schemes for some applications. This results from the use of the decoupling strategy, which facilitates faster training due to less training data required.

### 11.5.3.1.1 The TS fuzzy multiple-model paradigm

The Takagi–Sugeno (TS) fuzzy inference model is a simple description for a nonlinear dynamic system with locally linearised models, described by the set of rules in (11.77). Note that the system and observer descriptions are given here in the discrete-time form, although Wang *et al.* [113] considered the continuous-time form:

$$\textit{Rule } l: (l = 1, 2, \ldots, N) \quad \text{IF } \omega_k^l \text{ is } \mu^l \text{ THEN } \begin{cases} x_{k+1} = A^l x_k + B^l u_k, \\ y_k = C^l x_k, \end{cases} \quad (11.77)$$

$$\textit{Global model:} \quad x_{k+1} = \sum_{l=1}^{N} \alpha^l(\omega_k^l)(A^l x_k + B^l u_k),$$

$$y_k = \sum_{l=1}^{N} \alpha^l(\omega_k^l) C^l x_k, \quad (11.78)$$

where $l$ denotes the $l$th rule or submodel, $\mu^l$ is a fuzzy set, $\alpha$ is the rule firing strength based on the membership grade $\mu$ of the antecedent variable $\omega$ or the tensor product of grades of memberships, and if $\omega$ is a vector, $\alpha^l(\omega_k^l)$ satisfies the following constraints:

$$\sum_{l=1}^{N} \alpha^l(\omega_k^l) = 1, \quad 0 \le \alpha^l(\omega_k^l) \le 1 \,\, \forall l = 1, 2, \ldots, N. \quad (11.79)$$

### 11.5.3.1.2   Multiple-model structure in NFDFDS

In NFDFDS, sub-observers are based on locally recursive linear models with fault and disturbance as described below:

$$x_{k+1}^l = A_k^l x_k^l + B_k^l u_k^c + E_k^l d_k + F_{1k}^l f_k + w_{1k}^l,$$
$$y_{mk}^l = C_k^l x_k^l + w_{2k}^l, \quad (11.80)$$

where $u_k^c \in \mathbb{R}^r, x_k^l \in \mathbb{R}^n, y_{mk}^l \in \mathbb{R}^m$ are the input, state and output vectors for the $l$th consequent linear model, respectively. Each entry of $f_k \in \mathbb{R}^g$ corresponds to a specific fault while $d_k \in \mathbb{R}^q$ stands for the disturbance vector. $A_k^l$, $B_k^l$ and $C_k^l$ are system matrices with appropriate dimensions. The fault distribution matrix $F_{1k}^l \in \mathbb{R}^{n \times g}$ represents the effect of input and component faults on the system; $w_{1k}^l$ and $w_{2k}^l$ are supposed to be independent zero-mean white noise sequences with correlation matrices $Q_k$ and $R_k$, assumed to be known.

Note that a local recursive structure is employed here as compared to the global recursive one given in the equations (11.77) and (11.78). The global model can be described by fuzzy IF–THEN rules that represent local linear models of the nonlinear system in the following form:

$$\textit{Rule } l: (l = 1, 2, \ldots, N) \quad \text{IF } u_k^a \text{ is } M_l \text{ THEN } y_{mk} = y_{mk}^l. \quad (11.81)$$

The global output of the fuzzy model is obtained by defuzzification (a weighted sum of the outputs of the submodels) as

$$y_{mk} = \sum_{l=1}^{N} \alpha^l(u_k^a) y_{mk}^l. \quad (11.82)$$

In the above equation, $\alpha^l(\boldsymbol{u}_k^a)$ $\forall l = 1, \ldots, N$ is the firing strength of $l$th rule, which depends on the antecedent variable $\boldsymbol{u}_k^a$, and $M_l$ is a fuzzy set and $N$ is the total number of linear submodels.

The conventional fuzzy observer based on the system described by (11.77)–(11.82) involves an iterative procedure, selecting eigenvalues arbitrarily (by pole-placement) and solving the resulting inequalities using LMI to find a single positive definite Lyapunov matrix $\boldsymbol{P}$, common to all submodels. This common matrix ensures the stability of the global observer system. In this conventional approach, if such a matrix $\boldsymbol{P}$ does not exist, another set of arbitrary eigenvalues is selected and the whole procedure is repeated until a solution is found. As has been mentioned earlier, these computational limitations and stability constraints are alleviated in the proposed scheme. The improvements in design complexity can be attributed to the fact that the global state estimation is not necessary for FDI and hence there is no need for unnecessary global feedback. In NFDFDS, a method to generate locally optimal decoupled observers automatically from the NF model is proposed. This NF model operates in parallel to the local observers so that they can be modified or tuned online. In this way, the reasoning and approximation capabilities of NF networks can be exploited.

It should be noted here that for the FDI problem, the residual signal is the main requirement. Hence, global state estimation is obviated. However, each observer has a different state variable model from all other local observers. As is described in Reference 112, if all local observers are stable, the stability of the global observer is guaranteed. This represents a significantly relaxed stability requirement over that required by the TS-based observer scheme.

The NF model used here is a special form of the TS fuzzy model (also referred to as the multiple-model scheme) with linear models as fuzzy rule consequents. The model can be represented as a five-layer neural network where each layer corresponds to a function of the fuzzy logic inference system (layer 1: input, layer 2: input membership functions, layer 3: rules, layer 4: output membership functions, layer 5: defuzzification). This representation can be used for training or tuning model parameters online. An initial fuzzy model is created using a combination of CART [105] and GK clustering [115], and then structure simplification is carried out to fine-tune the model.

In order to design a robust observer, it is possible to use the unknown input approach by expressing the consequent linear models in the state–space form with faults [12]. The fault distribution matrix represents the effect of input and component faults on the system. The underlying problem is to determine the disturbance distribution matrix. This problem boils down to the determination of the disturbance term for the normal and faulty system operations, which corresponds to the possible solutions for disturbance and fault terms, respectively. An approach is suggested as part of the NFDFDS scheme to determine an individual distribution direction for each submodel. Moreover, in cases where this direction changes considerably with fault strength, a set of such directions may be determined for each submodel. The following three cases [112] can be considered in order to determine the disturbance terms. The first two cases enable disturbance vectors to be computed, while the third one considers the distribution matrices directly.

*Case 1.*  Consider the fact that the individual disturbances sources $d_{1k}$, $l = 1, \ldots, N$, correspond to each submodel.

*Case 2.*  It is now assumed that the overall disturbance in the system (5) can be decoupled by only the $i$th submodel corresponding to the most dominant rule.

*Case 3.*  The fault distribution matrix $E_{1k}^l$ is supposed to be constant over time for each fault scenario.

In NFDFDS, the output is estimated using the principle of disturbance decoupling in state estimation. The residual is then the output estimation error itself, and is robust against disturbance and satisfies a minimum variance condition (according to the Kalman filter theory). Any of the well-established hypothesis-testing procedures (generalised likelihood ratio (GLR) testing and sequential probability ratio testing (SPRT)) [116] can be utilised to examine the residuals for the likelihood of faults. The state estimation obtained by this method is an improvement over the results obtained using a standard Kalman filter, when disturbances are present in the system.

For fault isolation, the information is collected from the residuals $r^1, \ldots, r^M$ (where $M$ is the number of fault scenarios) in terms of the *fault effect in different operating points*. This information is collected in the form of fault symptoms $S_{i,j} \in [1, \ldots, N]$, where $i$ denotes the operating point and $j$ denotes the fault scenario. A set of fault symptoms for all observers forms a fault signature $S_j \in [1, \ldots, N]$. Different operating points are distinguished according to the most dominant rule (the one with the most firing strength) in that region. The information collected from this residual is a set of operating points in which the fault has an effect. It is expected that the fault will have a different effect in different operating points, as the system considered is nonlinear. When a fault occurs in the system, the signature is compared with the previously collected signatures for isolation.

## 11.6  Conclusions

In this chapter, selected aspects of the fault diagnosis problem for control systems have been considered with special attention paid to fault detection and, to a lesser extent, fault isolation. In particular, the robustness problem of FDI with respect to the requirement to maximise sensitivity to faults while minimising (or decoupling) the effect of uncertain effects through unknown inputs has been described. This severe robustness challenge accompanying analytical methods of FDI has led to the requirement of developing new methodologies for FDI in which analytical and CI techniques are combined to achieve good global and nonlinear modelling and robustness.

The chapter has outlined some of the recent research on FDI and fault diagnosis for dynamic systems, using this integrated approach. The examples have shown that by using evolutionary computing, neural networks and NF modelling structures realistic solutions are achievable. It is hoped that this direction of research will stimulate an increased adoption of real industrial application to make mode-based FDI more usual and effective in real process systems.

# References

1 GERTLER, J.: 'Fault detection and diagnosis in engineering systems' (Marcel Dekker, New York, 1998)

2 KORBICZ, J., KOŚCIELNY, J. M., KOWALCZUK, Z., and CHOLEWA, W. (Eds): 'Fault diagnosis. Models, artificial intelligence, applications' (Springer-Verlag, Berlin, 2004)

3 PATTON, R. J., FRANK, P., and CLARK, R. N. (Eds): 'Issues of fault diagnosis for dynamic systems' (Springer-Verlag, Berlin, 2000)

4 PATTON, R. J., and CHEN, J.: 'Robust fault detection and isolation (FDI) systems' in LEONDES, C. T. (Ed.): 'Dynamics and control, vol. 74, Techniques in discrete and continuous robust systems' (Academic Press, 1996), pp. 171–224

5 BILLINGS, S., and LEONTARITIS, I.: 'An input-output parametric model for non-linear systems. Part I: deterministic non-linear systems', *International Journal of Control*, 1985, **41**, pp. 303–28

6 KORBICZ, J., and CEMPEL, C. (Eds): 'Analytical and knowledge-based redundancy in fault detection and diagnosis', *Applied Mathematics and Computer Science, Special issue*, 1993, **3** (3)

7 PATTON, R. J., and CHEN, J.: 'Neural networks based fault diagnosis for non-linear dynamic systems'. AIAA Guidance, Navigation and Control Conference, Baltimore, MD, 7–10 August, 1995, paper AIAA-95-3219-CP

8 PATTON, R. J., and KORBICZ, J. (Eds): 'Advances in computational intelligence for fault diagnosis systems', *International Journal of Applied Mathematics and Computer Science, Special issue*, 1999, **9** (3), pp. 468–735

9 BROWN, M., and HARRIS, C. J.: 'Neuro-fuzzy adaptive modelling and control' (Prentice-Hall, NJ, 1994)

10 RUTKOWSKI, L.: 'Flexible neuro-fuzzy systems' (Kluwer Academic Publishers, Boston, 2004)

11 DAMADICS: 'Website of the research training network on *Development and application of methods for actuator diagnosis in industrial control systems*', 2002, http://diag.mchtr.pw.edu.pl/damadics

12 CHEN, J., and PATTON, R. J.: 'Robust model-based fault diagnosis for dynamic systems' (Kluwer Academic Publishers, London, 1999)

13 CHEN, J., PATTON, R. J., and ZHANG, H.: 'Design of unknown input observers and fault detection filters', *International Journal of Control*, 1996, **63**, pp. 85–105

14 CHEN, J., PATTON, R. J., and LIU, G. P.: 'Optimal residual design for fault diagnosis using multi-objective optimization', *International Journal of Systems Science*, 1996, **27** (6), pp. 567–76

15 KOWALCZUK, Z., and BIALASZEWSKI, T.: 'Genetic algorithms in the multi-objective optimisation of fault detection observers', in KORBICZ, J., KOŚCIELNY, J. M., KOWALCZUK, Z., and CHOLEWA, W. (Eds): 'Fault diagnosis. Models, artificial intelligence, applications' (Springer-Verlag, Berlin, 2004)

16   OBUCHOWICZ, A.: 'Evolutionary methods in designing diagnostic systems', in KORBICZ, J., KOŚCIELNY, J. M., KOWALCZUK, Z., and CHOLEWA, W. (Eds): 'Fault diagnosis. Models, artificial intelligence, applications (Springer-Verlag, Berlin, 2004)

17   MICHALEWICZ, Z.: 'Genetic algorithms + data structures = evolution programs' (Springer-Verlag, Berlin, 1996)

18   WITCZAK, M., and KORBICZ, J.: 'Observers and genetic programming in the indentification and fault diagnosis of non-linear dynamic systems', in KORBICZ, J., KOŚCIELNY, J. M., KOWALCZUK, Z., and CHOLEWA, W. (Eds): 'Fault diagnosis. Models, artificial intelligence, applications' (Springer-Verlag, Berlin, 2004)

19   KOZA, J. R.: 'Genetic programming: on the programming of computers by means of natural selection' (The MIT Press, Cambridge, 1992)

20   WITCZAK, M., OBUCHOWICZ, A., and KORBICZ, J.: 'Genetic programming based approaches to identification and fault diagnosis of non-linear dynamic systems', *International Journal of Control*, 2002, **75** (13), pp. 1012–31

21   WITCZAK, M.: 'Identification and fault detection of non-linear dynamic systems' (University of Zielona Góra Press, Zielona Góra, 2003)

22   FRANK, P. M., and KÖPPEN-SELINGER, B.: 'New developments using AI in fault diagnosis', *Engineering Applications and Artificial Intelligence*, 1997, **10** (1), pp. 3–14

23   KOSIŃSKI, W., MICHALEWICZ, Z., WEIGL, M., and KOLESNIK, R.: 'Genetic algorithms for preprocessing of data for universal approximators'. Proceedings of the seventh international symposium on *Intelligent information systems*, Malbork, Poland, 1998, pp. 320–31

24   CARSE, B., FOGARTY, T. C., and MUNRO, A.: 'Evolving fuzzy rule based controllers using genetic algorithms', *Fuzzy Sets and Systems*, 1996, **80** (3), pp. 273–93

25   KÖPPEN-SELINGER, B., and FRANK, P. M.: 'Fuzzy logic and neural networks in fault detection', in JAIN, L. C., and MARTIN, N. M. (Eds): 'Fusion of neural networks, fuzzy sets, and genetic algorithms' (CRC Press, New York, 1999), pp. 169–209

26   ISERMANN, R., and BALLÉ, P.: 'Trends in the application of model-based fault detection and diagnosis of technical processes', *Control Engineering Practice*, 1997, **5** (5), pp. 709–19

27   UCIŃSKI, D.: 'Optimal measurements methods for distributed parameter system identification' (CRC Press, New York, 2004)

28   MILANESE, M., NORTON, J., PIET-LAHANIER, H., and WALTER, E. (Eds): 'Bounding approaches to system identification' (Plenum Press, New York, 1996)

29   GUERNEZ, C., CASSAR, J. Ph., and STAROSWIECKI, M.: 'Extension of parity space to non-linear polynomial dynamic systems'. Proceedings of IFAC symposium on *Fault detection, supervision and safety of technical processes: SAFEPROCESS '97*, Hull, UK, 1997, vol. 2, pp. 861–6

30   YU, D. L., and SHIELDS, D.: 'Extension of the parity-space method to fault diagnosis of bilinear systems', *International Journal of Systems Science*, 2001, **32** (8), pp. 953–62

31   KRISHNASWAMI, V., and RIZZONI, G.: 'Non-linear parity equation residual generation for fault detection and isolation'. Proceeding of IFAC symposium on *Fault detection, supervision and safety of technical processes, SAFEPROCESS '94*, Espoo, Finland, 1994, vol. 1, pp. 317–22

32   SHUMSKY, A. Ye.: 'Robust residual generation for diagnosis of non-linear systems: parity relation approach'. Proceedings of IFAC symposium on *Fault detection, supervision and safety of technical processes, SAFEPROCESS '97*, Hull, UK, 1997, vol. 2, pp. 867–72

33   FRANK, P. M.: 'Fault diagnosis in dynamic system via state estimation – a survey' in TZAFESTAS, S., SINGH, M. and SCHMIDT, G., (Eds): 'System fault diagnostics, reliability and related knowledge-based approaches, vol. 1' (D. Reidel Press, Dordrecht, 1987), pp. 35–98

34   KORBICZ, J., FATHI, Z., and RAMIREZ, W. F.: 'State estimation schemes for fault detection and diagnosis in dynamic systems', *International Journal of Systems Science*, 1993, **24** (5), pp. 985–1000

35   ANDERSON, B. D. O., and MOORE, J. B.: 'Optimal filtering' (Prentice-Hall, NJ, 1979)

36   BOUTAYEB, M., and AUBRY, D.: 'A strong tracking extended Kalman observer for non-linear discrete-time systems', *IEEE Transactions on Automatic Control*, 1999, **44**, pp. 1550–6

37   THAU, F. E.: 'Observing the state of non-linear dynamic systems', *International Journal of Control*, 1973, **17** (3), pp. 471–9

38   HAC, A.: 'Design of disturbance decoupled observers for bilinear systems', *ASME Journal of Dynamic Systems, Measurement, and Control*, 1992, **114**, pp. 556–62

39   SHIELDS, D. N., and ASHTON, S.: 'A fault detection observer method for non-linear systems'. Proceedings of IFAC symposium on *Fault detection, supervision and safety of technical processes: SAFEPROCESS 2000*, Budapest, Hungary, 2000, vol. 1, pp. 226–31

40   SHILEDS, D., and YU, S.: 'Fault detection observers for continuous non-linear polynomial systems of general degree', *International Journal of Control*, 2003, **76** (5), pp. 437–52

41   YU, D., and SHIELDS, D. N.: 'Bilinear fault detection observer and its application to a hydraulic system', *International Journal of Control*, 1996, **64**, pp. 1023–47

42   HOU, M., and PUGH, A. C.: 'Observing state in bilinear systems: an UIO approach'. Proceedings of IFAC Symposium on *Fault detection, supervision and safety of technical processes: SAFEPROCESS '97*, Hull, UK, 1997, vol. 2, pp. 783–8

43   ALCORTA GARCIA, E., and FRANK, P. M.: 'Deterministic non-linear observer-based approaches to fault diagnosis', *Control Engineering Practice*, 1997, **5**, pp. 663–70

44   SELIGER, R., and FRANK, P.: 'Robust observer-based fault diagnosis in non-linear uncertain systems', in PATTON, R. J., FRANK, P., and CLARK, R. N. (Eds): 'Issues of fault diagnosis for dynamic systems' (Springer-Verlag, Berlin, 2000)

45   ZOLGHARDI, A., HENRY, D., and MONISION, M.: 'Design of non-linear observers for fault diagnosis. A case study', *Control Engineering Practice*, 1996, **4**, pp. 1535–44

46   CALADO, J. M. F., KORBICZ, J., PATAN, K., PATTON, R.J., and SÁ DA COSTA, J. M. G.: 'Soft computing approaches to fault diagnosis for dynamic systems', *European Journal of Control*, 2001, **7** (2–3), pp. 248–86

47   FRANK, P. M., and MARCU, T.: 'Diagnosis strategies and systems. Principles, fuzzy and neural approaches', in TEODORESCU, H.-N., MLYNEK, D., KANDEL, A., and ZIMMERMANN, H.-J. (Eds): 'Intelligent systems and interfaces' (Kluwer Academic Publishers, Boston, 2000)

48   GUPTA, M. M., JIN, L., and HOMMA, N.: 'Static and dynamic neural networks' (John Wiley & Sons, Hoboken, NJ, 2003)

49   OBUCHOWICZ, A.: 'Evolutionary algorithms for global optimization and dynamic system diagnosis' (Lubuskie Scientific Society, Zielona Góra, 2003)

50   RUTKOWSKA, D.: 'Neuro-fuzzy architectures and hybrid learning' (Physica-Verlag, Heidelberg, 2000)

51   RUTKOWSKI, L.: 'New soft computing techniques for system modelling, pattern classification and image processing' (Springer-Verlag, Berlin, Heidelberg, 2004)

52   AYOUBI, M.: 'Non-linear system identification based on neural networks with locally distributed dynamics and application to technical processes' (VDI-Verlag, Forstschritt-Berichte VDI, Reihe 8, No. 591, Düsseldorf, 1996)

53   KOIVO, H. N.: 'Artificial neural networks in fault diagnosis and control', *Control Engineering Practice*, 1994, **2** (7), pp. 90–101

54   KORBICZ, J., PATAN, K., and OBUCHOWICZ, A.: 'Neural network fault detection systems for dynamic processes', *Bulletin of the Polish Academy of Sciences, Technical Sciences*, 2001, **49** (2), pp. 301–21

55   SORSA, T., and KOIVO, H. N.: 'Application of artificial neural networks in process fault diagnosis', *Automatica*, 1993, **29** (4), pp. 843–9

56   HAYKIN, S.: 'Neural networks. A comprehensive foundation' (Prentice-Hall, Upper Saddle River, NJ, 1999)

57   MARCU, T., MIREA, L., and FRANK, P. M.: 'Development of dynamic neural networks with application to observer-based fault detection and isolation', *Applied Mathematics and Computer Science*, 1999, **9** (3), pp. 547–70

58   PATAN, K., and KORBICZ, J.: 'Artificial neural networks in fault diagnosis', in KORBICZ, J., KOŚCIELNY, J. M., KOWALCZUK, Z., and CHOLEWA, W. (Eds): 'Fault diagnosis. Models, artificial intelligence, applications' (Springer-Verlag, Berlin, Heidelberg, 2004), pp. 333–80

59   FARLOW, S. J. (Ed.): 'Self-organizing methods in modeling – GMDH type algorithms' (Marcel Dekker, New York, 1984)

60 IVAKHNENKO, A. G.: 'Polynomial theory of complex systems', *IEEE Transactions on Systems, Man and Cybernetics*, 1971, **SMC–1** (4), pp. 44–58

61 PHAM, D. T., and XING, L.: 'Neural networks for identification, prediction and control' (Springer-Verlag, London, 1995)

62 MUELLER, J. E., and LEMKE, F.: 'Self-organising data mining' (Libri, Hamburg, 2000)

63 WITCZAK, M., KORBICZ, J., MRUGALSKI, M., and PATTON, R. J.: 'GMDH neural network-based approach to robust fault detection and its application to solve the DAMADICS benchmark problem', Control Engineering Practice, CEP 2004 (accepted for publication to be published 2005)

64 LJUNG, L.: 'System identification. Theory for the users' (Prentice-Hall, NJ, 1987)

65 KOHONEN, T. K.: 'Self-organization and associative memory' (Springer-Verlag, NY, 1989)

66 MARCINIAK, A., and KORBICZ, J.: 'Diagnosis system based on multiple neural classifiers', *Bulletin of Polish Academy of Sciences, Technical Sciences*, 2001, **49** (4), pp. 681–702

67 KOŚCIELNY, J. M., and BARTYŚ, M.: 'Application of information system theory for actuator diagnosis'. Proceedings of the fourth IFAC symposium on *Fault detection, supervision and safety for technical processes, SAFEPROCESS*, Budapest, Hungary, 2000, vol. 2, pp. 949–54

68 AYOUBI, M.: 'Neuro-fuzzy structure for rule generation and application in the fault diagnosis of technical processes'. Proceedings of the American control conference, *ACC*, Washington, 1995, pp. 2757–61

69 CALADO, J. M. F., and SÁ DA COSTA, J. M. G.: 'An expert system coupled with a hierarchical structure of fuzzy neural networks for fault diagnosis', *Journal of Applied Mathematics and Computer Science*, 1999, **9** (3), pp. 667–88

70 JANG, J. S. R.: 'ANFIS: adaptive-network-based fuzzy inference systems', *IEEE Transactions on Systems, Man and Cybernetics*, 1993, **23**, pp. 665–85

71 JANG, J. S. R., SUN, C. T., and MIZUTANI, E.: 'Neuro-fuzzy and soft computing' (Prentice-Hall, NJ, 1997)

72 PATTON, R. J., LOPEZ-TORIBIO, C. J., and UPPAL, F. J.: 'Artificial intelligence approaches to fault diagnosis for dynamic systems', *International Journal of Applied Mathematics and Computer Science*, 1999, **9** (3), pp. 471–518

73 JANG, J. S. R., and SUN, C. T.: 'Functional equivalence between radial basis function networks and fuzzy systems', *IEEE Transactions on Neural Networks*, 1993, **4**, pp. 156–8

74 UPPAL, F. J., and PATTON, R. J.: 'Fault diagnosis of an electro-pneumatic valve actuator using neural networks with fuzzy capabilities'. Proceedings of the tenth European symposium on *Artificial neural networks, ESANN*, Bruges, Belgium, 24–26 April 2002

75 SIMPSON, P. K.: 'Fuzzy min–max neural networks – Part 1: classification', *IEEE Transactions on Neural Networks*, 1992, **3**, pp. 776–86

76   SIMPSON, P. K.: 'Fuzzy min–max neural networks – Part 2: clustering', *IEEE Transactions on Fuzzy Systems*, 1993, **1**, pp. 32–45

77   LIN, C. T., and LEE, C. S. G.: 'Neural network based fuzzy logic control and decision system', *IEEE Transactions on Computer*, 1991, **40**, pp. 1320–36

78   WANG, L. X., and MENDEL, J.: 'Generating fuzzy rules by learning from examples', *IEEE Transactions on Systems, Man, and Cybernetics*, 1992, **22**, pp. 1414–27

79   FARAG, W. A., QUINTANA, V. H., and LAMBERT-TORRES, G.: 'A genetic-based neuro-fuzzy approach for modelling and control of dynamical systems', *IEEE Transactions on Neural Networks*, 1998, **9** (5), pp. 756–67

80   BOSSLEY, K. M.: 'Neurofuzzy modelling approaches in system identification'. Ph.D. thesis, University of Southampton, UK, 1997

81   NAUCK, D., and KRUSE, R.: 'NEFCLASS-X – a soft computing tool to build readable fuzzy classifiers', *BT Technology Journal*, 1998, **16** (3), pp. 180–90

82   ANDREWS, R., and GEVA, S.: 'Rule extraction from local cluster neural nets', *Neurocomputing*, 2002, **47**, pp. 1–20

83   BERINJI, H. R., and KHEDKAR, P.: 'Learning and tuning fuzzy logic controllers through reinforcements', *IEEE Transactions on Neural Networks*, 1992, **3** (5), pp. 7254–740

84   TAKAGI, H., and HAYASHI, I.: 'NN-driven fuzzy reasoning', *International Journal of Approximate Reasoning*, 1991, **5** (3), pp. 191–212

85   HALGAMUGE, S. K., and GLESNER, M.: 'FuNe deluxe: a group of fuzzy-neural methods for complex data analysis problems'. European congress on Fuzzy and intelligent technologies '95, Aachen, Germany, 1995

86   TSCHICHOLD-GÜRMAN, N.: 'Generation and improvement of fuzzy classifiers with incremental learning using fuzzy RuleNet'. Proceedings of ACM symposium on *Applied computing*, Nashville, Tennessee, USA, 26–28 February 1995, pp. 466–70

87   KHAN, E., and VENKATAPURAM, P.: 'Neufuz: neural network based fuzzy logic design algorithms'. Proceedings of IEEE international conference on *Fuzzy systems*, San Francisco, CA, 1993, vol. I, pp. 647–54

88   HAYASHI, Y., CZOGALA, E., and BUCKLEY, J. J.: 'Fuzzy neural controller'. Proceedings of IEEE international conference on *Fuzzy systems*, San Diego, 1992, pp. 197–202

89   CARPENTER, G. A., and GROSSBERG, S.: 'Fuzzy ART: fast stable learning and categorization of analog patterns by an adaptive resonance system', *Neural Networks*, 1991, **4**, pp. 759–71

90   BROWN, M., and HARRIS, C. J.: 'The modelling abilities of the binary CMAC'. IEEE international conference on *Neural networks*, 1994, pp. 1335–9

91   ROAN, S. M., CHIANG, C. C., and FU, H. C.: 'Fuzzy RCE neural network'. Proceedings of IEEE international conference on *Fuzzy systems*, San Francisco, CA, 1993, vol. I, pp. 629–34

92   BENKHEDDA, H., and PATTON, R. J.: 'Information fusion in fault diagnosis based on B-spline networks'. Proceedings of IFAC symposium on *Fault*

*detection, supervision and safety for technical processes, SAFEPROCESS '97*, Pergamon Press, University of Hull, UK, 1997, pp. 681–7

93   UPPAL, F. J., and PATTON, R. J.: 'Application of B-spline neuro-fuzzy networks for fault detection and isolation'. Proceedings of the fourth IFAC symposium on *Fault detection supervision and safety for technical processes, SAFEPROCESS 2000*, Budapest, Hungary, 14–16 June 2000

94   UPPAL, F. J., and PATTON, R. J.: 'Application of B-spline neuro-fuzzy networks to identification, fault detection and isolation'. Proceedings of IEE international conference on *UKACC CONTROL 2000*, University of Cambridge, UK, 4–7 September 2000

95   TACHIBANA, K., and FURUHASHI, T.: 'A hierarchical fuzzy modelling method using genetic algorithm for identification of concise submodels'. Proceedings of second international conference on *Knowledge-based intelligent electronic systems*, Adelaide, Australia, 1994

96   OZYURT, B., and KANDEL, A.: 'A hybrid hierarchical neural network-fuzzy expert system approach to chemical process fault diagnosis', *Fuzzy Sets and Systems*, 1996, **83** (1), pp. 11–25

97   SUGENO, M., and KANG, G.: 'Structure identification of fuzzy model', *Fuzzy Sets and Systems*, 1988, **26** (1), pp. 15–33

98   TAKAGI, T., and SUGENO, M.: 'Fuzzy identification of systems and its applications to modelling and control', *IEEE Transactions on Systems, Man, and Cybernetics*, 1985, **15**, pp. 116–32

99   MAMDANI, E.: 'Advances in the linguistic synthesis of fuzzy controllers', *International Journal of Man–Machine Studies*, 1976, **8**, pp. 669–78

100  FÜSSEL, D., BALLÉ, P., and ISERMANN, R.: 'Closed-loop fault diagnosis based on a nonlinear process model and automatic fuzzy rule generation'. Proceedings of IFAC symposium on *Fault detection, supervision and safety for technical processes, SAFEPROCESS '97*, Pergamon Press, University of Hull, UK, 1997

101  PALADE, V., PATTON, R. J., UPPAL, F. J., QUEVEDO, J., and DALEY, S.: 'Fault diagnosis of an industrial gas turbine using neuro-fuzzy methods'. Proceedings of the 15th triennial IFAC World Congress, Barcelona, Spain, 21–26 July 2002

102  UPPAL, F. J., and PATTON, R. J.: 'Structure optimisation in neuro-fuzzy modelling for fault diagnosis'. Proceedings of the second design and application of methods for actuator diagnosis in industrial control systems (DAMADICS) workshop on *Neural networks methods for modelling and fault diagnosis*, DIST-University of Genova, Genova, Italy, 25–26 May 2001

103  UPPAL, F. J., PATTON, R. J., and PALADE, V.: 'Neuro-fuzzy based fault diagnosis applied to an electro-pneumatic valve'. Proceedings of the 15th triennial IFAC World Congress, Barcelona, Spain, 21–26 July 2002

104  SUGENO, M., and YASUKAWA, T.: 'A fuzzy logic based approach to qualitative modelling', *IEEE Transactions on Fuzzy Systems*, 1993, **1**, pp. 7–31

105  JANG, J. S. R.: 'Structure determination in fuzzy modelling: a fuzzy CART approach'. IEEE international conference on *Fuzzy systems*, Orlando, FL, 1994

106   DASGUPTA, D., and MICHALEWICZ, Z. (Eds): 'Evolutionary algorithms in engineering applications' (Spring-Verlag, New York, 1997)

107   BABUSKA, R.: 'Fuzzy modelling for control' (Kluwer Academic Publishers, Boston, MA, 1998)

108   NELLES, O., and ISERMANN, R.: 'Basis function networks for interpolation of local linear models'. IEEE conference on *Decision and control (CDC)*, Kobe, Japan, 1996, pp. 470–5

109   BEZDEK, C. J.: 'Pattern recognition with fuzzy objective function algorithm' (Plenum Press, New York, 1981)

110   GATH, I., and GEVA, A.: 'Unsupervised optimal fuzzy clustering', *IEEE Transactions on Pattern Analysis and Machine Intelligence*, 1989, **7**, pp. 773–81

111   KRISHNAPURAM, R., and FREG, C. P.: 'Fitting an unknown number of lines and planes to image data through compatible cluster merging', *Pattern Recognition*, 1992, **25** (4), pp. 385–400

112   UPPAL, F. J., and PATTON, R. J.: 'A hybrid fault diagnosis approach applied to electro-pneumatic valves in sugar process', *International Journal of Adaptive Control and Signal Processing*, accepted for publication in a special issue on *Condition monitoring*, 2005

113   WANG, H. O., TANAKA, K., and GRIFFIN, M. F.: 'Parallel distributed compensation of a non-linear systems by Takagi and Sugeno fuzzy models'. Proceedings of *FUZZ-IEEE/IFES '95*, 1995, pp. 531–8

114   TANAKA, K., TAKAYUKI, I., and WANG, H. O.: 'Design of fuzzy control systems based on relaxed LMI stability conditions'. Proceedings of the 35th CDC, Kobe, 1996, pp. 598–603

115   GUSTAFSON, D., and KESSEL, W.: 'Fuzzy clustering with a fuzzy covariance matrix'. Proceedings of IEEE CDC, 1979, San Diego, CA, USA, pp. 761–6

116   BASSEVILLE, M., and NIKIFOROV, I. V.: 'Detection of abrupt changes, theory and application' (Prentice-Hall, NJ, 1993)

*Chapter 12*

# Application of intelligent control to autonomous search of parking place and parking of vehicles

*F. Gómez-Bravo, F. Cuesta and A. Ollero*

## 12.1 Introduction

Parallel and diagonal parking of conventional vehicles is a non-trivial task which usually requires significant training when human drivers are involved. Nevertheless, intelligent navigation systems can manage this situation by taking advantage of knowledge of expert drivers and by using the kinematic constraints of the vehicles. These constraints are usually expressed by non-integrable equations on the generalised velocity vector, known as non-holonomic constraints. Control and planning motion of such systems should take into account these constraints.

Different approaches have been applied to non-holonomic autonomous vehicles manoeuvring [1–3]. There are also several approaches for parking manoeuvres based on planning techniques. On the one hand, some of these approaches consider environment geometrical constraints in the design of a discontinuous curvature collision-free manoeuvre by using arcs of circumference [4,5]. On the other hand, some developments, which are based on control theory, focus their attention on direct continuous curvature path generation [1–3,6,7]. A particular application of these techniques to parallel parking manoeuvre generation is presented in References 8 and 9.

The above strategies are based on the representation of the environment by means of models and the computation of a collision-free path to reach the goals [10]. These approaches are appropriated when an accurate model of the environment is known. Nevertheless, in applications where a suitable model of the environment is difficult to obtain, due to the sensors' capabilities, computational resources or dynamic environment, a reactive approach based on the ability to react to the continuous sensorial information, could represent a better option than planning a path [11]. Fuzzy logic has been widely applied to perform reactive navigation [12, and the references therein; 13–18]. Moreover, autonomous parking by using fuzzy controllers has been also reported [19–23].

However, in several situations a combination of both techniques, planned and reactive, could be the most convenient solution.

The method presented in this chapter is based on the autonomous search of the parking place, and the execution of the parking manoeuvre, combining reactive and planned navigation. This approach takes advantage of the reactive navigation robustness for some tasks, such as navigating looking for a parking place, stopping the vehicle over a selected point, etc., and applies planned navigation to consider the non-holonomic constraints and to perform an accurate navigation when high precision is required (manoeuvring in a constrained environment, for instance) [18,24–26].

The control architecture is based on the combination of multiple simultaneous behaviours working in a cooperative scheme. The behaviours' blending is performed by means of fuzzy logic weighting the behaviours by considering their applicability in the actual context of the vehicle. Furthermore, it incorporates a sequential controller which takes into account the different phases of the parking manoeuvre.

One of the most outstanding characteristics of this architecture is the incorporation of a parking manoeuvre selector based on a fuzzy inference system. In practice, it is interesting to select a parking manoeuvre from a solution set. Such selection should be made by considering the environment where the vehicle has to park, in the same way that a human driver selects the stopping point (to start the backward motion into the parking place) depending on the characteristics of the parking place. Fuzzy logic can be applied to consider the heuristic involved in this process in order to perform a more robust and safer parking manoeuvre. Moreover, the system is able to select continuous or discontinuous curvature parking manoeuvres to improve performance.

The application of this technique is illustrated by experiments with the electrically powered vehicles ROMEO-3R and ROMEO-4R (see Figure 12.14 and the videos at http://www.esi2.us.es/~fcuesta/videos/parking.htm) designed and built at the University of Seville for the experimentation of intelligent components and autonomous navigation strategies [27].

This chapter is organised as follows: the next section presents a general overview of the intelligent system for parking of vehicles. The main components and features of the system are also highlighted. Section 12.3, is devoted to introducing the proposed technique for planning of parking manoeuvres. An overview of some planning methods for non-holonomic vehicles is also included. Section 12.4 illustrates how the control system is able to select the most convenient manoeuvre according to the environment: First, a fuzzy system incorporating the knowledge of expert drivers selects the optimal manoeuvre. Second, the system selects to perform a continuous or a discontinuous curvature manoeuvre to improve performance. In Section 12.5, some experimental results with the ROMEO vehicles are shown. The chapter ends with the conclusions and references.

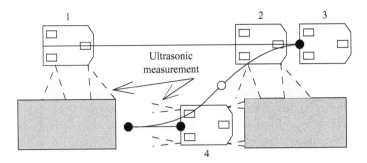

*Figure 12.1    Steps of the parking execution*

## 12.2    Intelligent system for parking of vehicles

The method presented in this chapter is based on the autonomous search of the parking place, and the execution of the parking manoeuvre, combining reactive and planned navigation. A mixed planned–reactive control architecture which takes advantage of the reactive navigation robustness and the planned navigation accuracy is proposed. Reactive navigation is applied when low precision is required for some task such as navigating looking for a parking place. At the same time, planned navigation allows to consider the non-holonomic constraints of the vehicle, and to perform an accurate navigation where high precision is required due to the characteristics of the manoeuvre.

Unlike other approaches, this architecture can be easily extended for parking more complex vehicles. For instance, parallel parking of tractor trailer vehicles has also been performed by means of this procedure [28].

Roughly speaking, the system works as follows: the vehicle is driven looking for a parking place (step 1 in Figure 12.1), paying attention to lateral ultrasonic measurement. In this phase three behaviours cooperate: line of vehicles following, obstacle avoidance and lateral parking manoeuvre. The parallel parking behaviour is only searching for a place large enough to park. The line of vehicles is followed navigating reactively. When the parking place is found (step 2 in Figure 12.1), the size of the parking place is estimated [22].

If the place is large enough then the lateral parking behaviour stops the vehicle (step 3 in Figure 12.1) and a collision-free manoeuvre is generated (see Section 12.3 for details). Different collision-free manoeuvres could be performed; however, a fuzzy logic system is used to select the best manoeuvre taking into account the environment (see Section 12.4.1 for details). Then, from the original manoeuvre a continuous curvature path can be generated [29]. Depending on the vehicle steering system, the new path could be followed whether or not stopping the vehicle. The system evaluates this issue and decides whether the continuous or the discontinuous curvature manoeuvre is followed (see Section 12.4.2 for details). Finally, the vehicle executes the selected manoeuvre, navigating from one stopping point to the next one, until the final point is achieved (step 4 in Figure 12.1). During this last procedure, rear and forward ultrasonic measurements are taken into account in order to stop the vehicle if any obstacle appears behind it. Then a sequential hybrid control is applied in order to navigate from a stopping point to the next one until the manoeuvre final configuration is achieved [26].

The autonomous navigation is based on the combination of behaviours to reach an objective or to execute a plan. In this strategy each behaviour working in parallel produces its own motion control command, named $c_i$. The output of each behaviour is weighted by a hybrid sequential controller. This sequential controller takes into account the situation the vehicle is in and determines the behaviour weights ($BW_i$) according to the context. The outputs of the behaviours are combined by the motion control law:

$$c = \frac{\sum_i BW_i \cdot c_i}{\sum_i BW_i}, \quad i = \text{flw, frw, oa}, \ldots.$$

On the other hand, several path tracking strategies have been implemented including Pure Pursuit, Generalized Predictive Path Tracking [30] and Direct Fuzzy Control [31]. For low speed manoeuvres, the results of the three methods are similar. Then, the simplest pure pursuit is usually the best option. It can be easily shown that the pure pursuit control law is given by:

$$\gamma = -\frac{2x_c}{D^2},$$

where $\gamma$ is the curvature commanded to the vehicle, $x_c$ is the lateral displacement of the vehicle with respect to a target point in the path and $D$ is the lookahead distance from the vehicle to this target point. The value of the lookahead $D$ is the key point in the algorithm. A large value of $D$ could bring out a bad tracking performance on curves. On the other hand, a small value could produce oscillations in the path tracking.

The following sections explain in detail the different procedures involved in the system.

## 12.3   Planning of parking manoeuvres for vehicles

### 12.3.1   *Manoeuvring in car and cart-like vehicles*

In recent years, many applications on autonomous vehicles have been developed from the adaptation of car and cart-like vehicles [22]. The kinematics of these types of vehicles can be expressed in the form [2]

$$\begin{bmatrix} \dot{x} \\ \dot{y} \\ \dot{z} \end{bmatrix} = \begin{bmatrix} \cos\theta & 0 \\ \sin\theta & 0 \\ 0 & 1 \end{bmatrix} \cdot \begin{bmatrix} v(t) \\ v(t) \cdot \rho(t) \end{bmatrix},$$

where $(x, y)$ are the coordinates of the reference point $i$ (see Figure 12.2), $\theta$ is the vehicle's heading, $v(t)$ is the linear velocity of point $i$ in the time $t$ and $\rho(t)$ is the curvature of the path described by point $i$.

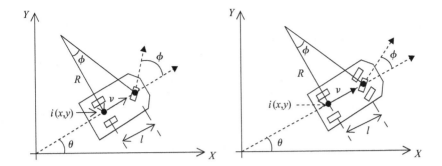

*Figure 12.2    Car and cart-like vehicles*

The non-holonomic constraint of this model is written as:

$$\dot{x} \sin \theta - \dot{y} \cos \theta = 0.$$

The curvature radius can be written as:

$$R(t) = \frac{1}{\rho(t)} = \frac{l}{\tan \phi}.$$

The value of $R(t)$ is usually bounded by $R_{\min}$, the minimum curvature radius. If $\phi$ is included in the state vector the model can be written in the form:

$$\begin{bmatrix} \dot{x} \\ \dot{y} \\ \dot{\theta} \\ \dot{\phi} \end{bmatrix} = \begin{bmatrix} \cos \theta & 0 \\ \sin \theta & 0 \\ \dfrac{\tan \phi}{l} & 0 \\ 0 & 1 \end{bmatrix} \cdot \begin{bmatrix} v(t) \\ \dot{\phi} \end{bmatrix}.$$

Now two non-holonomic kinematics constraints exist:

$$\dot{x} \sin \theta - \dot{y} \cos \theta = 0,$$

$$\dot{x} \sin(\phi + \theta) - \dot{y} \cos(\phi + \theta) - \dot{\theta} l \cos \phi = 0.$$

Generally speaking, the path planning problem of a non-holonomic vehicle consists of finding a path that connects the initial configuration to the final configuration and satisfies the existing non-holonomic constraints. Usually, collision-free path generation is also considered in this problem.

Different approaches have been proposed in the literature in order to solve this planning problem.

Several existing methods are based on the application of non-linear control theory. Applying this strategy to a non-holonomic system requires taking into account Brocket's condition. It can be shown that, for these systems, there is non-continuous smooth control law which stabilizes a point [32]. Different transformations in the kinematics models should be considered in order to accomplish Brocket's condition [33] or to allow open loop optimal controls, as in the case of chained form and sinusoidal inputs [1,6,34,35]. These methods do not consider explicitly collision avoidance; they

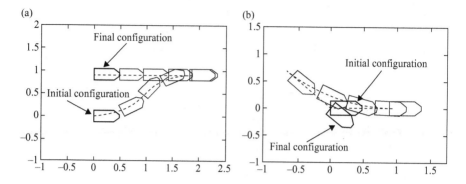

*Figure 12.3*　*Restricted manoeuvres changing (a) 'y' and (b) 'θ' coordinates*

usually apply techniques for collision detection once the path is generated, and obtain collision-free paths by means of an iterative algorithm.

Collision-free path generation for non-holonomic vehicles can be also considered by means of a searching algorithm. These methods decompose the configuration space into an array of small rectangloids and search for collision-free configurations in a graph whose nodes are these rectangloids [10,36]. There are other searching probabilistic methods such as RRT [37] which can be extended to solve this problem.

Car manoeuvring has been also analysed by means of fuzzy systems to incorporate expert knowledge into fuzzy sets. Genetic algorithms have been applied to optimise the rule base of the fuzzy controller [38].

Other approaches propose piece-wise path generation based on manoeuvres composed of circles and straight line segments following the ideas of Dubins [39] as well as Reeds and Shepp [40]. The problem can be solved in two steps [2]. First a collision-free path is constructed without considering the non-holonomic constraints. Second this path is transformed into a new path by linking the intermediate configurations in such a way that the non-holonomic constraints are accomplished.

Linking intermediate configurations can be performed in different ways. On one hand, shorter path can be obtained applying Reeds and Shepp's method [2]. On the other hand, canonical paths can be considered [7]. In both cases complete collision avoidance is obtained by applying a collision detection algorithm along the linked paths.

Finally, a new type of canonical manoeuvre, called restricted manoeuvre [25], can be also applied to the linking process. By this approach, geometric constraints can be incorporated in the manoeuvre generation process. Thus, collision-free paths can be obtained without any further collision test, which represents an outstanding advantage when real time processes are involved. A restricted manoeuvre represents a path in which some variables (independent variables) follow a closed trajectory; meanwhile, another variable (dependent variable) follows a trajectory which provides the desired increment in that variable. For instance, in Figure 12.3 car manoeuvres to change $y$ and $\theta$ are shown, respectively. As will be shown in the next subsection parallel parking

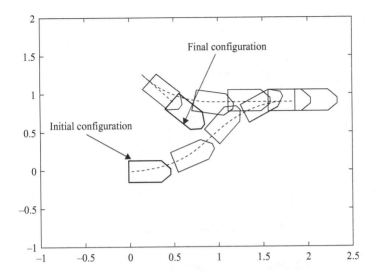

*Figure 12.4    Concatenation of several restricted manoeuvres*

can be performed by means of the application of manoeuvre of Figure 12.3(a) and diagonal parking by means of the application of manoeuvre of Figure 12.3(b).

Restricted manoeuvres can be concatenated in order to perform a more complex manoeuvre. Thus any final configuration can be achieved by a sequence of the afore-mentioned manoeuvres. Namely, Figure 12.4 shows a manoeuvre obtained by the combination and further simplification of three restricted manoeuvres: first changing $y$, second changing $x$ and finally a manoeuvre for changing $\theta$.

## 12.3.2    Parking manoeuvres

Manoeuvres for parallel and diagonal parking are particular cases of non-holonomic motion. Several of the aforementioned planning methods have been applied in a specific parking environment for solving parallel parking of different vehicles [6,10,35].

Some approaches allow parallel parking using successive sequential manoeuvres by means of sinusoidal paths in order to converge from a starting point to a goal point [8]. For each manoeuvre, the path is tested against collisions. Nevertheless, there is no way to know if there exists a collision-free solution, from a particular starting point. If the solution is not found the vehicle has to move to another starting point and repeat the whole process from the new point. In Reference 9, an offline computation stored in a look-up table is used in order to obtain a convenient start location.

Car parallel parking has also been solved by means of a fully reactive tech-nique [21]. In this approach, manoeuvres are generated by a fuzzy inference system from the perception of the onboard sensors. Environment and obstacles are modelled as a vector field, yielding a fuzzy perception model of the near environment.

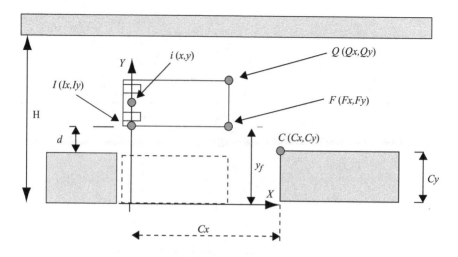

*Figure 12.5    Right-hand side parallel parking definition*

The parking manoeuvre is divided into three phases. Each one is associated with a set of rules. Activation of each phase is accomplished taking into account the fuzzy environment description [21]. However, this approach only considers as an obstacle the forward parked vehicle. There is no certainty about the manoeuvre length.

Several authors propose piece-wise manoeuvres composed of circles and straight line segments with collision avoidance based on geometrical issues [4,5]. However, they only deal with minimum constant value of the curvature, and consider minimal length paths, following Dubins as well as Reeds and Shepp results. Moreover, generation of continuous curvature manoeuvres is not considered.

The approach presented in this chapter is based on the concept of restricted manoeuvre. It provides not only one optimal manoeuvre but a set of collision-free manoeuvres and allows one to apply different optimal criteria.

Assume rectangular models of vehicles and obstacles, and that right-hand side parking is considered, as illustrated in Figure 12.5.

In order to park the vehicle, a restricted manoeuvre has to be designed in such a way that the initial and final values of variables $\theta$ and $x$ are the same, but the value of $y$ is changed. The key of the manoeuvre design consists of planning a collision-free path for a virtual vehicle leaving the parking place (see Reference 25 for details). This planning procedure provides a set of collision-free parking manoeuvres where two optimal solutions (minimum forward displacement and minimum lateral displacement) can be found; these manoeuvres are shown in Figures 12.6(a) and (b), respectively.

This method also provides the value of the minimum parking size for the existence of a direct collision-free parallel parking. This value represents an interesting parameter since it is used in order to determine whether the parking place is large and wide enough to perform a parking manoeuvre. Moreover, it is also applied in the selection of the final manoeuvre as will be shown in the next section.

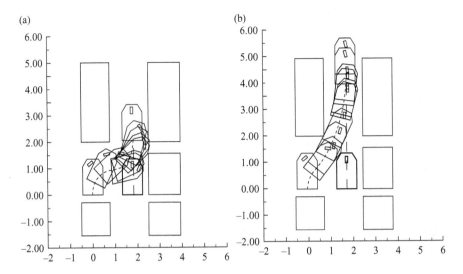

*Figure 12.6*   Optimal (extreme) manoeuvres for parallel parking with restricted
manoeuvres

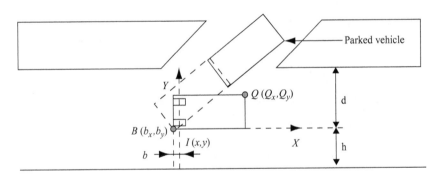

*Figure 12.7*   Left-hand side diagonal parking definition

Diagonal parking can be solved in the same way. Assume that a diagonal parking
to the left-hand side is defined as shown in Figure 12.7.

The problem of parking the vehicle can be solved in two steps: (1) planning a
restricted manoeuvre to modify the orientation of the vehicle; (2) following a straight
line to the end of the parking place. Application of restricted manoeuvres again
provides a solution set and two optimal solutions which are shown in Figure 12.8.

Observe that each solution, in both parallel and diagonal parking, is related with
a particular forward displacement (i.e. with the point where the vehicle starts the
backward motion). This fact will be used in the next section by an intelligent system
in order to select one manoeuvre from the solution set, according to criteria different
from the optimal ones presented in this section.

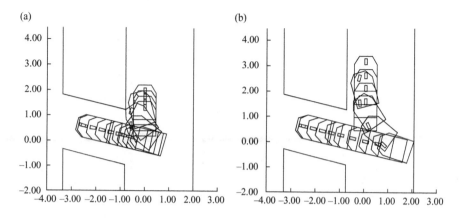

*Figure 12.8    Optimal (extreme) manoeuvres for diagonal parking with restricted manoeuvres*

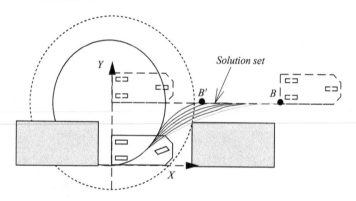

*Figure 12.9    Solution set of parallel parking manoeuvres*

## 12.4    Parking manoeuvre selection

In the previous section, different optimal manoeuvres have been introduced. Each of such manoeuvres is optimal only with respect to a simple given criterion: minimum lateral displacement, minimum forward displacement, minimum control effort and so on. These manoeuvres build up a solution set of collision-free manoeuvres, where the optimal manoeuvres are the extreme ones (see Figure 12.9). However, from a practical point of view, depending on the environment, the selection of one such extreme manoeuvre can be not a good option. Thus, for instance, a minimum forward displacement could imply a higher lateral displacement (the vehicle could navigate very close to the opposite parking line) and a poor path tracking performance due to the high curvature involved.

On the other hand, parking manoeuvres developed in this approach present curvature discontinuity. Curvature discontinuity could decrease the path tracking performance. Moreover, continuous curvature manoeuvres can be generated by means of $\beta$-splines [25,29,41]. Nevertheless, non-continuous curvature manoeuvres require smaller collision-free space and computational cost than continuous ones, whereas continuous curvature manoeuvres implies smoother control input and lower risk of error during path tracking. In general, when the parking place is large enough smoother manoeuvres are preferred, while non-continuous curvature manoeuvres are the best option when shorter parking places are involved.

Thus, depending on the situation the system should select the most appropriate type of manoeuvre.

### 12.4.1 Fuzzy manoeuvre selection

In practice, it is interesting to select a manoeuvre from the solution set. Such selection should be done by considering the environment where the vehicles have to park, and can be done by a fuzzy logic system in the same way that a human driver selects the starting point depending on the characteristics of the parking place.

Thus, fuzzy logic plays an important role in the practical application of this parking method, by considering the trade-off between optimal manoeuvre and environment conditions in order to perform a more robust and safer parking manoeuvre.

The fuzzy selection system will compute the best forward displacement based on the lateral distances, to the inner and outer parking lines, and the area of the parking place. It is interesting to note that the output of the fuzzy system should be considered as a point around which the vehicle should stop, but not as the exact point where the vehicle must stop. Indeed, once the vehicle is stopped, the parking manoeuvre is computed from the actual position (that will be collision-free since it starts from the solution set).

The inputs to the fuzzy system are: the inner distance (id), outer distance (od) and parking size (ps); the output is the forward displacement (fd). The membership functions of these variables are shown in Figure 12.10.

The knowledge of a human driver can be incorporated into the fuzzy selection system by means of the following rule base:

IF *outer distance* IS *closed* OR *inner distance* IS *far*
        THEN MAKE *forward displacement large*

IF *inner distance* IS *closed*
        THEN MAKE *forward displacement short*

IF *outer distance* IS *far enough* OR *inner distance* IS *medium*
        THEN MAKE *forward displacement medium*

IF *parking size* IS *small*
        THEN MAKE *forward displacement medium*

IF *parking size* IS *big*
        THEN MAKE *forward displacement short*

Figure 12.11 shows the application of the fuzzy selection system in different situations to obtain a convenient forward displacement, giving a robust manoeuvre

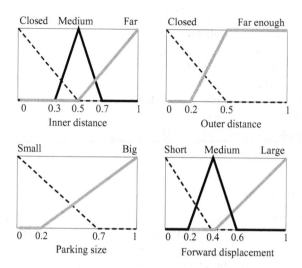

*Figure 12.10    Membership functions for the fuzzy selection system*

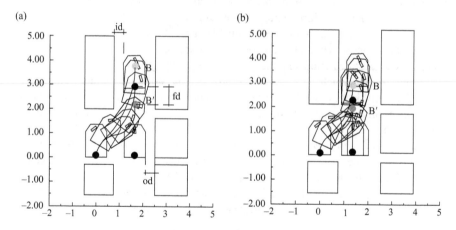

*Figure 12.11    Fuzzy selection of the optimal manoeuvre. (a) id = 0.45, od = 0.45, ps = 0.23, fd = 0.43. (b) id = 0.2, od = 0.7, ps = 0.23, fd = 0.19*

keeping away from the obstacles around. In the first case, inner and outer distances are 0.45 m, while the parking size is 0.23 with respect to the minimum parking size for a collision-free parallel parking (i.e. it is 23 per cent bigger than the minimum parking size, which takes a value of 2 for the given environment), yielding a medium forward displacement, around 0.43 (note that B and B' correspond to the extreme manoeuvres). In the second case, the vehicle is closer to the left line of vehicles (inner distance is 0.2 m) and a shorter forward displacement (namely around 0.19) is preferred.

In a similar way, the fuzzy system can be used to decide on the best diagonal parking manoeuvre, but then, width of the parking size can be neglected.

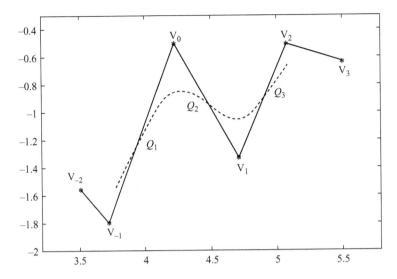

*Figure 12.12    Control vertices and β-spline curve*

### 12.4.2    Continuous versus non-continuous curvature manoeuvre selection

Parking manoeuvres developed in this approach present curvature discontinuity. The solution presented in this chapter to avoid such a discontinuity consists of applying a recursive algorithm to compute a new path, close enough to the first one, but with curvature continuity. Dynamic characteristics of the vehicle such as the steering system velocity are taken into account computing, if it exists, a feasible continuous curvature path. Thus, this algorithm is also used to select a feasible continuous or non-continuous manoeuvre. Curves with smooth properties, like β-splines, are applied [41,42].

A cubic β-spline is a parametric piece-wise curve which presents continuity of first and second derivatives at the joints between adjacent segments. This type of curve is specified by a set of points called control vertices. These vertices connected in sequence build up a control polygon. The curve tends to mimic the overall shape of the control polygon, Figure 12.12.

The idea of using β-spline curves for parking manoeuvres is based on placing the control vertices over the original manoeuvre, see Figure 12.13, and selecting a value for the distance between vertices ($\delta_f$) in such a way that the resulting β-spline is an admissible path [29]. A β-spline is admissible if:

(1)   it is a collision-free path,
(2)   the curvature is bounded along the curve,
(3)   the curve contains the starting and ending points defined by the original manoeuvre,
(4)   the change in curvature accomplishes velocity steering constraints.

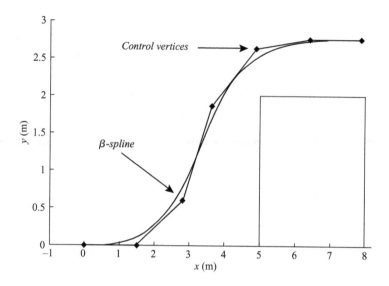

*Figure 12.13    Control vertices and continuous parking manoeuvre*

On the one hand, constraints 1 to 3 depend on the generated manoeuvre. On the other hand, constraint 4 depends both on the maximum steering velocity and the vehicle velocity.

Therefore, once the manoeuvre has been designed, in order to accomplish constraints 1 to 3, a value of $\delta_f$ is chosen taking into account several criteria based on the environment and the original manoeuvre (see Reference 29 for details). Then, for an initial velocity and taking into account the maximum steering velocity, the selected value of $\delta_f$ has to fulfil several criteria in order to accomplish constraint 4 (see Reference 29 for details). If this constraint is not accomplished, the vehicle velocity can be reduced to allow a correct path-following. If the obtained velocity is lower than a given minimum velocity value ($v_{thr}$), the $\beta$-spline is rejected and the original discontinuous curvature manoeuvre is performed.

The following algorithm has been implemented in order to tune the correct value of $v$ allowing the vehicle to accomplish the constraint:

*Parking place detection*
*Fuzzy Manoeuvre selection*
*Determine $\delta_f$*
continuous_manoeuvre = false
$v = v_{ini}$
**IF**  ($\delta_f$ accomplishes the admissibility condition 4 above)
    **THEN** *continuous_manoeuvre* = true
**ELSE** $v = v_2(\delta_f, maximum\_steering\_velocity)$
**IF** $v > v_{thr}$
    **THEN** *continuous_manoeuvre* = true
**END**

(a)  (b)

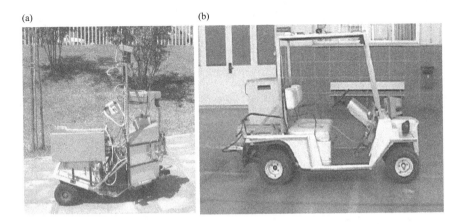

*Figure 12.14    ROMEO autonomous vehicles*

**IF** *continuous_manoeuvre = true*
    **THEN**

            generate $\beta$-spline
            follow $\beta$-spline
    **ELSE**
      follow the discontinuous curvature manoeuvre.
    **END**

Thus, the algorithm allows computing a feasible continuous curvature manoeuvre or selects to perform a non-continuous manoeuvre.

## 12.5    Experimental results

This section presents experiments with real autonomous vehicles performing parallel parking manoeuvres. Different environments have been considered in order to test the system in different parking situations.

Two types of vehicles have been used. On one hand, ROMEO-3R is the result of the adaptation of a conventional three-wheel electrical vehicle (see Figure 12.14(a)). The controller has been implemented in an industrial 486 PC system, with Lynx v2.2 real time operating system. On the other hand, ROMEO-4R is the result of the adaptation of a four-wheeled vehicle, with Ackermant steering (see Figure 12.14(b)). The controller has been implemented in a PC Pentium 133 MHz and Debian GNU/Linux 2.2.r4 operating system.

It should be highlighted that in both vehicles, only a minimal sensor layout is required to perform the experiments presented here.

Experiments presented in Figure 12.15 show two parallel parkings performed by ROMEO-3R. In both cases the system generates a manoeuvre built by a $\beta$-spline curve (the curvature changes continuously). Figure 12.16 presents different scenes from an experiment with the ROMEO-3R vehicle.

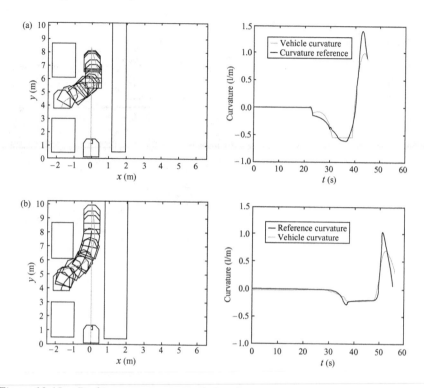

Figure 12.15    Parking manoeuvres with β-spline curves

Figure 12.16    ROMEO-3R performing a parallel parking

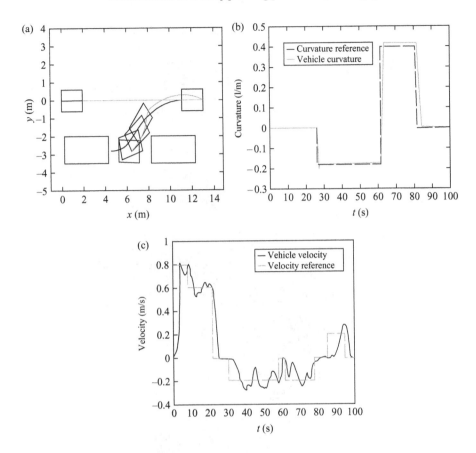

*Figure 12.17    Discontinuous curvature parallel parking*

The experiment shown in Figure 12.15(b) illustrates the effect of the vehicle being close to the opposite line of vehicles. The resulting manoeuvre is smoother than the first one (see Figure 12.15(a)). The forward displacement is larger and the curvature along backward motion presents lower values. Thus, the system provides the adaptation to the environment improving the manoeuvre performance.

The experiment presented in Figure 12.17 illustrates a parallel parking with ROMEO-4R. Unlike the previous experiments, the resulting manoeuvre is formed by a discontinuous curvature path. Thus the system has to stop the vehicle in order to change the steering angle when the curvature discontinuity appears. Observe the curvature discontinuity and the time instant when the vehicle stops and changes the steering angle.

Figure 12.17(c) also illustrates different phases of the parking manoeuvre. First the vehicle velocity is 0.8 m/s, while the vehicle is searching the parking place. Once the place has been found, the velocity reference is changed to 0.6 m/s. Finally during backward motion the reference velocity slows down to 0.2 m/s.

*Figure 12.18    ROMEO-4R performing a parallel parking*

Figure 12.18 presents different scenes from the experiment with the ROMEO-4R vehicle.

## 12.6    Conclusions

This chapter has presented an intelligent system for autonomous parking of vehicles, including intelligent navigation and searching of parking place. Thus, an intelligent control method has been implemented. This method uses soft-computing techniques to: (1) take advantage of knowledge of expert drivers, (2) manage imprecise information from sensors, (3) navigate searching for a parking place and (4) select and perform a parking manoeuvre.

Soft-computing techniques facilitate combining different strategies, providing the adaptation of the control motion to the environment and improving manoeuvre performance. The efficiency of the proposed method is demonstrated using the nonholonomic mobile robots ROMEO-3R and ROMEO-4R, developed at the University of Seville.

## Acknowledgements

This work was supported in part by the Ministerio de Educación y Ciencia under Contract DPI2002-04401-C03-03.

## References

1 MURRAY, R. M., and SASTRY, S. S.: 'Nonholonomic motion planning: steering using sinusoids', *IEEE Transactions on Automatic Control*, 1993, **38** (5), pp. 700–16

2 LAUMONT, J. P., JACOBS, P. E., TAIX, M., and MURRAY, M.: 'A motion planner for nonholonomic mobile robots', *IEEE Transactions on Robotics and Automation*, 1998, **10** (5), pp. 577–93

3 SEKHAVAT, S., LAMIRAUX, F., LAUMONT, J. P., BAUZIL, G., and FERRANT, A.: 'Motion planning and control for Hilare pulling a trailer: experimental issues'. Proceedings of the IEEE international conference on *Robotics and automation*, 1997, pp. 3306–11

4 NEFF PATTEN, W., WU, H., and CAI, W.: 'Perfect parallel parking via Pointryagin's principle', *Journal of Guidance, Control and Dynamics*, 1994, **116**, pp. 723–9

5 JIANG, K., and SEREVIRATNE, D.: 'A sensor guided autonomous parking system for nonholonomic mobile robots'. Proceedings of the 1999 IEEE international conference on *Robotics and automation*, pp. 311–16

6 SEKHAVAT, S., and LAUMONT, J. P.: 'Topological property for collision-free nonholonomic motion: the case of sinusoidal inputs for chained systems', *IEEE Transactions on Robotics and Automation*, 1998, **14** (5), pp. 671–80

7 LAMIRAUX, F., and LAUMONT, J. P.: 'Smooth motion planning for car-like vehicles', *IEEE Transactions on Robotics and Automation*, 2001, **17** (4), pp. 498–502

8 PAROMTCHIK, I., and LAUGIER, C.: 'Motion generation and control for parking an autonomous vehicle'. Proceedings of the IEEE international conference on *Robotics and automation*, 1996, pp. 3117–22

9 PAROMTCHIK, I., LAUGIER, C., GUSEV, S. V., and SEKHAVAT, S.: 'Motion control for an autonomous vehicle'. Proceedings of the international conference on *Control, automation, robotics and vision*, 1998, vol. 1, pp. 136–40

10 LATOMBE, J. C.: 'Robot motion planning' (Kluwer Academic Publishers, Norwell, MA, 1991) pp. 424–5

11 BORENSTEIN, J., and KOREN, Y.: 'The vector field histogram-fast obstacle avoidance for mobile robots', *IEEE Transactions on Robotics and Automation*, 1991, **7** (3), pp. 278–88

12 SAFFIOTTI, A.: 'Fuzzy logic in autonomous robot navigation: a case study', *Soft Computing*, 1997, **14**, pp. 180–97

13 BRAUNSTINGL, R., SANZ, P., and EZQUERRA, J. M.: 'Fuzzy logic wall following of a mobile robot based on the concept of general perception', *Proceedings of the ICAR '95*, 1995, **1**, pp. 367–76

14 TUNSTEL, E.: 'Mobile robot autonomy via hierarchical fuzzy behaviour control'. Second World Automation Congress, Montpellier, 1996

15 GASOS, J., and ROSETTI, A.: 'Uncertainty representation for mobile robots: perception, modeling and navigation in unknown environments', *Fuzzy Sets and Systems*, 1999, **107**, pp. 1–24

16 LUO, R. C., and CHEN, T. M.: 'Autonomous mobile target tracking system based on grey-fuzzy control algorithm', *IEEE Transactions on Industrial Electronics*, 2000, **47** (4), pp. 920–31

17 SONG, K.-T., and SHEEN, L.-H.: 'Heuristic fuzzy-neuro network and its application to reactive navigation of a mobile robot', *Fuzzy Sets and Systems*, 2000, **110** (3), pp. 331–40

18 CUESTA, F., and OLLERO, A.: 'Intelligent mobile robot navigation', Springer tracks in advanced robotics (Springer-Verlag, Berlin, 2005)

19 SUGENO, M., and MURAKAMI, K.: 'Fuzzy parking control of model car'. Proceedings of the IEEE 23rd conference on *Decision and control*, Las Vegas, NV, 1984

20 YASUNOBU, S., and MURAI, Y.: 'Parking control based on predictive fuzzy control'. Proceedings of IEEE international conference on *Fuzzy systems*, 1994, vol. 2, pp. 1338–41

21 BRAUNSTINGL, R., and VOESSNER, S.: 'Perception based fuzzy parking'. Third World Automation Congress, ISORA 030, Anchorage, USA, May 1998

22 OLLERO, A., CUESTA, F., BRAUNSTINGL, R., ARRUE, B. C., and GÓMEZ-BRAVO, F.: 'Perception for autonomous vehicles based on proximity sensors. Application to autonomous parking'. *14th IFAC World Congress*, 1999, vol. B (B-1e-02-4), pp. 451–6

23 BATURONE, I., MORENO-VELO, F. J., SANCHEZ-SOLANO, S., and OLLERO, A.: 'Automatic design of fuzzy controllers for car-like autonomous robots', *IEEE Transactions on Fuzzy Systems*, 2004, **12**, pp. 447–65

24 CUESTA, F., GÓMEZ-BRAVO, F., and OLLERO, A.: 'A combined planned/reactive system for motion control of vehicles manoeuvres'. MC98, France, September 1998, pp. 303–8

25 GÓMEZ-BRAVO, F., CUESTA, F., and OLLERO, A.: 'Parallel and diagonal parking in nonholonomic autonomous vehicles', *Engineering Applications of Artificial Intelligence*, 2001, **14** (4), pp. 419–34

26 GÓMEZ-BRAVO, F., CUESTA, F., and OLLERO, A.: 'A new sequential hybrid control structure for car and tractor-trailer parallel parking'. Proceedings of seventh international IFAC symposium on *Robot control*. SYROCO'03, 2003

27 OLLERO, A., ARRUE, B. C., FERRUZ, J., *et al.*: 'Control and perception components for autonomous vehicle guidance. Application to the Romeo vehicles', *Control Engineering Practice*, 1999, **07** (10), pp. 1291–9

28  CUESTA, F., GÓMEZ-BRAVO, F., and OLLERO, A.: 'Parking manoeuvres of industrial-like electrical vehicles with and without trailer', *IEEE Transactions on Industrial Electronics: Special Session on Automotive Electronic Systems*, 2004, **51** (2), pp. 257–69

29  GÓMEZ-BRAVO, F., CUESTA, F., and OLLERO, A.: 'Continous curvature path generation for parallel parking manoeuvres based on $\beta$-spline curves'. Proceedings of IFAC international conference on *Intelligent control systems and signal processing*, ICONS-03, 2003

30  OLLERO, A., GARCÍA-CEREZO, A., and MARTÍNEZ, J.: 'Fuzzy supervisory path tracking of mobile robots', *Control Engineering Practice*, 1994, **2** (2), pp. 313–19

31  GARCÍA-CEREZO, A., OLLERO, A., and MARTÍNEZ, J. L.: 'Design of a robust high-performance fuzzy path tracker for autonomous vehicles', *International Journal of Systems Science*, 1996, **27** (8), pp. 799–806

32  BROCKET, R. W.: 'Asymptotic stability and feedback stabilization', in BROCKET, R. W., MILLMAN, R. S., and SUSSMAN, H. J. (Eds): 'Differential geometric control theory' (Birkhauser, Boston, MA, 1998)

33  AICARDI, M., CASALINO, G., BICCHI, A., and BALESTRINO, A.: 'Closed loop steering of unicycle-like vehicles via Lyapunov techniques', *IEEE Robotics and Automation Magazine*, 1995, vol. 2, no. 1, pp. 27–35

34  MURRAY, R., and SASTRY, S. S.: 'Steering nonholonomic system using sinusoids'. Proceedings of the IEEE international conference on *Decision and control*, 1990, pp. 1136–41

35  TILBURY, D.: 'Exterior differential systems and nonholonomic motion planning'. Ph.D. thesis, University of California at Berkeley, 1994

36  BARRAQUAND, J., and LATOMBE, J. P.: 'On nonholonomic mobile robots and optimal manoeuvering', *Revue d'Intelligence Artificiale*, Hermes, Paris, 1989, **3** (2), pp. 77–103

37  LaVALLE, S. M., and KUNER, J. J. Jr.: 'Randomized kinodynamic planning', *International Journal of Robotics Research*, 2001, **20** (5), pp. 378–400

38  LEITCH, D., and PROBERT, P.: 'Context dependent coding in genetic algorithms for the design of fuzzy systems'. Proceedings of the IEEE international conference on *Fuzzy logic, neural net and genetics algorithms*, Japan, August 1994

39  DUBINS, L. E.: 'On curves of minimal length with a constraint on average curvature and with prescribed initial and terminal position and tangents', *American Journal of Mathematics*, 1957, **79**, pp. 497–516

40  REEDS, J. A., and SHEPP, R. A.: 'Optimal paths for a car that goes both forward and backward', *Pacific Journal of Mathematics*, 1990, **145** (2), pp. 367–93

41  MUÑOZ, V., MARTÍNEZ, J., and OLLERO, A.: 'New continuous curvature local path generators for mobile robot', *SICICA'92*, 1992, vol. 1, pp. 551–6

42  BARSKY, B.: 'Computer graphics and geometric modelling using beta-splines' (Springer-Verlag, Berlin, 1987)

*Chapter 13*

# Applications of intelligent control in medicine

*M. F. Abbod and D. A. Linkens*

## 13.1 Introduction

The first part of the chapter describes the results of two literature surveys on intelligent systems allied to medicine which were commissioned by the European Networks of Excellence ERUDIT and EUNITE. The first survey covered the use of fuzzy technology across the whole of medicine divided into ten sub-disciplines, ranking from surgery to many branches of internal medicine. The second survey covered the field of intelligent adaptive systems in medicine, again applied to the same ten subdivisions as above, but reduced to five sub-areas in the analysis of the results. In this case a wide range of intelligent techniques was considered including fuzzy inference, neural networks and model-based reasoning. Both surveys attempted to show the merits and limitation of the current work and to give signposts for future developments in the field.

The second part of the chapter contains two relevant case studies. The first concerns intelligent systems in anaesthesia monitoring and control. The theme is unconsciousness management in operating theatres, which is a particularly challenging application for AI techniques. The second case study concerns the developments of nonlinear models based on ANN and neuro-fuzzy techniques for both classification and prediction in cancer survival studies. The particular application is that of bladder tumour prognosis using gene expression markers together with demographic data and social conditions such as smoking.

## 13.2 Fuzzy technology in medicine

The complexity of biological systems makes traditional quantitative approaches of analysis inappropriate. There is an unavoidable substantial degree of fuzziness in the description of the behaviour of biological systems as well as their characteristics.

The fuzziness in the description of such systems is due to the lack of precise mathematical techniques for dealing with systems comprising a very large number of interacting elements or involving a large number of variables in their decision tree [1]. Fuzzy sets are known for their ability to introduce notions of continuity into deductive thinking [2]. Practically, this means that fuzzy sets allow the use of conventional symbolic systems (specified in the form of tabulated rules) in continuous form. This is essential since medicine is a continuous domain. Many practical applications of fuzzy logic in medicine use its continuous subset features such as: fuzzy scores, continuous version of conventional scoring systems and fuzzy alarms. The best developed approach is for fuzzy control, providing the most successful application to date in which a rule-based mapping from input to output variables effectively implements a continuous control law.

In medicine it is not necessary to deal with micro-phenomena and micro-objectives to encounter the problem of incompleteness, uncertainty and inconsistency. The lack of information and its imprecise, and sometimes contradictory, nature is much more a fact of life in medicine than in, for example, the physical sciences. These problems have to be taken into account in every medical decision, where they may have important, even vital, consequences for the object of medical attention.

The inherited sources of inaccuracy can be classified as:

1. Information about the patient, which can be divided into a number of categories, all of which have uncertainty.
2. Medical history of patients, which is supplied by the patient, and is usually highly subjective and may include simulated, exaggerated or understated symptoms; ignorance of previous diseases; failure to mention previous operations and general recollection often leads to doubts about the patient medical history.
3. Physical examinations, which the physicians conduct to obtain objective data. They are subject to mistakes and overlooking important indications, or may even fail to carry out a complete test. Furthermore, they may misinterpret other indications because the boundary between normal and pathological status is not always clear.
4. Results of laboratory tests are objective data, which depend on the accuracy of the measurements, organisational problems (mislabelling sample, wrong laboratory etc.), and even on the improper behaviour of the patient prior to the examination.
5. Results of histological, X-ray, ultrasonic and other clinical investigations, which depend on correct interpretation by medical staff.

This section surveys the use of fuzzy logic in medicine based on searches in medical databases such as MEDLINE, INSPEC and edited books. The purpose of the study was to establish a 'Roadmap' which may help to forecast the future developments of fuzzy technology in medicine and healthcare. A simple search of the word 'fuzzy' obtained many cited papers, in half of which the term was used to describe an unsharp border of structure or situation. For the rest, most of them recent ones, the word 'fuzzy' was used as part of fuzzy sets or fuzzy logic [3]. The main topics which were cited can be classified into ten fields (each of which can be further classified into subheadings) as follows: conservative medicine, invasive medicine, regionally defined medical disciplines, neuromedicine, image and signal processing, laboratory, basic science, nursing, healthcare and oriental medicine.

### 13.2.1  Bibliographic materials

The first comprehensive bibliographic paper covering all fields of fuzzy logic was compiled by Kandal and Yager [4]. Another survey that covered 20 years of fuzzy set theory application in medicine was conducted by Maiers in 1985 [5]. It summarises the medical applications of fuzzy set theory, assesses the value of fuzzy sets in medical applications, and suggests their future potential. An introduction to fuzzy sets, approximate reasoning and fuzzy rules as a tool for modelling sets with ill-defined or flexible boundaries with uncertainties has been presented by Dubois, Prade and co-workers [6–12]. Other papers focused on surveying special fields like virtual reality and robotics in medicine [13], general application of fuzzy logic in medicine [14], and a case study of different medical applications domain [15]. A good source of references and an excellent book in the field of fuzzy logic is by Zimmermann [16]. A book edited by Cohen and Hudson [17] contains a collection of papers that deal with medical decision-making relating to heart disease. Different approaches were reported: knowledge-based systems, statistical approaches, modelling and hybrid systems. Finally, a more recent book that has an excellent survey of fuzzy logic in medicine was edited by Teodorescu, Kandel and Jain [18]. It also covers recent applications in modelling of the brain, tumour segmentation, dental developments, myocardial ischemia diagnosis, heart disease diagnosis and usage in the ICU. Neuro-fuzzy system applications were also reported in the book. Also, in 1989, a group of Japanese scientists established a learned society for fuzzy systems and biomedical applications: Japanese Bio-Medical Fuzzy Systems Association (BMFSA) which has an international journal 'Biomedical Soft Computing and Human Sciences' published in Japanese with a yearly issue in English.

### 13.2.2  Discussion on intelligent fuzzy technology in medicine

Based on this study, future developments of fuzzy technology in medicine and healthcare can be tentatively forecast. The sectors of medical activities can be brought together in a hierarchical scheme according to the mode of the medical procedure. This means that significant methodologies, relationships and demands are correlated. This scheme substantiates the hypothesis that a successful application in one sector should lead to a successful application in neighbouring sectors. All information gathered in this survey retains its time value. The papers are arranged on a timescale. This means that not only the actual state of the art, but also the dynamic process of the information spread in each sector can be determined. This two-dimensional representation is useful in the prediction of future events. It suggests that in two to three years the spread of fuzzy logic applications will follow the tendencies of recent years. Table 13.1 summarises the number of applications of fuzzy logic for each criterion classified on a yearly basis based on two databases: MEDLINE and INSPEC. Each database is different, since MEDLINE mainly reviews medical journal papers,

*Table 13.1   Number of applications of fuzzy logic in each year for the specified medical sub-topics*

| | Publication year | | | | | | | | | | |
| --- | --- | --- | --- | --- | --- | --- | --- | --- | --- | --- | --- |
| | <89 | 90 | 91 | 92 | 93 | 94 | 95 | 96 | 97 | 98 | Total |
| **Conservative disciplines** | | | | | | | | | | | |
| Internal medicine | 35 | 3 | 2 | — | 2 | 9 | 7 | 5 | 2 | 1 | 66 |
| Cardiology | 9 | — | 3 | 2 | 1 | 7 | 2 | 3 | 1 | 2 | 30 |
| Invasive care | 3 | — | — | 1 | 2 | 2 | — | 2 | 2 | 1 | 13 |
| Paediatry | 1 | — | — | — | — | 4 | — | 3 | — | — | 8 |
| Endocrinology | 3 | 1 | — | — | 1 | — | 1 | — | 1 | — | 7 |
| Oncology | 1 | — | — | — | — | 1 | — | — | 7 | — | 9 |
| Gerontology | 1 | — | 7 | 4 | 3 | 3 | 6 | 4 | 4 | — | 32 |
| General practice | 2 | — | — | — | 1 | — | — | — | — | — | 3 |
| **Invasive medicine** | | | | | | | | | | | |
| Surgery | — | — | — | — | — | 1 | — | — | 1 | 1 | 3 |
| Orthopaedics | — | — | — | — | — | 2 | — | — | — | — | 2 |
| Anaesthesia | 6 | 1 | 1 | 3 | 1 | 8 | 3 | 6 | 9 | 3 | 41 |
| Artificial organs | — | — | — | — | — | 1 | — | 2 | 3 | 1 | 7 |
| **Regionally defined medical disciplines** | | | | | | | | | | | |
| Gynaecology | — | 2 | — | — | 1 | — | — | 2 | 3 | — | 8 |
| Dermatology | — | — | — | — | — | 2 | — | — | — | 1 | 3 |
| Dental medicine | — | — | — | — | 1 | 1 | — | — | 1 | — | 3 |
| Ophthalmology | 1 | — | 1 | — | — | 1 | — | 1 | 1 | — | 5 |
| Otology, rhinology etc. | — | — | — | 1 | — | — | 2 | 1 | — | — | 4 |
| Urology | 1 | 1 | — | — | 1 | 2 | 1 | 1 | 3 | — | 10 |
| **Neuromedicine** | | | | | | | | | | | |
| Neurology | 5 | — | — | — | — | 1 | 2 | 2 | 7 | — | 17 |
| Psychology | 5 | — | 1 | 3 | 1 | 3 | 4 | 6 | 4 | 4 | 31 |
| Psychiatry | 4 | — | 1 | — | — | 1 | 3 | 1 | 1 | 2 | 13 |
| **Image and signal processing** | | | | | | | | | | | |
| Signal processing | 4 | 3 | 3 | 2 | — | 3 | 1 | 3 | 5 | 3 | 27 |
| Radiation medicine | 3 | 1 | — | — | — | — | 2 | 2 | 2 | 3 | 13 |
| Radiology | 3 | 3 | — | — | 2 | 4 | 5 | 7 | 10 | — | 34 |
| **Laboratory** | | | | | | | | | | | |
| Biochemical and tests | 1 | 1 | 3 | 2 | 1 | 1 | 2 | 4 | 2 | 6 | 23 |
| **Basic science** | | | | | | | | | | | |
| Medical information | 29 | 2 | 3 | 5 | 7 | 6 | 16 | 11 | 9 | 1 | 89 |
| Anatomy, pathology etc. | 2 | 1 | 1 | 4 | 2 | 1 | — | 2 | 2 | 1 | 16 |
| Physiology | — | — | — | — | — | 1 | 2 | 1 | — | — | 4 |
| Pharmacology | 2 | 1 | — | 1 | 2 | 6 | 7 | 4 | 5 | — | 28 |
| Education | 1 | — | — | — | 1 | 1 | 1 | 2 | 2 | 1 | 9 |
| **Nursing** | 1 | — | — | — | — | — | 1 | — | 1 | 2 | 5 |
| **Healthcare** | 2 | — | — | — | 2 | 1 | 3 | 1 | 6 | 5 | 20 |
| Year Totals | 125 | 20 | 26 | 28 | 32 | 73 | 71 | 76 | 94 | 38 | 583 |

whereas INSPEC reviews computer science journals and conference papers. It is noticeable, however, that there is a surprisingly limited intersection between the two databases which grows as the number of publications increases. A notable exception is Adlassnig who is well known in both fields. Nevertheless, the two databases are correlated as there is an increase in the number of publications as the year scale progresses.

From such an analysis, some tentative conclusions are:

1. Internal medicine, anaesthesia, radiology, electrophysiology, pharmacokinetics and neuromedicine already use fuzzy logic methods to a considerable degree. Such techniques are: fuzzy expert systems, fuzzy control, fuzzy signal and image processing, fuzzy modelling and fuzzy neural simulation.
2. In the field of surgical disciplines, dental medicine, general practice and nursing, there are no specific applications.
3. Most regionally defined medical disciplines are developed from methods used in neighbouring disciplines.
4. Papers in the field of medical reasoning and decision support sciences are growing rapidly in number. This will keep developments in other sectors growing.

Although the literature search was not exhaustive, it is sufficient to illustrate the main features governing the penetration of fuzzy technology into the medicine and healthcare aspects of the life sciences. It is hoped that the survey will prompt the identification of matching techniques and subject areas where further exploitation of fuzzy systems may be both feasible and timely. Further details on the survey are given in Reference 19 which contains 202 references.

## 13.3   Intelligent adaptive systems in medicine

Adaptivity is a very common feature of everyday life, essential for our existence but difficult to define, analyse and synthesise. Generally speaking, adaptive systems are designed to cope with changing environmental conditions while maintaining performance objectives. Over the years, the theory of adaptive systems has evolved from relatively simple and intuitive concepts to a complex multifaceted theory dealing with stochastic, nonlinear and infinite dimensional systems.

In order to classify different forms of adaptivity the following good definitions have been adopted by EUNITE (European Network for Intelligent Technologies) which is concerned with the topic areas of Smart and Adaptive systems:

Level I   :   Adaptation to a changing environment
Level II  :   Adaptation to a similar setting without explicitly being ported to it
Level III:   Adaptation to a new/unknown application.

In the first case the system must adapt itself to a drifting environment (over time, space etc.), applying its intelligence to recognise the changes and react accordingly. This is probably the easiest concept of adaptation for which examples abound: customer preferences in electronic commerce systems, control of non-stationary systems (e.g. drifting temperature), telecommunication systems with varying channel or user characteristics. In the second case, the accent is more on the change of the environment itself than on a drift of some features of the environment. Examples are: systems that must be ported from one plant to another without explicitly changing their main parameters, a financial application that must be ported from a specific market to a similar one (e.g. a different geographical location). Other interesting problems are the porting of compiled software from one process to another (obviously without recompilation) or the porting of algorithms, usually developed in software, to dedicated (and often resource-limited) hardware for embedded systems. The third level is the most futuristic one, but its open problems have been addressed already by a number of researchers, especially in the Machine Learning field, where starting from very little information on the problem it is possible to build a system through incremental learning. Note that the boundaries between the three levels are very fuzzy: an environment can change in such a way as to result in a completely different setting or even an entire new kind of problems to be solved.

In terms of development of a theoretical basis for adaptive systems together with practical engineering implementations, much research has been undertaken in recent decades. In particular, early applications were in aircraft and missile flight control, while more recently emphasis has been on the process and manufacturing industry. Early flight adaptive systems soon ran foul of the resultant lack of proven stability theory together with case studies which demonstrated dangerous instabilities. Thus, in industrial applications the major attempt concentrated on linearised processes with several avenues of self-adaptive theory being developed. These included model reference adaptive control (MRAC), self-tuning control (STC) and model-based predictive control (MPC) with special emphasis on generalised predictive control (GPC). Many examples now exist where such schemes have been successfully implemented. However, the extension of these approaches to nonlinear systems remains an active yet elusive research objective.

In the realm of medicine, as it relates to the human body, one is immediately faced with endemic problems of nonlinearity and non-stationarity. Further, the degree of complexity in physiological systems is very large. Many natural systems (e.g. brains, immune systems, ecologies, societies) and many artificial systems (e.g. parallel and distributed computing systems, artificial intelligence systems, artificial neural networks, evolutionary programs) are characterised by apparently complex behaviours which emerge as a result of nonlinear spatio-temporal interactions among a large number of component systems at different levels of organisation. Consequently, researchers in a number of distinct areas (e.g. computer science, artificial intelligence, neural networks, cognitive science, computational economics, mathematics, optimisation, complexity theory, control systems, biology, neuroscience, psychology and engineering) have begun to address the analysis and synthesis of such systems through a combination of basic as well as applied techniques. Even more advanced types of adaptive system require evolving complex systems which have the following characteristics: self-maintenance, adaptivity, information preservation and spontaneous increase of complexity. Living systems are an obvious subset but also there are autocatalytic chemical reactions with the same properties.

In control systems, adaptation is required to keep the controller in tune as the parameters of the process itself vary. Types of adaptation vary from common sense approaches to much more mathematical ones. The extra complexity of the mathematical calculations is often justified by less hardware requirements and more acceptable operation. One approach is for the controller to set up its own parameters to suit the changing process parameters by using techniques such as gain-scheduling, STC and MRAC. In this approach, the controller uses online parameter estimation with the general structure of the plant model being predetermined and remaining constant. In practice, it is often desirable to be able to change the structure of the control strategy to meet changing circumstances, which requires a second type of adaptive controllers. One particular example of this is the sliding mode control strategy where the controlled nonlinear system is made to slide along a desired phase plane trajectory.

Thus there is a strong driving force for the study and implementation of adaptive systems in biomedicine. The purpose of this survey is to investigate from the published literature the penetration and likely future of such systems. The following hypotheses were considered:

Hypothesis 1: Adaptivity ought to be an important component in biomedical research.

Hypothesis 2: Some technologies should be particularly appropriate for adaptive studies.

Hypothesis 3: The medical sub-specialisms should indicate where penetration is already taking place.

Hypothesis 4: Future trends, including particular limitations, may become apparent with indications of where to expect significant advances.

The survey sought to classify the application areas into a number of broadly based medical specialisms. In addition, the whole field was considered under the basic modalities of diagnosis (including monitoring technologies), therapy (including control/decision support), and imaging (as an important sub-branch of diagnosis). It surveyed the use of adaptive systems in these areas of medicine. It was based on extensive searches in medical and engineering databases such as, MEDLINE, INSPEC, Biological Abstracts, EMBASE, Web of Science and edited books. The keywords search was based on the logical linguistic pattern '(adaptive or smart or intelligent) and (biomedical or medicine) and (engineering systems)'. Very many of the search entries were found to be unsuitable for the purpose of the survey, in that they did not describe work which had significant smart or adaptive contributions. The eventual purpose of the study was to establish a 'Roadmap' which may help to forecast the future developments of smart technology in medicine and healthcare, where the word 'smart' may include intelligent or adaptive capability.

### 13.3.1 Discussion on intelligent adaptive systems in medicine

The main topics which were cited in the survey were classified into five broad fields as follows: emergency and intensive care unit (ICU), general medicine, surgical

*Table 13.2*  *Number of applications of smart and adaptive systems in each year for the specified medical sub-topics*

| | Publication year | | | | | | | | | | | | |
| --- | --- | --- | --- | --- | --- | --- | --- | --- | --- | --- | --- | --- | --- |
| | <90 | 91 | 92 | 93 | 94 | 95 | 96 | 97 | 98 | 99 | 00 | 01 | Total |
| Emergency and ICU | 2 | — | — | 1 | 1 | 1 | 1 | — | 1 | — | — | 2 | 9 |
| General medicine | | | | | | | | | | | | | |
|   Neurology | — | 1 | — | 1 | — | 1 | — | 4 | 1 | 5 | 2 | 1 | 16 |
|   Cardiology | 2 | — | — | 1 | 2 | 2 | 2 | 4 | 1 | 3 | — | 3 | 20 |
|   Ear-nose-throat | — | — | — | — | — | — | — | 1 | 1 | — | 1 | — | 3 |
|   Endocrinology | — | — | — | — | 1 | — | — | — | — | 1 | — | 1 | 3 |
|   Gynaecology | 1 | — | — | — | — | — | — | — | — | 1 | — | — | 2 |
|   Internal medicine | 1 | — | — | — | — | — | — | — | — | — | — | 1 | 2 |
|   Respiratory medicine | 2 | — | — | 1 | 1 | 2 | 1 | — | — | — | 2 | 1 | 10 |
|   Paediatric medicine | 1 | — | — | — | — | — | — | 1 | — | — | 1 | — | 3 |
|   Ophthalmology | 1 | — | — | — | — | — | 2 | — | 1 | — | — | — | 4 |
| Surgical medicine | | | | | | | | | | | | | |
|   Surgery and | | | | | | | | | | | | | |
|     anaesthesia | 6 | 1 | 1 | 2 | 1 | 1 | — | 4 | 4 | 3 | 2 | — | 25 |
|   Orthopaedics | 1 | — | — | — | — | — | 1 | — | — | — | — | — | 2 |
| Image and signal processing | | | | | | | | | | | | | |
|   Signal processing | — | — | — | — | 1 | — | — | 2 | 1 | — | 2 | — | 6 |
|   Radiology | 1 | — | — | 2 | 1 | 3 | — | 7 | 5 | 2 | 3 | 3 | 27 |
| Pathology | — | — | — | — | — | — | — | — | 1 | — | — | — | 1 |
| Year totals | 18 | 2 | 1 | 8 | 8 | 10 | 7 | 23 | 16 | 15 | 13 | 12 | 133 |

medicine, pathology and medical imaging. Each of these fields can be further sub-classified.

In examining the hypotheses stated in Section 13.2, it is clear from the number of papers (133) cited in the full paper [20] that there have been many contributions to adaptive technology in biomedicine in the past ten years. Thus Hypothesis 1 can be demonstrated to be true. Although the literature search was not exhaustive, it is sufficient to illustrate the main features governing the penetration of smart and adaptive technology into medicine and healthcare aspects of the life sciences. Table 13.2 summarises the number of applications of intelligent and adaptive smart systems for each sub-discipline classified on a yearly basis.

In terms of the technologies employed as suggested under Hypothesis 2, there are many different approaches and combinations of hybrid forms. Unsurprisingly, because of the endemic nonlinearity in human systems the theme of artificial neural networks (ANN) often occurs. What is interesting is the wide variety of usage to which ANN are being put, particularly bearing in mind that only adaptive forms of ANN are surveyed here. This means that only dynamic versions of ANN are considered, that is, ones that update themselves to changing data input and not merely learn their weights in 'batch-like' mode. Thus, in the papers cited there are examples of ANN used in diagnosis via classification approaches. Also, they can be used for (nonlinear) predictive modelling for decision support. More ambitiously, they are being used for online control which raises important questions about stability when inverse modelling approaches are used. In addition to supervised learning, ANN are being used extensively in unsupervised mode (especially Kohonen self-organising map (SOM) techniques), often in pattern recognition for image analysis.

Another technique which has been largely ignored recently is that of self-organising fuzzy logic control (SOFLC). This has the potential to achieve Level II adaptivity as defined earlier. This has been used extensively in online drug administration trials in muscle relaxation and depth of anaesthesia control. A simpler system of gain-scheduling has been used in several sub-disciplines and can achieve Level I adaptivity. This latter technique is being employed in aerospace and process industries with considerable success because of its simple structure and basically open-loop/supervisory nature.

While ANN are basically quantitative and have self-learning capability, Fuzzy Logic is mainly qualitative but offers transparency through its rules. Neuro-fuzzy techniques such as adaptive network based fuzzy inference system (ANFIS) merge both approaches and are being used in several sub-disciplines. They are being used for model identification, prediction and quite commonly in adaptive 'noise' cancellation. 'Noise' here includes the separation of wanted from unwanted monitoring signals, for example, foetal from maternal ECG.

Model-based predictive control appears in several areas, including that of diabetes management. While fixed models can be utilised in MPC, interest heightens when they are both nonlinear and adaptive. This is occurring in ventilator management and anaesthesia control. The ideal is to embed a physiologically realistic model in the control strategy and this is happening in some of the cited research. However, there is also an important place for just modelling the biological sub-systems under study and then using this in simulation conditions for the offline design of less complex controllers. Examples of this include several eye mechanism models and body motion dynamics for functional nerve simulation. Blood pressure regulation via the use of medium complexity cardiovascular models is another promising example.

Signal processing encompasses many facets for one-, two- and three-dimensional monitoring. Adaptive techniques are commonly being applied in conjunction with feature extraction algorithms such as Wavelets (in this case the frequency domain). Time series processing can include recurrent ANN which allows for dynamics in the underlying processes being analysed. In biomedicine, the use of intelligent sensor fusion is obvious because of the system complexity and the inferential nature of the signals at their point of measurement (often required non-invasively).

Table 13.3 shows the distribution of the cited publications between the three complementary areas of diagnosis, therapy and imaging. Some tentative conclusions can be drawn from these tables regarding Hypothesis 3. From Table 13.2 the overall totals suggest considerable disparity between the sub-disciplines. Thus, neurology,

*Table 13.3   Sub-classification   of   adaptive   and   intelligent applications into medicine*

|                              | Diagnosis | Therapy | Imaging | Total |
|------------------------------|-----------|---------|---------|-------|
| Intensive care unit          | 5         | 4       | —       | 9     |
| General medicine             |           |         |         |       |
|   Neurology        | 1         | 13      | 2       | 16    |
|   Cardiology       | 6         | 14      | —       | 20    |
|   Ear-nose-throat  | —         | 3       | —       | 3     |
|   Endocrinology    | —         | 3       | —       | 3     |
|   Gynaecology      | 2         | —       | —       | 2     |
|   Internal medicine| 2         | —       | —       | 2     |
|   Respiratory medicine | 4     | 6       | —       | 10    |
|   Paediatric medicine | 1      | 2       | —       | 3     |
|   Ophthalmology    | 2         | 2       | —       | 4     |
| Surgical medicine            |           |         |         |       |
|   Surgery and anaesthesia | 1  | 24      | —       | 25    |
|   Orthopaedics     | —         | 2       | —       | 2     |
| Image and signal processing  |           |         |         |       |
|   Signal processing | 5        | 1       | —       | 6     |
|   Radiology        | 1         | —       | 26      | 27    |
| Pathology                    | —         | —       | 1       | 1     |
| Total                        | 30        | 74      | 29      | 133   |

cardiology, surgery and anaesthesia, and radiology are well represented. In contrast, there is practically no activity in the areas of gynaecology, pathology and internal medicine. In the latter case this seems surprising, since expert systems were applied in this field at an early stage. Over the timescale of the survey there has been a general increase in activity since 1991. There was a peak in 1997, which gives a warning that interest and activity in adaptive systems may not have been consolidated. This provides a major challenge. In the individual sub-fields, there has been an increase over the period in cardiology and neurology applications. In respiratory medicine there has been a small increase, while in radiology there has been a significant increase. In surgery and anaesthesia there has been a steady output throughout the period of the survey.

From Table 13.3 it can be seen that there is a fairly even distribution over the sub-fields in the theme of diagnosis. One would have expected contributions in internal medicine for this area. The theme of therapy is non-evenly distributed across the sub-fields, which is not surprising because of the diverse nature of the specialisms. However, in adjacent areas of paediatrics and orthopaedics, where therapy is particularly important, one would expect contributions to be possible. The theme of imaging has a very peaky distribution, possibly because other sub-disciplines are implicated within the area of radiological imaging. However, there is adjacency of concept for intelligent imaging in areas such as gynaecology and ophthalmology which could benefit from the published work in radiology. The limitations arise mainly from resource and motivation aspects rather than technological barrenness. However, in terms of Hypothesis 4 on trends in the theme of smart, adaptive systems the survey has demonstrated that there are certain 'white spots' of medical specialisms which could benefit from development in adjacent sub-areas. This accords with the equivalent survey of the use of fuzzy technology in medicine [19], which was conducted within the European Network of Excellence ERUDIT and incorporated into its 'Roadmap' (see Section 13.2). In that Roadmap important issues of validation (including safety aspects), evaluation (including cost/benefits analysis), litigation concerns and balance between technical 'push' and clinical 'pull' were highlighted as challenges. In addition, the following needs and possible limitations were identified: non-serendipitous linkage between applications in the sub-areas (systems scientists and engineers should have an important role here); a critical mass of focused clinicians to provide take-up of the technology; sufficient funding and career prospects for bio-engineers to maintain impetus in developing the particular applications; and globalisation of scarce resources via journals, conferences and networking.

In summary, examples of Level I and Level II adaptation can be found in medical applications, but Level III has not yet been achieved. This level implies a degree of autonomy which medical applications are unlikely to allow in the foreseeable future.

## 13.4   Unconsciousness monitoring and control

Depth of anaesthesia (DOA) during surgical operation is assessed by anaesthetists via observation of blood pressure, heart rate, lacrimation pupil response, diaphoresis etc. These signs are known as 'clinical signs of DOA' since they depend on skeletal muscle activities. Modern anaesthesia utilises a number of drugs in addition to narcotic anaesthetic drugs, such as neuromuscular blockers and pain relieving drugs. All of these tend to obscure these signs of anaesthetic depth. Anaesthetic depth has been defined in terms of mean alveolar concentration (MAC). The MAC of an inhalational agent is the concentration at which 50 per cent of the patient population will not move to a painful stimulus.

One of the modern methods for monitoring DOA is the use of auditory evoked responses (AER). It consists of producing auditory signals caused by clicks which are delivered to the patient via earphones. The electrical signals produced by the central nervous system in response to the clicks are recorded by electrodes placed on the scalp. The waveform represents the passage of electrical activity from the cochlea to the cortex. The form of an AER signal is presented by a number of peaks and can be characterised by its amplitude and latency. The brainstem auditory evoked potential (0–10 ms) is associated with the primary hearing system. The early cortical response (10–60 ms) is associated with the primary hearing cortex. The amplitude of the peaks decreases and the latencies increase with increase in the anaesthetic depth.

Previous research on closed-loop control has indicated that patients can be modelled and controlled well with artificial intelligence (AI) techniques such as fuzzy logic and neural networks [21,22]. Muscle relaxation has been controlled using a fuzzy logic controller. DOA is, however, more difficult to measure. Recent research on the measurement of DOA has concentrated on use of the AER signal [23,24]. More recently, neural networks and fuzzy logic have been used to analyse the signal and measure the DOA [25]. In recent work by the authors, neural networks and fuzzy logic [26] have been used to control the amount of infused drug using a rule-based fuzzy logic controller.

This section describes the structure of a real-time measuring system based on fuzzy logic. The system uses neuro-fuzzy and multiresolution wavelet analysis for monitoring the DOA based on AER signals. The measuring system uses a DSP signal processing chip hosted in a PC, providing averaging and analysis using multiresolution wavelet analysis. The analysed signal is fed to a neuro-fuzzy system where the inference takes place to obtain a measure of the DOA. Another measure for DOA is based on the cardiovascular system status using a rule-based fuzzy logic classifier. The two measures are merged together using rule-based fuzzy logic data fusion to decide the final DOA. Based on the classified DOA, a target concentration is decided by a rule-based fuzzy logic controller which feeds the target to a target controlled infusion (TCI) algorithm.

### 13.4.1   Assessments of DOA

#### 13.4.1.1   Auditory evoked responses for DOA assessment (AER_DOA)

Depth of anaesthesia is measured via the evoked brain potentials which are measured using one or more scalp electrodes. The AER are the responses in the EEG to clicks applied to the ear. They can be recorded in operating theatre and extracted from the brain by computer averaging. The signal is recorded using a computer fitted with a DSP card which produces the auditory stimulus in the patient's headphone, and samples the EEG at 1 kHz for a sweep period of 120 ms after each click. The computer produces an average of 188 data sweeps plus calculation time, each 30 s. A typical AER signal can be classified into three sections: brain stem (0–15 ms), early cortical (15–80 ms) and late cortical (80–1000 ms). The response features comprise a number of peaks with different amplitudes and latencies. The peaks appearing in the early cortical section are labelled as Na, Pa, Nb and Pb. A Pathfinder electro-diagnostic system was used to record AEP signals using a 0.1 ms click duration at a 6.122 Hz stimulating frequency. A study carried out by Schwender and others [24] measured latencies of the peaks Na, Pa, Nb and Pb (ms) and amplitudes Na/Pa, Pa/Nb and Nb/Pb ($\mu$V). The AER signals were recorded for patients awake and asleep. A grand average was calculated for different levels of anaesthesia for constant level of anaesthetic drug, showing the relationships which exist between DOA and latencies.

One of the methods for analysing the AER signal is that of Fourier transform (FT). Schwender and others [27] have used it to find the region corresponding to the maximum power in the frequency spectrum as a means to identify the major oscillatory components in the signal. They showed that an increase in the DOA yields a shift in the power spectrum of the oscillatory component along with an attenuation in the response. The shortcoming of FT is that it cannot represent the transient nature of the peaks in the signal. However, the wavelet transform (WT) has the ability to deal with non-stationary signals, as well as the ability to decompose the signal into its frequency components, allowing the identification of major oscillatory components.

The DOA was based on the wavelets analysis of the AER signal and classified using an ANFIS. The ANFIS architecture is based on a fuzzy inference system implemented in a framework of an adaptive network [28]. Using a hybrid learning procedure, ANFIS can learn an input–output mapping based on human knowledge (in the form of if–then fuzzy rules). The ANFIS architecture has been employed to model nonlinear functions, identify nonlinear components online in a control system, and predict a chaotic time series etc. It performs the identification of an input–output mapping, available in the form of a set of N input–output examples, with a fuzzy architecture, inspired by the Takagi–Sugeno modelling approach [29]. The fuzzy architecture is characterised by a set of rules, which are properly initialised and tuned by a learning algorithm. The rules are in the form:

1. if input 1 is A11 and input 2 is A12 then output $=$ f 1(input1,input2),
2. if input 1 is A21 and input 2 is A22 then output $=$ f 2(input1,input2),

where A$ij$ are parametric membership functions.

Multiresolution Wavelet Analysis was used to obtain the coefficients of the Detail components D1–D4 of the AER. A 5-point weighted moving average was used to smooth the data; it was observed that this weighted moving average did not add any noticeable delay in the signal [30,31]. These wavelet coefficients were then used to calculate the average energy in each of D1, D2, D3_1, D3_2, D3_3 and D4. The notes taken down during the course of the surgical procedure were then used to label the features. These features were furthermore labelled with the anaesthetist's expert opinion (based on the clinical signs of anaesthesia and his experience) on the DOA at different stages of the surgical procedure. Thus, labelling of the data could be carried out. Six labels, corresponding to distinct anaesthetic depths were used: awake, light, ok_light, ok, ok_deep and deep. The data outliers were removed and the remaining data were normalised to zero mean and unit variance for each of the six categories. The data from nine patients were used to create a training and a validation set. One thousand one hundred and thirty one data points were used in the training and validation sessions.

### 13.4.1.2   Fuzzy logic rule-based classifier for CV based DOA (CV_DOA)

The CV_DOA classifier is based on two inputs, namely heart rate (HR) and systolic arterial pressure (SAP) and the output is the DOA. The input variables can be classified into three linguistic levels for each variable, low, medium and high. Medium means the input variable is within the normal range, low and high mean below or above the normal range respectively. In contrast, the output variable (DOA) has six linguistic levels, deep, ok_deep, ok, ok_light, light and awake as shown in Table 13.4.

*Table 13.4   CV_DOA classifier rule-base*

|  | CV_DOA | SAP | | |
|---|---|---|---|---|
|  |  | High | Medium | Low |
| HR | High | Awake | Ok_light | Ok |
|  | Medium | Light | Ok | Ok_deep |
|  | Low | Ok | Ok_deep | Deep |

*Table 13.5   F_DOA classifier rule-base*

|  | F_DOA | AER_DOA | | | | | |
|---|---|---|---|---|---|---|---|
|  |  | Awake | Light | Ok_light | Ok | Ok_deep | Deep |
| CV_DOA | Awake | Awake | Awake | Light | Light | Ok_light | Ok |
|  | Light | Awake | Light | Ok_light | Ok_light | Ok | Ok |
|  | Ok_light | Light | Ok_light | Ok_light | Ok | Ok | Ok_deep |
|  | Ok | Ok_light | Ok | Ok | Ok | Ok | Ok_deep |
|  | Ok_deep | Ok_light | Ok | Ok | Ok | Ok_deep | Deep |
|  | Deep | Ok | Ok | Ok_deep | Ok_deep | Deep | Deep |

Details of the rule-base derived from anaesthetists' experience have been reported in a previous study by Linkens [26].

### 13.4.1.3   Fuzzy logic rule-based data fusion for merging AER_DOA and CV_DOA

The last junction is to merge the classified DOA measured from the previous agents (AER_DOA and CV_DOA) to produce a final DOA(F_DOA) using rule-based fuzzy logic data fusion. As mentioned in the last two sections, each variable has six linguistic levels. The same structure is used for the output measure of DOA. The rule-base for merging the two measures is shown in Table 13.5.

### 13.4.2   Target controlled infusion (TCI)

The TCI system was first introduced by Alvis and co-workers [32] who controlled fentanyl infusion during anaesthesia. The idea behind the system is to keep the plasma concentration level constant. In order to do that, the infusion rate has to be determined. Once the desired level is reached, the system maintains the concentration in an open-loop manner. The TCI system used in this study is based on a three-compartment patient model with the pharmacokinetic parameters obtained from Glass *et al.* [33].

The pharmacokinetic model of the patient as well as the pharmacokinetic parameters (in $min^{-1}$) describe the flow rate of the drug between the various compartments. The three-compartment model is a linear model since the pharmacokinetic parameters describing the model are constant. The rate of transfer of the drug from one compartment to another is proportional to the amount of drug present in the first compartment.

### 13.4.3   Closed-loop system for control of DOA

A schematic block diagram of the system is shown in Figure 13.1 where the system is implemented on two computers, one acting as the controller and the other acting as the patient simulator. The controller computer is used as a host for the DSP chip card for recording the AER signal. In addition, it is used for analysing the signal, recording heart rate and blood pressure, and the associated calculation regarding DOA. The TCI system is also implemented on the controller computer for calculating the amount of the infused drug. In some of the latest syringe driver systems, a TCI system is installed on board the device, in which case the TCI system would not be required. The DSP card records the AER signal from the amplifier board as well as providing the auditory trigger signal at 6.122 Hz rate. One hundred and eighty eight sweeps each of 120 ms are collected, and by the end of the collection the data are averaged and filtered to produce the final AER signal every 30 s. Other measurements are recorded such as the heart rate and systolic blood pressure each 30 s, therefore the whole system sampling period is 30 s.

On the simulator computer, the patient model is represented using a fourth order Runge–Kutta algorithm in two parts, the pharmacokinetic model and the pharmacodynamic model. The output of the pharmacokinetic model is the plasma concentration which is fed to two paths; the first path is to the pharmacodynamic model block (effect compartment) then to the nonlinear effect block (Hill equation). This output is the mean arterial pressure which is fed in turn to two more blocks, to calculate the associated systolic pressure and the heart rate. The second path is to the AER signal simulation block which consists of a look-up table for the AER signal classified on the basis of the plasma concentrations from nine patients. According to the plasma concentration calculated by the model, a certain group is selected which is related to the present plasma concentration, then a single AER trace is selected randomly from ten stored traces. Each trace consists of 120 data points sampled at 100 ms. The selected trace is used as an output AER signal via a D/A converter. The AER signal is released on the reception of the trigger signal from the AER system amplifier which is used in real-time for the earphone trigger. The input to the simulator is the infused drug calculated by the controller computer via an A/D converter. The final output of the simulator is SAP and HR at 30 s update intervals. The AER signal is sampled at 120 ms for each trace.

After collecting the AER signal, it is fed to the wavelet analysis block to determine the required components of the signal, then it is fed to the ANFIS block where the AER_DOA is calculated. In the meantime the recorded HR and SAP are fed to the fuzzy logic rule-base for analysis of CV_DOA. Prior to that, both SAP and HR

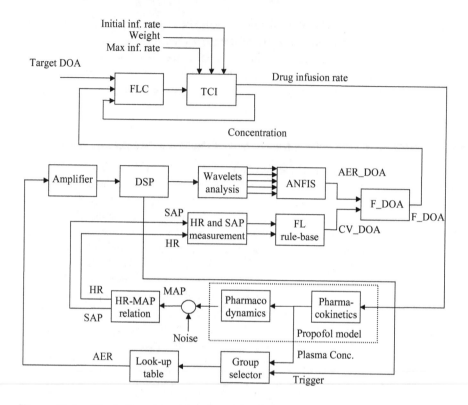

*Figure 13.1    Block diagram of the closed-loop control system*

are filtered using a weighted moving average filter. The last stage in DOA analysis is to feed CV_DOA and AER_DOA to the final DOA analysis block. The calculated F_DOA is fed to the next block in the system which is the plasma concentration controller block. Based on the current target concentration and the measured DOA, a new target concentration is calculated using a fuzzy logic rule-based controller. The target concentration controller consists of three rule-bases. One rule-base is selected depending on the required DOA level. The three levels are ok_light, ok and ok_deep.

The plasma target concentration is fed to the TCI block which is based on the initial information supplied. It calculates the drug infusion rate and feeds it to the syringe driver which in this case will be the simulator via a DA converter. The initial settings for the TCI algorithm are based on the patient weight, maximum infusion rate and the initial infusion rate.

## 13.4.4    Simulation and results

In this section, results are presented for different types of simulation runs: a nominal patient and a sensitive patient. The type of AER signal used in the simulator, and its random selection, gives the phenomena of a random stimulus component to the AER signal, together with inherited noise through the AER amplifier.

The following parameters were used during the simulation runs reported here:

- Patient weight            70 kg
- Propofol concentration    10,000 ng/ml
- Maximum infusion rate     6000 mg/h
- HR base line              70 BPM
- SAP base line             140 mmHg
- Patient sensitivity (gain)    Normal   6.58
                                Sensitive 9.32
- Moving average filter used for *HR*, *SAP*, *AER_DOA*, *CV_DOA* and *F_DOA*.

In the simulation run graphs (Figures 13.2 and 13.3), each figure is divided into four subplots: the target, central compartment concentration, and the infused drug are shown in the first plot (a). The second plot (b) shows the heart rate and blood pressure, the third plot (c) shows the classified AER_DOA and CV_DOA, while the last plot (d) shows the final DOA as the two intermediate DOAs are merged together.

### 13.4.4.1  Normal sensitivity patient

The simulation run for a nominal patient is shown in Figure 13.2. As seen in the figure the controller performance is not steady. This is due to the fact that the selection of AER signals is chosen randomly, and since the signals contain noise and different operation stimuli, the random selection factor appears to have affected the controller behaviour. This will give the model the facility to behave differently each time the simulator runs, which makes it look more like a real set of patients. The simulation of the patient was in steady state for the whole period. AER_DOA indicated a deeper level of anaesthesia than CV_DOA, but the final level after merging was within the ok level as required by the input command signal. The average concentration during the maintenance stage was about 4000 mg/ml.

### 13.4.4.2  High sensitivity patient

A sensitive patient is known to consume a lower amount of drug to reach the same level of DOA as the nominal patient. This is due to the high sensitivity to the drug. Such phenomena can be simulated by altering the patient's model gain to a higher level. For a nominal patient a gain of 6.58 was used, while the sensitive patient gain was increased to 9.32. This is an increase of 41 per cent which means 28.5 per cent reduction in the drug consumption. The simulation run is shown in Figure 13.3. In the first part where the target TCI and plasma concentration is shown, the target TCI appears lower than the plasma concentration, this being due to the high sensitivity of the patient. In contrast, in Figure 13.2, the plasma concentration follows closely the target TCI which is part of the TCI algorithm. In the case of the sensitive patient, the model used in the TCI algorithm is the nominal parameters model, accordingly the TCI target was kept below the nominal target. However, due to the correction by the feedback loop, the actual plasma concentration was controlled near to 3000 ml/mg.

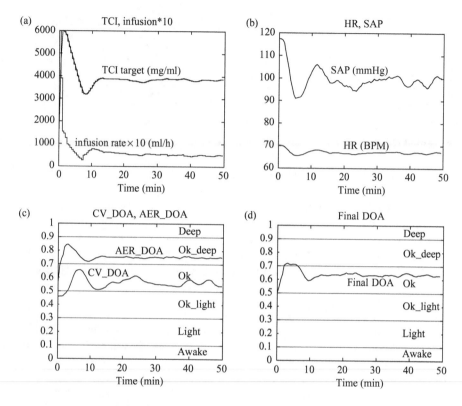

*Figure 13.2    Simulation run for a nominal 70 kg patient*

## 13.4.5   Clinical trials

The system has been used online in the operating theatre for clinical trials in the Royal Hallamshire Hospital, Sheffield. This procedure required connecting the system to a DATEX device for recording HR and BP, as well as a Graseby 3400 syringe pump via the RS232 serial port.

The first clinical trial carried out using the system was based on an injection of a 100 mg bolus of morphine and 100 μg of fentanyl. Then the propofol drug infusion was started with a TCI level of 4000 mg/ml. The patient data are shown below:

| | |
|---|---|
| Patient initials: | CJ |
| Sex: | Male |
| Weight: | 96 kg |
| Date of Birth: | 24-12-1961 (Age: 37) |
| Blood Pressure: | 150/80 |
| Heart Rate: | 91 |
| Operation Type: | Hernia repair |
| Anaesthesia: | start with 10 mg morphine, 100 μg fentanyl |

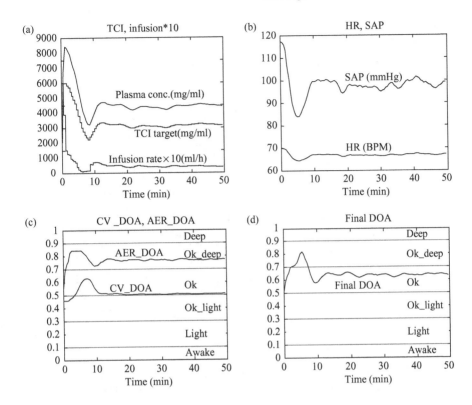

*Figure 13.3    Simulation run for a sensitive 70 kg patient*

The time response as shown in Figure 13.4 has four parts: (a) shows the TCI level and the drug infusion rate, (b) shows the systolic blood pressure and the heart rate, (c) shows AER_DOA and CV_DOA and finally (d) shows F_DOA by merging both measures of DOA. It can be seen that the drug TCI level started with a small value, increased during the middle of the operation, then decreased towards the end of the operation. This type of profile is usually adopted by anaesthetists during a typical operation cycle. During the operation, CV_DOA varied between ok_light and ok but did not reach ok_deep due to the effect of the muscle relaxant drug. AER_DOA was indicating ok_deep but also was affected by surgical interference from X-ray equipment and diathermy. The final fusion of the two measures indicated that DOA was in the ok region during most of the operation.

### 13.4.6   Discussion on DOA monitoring and control

A closed-loop control system for measuring and controlling depth of anaesthesia has been developed based on two different measures, the AER system and the cardio-vascular system. The system is aimed at providing more reliable measurement of DOA using data fusion principles. The system has been tested under three different

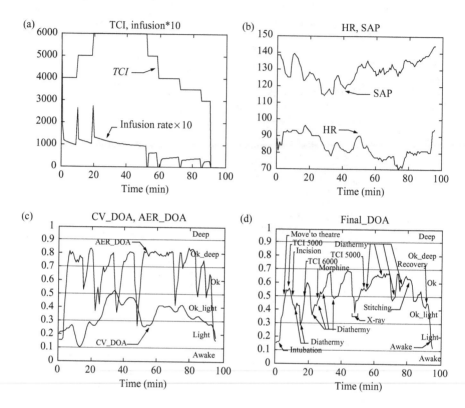

*Figure 13.4　Time response for a clinical trial (patient CJ)*

patient sensitivities: nominal, sensitive and insensitive. The system has been tested on a patient simulator where a second computer acts as the patient. A numerical model is simulated using a Runge–Kutta integration method. The results obtained from the simulations carried out have shown that the closed-loop system effectively maintains the patient at the clinically acceptable anaesthetic depth even when the stimuli reflect those of a real surgical procedure.

The system has been used online in the operating theatre for clinical trials at the Royal Hallamshire Hospital, Sheffield. This procedure has required connecting the system to a Datex AS/3 device for recording HR and BP in addition to the AER monitor, as well as a Graseby 3400 syringe pump. Initial results are promising, indicating reliable and robust measurement of DOA via data fusion techniques.

## 13.5　Intelligent systems in bladder cancer prediction

New techniques to enable the prediction of individual cancer behaviour are needed to improve patient care. The discovery of robust biomarkers combined with improved methods of data modelling is likely to deliver this ability. Like many forms of malignancy, the incidence of bladder cancer is increasing, with over 11,000 new cases in 1994 in England and Wales. At presentation, 70 per cent of transitional cell carcinomas of the bladder (TCC[3]) are superficial and non-invasive. They can be managed by a combination of local endoscopic resection and intra-vesical chemotherapy. Following treatment, 50 per cent will recur as similar non-invasive lesions, and a smaller percentage (20 per cent) will progress to muscle invasion. Muscle invasive tumours have a poor prognosis (50 per cent five-year survival rates) and require radical therapy if cure is to be achieved.

To date, the most reliable predictors of tumour behaviour are the pathological stage and grade at diagnosis (TNM classification). Specific tumours may also have additional prognostic information, for example, for superficial TCCs, the presence of carcinoma *in situ* and the state of the bladder three months after surgery are important prognostic factors. While these parameters are useful for stratifying patients into subgroups, it is still impossible to predict the behaviour of *individual* tumours. The development of molecular medicine has yielded many new molecules that may be useful as biomarkers, to help the prediction of cancer behaviour. Some of the most biologically promising are the p53 and mismatch repair (MMR) proteins hMSH2 and hMLH1. The p53 gene is estimated to be mutated in over 50 per cent of human cancers [34] and has been shown to predict recurrence and survival in bladder cancer [35]. Loss of the MMR proteins (most frequently hMSH2 and hMLH1) results in microsatellite instability and subsequent carcinogenesis in hereditary [36] and sporadic colorectal cancer [37]. The resultant tumours have a distinctive phenotype with chemoresistance [38] and better than expected (for their stage) clinical outcomes [39]. However, despite optimistic reports for these and other molecular biomarkers, to date no single molecule is sufficiently robust to use in routine clinical practice. It appears likely that the best results will be achieved with a panel of molecular markers.

An alternative solution to the problem of accurately predicting tumour behaviour lies within the interpretation of data. Traditional statistical methods (e.g. logistic regression) are limited by their need for linear relationships between variables, for large datasets and their poor performance when large amounts of 'noise' (inherent variation) contaminate a dataset. Thus predictions are only accurate in 70 per cent of tumours using the TNM classification [40]. By using AI methods, such as neuro-fuzzy modelling (NFM) and ANNs, complex nonlinear relationships between dependent and independent variables, within a population whose distribution may not be normal, can be identified. The most commonly used ANN is the multilayer perceptron and ANNs have been applied to clinical medicine since 1989 [41]. While individual authors have shown that ANNs are superior to standard statistical analysis using the TNM staging system to predict breast, colorectal and bladder cancer outcomes [40,42], a recent review concluded that to date evidence suggests that ANNs may only outperform statistical methods in small datasets [43]. Also, ANNs are not without problems. They can be 'overtrained' to learn the inherent variation of a sample population and are non-robust. They do not generalise across the specific problem range of variables, for either interpolation or extrapolation. More importantly, the network is hidden within a functional 'black box'. Thus, it is difficult to gain insight into the model obtained from the data and ensure that clinical and statistical sense prevails. As a result, statisticians are reluctant to believe in the validity of ANN [44]. In addition, the weights attached to different variables are uninterpretable making the interrogation of new variables difficult.

Neuro-fuzzy modelling is an alternative AI method, without many of these drawbacks of ANN. In contrast to ANN, which relies entirely on quantitative (i.e. numerical) ideas, NFM combines both quantitative and qualitative (i.e. linguistic) concepts [45]. The model obtained from the data comprises a number of rules in parallel. Each rule is in the form 'IF Input 1 is $X_1$ AND Input 2 is $X_2$... THEN output 1 is $Y_1$...' where $X$ and $Y$ are qualitative fuzzy labels. The labels are quantified to represent the strength or certainty of the particular input or output. The parallel rules are merged using fuzzy reasoning to produce the necessary quantitative output. While the basic structure of an NFM model is similar to that for ANN, the model rules are entirely transparent, facilitating interpretation and validation.

In the following sections, the predictive abilities of two AI methods, NFM and ANN, are compared with traditional statistical methods using a cohort of TCCs. For each method, the results using both conventional clinicopathological data and experimental findings using putative molecular biomarkers are compared. To date, this was the first study to report the use of NFM in cancer prediction.

### 13.5.1    Patients and methods

*Patients.* One hundred and nine patients with primary TCC of the bladder were studied. These represented a typical UK population of affected patients. The median age at diagnosis was 70 years (range 34–90 years) and the majority of patients were male ($n = 77$ (69 per cent)). Each TCC had been treated at the Royal Hallamshire Hospital, Sheffield, UK, and has a clinical follow up of at least 5 years (median 6, range 5–16). Standard immunohistochemistry was performed using commercially available antibodies. Abnormal expression of hMLH1 (17 per cent), hMSH2 (17 per cent), either (24 per cent) or both (8 per cent) and p53 (75 per cent) was seen in the tumours respectively [46]. Abnormal (reduced) MMR expression was significantly associated with tumours of a more advanced stage ($p = 0.01$ for hMLH1 and $p = 0.03$ for hMSH2) and a worse differentiation ($p = 0.03$ for both hMLH1 and hMSH2), when compared to tumours with normal expression. By five years, tumours with reduced expression of either MMR protein had significantly fewer relapses ($p = 0.03$) than tumours with normal MMR expression. Abnormal (increased) p53 immunohistochemical expression was not significantly related to disease stage, grade or subsequent relapse. Complete methodological and patient details are described elsewhere [47].

*Artificial intelligence modelling.* Two models were developed using both NFM and ANN as follows:

- A Classifier predicted the likelihood of a tumour relapse (yes or no).
- A Predictor predicted the timing of this relapse (months after surgery).

These two models were combined together in series; thus predicting if and when a relapse would occur, to produce the most clinically useful model. To discover the value of the putative molecular biomarkers, the data were analysed twice in each model, as follows;

- In Analysis (A) only the conventional clinicopathological data were studied.
- In Analysis (B) both the conventional clinicopathological data and the additional three molecular putative biomarkers were studied.

For both NFM and ANN, the models were trained on 90 per cent of the patients, before testing on the final 10 per cent. This was then repeated until all patients had been used to test the model, via so-called 'ensembling' and cross-validation. While this was necessary for ANN because of the small dataset, it was also used for NFM in order to standardise the methods. In addition, NFM was modelled without cross-validation, to assess the performance with small datasets.

*Neuro-fuzzy modelling.* The NFM analyses were performed via an extensive in-house suite of software [48] developed using Matlab. The modelling procedure involves a number of iterative loops subsequent to careful data preparation and initialisation of the starting model structure and parameters. These loops refine the model parameters, simplify its structure and component terms to the minimum complexity consistent with the model (i.e. parsimonious modelling), and validate the results.

*Artificial neural network.* The ANN models were produced using the Matlab Neural Network toolbox. For each session, ten models were generated and the best was selected. Using this 'best fit' model, the total dataset was then re-tested to produce the final results. Training was performed using 50 iterative loops, after which number the model's accuracy deteriorated.

*Statistical analysis.* To obtain a probability of tumour relapse, using traditional statistical analysis, logistic regression (LoR) was performed using statistical package for the social sciences (SPSS). To obtain a statistical prediction of the timing of tumour relapse, linear regression (LiR) was performed using SPSS. Statistical comparisons of the accuracy of ANN, NFM and LoR or LiR were performed using a $t$-test. The relationship of the clinical and molecular variables was also assessed with a $t$-test. A $p$-value of $<0.05$ was taken to be significant.

### 13.5.2   Results for bladder cancer prediction

*Artificial intelligence modelling.* The results of the Classifier and Predictor models generated using NFM, ANN, LoR and LiR are described in Tables 13.6 and 13.7. For both models, AI performs better than statistics, with NFM more accurate than ANN in all but one case (Classifier, analysis B). The AURoC (area under receiver operator characteristic curve) for NFM (0.98) is superior to that for both ANN (0.91–0.88) and LoR (0.47–0.49). In Table 13.7, the differences between the predicted and actual timing of relapse are shown as root mean squared (RMS) values. As can be seen, NFM is superior to both ANN and LiR, with lower RMS values. When these are compared, both AI methods are significantly superior to LiR ($p < 0.0006$) and NFM is significantly better than ANN at predicting tumour relapse ($p = 0.01$) in the testing phase, but for the best fit model the $p$-value falls just short of significance ($p = 0.079$). If analyses (A) and (B) are compared individually, there is

*Table 13.6   Presence of tumour relapse: the characteristics of each* Classifier *predictive model*

|     | Sensitivity | Specificity | PPV | NPV | Accuracy | AURoC |
|-----|-------------|-------------|-----|-----|----------|-------|
| NFM |             |             |     |     |          |       |
| A   | 0.92        | 0.90        | 0.98 | 0.72 | 0.92    | 0.98  |
| B   | 0.90        | 0.80        | 0.92 | 0.74 | 0.88    | 0.98  |
| ANN |             |             |     |     |          |       |
| A   | 0.90        | 0.89        | 0.98 | 0.64 | 0.90    | 0.88  |
| B   | 0.94        | 0.96        | 0.99 | 0.84 | 0.95    | 0.91  |
| LoR |             |             |     |     |          |       |
| A   | 0.77        | 1.0         | 1.0  | 0.00 | 0.77    | 0.47  |
| B   | 0.72        | 0.40        | 0.96 | 0.07 | 0.71    | 0.49  |

*Table 13.7   Timing of tumour relapse: the accuracy of each method's* Predictor *model is shown as an RMS value (difference between the timing of the predicted and actual relapse)*

|     | Training | Testing | Best fit (ANN, NFM) |
|-----|----------|---------|---------------------|
| NFM |          |         |                     |
| A   | 7.8      | 9.1     | 7.9                 |
| B   | 6.8      | 4.8     | 6.7                 |
| ANN |          |         |                     |
| A   | 9.1      | 12.3    | 8.8                 |
| B   | 7.1      | 11.7    | 7.6                 |
| LiR |          |         |                     |
| A   | —        | —       | 13.1                |
| B   | —        | —       | 13.1                |

no difference between NFM and ANN for analysis (A) ($p = 0.87$), while for analysis (B), NFM is superior to ANN ($p < 0.001$). The predictions of all three methods are shown graphically in Figure 13.5. When cross validation is omitted from the NFM model, the RMS values are: for analysis (A) 7.6 (training), 10.3 (testing) and 8.5 (best fit), and for analysis (B) 7.2 (training), 7.6 (testing) and 7.3 (best fit).

## 13.5.3   Discussion on intelligent systems for bladder cancer prediction

The accurate prediction of an individual patient's tumour response to treatment is the Holy Grail of oncology. Indeed, recent discoveries in molecular medicine [38] and improvements in clinical treatments have made it now more important than ever

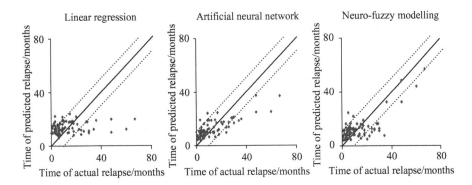

*Figure 13.5   Scatter plots showing the actual and predicted times of tumour relapse.*
*The actual (x-axis; 0–80 months after surgery) and predicted times*
*(y-axis) of tumour relapse are plotted against each other. The proximity*
*of each point to the line of perfect match (continuous line), indicates the*
*prediction's accuracy. Dashed lines indicate the ±10 per cent range.*
*The data shown are from analysis (A)*

to predict tumour behaviour. It has been shown that AI methods can predict tumour behaviour with greater accuracy than traditional statistical methods, and that NFM is superior to ANN in this section. Until the advent of AI, the best method of predicting tumour behaviour was logistic regression. However, the poor predictive accuracy of this method (70–77 per cent in this study and 69–73 per cent previously) and the fact that a general probability is not applicable to an individual patient, show the need for improvements. Using Figure 13.5, if LiR were applied in clinical practice, those patients with late relapsing tumours (over 40 months) would have had their most intensive cystoscopic surveillance too early for their actual relapse. On an individual basis, those two patients with relapse after 60 months (both predicted to recur by 20 months) may have been falsely reassured and discharged.

It has been confirmed that ANN provides a powerful and accurate predictive method, and, as with other reports, that it is superior to traditional statistical methods. Despite optimistic reports on the application of ANN, the functional layers of the ANN are hidden, with the uninterpretable weights attached to individual variables. This opacity remains an obstacle to the widespread introduction of ANN. Unlike previous studies a comparison has been made of ANN with NFM, which has transparent functional layers. The study has shown that NFM produces a significantly more accurate prediction than LoR and LiR, and is equivalent to or better than ANN. NFMs are particularly good when working with small datasets, as shown by the relative improvement of NFM over ANN when using analysis (A), and the small improvement with the use of cross-validation for NFM.

While the accuracy of NFM is important, it has many other benefits over ANN that promise to make it more acceptable to the clinical and scientific community. Unlike the 'black-box' approach of ANN, the NFM approach is transparent. The problem specific qualitative modelling representation can be easily translated into understandable medical terms. NFM uses the modelling abilities of fuzzy logic to complete a profile for each variable. This produces a set of parallel rules, which are summated in series and interpreted to produce a quantitative output. An example is shown in Figure 13.6(a). The top line, Rule 1, shows that a poorly differentiated (grade 3), superficially invasive tumour (stage T1), in a 70-year old male current smoker (30 cigarettes per day), with abnormal p53 staining, but otherwise normal variables, will have a short time to tumour relapse.

Once trained on a dataset, the NFM suite is then able to reduce the inputs into those that have most influence. This is likely to assist clinicians in day-to-day clinical practice, where insufficient data may be present in a patient's casenotes. In Figure 13.6(b), the NFM suite of programs has automatically reduced the nine inputs to the four most effective ones. These can be interpreted, Figure 13.6(c), to confirm that these are clinically sensible inputs (tumour grade, patient age, smoking history and p53 expression). The automatic selection of these inputs can easily be validated; for example, authors have used multivariate analysis of large TCC series to show that grade is more important than stage at predicting relapse [49]. This ability to prioritise the inputs will prove very useful in numerous clinical areas. Furthermore, because the fuzzy representation allows the predictive modelling rules to be understood, expertise can be incorporated into the selection of the inputs and manipulation of the model's rules. As a result, non-sensible variables which may be incidentally over-represented in the training dataset can be removed, or have their importance reduced. In addition, if there is medical knowledge available relevant to the NFM rule set, this can be added to the model easily. This will then enable the model to extrapolate and be more robust than other AI approaches.

According to the literature, this was the first report of the use of NFM to predict the behaviour of cancer. It has been demonstrated that NFM can successfully predict the occurrence and timing of tumour relapse, with similar or greater accuracy than ANN and LoR. In addition, NFM modelling appears to be superior to ANN and LoR in its transparency, its ability to incorporate expertise, its superior performance with sparse data, its ability to select the most useful input criteria (parsimony) and to allow predictions of outcome that result from changes in the value of individual inputs. These features suggest that NFM could be used as a valuable and versatile tool to address numerous clinical situations. While the predictions have been performed using bladder cancer data, these methods are transferable to all other human malignancies. The full paper is presented in Catto *et al.* [47].

## 13.6    Conclusions

With its vast uncertainties and lack of precise measurements the field of medicine and allied topics clearly provides a challenging scope for utilisation of intelligent systems technology. The two surveys show that there is both potential and awareness of this fact. In the realm of fuzzy technology there are many successful applications within medicine and this is likely to continue. However, the transfer of technology into routine clinical management is, as always, slower than could be wished.

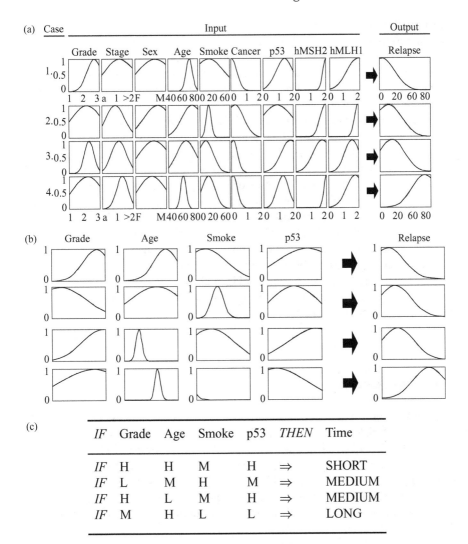

*Figure 13.6* *(a) Neuro-fuzzy modelling output, after defuzzification. The above diagram represents the modelling rules used by NFM. (b) Neuro-fuzzy modelling using the most useful input variables. NFM has rationalised the nine inputs, into those that are most informative: grade of tumour, age of patient, smoking exposure and p53 immunohistochemical status of the tumour. In the first case (top line), we can see that a grade-3 tumour, in a 70-year old smoker, with abnormal p53 immunostaining is predicted to relapse at an early stage. (c) A Linguistic translation of the NFM output shown in (b). In each box a word has replaced the output graph seen in (b). It should be emphasised that these are fuzzy words and make sense within the model's rules, not necessarily in isolation. For example, the term MEDIUM when applied to p53 does not make sense in isolation*

Other intelligent techniques such as ANN, neuro-fuzzy and evolutionary computing (for optimisation) are also well represented in medical applications. The theme of adaptivity, however, is not so clearly researched largely because of lack of clarity in its meaning and concerns about robustness and stability. This continues to be a major matter for research endeavour before there will be the necessary technology transfer into clinical applications. There are, however, numerous examples of lower level adaptivity which have been successfully documented and implemented.

The case studies are very different in nature. The first one on unconsciousness monitoring and control presents a big challenge in measurement via highly inferential methods, together with many aspects of intelligent technology to achieve closed-loop control. The successful completion of the project entailed the strong integration of several techniques plus research into novel concepts in the field of anaesthesia. The second case study on bladder cancer prediction was very different and relied on a single technology of either ANN or neuro-fuzzy models on sparse data sets incorporating novel genetic markers. The success of these initial studies in prediction was augmented significantly by the resultant interpretability of the neuro-fuzzy model particularly for prognosis relating to smoking habits.

Obviously, there is a long way to go, but the future of intelligent systems in medicine is assured. In the field of biology, these remarks can be amplified since vast studies in measurements at genetic, proteomic and cellular levels are currently being made. The future is exciting!

## Acknowledgements

The collaboration of a number of co-authors in the main papers representing the source of materials from which this chapter has been produced is acknowledged. These include M. Mahfouf and M. Chen from the Department of Automatic Control and Systems Engineering, J. W. Catto and F. C. Hamdy from the Academic Unit of Urology, The University of Sheffield, D. G. Keyserlingk, Institute fur Anatomie, RWTH Aachen and G. Dounias, Business School, University of Aegean. Funding for the work on anaesthesia monitoring and control was provided by the UK EPSRC.

## References

1 ZADEH, L.: 'Biological application of the theory of fuzzy sets systems'. Proceedings of International Symposium on *Biocybernetics of the Central Nervous System*, Little, Brown and Co., Boston, MA, 1969, pp. 199–212

2 MARTIN, J. F.: 'Fuzzy control in anaesthesia', *Journal of Clinical Monitoring*, 1994, **10**, pp. 77–80

3 VON KEYSERLINGK, D. G.: 'Roadmap for ERUDIT', **EUNITE** – European Network on Intelligent Technologies for Smart Adaptive Systems, Aachen, Germany, February 1998

4 KANDAL, A., and YAGER, R. R.: 'A bibliography on fuzzy sets, their applications, and related topics', in GUPTA, M. M., RAGADE, R. K., and

YAGER, R. R. (Eds): 'Advances in fuzzy set theory and applications' (North-Holland, Amsterdam, Netherlands, 1979)

5  MAIERS, J. E.: 'Fuzzy set theory and medicine: the first twenty years and beyond'. Proceedings of the ninth Annual Symposium on *Computer Applications in Medical Care*, IEEE Computer Society Press, Washington, DC, 1985, pp. 325–9

6  DUBOIS, D., and PRADE, H.: 'Possibility theory – an approach to computerised processing of information' (Plenum Press, New York, 1988)

7  DUBOIS, D., and PRADE, H.: 'Representation and combination of uncertainty with belief function and possibility measures', *Computational Intelligence*, 1988, **4**, pp. 244–64

8  DUBOIS, D., and PRADE, H.: 'Weighted fuzzy pattern matching', *Fuzzy Sets and Systems*, 1988, **28**, pp. 313–31

9  DUBOIS, D., LANG, J., and PRADE, H.: 'Fuzzy sets in approximate reasoning, part 2: logical approaches', *Fuzzy Sets and Systems*, 1991, **40**, pp. 203–44

10  DUBOIS, D., and PRADE, H.: 'What are fuzzy rules and how to use them', *Fuzzy Sets and Systems*, 1996, **84**, pp. 169–85

11  DUBOIS, D., and PRADE, H.: 'An introduction to fuzzy systems', *Clinica Chimica Acta*, 1998, **270** (1), pp. 1–29

12  BEN FERHAT, S., DUBOIS, D., and PRADE, H.: 'Practical handling of exception-tainted rules and independence information in possibilistic logic', *Applied Intelligence*, 1998, **9**, pp. 101–27

13  BURDEA, G. C.: 'Virtual reality and robotics in medicine'. Proceedings of fifth IEEE International Workshop on *Robot and Human Communication* RO-MAN '96, Tsukuba, IEEE, New York, 1996, pp. 16–25

14  STEIMANN, F.: 'Fuzzy set theory in medicine', *Artificial Intelligence in Medicine*, 1997, **11**, pp. 1–7

15  MOLLER, D. P. F.: 'Fuzzy logic in medicine'. SO: Fourth European Congress on *Intelligent Techniques and soft computing Proceedings*, EUFIT '96, Verlag Maniz, Aachen, Germany, 1996, vol. 3, pp. 2036–45

16  ZIMMERMANN, H. J.: 'Fuzzy set theory and its applications' (Kluwer Academic, London, Boston, MA, 1985)

17  COHEN, M. E., and HUDSON, D. L.: 'Comparative approaches to medical reasoning. Advances in fuzzy systems – applications and theory, vol. 3' (World Scientific Publishing, Singapore, 1995)

18  TEODORESCU, H. N., KANDEL, A., and JAIN, L. C.: 'Fuzzy and neuro-fuzzy systems in medicine' International Series on Computational Intelligence, vol. 2 (CRC Press, Boca Raton, London, NY, Washington DC, 1999)

19  ABBOD, M. F., VON KEYSERLINGK, D. G., LINKENS, D. A., and MAHFOUF, M.: 'Survey of utilisation of fuzzy technology in medicine and healthcare', *Fuzzy Sets and Systems*, 2001, **120**, pp. 331–49

20  ABBOD, M. F., LINKENS, D. A., MAHFOUF, M., and DOUNIAS, G.: 'Survey on the use of smart and adaptive engineering systems in medicine', *Artificial Intelligence in Medicine*, 2002, **686**, pp. 1–31

21  LINKENS, D. A., MAHFOUF, M., and PEACOCK, J. E.: 'Propofol induced anaesthesia: a comparative control study using a derived

pharmacokinetic-pharmakodynamic model'. Internal report, Department of Automatic Control and Systems Engineering, University of Sheffield, 1993

22　LINKENS, D. A., SHIEH, J. S., and PEACOCK, J. E.: 'Machine learning rule-based fuzzy logic control for depth of anaesthesia', *IEE Control '94*: International Conference on Control, Coventry, UK, vol. 1, 21–24 March 1994, pp. 31–6

23　NEWTON, D. E. F., THORNTON, C., KONIECZKO, K. M. *et al.*: 'Auditory evoked response and awareness: a study in volunteers at sub-Mac concentration of isoflurane', *British Journal of Anaesthesia*, 1992, **69**, pp. 122–9

24　SCHWENDER, D., KLASING, S., CONZEN, P., FINSTERER, U., POPPEL, E., and PETER, K.: 'Midlatency auditory evoked potentials during anaesthesia with increasing endexpiratory concentrations of desflurane', *Acta Anaesthesiologica Scandinavica*, 1996, **40** (2), pp. 171–6

25　WEBB, A., ALLEN, R., and SMITH, D.: 'Closed-loop control of depth of anaesthesia', *Measurement and Control*, 1996, **29**, pp. 211–15

26　LINKENS, D. A., ABBOD, M. F., and BACKORY, J. K.: 'Fuzzy logic control of depth of anaesthesia using auditory evoked responses', IEE Colloquium on *Fuzzy Logic Controller in Practice*, IEE Savoy Place, London, 15 November 1996, pp. 4/(1–6)

27　SCHWENDER, D., KAISER, A. D., KLASING, S., PETER, K., and POPPEL, E.: 'Midlatency auditory evoked potentials and explicit and implicit memory in patients undergoing cardiac surgery', *Anaesthesiology*, 1994, **80**, pp. 493–501

28　JANG, J. R.: 'ANFIS: Adaptive-network based fuzzy inference system', *IEEE Transactions on Systems, Man, and Cybernetics*, 1993, **23** (3), pp. 665–85

29　BERSINI, G., BONTEMPI, C., and DECAESTECKER, C.: 'Comparing RBF and fuzzy inference systems on theoretical and practical basis', in FOGELMAN-SOULIE, F., and GALLINARI, P. (Eds): ICANN '95, International Conference on *Artificial Neural Networks*, Paris, 1995, vol. 1, pp. 169–74

30　LINKENS, D. A., ABBOD, M. F., and BACKORY, J. K.: 'Closed-loop control of anaesthesia using a wavelet-based fuzzy logic system'. EUFIT '97, 1997, Aachen, Germany, September 8–12

31　LINKENS, D. A., ABBOD, M. F., and BACKORY, J. K.: 'Auditory evoked responses for measuring depth of anaesthesia using a wavelet fuzzy logic system', *Control Engineering Practice*, 1997, **5** (12), pp. 1717–26

32　ALVIS, J. M., REVES, J. G., SPAIN, G. A., and SHEPPARD, L. C.: 'Computer assisted continuous infusion of the intravenous analgesic fentanyl during general anaesthesia – an interactive system', *IEEE Transactions on Biomedical Engineering*, 1985, **BME-32** (5), pp. 323–9

33　GLASS, P. S., GOODMAN, D. K., GINSBERG, B., REEVES, J. G., and JACOBS, R. G.: 'Accuracy of pharmacokinetic model-driven infusion of propofol', *Anesthesiology*, 1989, **71** (3A), p. A277

34　HOLLSTEIN, T., SIDRANSKY, D., VOGELSTEIN, B., and HARRIS, C. C.: 'p53 mutations in human cancers', *Science*, 1991, **253**, p. 49

35　ESRIG, D., ELMAJIAN, D., GROSHEN, S. *et al.*: 'Accumulation of nuclear p53 and tumour progression in bladder cancer', *New England Journal of Medicine*, 1994, **331**, p. 1259

36 AALTONEN, L. A., PELTOMAKI, P., LEACH, F. S. *et al.*: 'Clues to the pathogenesis of familial colorectal cancer', *Science*, 1993, **260**, pp. 812–16

37 LIU, B., NICOLAIDES, N. C., MARKOWITZ, S. *et al.*: 'Mismatch repair gene defects in sporadic colorectal cancers with microsatellite instability', *Nature Genetics*, 1995, **9**, pp. 48–55

38 DRUMMOND, J. T., ANTHONEY, A., BROWN, R., and MODRICH, P.: 'Cisplatin and adriamysin resistance are associated with MutLA and mismatch repair deficiency in an ovarian tumour cell line', *Journal of Biological Chemistry*, 1996, **271**, pp. 19645–8

39 GRYFE, R., KIM, H., HSIEH, E. T. *et al.*: 'Tumor microsatellite instability and clinical outcome in young patients with colorectal cancer', *New England Journal of Medicine*, 2000, **342**, pp. 69–77

40 BURKE, H. B., GOODMAN, P. H., ROSEN, D. B. *et al.*: 'Artificial neural networks improve the accuracy of cancer survival prediction', *Cancer*, 1997, **79**, pp. 857–62

41 BAXT, W. G.: 'Use of an artificial neural network for data analysis in clinical decision-making: the diagnosis of acute coronary occlusion', *Neural Computation*, 1989, **2**, pp. 480–9

42 QURESHI, K. N., NAGUIB, R. N. G., HAMDY, F. C., NEAL, D. E., and MELLON, J. K.: 'Neural network analysis of clinicopathological and molecular markers in bladder cancer', *Journal of Urology*, 2000, **163**, pp. 630–3

43 SARGENT, D. J.: 'Comparison of artificial neural networks with other statistical approaches', *Cancer*, 2001, **91**, pp. 1636–42

44 SCHWARZER, G., VACH, W., and SCHUMACHER, M.: 'On the misuses of artificial neural networks for prognostic and diagnostic classification in oncology', *Statistics in Medicine*, 2000, **19**, pp. 541–61

45 NIE, J., and LINKENS, D. A.: 'Fuzzy-neural control' (Prentice Hall, London, 1995)

46 CATTO, J. W. F., XINARIANOS, G., BURTON, J. L., MEUTH, M., and HAMDY, F. C.: 'Differential expression of hMLH1 and hMSH2 is related to bladder cancer grade, stage and prognosis, but not microsatellite instability', *International Journal of Cancer*, 2003, **105**, pp. 484–90

47 CATTO, J. W. E., LINKENS, D. A., ABBOD, M. F. *et al.*: 'Artificial intelligence in predicting bladder cancer outcome: a comparison of neuro-fuzzy modelling and artificial neural networks', *Clinical Cancer Research*, 2003, **9**, pp. 4172–7

48 CHEN, M., and LINKENS, D. A.: 'A systematic neurofuzzy modelling framework with application to material property prediction', *IEEE Transactions on Systems, Man, and Cybernetics (SMC). Part B: Cybernetics*, 2001, **31** (5), pp. 781–90

49 MILLAN-RODRIGUEZ, F., CHECHILE-TONIOLO, G., SALVADOR-BAYARRI, J., PALOU, J., and VICENTE-RODRIGUEZ, J.: 'Multivariate analysis of the prognostic factors of primary superficial bladder cancer', *Journal of Urology*, 2000, **163**, pp. 73–8

# Index

Printed in the USA
CPSIA information can be obtained
at www.ICGtesting.com
JSHW011508221024
72173JS00005B/1238

9 780863 414893